N. Schmidt-Kittler · K. Vogel (Eds.)

Constructional Morphology and Evolution

With 147 Figures

Springer-Verlag
Berlin Heidelberg New York
London Paris Tokyo
Hong Kong Barcelona
Budapest

Prof. Dr. NORBERT SCHMIDT-KITTLER
Institut für Geowissenschaften
Johannes Gutenberg-Universität
Saarstr. 21
6500 Mainz, FRG

Prof. Dr. KLAUS VOGEL
Geologisch-Paläontologisches Institut
J. W. Goethe-Universität
Senckenberganlage 32 – 34
6000 Frankfurt a. M., FRG

Cover illustration (drawn by H. Schäfer): The cover illustration symbolizes a basic principle of constructional morphology: a body surrounded by a flexible envelope and tightly filled with fluid or viscid material automatically assumes a *spherical shape*. This is true for protists as well as for multicellular organisms. Every deviation must be actively generated by a bracing system. Ontogeny and phylogeny are governed by this principle. This holds also for skeleton bearing organisms such as *trilobites*. Skeletal growth requires preformed hydraulic bodies. In trilobites as in other arthropods this reciprocal effect of hydraulics and bracing systems also maintains the shape during molting processes.

ISBN-13: 978-3-642-76158-4 e-ISBN-13: 978-3-642-76156-0
DOI: 10.1007/978-3-642-76156-0

Softcover repint of the hardcover 1st edition 1991

31/3145-543210 – Printed on acid-free paper

Preface

This book contains the results of an international symposium held in Bad Homburg, FRG, from the 17th to the 21st April 1990 under the title "Constructional Morphology and Biomechanics: Concepts and Implications". Twenty-five biologists and paleontologists from different countries working in this field were invited to present review papers and case studies on the main fields of constructional morphology. The list of participants is practically identical with the list of contributors given on page XI with the exception that in co-authored works only the first author attended.

The contributions of the participants were submitted before the meeting took place so that the symposium could mainly be dedicated to discussion. All contributions were reconsidered by their authors after the symposium in response to discussion remarks and after having been reviewed by one or two other participants of the meeting.

The articles are divided into four main topic fields: Papers dealing with conceptual and methodical questions of constructional morphology form the first part. It is followed by a group of papers focussing on the operation of bioconstructions and behavioural control within the environment. Morphogenetic aspects of bioconstructions and organic patterns are the subject of the third part. The last part focusses on the evolution of bioconstructions and body plans. Some of the contributions, due to their more general approach or the combination of treated subjects, could have been assigned to more than one topic field. Thus, their arrangement in one part of the book must seem somewhat arbitrary, however, a decision had to be made.

The symposium was financed by the Werner Reimers-Stiftung in Bad Homburg and the Siftung Volkswagenwerk in Hannover, both of which are thanked for their support. We are also grateful to the President of the University of Mainz for his financial contributions to the organization of the symposium.

The very pleasant and relaxing atmosphere of the facilities and grounds of the Reimers-Stiftung in Bad Homburg was an ideal meeting place for scientific encounters of a small working group.

March 1991

Norbert Schmidt-Kittler, Mainz
Klaus Vogel, Frankfurt

Contents

List of Contributors

You will find the addresses at the beginning of the respective contribution

Introduction: A New Look at Morphology

N. SCHMIDT-KITTLER[1] and K. VOGEL[2]

Is "constructional morphology", as used in the title of this book, really the appropriate term for the treatment of morphological phenomena exemplified by the contributions of the participants of this Symposium? Is this not the only kind of approach that all biologists and paleontologists should choose when dealing with organismic form and patterns? Should not the traditional descriptive and comparative morphology be better called "morphography" leaving the term "morphology" to the attempt to analyze those processes that enforce and determine organismic form, that make it develop and evolve and that are produced by it? Or, should we not speak of "morphodynamics" (as proposed by Seilacher during the Symposium)?

The editors rejected both alternatives. They preferred to mark the difference in scientific treatment of organisms distinctly by an own term, but by a term that already has a tradition and not by creation of another new term such as morphodynamics.

"Constructional morphology" is the literal translation of the German word "Konstruktionsmorphologie" introduced by Weber (1955) to characterize a morphological approach that aims at the operational interconnections between form and function (Vogel this Vol.). Is the meaning of this word changed when translated into English? Bock (this Vol.) thinks it would be. Other English-speaking participants do not see translation problems, but would have difficulties pronouncing the German word.

The editors decided to speak of "constructional morphology". They propose the following definition: Constructional morphology is the study of organisms and their parts as coherent systems and subsystems that obey statically and dynamically (at rest and in movement) physical principles and processes. The choice of this term also focuses attention on potentials and limitations set on these systems by ontogenetic growth and evolutionary history. It includes exploration of how mechanical and chemical forces and processes influence each other and how construction, behavior, and environment act and react on each other.

There are participants who object to such a broad definition. Thus, for instance, Bock (this Vol.) argues that our term should be restricted to nomological-deductive explanation and should not include historical-narrative explanations, which would best be termed evolutionary morphology.

[1] Institut für Geowissenschaften, Paläontologie, Universität Mainz, Saarstr. 21, D-6500 Mainz, FRG
[2] Geologisch-Paläontologisches Institut, Universität Frankfurt, Senckenberganlage 32–34, D-6000 Frankfurt a. M., FRG

N. Schmidt-Kittler and K. Vogel (Eds):
Constructional Morphology and Evolution

This brings us to the main sources of controversies which occurred during the Symposium. They can ultimately be assigned to two categories of problems: either to questions of scope or to normative difficulties. Already in the entrance discussion and also in the following argumentations, the distinction between nomological-deductive explanations and historical-narrative explanations became relevant. This distinction goes back to Nagel (1961), and was introduced in order to distinguish clearly between the different methodical situations in the so-called exact natural sciences and in historical sciences. The fact that biology, by the very reason of its research objects, the organisms, comprises both approaches, is the main source from which the different conceptual problems derive. Gould (1980) already showed that a great deal of the differences in research strategy in evolutionary science are ultimately not really controversial, since they reduce to differences in scope and not in ideas. A closer look at the various standpoints defended in the discussions and also mirrored in the points defended in the chapters presented in this book reveals that within both the nomothetic and the idiographic approach (as they were originally called by Nagel) further methodical distinctions can be made. Within the idiographic or historical-narrative approach, beyond data-gathering through comparative anatomical studies of Recent organisms and paleontological research, explanations can be based on the multifariously defended (Rudwick 1964, and, e.g., Paul 1975) but also criticized paradigm method (Grant 1972; Signor 1982). They are thus ultimately acquired through morphoconstructional arguments which gain their power from nomological-deductive reasoning (as discussed by Bock in the present Vol.). Explanations, however, are also frequently derived from the environmental scenario (Skelton this Vol.).

Within the nomological-deductive approach two distinct main research strategies are followed, which differ in the crucial point of whether or not the synthetic theory of evolution forms part of the basic assumptions required for the respective methodology. In fact, morphoconstructional studies in a very puristic sense can be based exclusively upon physicochemical analysis of organic structures and processes. This methodical program is most consequently followed by Nachtigall (1972, see also Wisser and Nachtigall this Vol.). The evolutionary dimension, though not questioned by the holders of this position, is not integrated directly in their research field. Nomological-deductive concepts, which are inclusive of the dimension of structural change of organisms with time, have in common that assumptions about economization processes play an important role in addition to the puristic methodology and form the main argument in the deduction of possible evolutionary transformations of organic systems (Peters and Gutmann 1971; Gutmann 1972). This is only possible if the economy aspect is consistently derived from synthetic theory.

It does not make much sense to argue about different scopes of research as long as they are methodically consistent, because they ultimately mirror the sometimes quite diverging scientific interests of their protagonists. However, part of the differences of standpoints traces back to reasons which, though interpreted individually, are caused by more objective difficulties. These are the normative questions of how much evidence is required or must be made available in order to give reasonable support to the statements made in the different methodical approaches. To which extent can we, for instance, infer possible directions of constructional transformation by assuming environmental factors as given theoretically? And if we can successfully do that in the case of highly environment-independent organic performances, how are the chances of applying this deductive strategy successfully in the case of more environment-dependent performances? On the other hand, we can ask how much evidence drawn from paleontological and geological data is needed in order to support actual assumptions about environmental influences upon organic evolution. Answers to these questions can frequently not be given very clearly and are sometimes extremely divergent. And with this, notwithstanding the fact that the basic assumptions held by different workers are very close to each other, the scientific relevance of statements defended on the basis of the various methodologies is judged quite differently.

It is evident that the different scopes in morphoconstructional research and different views about priorities in research strategies are severe obstacles on the way to agreements about concepts and general headings for research programs.

In the following we present various impressions about the diversity of approaches used in the different fields of constructional morphology. Thus, the topic field of *Bioconstruction: operational interconnections, behavior, environment* (see below) was approached experimentally (Bereiter-Hahn; Wisser and Nachtigall), by paradigm comparison (Gutmann), by description and comparison between different animal groups or representatives of the same group, or by comparison of skeletal elements (Alexander; von Koenigswald and Pfretzschner; Paul; Preuschoft; Thomas), or by treatment of pathological cases and their meaning for selection and evolution (Peters; cf. also Bandel).

The topic field of *Bioconstruction: morphogenesis* was treated experimentally (Tautz) or by paradigm comparison (Gutmann), by a theoretic attempt to reconstruct morphogenetic gradients (Weishampel; cf. also Seilacher), or by comparative morphology more or less supported by biomechanical arguments (Bandel; Liem; Seilacher).

The same methodological diversity was used to approach the topic field *Bioconstruction: evolution.* Phylogenetic reconstruction on molecular basis (Valentine), by paradigm comparison or transformation series of models (Edlinger), by organism comparison (Duncker) and – surprisingly chosen only once despite the relatively great number of participating researchers in geology and paleontology – by geologic-paleontological documentation (Skelton).

In spite of this methodological diversity, the participants agree on one essential point: we will never understand organisms without an integrative approach. This follows from the systemal organization of the organism, or from their physicochemical coherence, or from their operational closedness, however this frame condition of organismic beings is called. The more and more detailed experimental exploration is an unrenouncable but not sufficient opening to life. The relevance, for instance, of biochemical processes for the organism results only from their functional position within the coherent system. Thus, in a very impressive way, it can be shown that not only is cell mechanics influenced by cell chemistry but, vice versa, cell chemistry (including genetic information) by cell mechanics (Bereiter-Hahn)!

In the context of the history of science, it seems remarkable that at this point biology should match the development in physics insofar as in this subject insight

References

Gould SJ (1980) The promise of paleobiology as a nomothetic, evolutionary discipline. Paleobiology 6:96–118
Grant RE (1972) The lophophore and feeding mechanism of the Productina (Brachiopoda). J Paleontol 46:213–249
Gutmann WF (1972) Die Hydroskelett-Theorie. Aufs Red Senckenb Naturforsch Ges 21:1–91
Nachtigall W (1972) Biologische Forschung. Aspekte, Argumente, Aussagen. Quelle & Meyer, Heidelberg
Nagel E (1961) The structure of science. Harcourt, Brace, New York
Paul CRC (1975) A reappraisal of the paradigm method of functional analysis in fossils. Lethaia 7:15–21
Peters DS, Gutmann WF (1971) Über die Lesrichtung von Merkmals- und Konstruktionsreihen. Z Zool Syst Evolutionsforsch 9:237–263
Rudwick MJS (1964) The inference of function from structure in fossils. Brit J Philos Sci 15:27–40
Signor PW (1982) A critical re-evaluation of the paradigm method of functional inference. N Jahrb Geol Paläontol Abh 162:59–63
Weber H (1955) Stellung und Aufgaben der Morphologie in der Zoologie der Gegenwart. Zool Anz Suppl 18:137–159

Part I Concepts of Constructional Morphology

Explanations in Konstruktionsmorphologie and Evolutionary Morphology

WALTER J. BOCK[1]

Summary

An analysis of the different uses of Konstruktionsmorphologie (KM) is presented in the light of nomological-deductive (N-D E) and historical-narrative explanations (H-N E) in science. It was argued that the term KM should be restricted to N-D approaches, and that the proper English translation is engineering, not constructional, morphology. H-N approaches to KM are best termed evolutionary morphology. Studies of evolutionary morphology must be based on accepted causes and processes of evolutionary change (N-D Es). Constraints are initial and boundary conditions in explanations of evolutionary processes; they do not serve as causes of evolutionary change. Explanations of evolutionary change based only internal causes (e. g., causes completely internal to living organisms) are unacceptable, as not all of these causes have been demonstrated empirically. Complete morphological explanations require both N-D and H-N explanations.

1 Introduction

Over the past three decades, many morphologists, systematists, and paleontologists have advocated Konstruktionsmorphologie (= KM) as a research strategy for morphological analyses, yet clear statements of underlying assumptions and nomological statements are lacking, and no analysis has been provided of the types of explanations included under KM. It is not sufficient to state that KM is a methodology or a research program characterized by a triangle of interacting components (Thomas 1979; Reif et al. 1985; Seilacher 1989) with observed structures explained by the dual constraints of function and fabrication. In spite of the considerable use of KM over the past three decades, a philosophical inquiry of this approach is lacking. The result is that almost everything goes. Different biologists attempt radically different explanations under the heading of KM. The result is that morphological explanation under KM is at an impasse. Hence the timeliness of the current conference.

In science, explanation is everything. After scientists make empirical observations of some phenomena in nature, thereby having a basis for analyses, their interests lie strictly in explanations, be they partial or complete. For certain aspects of science, a partial explanation of the phenomena may be sufficient, perhaps only a "how" question and not a "why" question, or "how does it work" as opposed

[1] Department of Biological Sciences, Columbia University, New York, NY 10027, USA

N. Schmidt-Kittler and K. Vogel (Eds):
Constructional Morphology and Evolution

to "how did it originate". Nevertheless, scientific inquiry stops only when explanations are complete, meaning that all aspects of the phenomenon, including its historical development, have been explained. In biology, including morphology, complete explanations must be historical-narrative in addition to nomological-deductive. The major questions which I will address are the nature of explanations used in KM and the requirements for full explanations in morphology.

2 Explanations in Science

Explanations in science are not monolithic, but vary considerably. Nevertheless, scientific explanations form definite patterns. A number of classifications for explanations have been proposed (Nagel 1961), most of which have real merit. I will consider a simple system of explanations without any pretence that it is complete or even adequate for all purposes. However, it does cover essential aspects of explanations in sciences such as biology which possess a major historical aspect (Bock and von Wahlert 1963; Bock and Caplan in prep.). Explanations will be divided into nomological-deductive (N-D E), the typical law-like approach, and historical-narrative (H-N E) explanations. Nomological-deductive explanations are the standard form of explanations in science. Although historical-narrative explanations have scarcely been considered by philosophers, they represent the major class of scientific explanation.

2.1 Nomological-Deductive Explanations

N-D Es have the following form. Given a set of facts (e.g., initial and boundary conditions) and a set of laws which together form the explanatory statement, or *explanans,* a particular conclusion, or *explanandum*, must follow. N-D Es answer the question: why or how has a particular phenomenon [explanandum] occurred? Or, to put it in the opposite form, given a particular phenomenon, can it be explained with a set of causes and initial plus boundary conditions. N-D Es apply to universals (nonlimited sets of phenomena), do not depend on the past history of the objects or the phenomena being explained, and their premises (the nomological statements) are assumed to be always true. Thus N-D Es are temporally-spatially unrestricted. Many workers wrongly assume this means that N-D Es must hold everywhere in the known universe, and thereby claim that biological explanations are not true N-D Es because we do not know whether they hold everywhere in the universe. This position pushes the concept of nomological explanation to ridiculous extremes, and even renders its application in physics and chemistry dubious. Universally applicable does not mean applicable everywhere in the known universe, but to the relevant universe of the particular N-D E. It is thus necessary to provide a reasonable restriction of the realm of application) for each N-D E. For biology, the appropriate universe is the earth, and more specifically its surface – the upper part of its crust. If the explanandum did not always result from the conjunction of the set of facts invoked (the initial and boundary conditions) and the set of general laws, then the N-D E is not valid (it has been falsified), and one must search for the basis of its falsification. Falsification, i.e., the N-D-E being invalid, means that the explanandum (i.e., predictions generated from the initial and boundary conditions and the set of law-like statements) does not agree with objective, noncircular, empirical observations. Falsification does not automatically imply that the general laws used in the explanation are in error. Possibly, the set of initial and/or boundary conditions used in the empirical test was in error. Or possibly the predictions were incorrectly generated. Examples of N-D Es include the law of gravity, the statement of natural selection (nonrandom, differential reproduction of genes), the theory of plate tectonics, the theory that all organisms descended with modification from a single origin of life, etc. Accepting the existence of N-D E in biology necessitates acceptance of the belief that laws exist in biology.

2.2 Historical-Narrative Explanations

H-N E provide an understanding of the existing attributes of a particular set of objects at a definite point in time; they deal with outcomes, depend on the past history of these objects, and must use pertinent N-D Es. Objects explained by an H-N E are particulars (unique), not universals, in that they constitute a finite set having definite spatial-temporal positions. H-N Es are stated on a nondeductive basis with the hope of reaching the most reasonable and probable explanation for the objects studied. Five aspects of the H-N E explanation should be stressed, the first being the most important: (1) These explanations are historical, which means that earlier events affect later ones. Consequently, in any H-N E, special attention must be given to formulating the analysis within the correct chronological order of events and changes. (2) Explanations should be given on a probability basis of being correct (Nagel 1961, p. 26). This is necessary because H-N Es frequently involve a number of conflicting N-D Es and uncertain initial and boundary conditions. (3) All H-N E must be based on pertinent N-D Es. These N-D Es form part of the argument chain used in testing H-N Es. (4) H-N Es are unique and cannot be generalized in that a successful explanation for one phenomenon (e.g., origin of homoiothermy in mammals) need not hold for a similar phenomenon (e.g., origin of homoiothermy in birds). (5) Because of their complexity, because of the possible confusion between conflicting explanations and because of the difficulty in identifying valid empirical tests, H-N Es must be stated clearly and unambiguously. Failure to do this can preclude formulation of meaningful tests and/or appraisal of rival H-N Es. Examples of H-N E are the phylogeny and the evolutionary history of organisms, biological classification, historical biogeography, and geological history of the earth; see Bock (1986b) for a detailed H-N analysis of the arboreal origin of avian flight.

Both N-D and H-N Es are scientific under the criterion of demarcation for scientific explanations advocated by Popper in that both are available for testing by falsification against objective, empirical observation. But the two modes of explanations differ in how they are expressed, tested, and used to test other theoretical statements. N-D Es are tested strictly by attempts to falsify the explanation using deductive approaches. Although H-N Es can be tested in this way, these explanations are generally tested by verification using inductive methods.

2.3 *Types of Nomological Statements*

N-D E are characterized by the existence of law-like statements (scientific laws) which provide the explanatory foundation. But law-like statements are not all identical; rather, they can be arranged into three categories – causes, processes, and outcomes. The use of vague terms such as "agents", "principles", and even "mechanisms" must be avoided at all costs because they can apply to quite diverse nomological statements. Principle is equivalent to lawlike statement and could be a cause, process, or outcome. Agent, as usually used, can refer to cause or to initial and/or boundary conditions. It is simply not possible to evaluate the meaning of scientific papers which rely on an extensive use of these ambiguous general terms. An example is Gould and Lewontin's (1979) analysis of the adaptational program, which is rendered unintelligible because of their constant use of the phrase "agent of evolutionary change" to cover both causes and initial plus boundary conditions.

1 Causes. Causes (= efficient causes, which is the only proper use of cause in science) bring about changes in the state of an object over time. That is, a cause results in an object changing from an initial state A at time 1 to a different state B at a later time 2. A physical force is a cause, and could result in a mass changing its acceleration, and hence its velocity over time. A selective demand is a cause and could result in change in the pattern of genetic variation of a population over time. A major goal of workers in any science is the elucidation of pertinent causes operating in that science. Physicists have identified four fundamental forces operating on physical objects, and much debate exists at to whether these four forces are sufficient to explain all state changes in these objects or whether additional forces exist. Evolutionary biologists have been much remiss in not clarifying and codifying the fundamental causes operating in evolution. Because forces are well known causes in physics, some workers have labeled causes of evolutionary changes as forces. This is unfortunate and unnecessary as the term evolutionary cause or simply cause is available and unambiguous.

2 Processes. A process is the actual time course of modification shown by an object as it changes from its initial state A at time 1 to its modified state B at time 2. Involved in any process is at least one cause, a set of initial conditions describing the nature of the object at the onset of the process, and a set of boundary conditions describing the surrounding environment during the time course of the process. The initial and boundary conditions influence the exact consequence of the causes acting on the object and hence the nature of the modified state B reached by the object at time 2. The path taken by an artillery shell acted on by the detonation of the power charge is a process. The cause is the force exerted on the shell by the detonating power charge, the initial conditions are the mass of the shell, the structure, direction and elevation of the cannon, and the boundary conditions are the density and movement of the air, and gravity. In evolutionary biology, the important processes are phyletic evolution (change along a phyletic lineage) and speciation (splitting of a phyletic lineage; Bock 1979, 1986a).

3 Outcomes. An outcome (result) is the final modified state of an object at the termination of the process initiated by the cause. The particular outcome, or the modified state of the object, is determined by the nature of the cause, the set of initial conditions including the object, and the set of boundary conditions. The final resting place of the artillery shell is the outcome of its trajectory (process) and is determined by the combination of the cause and initial plus boundary conditions described above. If the shell did not hit the desired target, then the causes, initial or boundary conditions must be altered to achieve the desired result. In evolutionary biology, the altered pattern of genetically based phenotypic variation in a population is the result of phyletic evolution, and two or more species is the result of speciation. The statement of natural selection as nonrandom differential reproduction of genes is an outcome and depends on the interaction of a set of causes and initial and boundary conditions. Natural selection as so defined is not a cause of evolutionary change, contrary to the belief of most biologists.

3 Functional and Historical Biology

Inquiry in biology has been divided into two major domains by Mayr (1982, pp. 67–76), who recognized the areas of functional biology and evolutionary biology, which differ in their research interests, methods of study, and causes, namely proximal versus ultimate. Although Mayr's use of the term cause is considerably broader than that of efficient cause, his concept of proximal versus ultimate causes (Mayr 1961, 1976) is clear within the framework of his discussion. Dullemeijer (1972, 1974, 1980, 1985; Dullemeijer and Barel 1977) has also been long interested in the nature of explanations morphology and has given particulat attention to the distinctions between the fields of functional morphology and evolutionary biology. His delimitation of functional and evolutionary biology is close to that of Mayr. Functional biology is characterized by questions of "how does it work?", while evolutionary biology deals primarily with questions of "how did it come into being?" (see Bock 1988). The distinction between functional and evolutionary biology is most useful in analyses of the history of biology, but becomes less useful for the clarification of explanatory systems.

Functional biology includes numerous fields from physiology, biochemistry, development, genetics, to morphology and ecology, and deals largely with questions such as how are biological features put together in an organism?, how did they develop ontogenetically?, how do they operate?, how do different features in an organism interact mutually?, how do organisms interact with each other and with the demands of their external environment?, etc. The basic approach is that of N-D E, using direct observations, experiments, and comparisons to generate the empirical observations needed to test scientific hypotheses.

Evolutionary biology covers the basic questions of how features of organisms and groups of organisms came into being and developed their existing characteristics, including their ecological relationships, pattern of community structure, biogeographic distribution, etc. It is not correct to claim that evolutionary biology is equivalent to H-N E, as this domain of biology also includes a large segment of typical N-D Es such as clarifying the causes and processes of evolutionary change. Nor is it valid to conclude that evolutionary biology uses only comparative methods to test theoretical statements.

Mayr (1982, pp. 67–76) suggests that the dichotomy of functional versus evolutionary biology is correlated with the dichotomy of proximal versus ultimate causes (Mayr 1961). From his initial presentation of these ideas, it appears that Mayr used proximal versus ultimate causes in the sense of explanations, with proximal cause equal to functional explanation, and ultimate cause to evolutionary explanation. I will accept this interpretation and will not use proximal versus ultimate causes independently of functional versus evolutionary explanations.

Further, both Mayr and Dullemeijer used the dichotomy of functional versus evolutionary not only to distinguish between two major domains in biology study, but to distinguish between two major approaches to explanations in biology (Bock 1988). Although this dichotomy of explanations in biology appears reasonable and has much support, no valid characterization of these explanatory types has ever been offered. This dichotomy of explanations is not equal to N-D E versus H-N E, although many workers assume this relationship.

Explanations in functional biology are usually, if not always, N-D Es. It is not clear whether any explanations in the domain of functional biology are true H-N Es. One must exclude simple time factors in the phenomena being explained. Perfectly good processes can require a long time period for completion; hence, long time duration for an event does not automatically mean that it falls under the mode of H-N E.

Explanations within the domain of evolutionary biology can be either N-D E or H-N E, depending on what is being analyzed (Bock and von Wahlert 1963; Bock and Caplan in prep.). Explanations such as mechanisms of populational genetics, competition between individuals of different species, the analysis of adaptations (state of being) in organisms, the action of selective demands arising from the external environment on individual organisms, etc. are all part of evolutionary biology, but are strict N-D Es. Other evolutionary subjects such as elucidation of the phylogeny of a group, the biogeographical history of a group, clarification of the evolution of flight in birds, analysis of convergence of features, the analysis of multiple pathways of evolution, etc. are all proper H-N Es.

Great care must be taken to characterize explanations within evolutionary biology properly as N-D or H-N. The establishment of the evolutionary history (phylogeny) of a set of species based on a knowledge of their attributes is a H-N E and requires an analyses of the homology and other taxonomic properties of features (Bock 1989c) as well as the necessary N-D Es of evolutionary theory. Functional and adaptive analyses of the features are needed to reach convincing conclusions about taxonomic properties (Bock 1981; Szalay and Bock 1991). Yet many studies labeled as phylogenetic reconstruction (e.g., Gutmann in this Vol.) are clearly N-D Es, not H-N-Es! These studies do not establish a phylogeny of a group of species based on analysis of their attributes. Rather they begin with a particular phylogeny or a few alternative phylogenies and analyze the probability of the postulated evolutionary changes using morphological N-D Es and assumptions such as increased efficiency and optimization. Such approaches to phylogenetic reconstruction do not require use of homology as particular phylogenies are assumed.

Although the dichotomy of evolutionary versus functional biology is valuable in the analysis of the biological history, the usefulness of evolutionary versus functional explanations is far less clear. An initial difficulty would be formulating an acceptable set of characteristics of each type of explanation which is not equal to N-D E and H-N E. Perhaps it is best to recognize these two broad domains within biology for historical analyses, but not to attempt to extend this particular dichotomy to explanations in biology. I will, therefore, accept only the dichotomy of N-D E and H-N E in this treatment of explanations in morphology.

4 Words, Concepts, and Translations

Quite to the contrary of most other participants to this conference, I chose to use the term Konstruktionsmorphologie rather than the English constructional morphology. This decision was not made to be perverse, but because I believe strongly that words are important and must be used carefully. I reject the argument of those who claim that it does not matter what objects and ideas are called. Biology and especially evolutionary biology has suffered greatly because of the indifferent attitude and sloppiness of biologists in choosing and defining terms used to express their ideas. Contrary to claims made by Pranger (1990), pluralistic definitions of scientific terms are most undesirable.

4.1 Words and Concepts

Words stand for particular concepts and must be defined carefully. One distinction between scientific and general language is that scientists attempt to restrict their terms to single unambiguous concepts. For example "elastic" in physics refers to a material that returns to its original configuration after a deforming force is removed. In general English, "elastic" is used only roughly in this sense, but more in the sense of a material which deforms greatly under a deforming force (= compliant in physical terminology) and returns to its original shape after that force is removed. Steel balls are not considered elastic in general English usage in spite of the fact that they are indeed highly elastic. Considerable problems exist in morphological discussions if elastic is not used in the proper physical sense when discussing morphological structures. One can read frequently that bones and collagenous bands are not "elastic" (meaning that they do not deform greatly under deforming forces; e.g., they are highly incompliant), when in truth these structures are highly elastic.

Moreover, one is simply not free to define any word within scientific usage in an arbitrary way. Severe restrictions exist as words used in science still possess associations from general usage. If a scientific concept has absolutely no connections with general ideas, then it is possible to coin terms freely. Thus, subparticle physics uses a series of terms as "quarks", "flavor" and "color" which are so foreign to any common usage that no confusion results. But in scientific usage, many words still possess connotations and hidden meanings, in addition to the open or defined meaning. Often these hidden meanings are far more important than the defined meanings. In science, words should be used as closely as possible to their general meanings in the common language. One can rightly object if in science the word "white" is used to designate the absence of reflected light, the opposite of its general meaning.

4.2 *Translation Difficulties*

Science is international, but even today science is communicated in many different languages. Over the past three to four decades, English has become the international language of science, a trend with many advantages, including one not usually discussed. With the use of an international language in science, one avoids problems resulting from difficulties in translating many words. The greatest problems exist for words which are similar or identical in structure (spelling) in different languages. One such word is "zu kontrollieren"; "Konstruktion" is another. These words are usually incorrectly translated in translation dictionaries, and it is often not possible to comprehend their real meanings until learning their incorrect usage, e. g., a German using English equivalents for "zu kontrollieren". In this case, "zu kontrollieren" means "to check" or "to record", not "to exercise restraining or directing influence over" which is the usual meaning of "to control" in English. Native English-speaking scientists are startled when hearing a German biologist tell of "controlling the independent variable" even when knowing the difference in meanings of "to control" and "zu kontrollieren". Yet such problems are relatively simple compared to words of varying meanings in different languages. Although the meanings of "Morphologie" and "morphology" in German and English are the same, this word has very different connotations when used to designate a biological discipline because traditional interests of German morphologists are quite different from those of Anglo-American workers. These differences are little appreciated without an extensive and intensive understanding of research interests of both German- and English-speaking morphologists.

4.3 *English Words Ending in "tion"*

This class of English words represents special problems in translation as well as in English usage. Most, if not all of the words ending in "tion", such as adaptation, function, construction, have the duel meanings of expressing "a state of being" as well as "a process of becoming". The meaning can be determined only by use of the word, or by the construction of the sentence. Indeed even within English, workers have argued against the duel use of words such as adaptation in science because of their potential confusion. As an example, adaptation, which is used in both senses (Bock and von Wahlert 1965; Bock 1980), has the meanings within evolutionary biology of: (1) The properties of a phenotypic feature of an organism relative to existing selective demands arising from the external environment, and (2) the process of change in a feature over evolutionary time under the control of selective demands arising from the external environment. Clearly, analysis of a feature as an adaptation (the state of being) is a N-D E and as such is completely independent of how this feature evolved (that is, how the adaptation came into being), contrary to the position advocated by Gould and Vrba (1982).

Construction is also used in the sence of the state of being and the process of becoming. Thus construction has the meaning of: (1) The actual building or putting together of an object – how a particular object has come into being; and (2) the actual configuration of an object. The latter is in the meaning of the physical structure, its properties of form – the material composition and how the materials are put together. Construction, as the state of being, generally does not have the meaning of how the object works or operates because of its structural properties. For this meaning, one would have to use a term such as functional or engineering design.

4.4 Konstruktion and zu Konstruieren

With the realization of the problems of meanings of similar words in German and English, I made a survey of the meaning of "Konstruktion" by writing to a number of German colleagues (not only morphologists) and asking their understanding of this word. My questions were placed within the frame of comprehending the meaning of KM, but I phrased my inquiries carefully as to the meaning of "Konstruktion", not KM. Most of my correspondents discussed the general meaning of Konstruktion in German, some also covered KM. Although some differnence of opinions existed in the replies I received, most persions stated firmly the "Konstruktion" in Geman had the meaning of the operation, actions, functions of an object as a consequence of its structure (material composition and arrangement). This would include the strength of the object relative to external forces, how it could move, etc. They pointed out that "Konstruktion" is not used in the sence of actually building or forming (to build = zu bauen) the object. It is not used in the sence of a construction site (= Bauplatz). Some correspondents differed on this point. Some indicated that it was proper to extend the meaning of a compound word such as "Konstruktionsmorphologie" to include the processes by which morphological features came into being, both in an evolutionary sense and in an ontogenetic sense. These differences may reflect valid differences in the use of "Konstruktion" within German, or valid extensions of its usage within scientific German. However, the basic problems still exist in the use of KM and its translation into English.

4.5 Translation of "Konstruktion" into English

No simple English translation exists for the German "Konstruktion". It is not correct to translate the German "Konstruktion" into the English "construction", or the German "zu konstruieren" to the English "to construct". Hence one cannot simply translate the German KM into English as "constructional morphology", and expect the same meaning. If KM is translated into English, the best term might be "engineering morphology" as engineers are concerned with functional properties arising from the form of the designed system. This term reflects much of the meaning intended under KM and reflects the important aspect of living organisms that they are functioning machines. It might, however, be best not to translate KM.

This problem of translation is one of the reasons why I have great problems understanding all concepts included in KM. Even with a comprehension of translational problems, I still do not understand what is included under the KM by all workers using this term. Moreover, I doubt that the correct meaning of KM has survived its translation into the English constructional morphology for most, if not all, workers. Examination of the use of "constructional morphology" by most English-speaking workers provides support for this conclusion.

5 Konstruktionsmorphologie

5.1 Meanings

An analysis of the literature suggests that KM, and the English "constructional morphology" has been used in at least two major different meanings; namely: (1) Explanation of the operation, use, action of an object (its properties of function) from an analysis of its composition and the arrangement of its parts (its properties of form; Weber 1955, 1958). This would be in the sense of understanding functional properties of morphological features from an analysis of their properties of form (Bock and von Wahlert 1965). And (2) Explanation of the formation of an object considering all influences on it during its evolutionary and ontogenetic developments, but also including all influences of surrounding features and the external environment, etc. (Seilacher 1970, 1972, 1979, 1989; Reif et al. 1985; Thomas 1979). However, it is not totally clear whether all proponents of the second meaning of KM also include the first engineering meaning.

5.2 Explanations Included Under Konstruktionsmorphologie

With the diverse meanings of the term KM, the types of explanations included under this term vary considerably. Workers using KM have not been clear as to which type of explanations are to be covered. It is clear that many uses of KM include both N-D and H-N Es. Yet workers have not distinguished between them and have not indicated what types of tests are to be used, etc., in dealing with these diverse explanations. Because of the predictions generated, their test against empirical observations, etc., N-D E and of H-N E differ considerably and are sometimes in direct conflict. Hence, it is simply not possible to lump all explanations in KM under the same heading and to use the same research approaches for all analyses within this field without generating major confusion.

Considering the two major meanings of KM given above, it is clear that its first meaning of engineering morphology is a pure N-D E, and explanation under this concept of KM must be approached strictly with N-D methods. The second meaning of ontogenetic and evolutionary development of morphological features involves a combination of N-D and H-N Es, depending on exactly which aspect of the development of the feature is being considered.

5.3 N-D E Explanations in Konstruktionsmorphologie

These apply only to nonhistorical explanations of KM. Within the standard meaning of Konstruktion in German, the term KM has a clear and fully understandable meaning, and can be defined as: "explanation of the functional properties of an individual feature or the whole organism based on its properties of form, the organism being considered as an integrative system". These studies attempt to explain how the feature or the organism operates based on an understanding of its existing morphology. The starting point for analysis is the existing properties of form of the feature or organism, that is, the existing morphological form.

The introduction of the concept of KM by Weber (1955, 1958) was to propose a substitute for morphological approaches covered by the German term Bauplan. "Bauplan" dealt with the structure of organisms under a strict concept of form – a science of pure form – which had long been sterile as a research strategy in morphology and evolutionary biology, contrary to the position taken by Gould (1970). Weber was interested in a morphology which included a combined concept of morphological form and function. He stressed the importance of understanding functional properties as well as properties of morphological form, and more importantly, the relationships between the two. Weber's ideas were a continuation of the those advocated in the 1920's and 1930's by Böker (1935–1937), and earlier by Hesse and Doflein (1910–1914). They constituted an important advance from the approach of a pure morphology of form characterized by the Bauplan to functional and ecological morphology (Bock 1990). Similar developments were occurring elsewhere under the heading of functional morphology (Davis 1958; Dullemeijer 1974; Wake 1982; Gans 1985; Liem and Wake 1985).

Thus, this concept of KM would include all aspects of the explanation of the form-function complex of features and whole organisms (Bock and von Wahlert 1965; Bock 1980). It would include analysis of morphological form done by workers such as Dullemeijer (1974) and his colleagues in Leiden (see Osse et al. 1989), by Gutmann (1966, 1971, 1972, 1988, 1989) in a large series of papers, such as his analysis of the hydraulic principle, the hydrostatic skeleton, etc., and similar papers by other workers in the Senckenberg Museum, by Homberger (1986, 1988) in her analyses of avian morphology and concepts of models in morphology, by Bock (1974, 1988, 1989a, b) in a series of papers on nonevolutionary explanations in morphology (see also the symposia in The American Zoologist, Bock and Homberger 1988, and Kraemer and Bock 1989). One should also include N-D evolutionary explanations such as the analysis of the adaptive significance of features (Bock 1979, 1980). This list is far from complete as there are many other important papers using a nomological deductive approach to KM, regardless of whether this term was used.

I have not discussed those KM approaches based on N-D E of ontogenetic development (see discussions in Wake and Roth 1988, for example), partly because the goals of these studies are not always clear. I agree strongly with workers who argue that analysis of adult morphological phenotypes must begin with ontogenetic analyses as stressed by Gutmann (1978, 1987a, b; Gutman and Bonik 1981). Such analyses fall completely within N-D E. However, many workers considering ontogenetic aspects of morphology do so in the hopes of clarifying causes and processes of phyletic evolution. It must be stressed that the causes associated with the production of the phenotype of individual organisms do not represent, by themselves, causes of evolutionary change.

5.4 H-N (Evolutionary) Explanations in Konstruktionsmorphologie

Herein are included all approaches advocated by diverse workers under the heading of KM which are historical evolutionary explanations, be they analyses of the phylogeny of features or organisms, the evolutionary history of features or organisms, the classification of organisms, etc. These approaches cover modification of features over evolutionary time, and hence the question exists whether such analyses are properly covered by the term Konstruktion. H-N approaches of KM include those advocated by Seilacher (1970, 1972, 1979, 1989), Reif (1983; Reif et al. 1985), Thomas (1979), Vogel (1989), and many other workers interested in explaining the phyletic history of organisms and their morphological features. This approach represents a valid, and indeed a most important, concern for morphologists, systematists and paleontologists. But it differs significantly from nonhistorical N-D KM advocated by Weber, Dullemeijer, Gutmann, etc. The frames of reference, questions to be answered, and even the permissible empirical tests differ sharply in N-D and H-N KM.

Problems in understanding H-N approaches of KM result from basic ideas not being set forth in a clear and sufficient manner. It is difficult to know what is included, the nature of explanations, how predictions are tested, etc. Most papers contain a mixture of N-D and H-N Es associated with morphological features, and it is difficult to impossible to separate these explanations.

Many serious problems arise when H-N KM is advocated by English-speaking workers, suggesting difficulties in understanding ideas presented in German, problems in translation of key words, misunderstanding of central concepts, etc. Thus, in advocating the ideas of KM adopted from Seilacher, Gould (1970), and Gould and Lewontin (1979) argue that these concepts fall within the Bauplan tradition. Their discussions are either within the framework of a science of "pure form" or one that does not include functional analyses in a central role. These ideas contrast sharply with the concept of KM as advocated by workers such as Böker, Weber, Dullemeijer, Gutmann as well as Seilacher, Reif, and others, who developed KM as an escape from the restrictions of the Bauplan approach, including concepts of pure morphological form. Weber insisted on explaining functional properties in proper morphological explanations, going far beyond any Bauplan approach.

The second and more serious problem of the various H-N approaches to KM is that some unacceptable concepts of evolutionary change – usually some evolutionary cause – appear to be incorporated into historical KM. Sometimes these evolutionary causes are clearly stated, but at other times they are hidden. A number of these ideas appear to stem from the teleomechanistic tradition developed in central Europe during the 19th century (Lenior 1982). Gould and Lewontin (1979) argue that one set of causes (= agents) for evolutionary change are constraints. Constraints have an important role in the processes of evolutionary change, but they are not causes. Frequently, the constructional system itself is regarded as a cause for further evolutionary change. But a morphological construction (either in terms of the property of form alone or in terms of properties of the form-function complex) does not constitute a cause; it is an initial condition. Some workers imply the existence of concepts of evolutionary change arising from developmental mechanisms (see discussions in Wake and Roth 1989). That is, the claim is made that causal connections exist between known ontogenetical causes or processes and evolutionary causes, an old idea within evolutionary biology dating back well over 100 years. Or, evolutionary causes are advocated which are strictly internal to the organism; e.g., the complete set of causes needed for evolutionary change are internal to organisms. To be sure, some evolutionary causes are internal. Most of the causes involved in the production of individual genetic variation generation by generation are internal to the organism, although it is also well known that external factors such as radiation can increase mutation rate. However, I know of no set of completely internal causes which are, by themselves, sufficient for phyletic evolutionary change (Bock 1980). At least one external cause, e.g., selective demands arising from the external environment, must be present for evolutionary change to take place.

5.5 Conclusions

Two very different types of inquiry have been included under the heading of KM, which can be designated the N-D and the H-N approaches. N-D KM corresponds closely to the initial concept of KM as advocated by Weber; I would advocate that KM should be restricted to this approach. If KM is translated into English, the best term may be "engineering morphology" which may be redundant to the common use of "functional morphology". H-N KM deals with equally important inquiries in morphology, but the questions posed, the basis assumptions to be used, tests against empirical observations needed, etc., are all quite different from N-D KM. I would argue that including both areas under the same heading of KM leads to excessive misunderstanding and confusion. Hence, I propose that H-N Es be excluded from KM, and placed under a different heading. Indeed the suitable heading of evolutionary morphology has been available since the 1950's, if not before (see Davis 1958, 1964).

6 Evolutionary Morphology

Space does not permit a full treatment of the goals and approaches of morphological inquiry under the heading of evolutionary morphology. Only the briefest sketch can be presented. Evolutionary morphology includes all analyses of the evolution of morphological features either as unit structures or complex systems up to the organismic level. Evolutionary morphology may be considered as the most inclusive discipline of morphology and includes descriptive, functional, integrative (including connectivity analyses, Bock 1989b), and ecological morphology (Bock 1977, 1990) as subdisciplines. Good analyses in all these subdisciplines are basic to proper explanation in evolutionary morphology. Moreover, evolutionary morphology depends on a foundation of well-established N-D causes and processes of evolutionary change (see above for a discussion of causes, processes, and outcomes).

6.1 Evolutionary Processes

It is easiest to discuss evolutionary processes first and then consider the underlying causes; this will be done only for sexually reproducing organisms. Evolutionary change in sexually reproducing organisms is always modification in a population of reproducing individuals, never of individual organisms. And evolutionary change is that observed between phenotypic characteristics in an initial generation and a descendent one. Discussion of causes and processes of organic evolution must always be based on a full understanding of population biology (including genetics) and ecology.

Similar processes and causes exist for phyletic evolution in asexually reproducing organisms, but their expression must be presented more carefully and requires more space. The process of speciation does not exist for asexually reproducing organisms, as these organisms are not arranged into species. Two major evolutionary processes exist, namely those of:

Phyletic evolution, which is evolutionary change of a species over time, or evolutionary change along a single phyletic lineage (see Bock 1979, 1986a for definitions of species and phyletic lineages). The result of phyletic evolution is change in the characteristics of the species, but it does not increase (multiply) the number of species. Phyletic evolution can occur without speciation.

Speciation or multiplication of species (splitting of a phyletic lineage), which is the division of a single species into two or more species, or in other terms, the splitting of a phyletic lineage into two or more. The result is increase in the number of species. Speciation requires phyletic evolution in at least one of the phyletic lineages.

6.2 Causes Underlying Evolutionary Processes

Although two distinct processes of evolutionary change exist, only that of phyletic evolution has distinct causes associated with it. The process of speciation has only the causes of phyletic evolution plus a distinctive boundary condition, namely an external isolating barrier separating two segments of the ancestral species for a sufficiently long period of time to permit evolution of intrinsic isolating mechanisms. The number of causes involved in phyletic evolution differs, depending on whether one begins with individual genetic variation or individual phenotypic variation in a population; I will start with the former for a more complete treatment. Causes of phyletic evolution operate generation by generation; they cannot be averaged over a number of generations. And they act only on individual organisms, although the processes of evolutionary change in sexually reproducing organisms are always population phenomenon. No N-D causes of phyletic evolutionary change have been demonstrated for populations, species, or any other groups (classes; see Caplan and Bock 1988). The N-D causes for phyletic evolutionary change are:

1. Production of Individual Genetic Variation Generation After Generation. These causes are: (a) Recombinations of the existing genetic material. New combinations are formed, but without change in the genetic material. (b) Gene flow resulting from migration of individual organisms and breeding with members of the new population, resulting in the appearance of new genetic material for the population, but not for the species. (c) Mutations of all sorts from point mutations to duplications of the entire set of chromosomes providing new genetic material for the population and the species. (d) Genetic drift, which is modification in gene frequencies of the population as a result of stochastic events acting in small populations during gamete and zygote formation. Many evolutionists regard genetic drift and other chance-based events during reproduction to be a distinct evolutionary cause, but these are best considered under the general heading of causes producing individual genetic variation.

The several causes underlying production of individual genetic variation in a population generation after generation are stochastic with respect to selective demands acting on the individual organisms. Hence the phenotypic features which arise from the action of these causes are accidental with respect to these selective demands. These causes producing individual genetic variation result in the chance-based or accidental aspect of evolutionary change (Mayr 1961), and in the phenomena of paradaptation and multiple pathways of adaptive evolution (Bock 1959, 1967). Several adaptive answers may evolve in response to the same selective demand acting on different populations of organisms.

2. Production of the Phenotype of the Individual Organism from its Genotype. This set involves all causes governing the process of ontogenetic development as well as all processes of physiological adaptation (= somatic modifications, Bock and von Wahlert 1965). These developmental causes are all genetically based. The phenotype results from the genotypic information plus influence of the external environment acting on the organism during its ontogeny. Physiological adaptations can also result from diverse interactions between parts of the organism, e.g., between bones and muscles, in addition to influences of the external environment.

Causes of physiological adaptation arise from connectivity between parts of the organism. It is important to remember that connectivity begins at the origin of the organism as a zygote, as stressed by Gutmann (1978). Theoretical analysis of connections between parts of an organism cannot be approached by adding or subtracting features of the developed organism.

Because different features may be linked together genetically or developmentally (pleiotropic connections), selection for one feature and its resulting adaptive modification could result in evolutionary modification in other features. These pleiotropic evolutionary changes would not be adaptive evolutionary processes although the modified features could be new adaptions. Because of the potential appearance of different phenotypes from the same genotype, evolutionary change can occur in features without accompanying genetic change. Hence, the general definition of evolutionary change need not be stated in terms of heredity or genetics, especially when considering modification of individual features.

3. Selective Demands Arising from the External Environment. This evolutionary cause is usually called natural selection or simply selection, but the definition of natural selection in Darwin's *On the Origin of Species* (1859, p. 61), as well as the generally accepted definition of Natural Selection in modern population genetics and evolutionary theory (= nonrandom differential reproduction of genes), is an outcome, not a cause. Hence I will use the term selective demand for this evolutionary cause and will use natural selection (= selection) strictly in its outcome definition. Selective demands arise only from the external environment, hence this cause is external to the organism. Selective demands are determined by interaction between environmental demands acting on the organism and how the organism responds to them (Bock 1980); selective demands do not arise unilaterally from the external environment. Thus the same environmental demand acting on several species may result to quite different selective demands depending on how individuals of each species react to the environmental demand. The concept of internal selection, or selective demands arising from the internal environment, is commonly advocated but without the necessary supporting evidence. This concept appears to be a misnomer for regulatory interactions between parts of the organism, and is largely, if not completely, covered by the concept of physiological adaptation (Bock and von Wahlert 1965; Bock 1980). Selective demands act on the phenotype of the whole organism, but their consequences depend on the phenotypic characteristics of features interacting directly with them and on the heritability of these features. Selective demands do not act directly on the genotype or on individual genes. Yet, the consequence of selective demands after the favored individuals reproduce could be natural selection (= modification in the pattern of individual genetic variation in the population).

Given a pattern of genetically based phenotypic individual variation in a population, directional selective demands are the major cause modifying this pattern of variation over time. In the absence of directional selective demands, differential reproduction of individuals possessing different genotypes may result in change, but only during the briefest periods while the population is below the carrying capacity of its environment. Some workers have claimed that other evolutionary causes (i.e., constraints, mechanisms of ontogenetic development, internal selection, etc.) can change the pattern of individual genetic variation in a population over time, but none of these proposed causes has ever been demonstrated empirically. The role of genetic drift as a cause of phyletic evolution of phenetic features is still controversial.

In the absence of other known causes, phyletic evolutionary change of populations would be under the control of selective demands. Hence, phyletic evolution is panselective. Care must be taken whether statements about phyletic evolution imply modification in a population of individual organisms or change in a particular feature of these organisms. Both approaches are completely valid. However, simply because phyletic evolution of populations is panselective, this does not mean that every attribute of the organism is an adaptation or that the evolution of every feature is adaptive. Because diverse features of organisms may be linked together pleiotropically, it is quite possible that adaptive evolution of one feature can result in nonadaptive pleiotropic evolution of other features. Pleiotropic evolutionary change had been well known to evolutionists for many decades, but some workers have used it to argue for the existence of causes of phyletic evolutionary changes other than selective demands.

The conclusion that all phyletic evolution is panselective holds only for evolution of populations, but not necessarily for individual features. Nevertheless evolutionary morphological investigations must begin by assuming that features are adaptive. The adaptationist program (= a research strategy) must be used, and features must be investigated thoroughly to ascertain their possible adaptive significance and mode of evolutionary change. Postulating nonadaptiveness of a feature is feasible only after failing to demonstrate its adaptiveness following an intensive investigation. Moreover, detailed studies of integrative morphology (connectivity analyses) are needed to demonstrate interconnections between features – structural, functional, genetic and developmental – as the basis of possible pleiotropic evolutionary change. Hence the need for proper N-D studies of KM as a foundation for evolutionary morphology and for complete explanations of morphological features.

6.3 The Role of Constraints in Phyletic Evolution

A great deal has been written about constraints of all types in biology and their role in evolution. Unfortunately much of this discussion has been overstated, including the claim by Gould and Lewontin (1979) that constraints are causes (= agents) of phyletic evolutionary change. The role of constraints in phyletic evolution is not new, but has been well understood since the days of the great German evolutionary biologist August Weismann (Mayr 1985).

1. Types of Constraints. Many different forms of constraints exist, and can be divided into two major types.

a) Physical-chemical constraints are the result of the form-functional properties of materials and are a general attribute of all materials. Thus bone has particular strengths against different types of stresses depending on the exact nature of bony material and how the bony material is arranged into structural units. Skeletal muscle has limits in the amount of force generated and the distance over which the muscle can shorten. These constraints of the skeleto-muscular system restrict the strength of vertebrates, and therefore the possibilities of vertebrate action. Limits exist on the maximum size of flying vertebrates because of the strength of the skeleto-muscular system, similar to restrictions on the size and speed of airplanes because of constraints of materials available for their construction. Materials used in machines and in living organisms can be changed. Modification of biomaterials will alter the physical constraints of organisms, as argued persuasively by Gutmann (1966, 1972, 1988) in his treatment of the physical properties of diverse materials used in skeletal systems.

b) Developmental (ontogenetic) constraints restrict the possible phenotypic expressions of genotypes. These developmental restraints are a combination of physicochemical constraints and historical constraints, and possess a genetic control. Developmental constraints are important factors in understanding ontogenetic development and hence the production of individual phenotypes. But they neither act as an evolutionary cause nor are they absolute, that is, incapable of being modified. Developmental constraints have been discussed under the heading of controls by some authors (Wake and Roth 1989).

c) Historical constraints are a standard attribute of historical explanations in which the definition of historical change is modification which depends, in part, on past events and hence on characteristics of the subject during its past. Because evolutionary modifications are historical, they depend on historical constraints which are existing genetic and phenotypic variation found in populations of interbreeding organisms. They are frequently termed phyletic constraints (Gans 1989). Evolutionary change in a population of elephants depends on the existing structure of elephants, and further evolutionary change in a population of hummingbirds depends on the existing structure of hummingbirds. These historical constraints are real, but they are not absolute in that they alter as the population evolves. It is entirely possible for evolutionary change beginning with a population of elephants to terminate in a population of creatures possessing characteristics of present-day hummingbirds. Modification of historical constraints during phyletic evolution was charmingly and thoroughly demonstrated by Stümpke (1961) in his epic monograph on the Rhinogradentia.

2. The Role of Constraints in Evolutionary Explanations. Recall that a process is a combination of a set of causes, the object on which the causes are acting, the initial condition of the object, and the boundary conditions extant during the process. Constraints of all kinds constitute the initial and boundary conditions acting during evolutionary processes. Constraints are not causes of evolutionary change. They influence the details of the particular evolutionary process, and hence its outcome, but they do not bring about change in a population from the ancestral state to its descendent state. Thus in the process of speciation, the presence of an external isolating barrier separating the original species into two isolated segments constitutes an essential boundary condition. But the existence of this boundary conditions will never lead to speciation – it is not a cause, and by itself will never bring about the needed modifications of these isolated populations. The evolutionary causes acting during speciation are those of phyletic evolution. They can effect modifications in at least one of the isolated populations and hence in the possible appearance of intrinsic isolating mechanisms required for the two populations to co-exist as distinct species.

Constraints have a most important role in evolutionary explanations in that they provide information about the initial and boundary conditions present during a particular evolutionary process. This distinction between causes and initial plus boundary conditions operating during a process has been confused in discussions of evolutionary theory and in explanations of evolutionary morphology with disastrous results. Constraints are frequently cited as evolutionary agents (= causes) and no distinction made between diverse factors vaguely termed "agents of evolution" (e. g., Gould and Lewontin 1979). Unless the correct role of constraints as initial and boundary conditions in evolutionary processes is understood and unless evolutionary factors are correctly identified as causes or constraints, analyses under KM and evolutionary morphology are doomed to failure.

7 Conclusion

As an evolutionary biologist and morphologist, my interests are in complete explanations of morphological features, which must include all aspects of N-D and H-N Es. Proper explanations can be achieved only if these two types are clearly recognized and distinguished from one another. Further, it is necessary to differentiate between law-like statements that are causes, processes, and results, and to understand the nature and role of initial plus boundary conditions in all types of explanations. These distinctions have not been made in earlier morphological explanations, resulting in ambiguous conclusions. Morphological features do not exist as separate entities, but are always found in complex systems possessing tight organization (LIMAs, Dyke 1988) known as individual organisms which in turn do not exist as isolated entities but always in relationship to their external environment. The integration of features in individual organisms appears and develops ontogenetically from their origin as zygotes. The individual organism from its appearance as a zygote always interacts with its external environment from the beginning of its ontogenetic development. Full explanation of complex morphological systems demands multiple approaches, including descriptive, functional, integrative, ecological, and evolutionary morphology. KM, or engineering morphology, as conceived by Weber and developed by Dullemeijer, Gutmann, and many other workers, is best restricted to N-D Es associated with N-D descriptive, functional, integrative, and ecological morphology. A critical conceptual advance was achieved by the postulation of KM as a replacement of the idealistic concept of the Bauplan, in which the concept of organisms as integrated, functional machines was substituted for the sterile notion of organisms as systems of pure morphological form. Subsequent extension of KM to evolutionary explanations, and especially to H-N Es, has been less successful because the proponents of this extension failed to distinguish between the N-D and H-N Es. Clarification in morphological analyses would be improved if KM was restricted to N-D Es, and if H-N analyses of morphology were considered under the heading of evolutionary morphology.

Acknowledgments. I would like to express my sincere thanks to the many friends with whom I have discussed facets of KM, and especially those who have patiently instructed me in some of the subtleties of the German language.

References

Bock WJ (1959) Preadaptation and multiple evolutionary pathways. Evolution 13:194–211
Bock WJ (1967) The use of adaptive characters in avian classification. In: Snow D (ed) Proc 14th Int Ornithol Congr Oxford. Blackwell, Oxford, pp 61–74
Bock WJ (1974) The avian skeletomuscular system. In: Farner DS, King JR (eds) Avian biology, Vol 4. Academic Press, New York London, pp 119–257
Bock WJ (1977) Toward an ecological morphology. Vogelwarte 29:127–135
Bock WJ (1979) The synthetic explanation of macroevolutionary change – a reductionistic approach. Bull Carnegie Mus 13:20–69
Bock WJ (1980) The definition and recognition of biological adaptation. Am Zool 20:217–227
Bock WJ (1981) Functional-adaptive analysis in evolutionary classification. Am Zool 21:5–20

Bock WJ (1986a) Species concepts, speciation, and macroevolution. In: Iwatsuki K, Raven PH, Bock WJ (eds) Modern aspects of species. Univ Press, Tokyo, pp 31–57

Bock WJ (1986b) The arboreal origin of avian flight. Mem Cal Acad Sci 8:57–72

Bock WJ (1988) The nature of explanations in morphology. Am Zool 28:205–215

Bock WJ (1989a) Principles of biological comparisons. Acta Morphol Neerl Scand 26:17–32

Bock WJ (1989b) Organisms as functional machines: a connectivity explanation. Am Zool 29:1119–1132

Bock WJ (1989c) The homology concept: its philosophical foundation and practical methodology. Zool Beitr NF 32:327–353

Bock WJ (1990) From biologische Anatomie to ecomorphology. Neth J Zool 40:254–277

Bock WJ, Caplan AL (in preparation) Explanations in evolutionary biology

Bock WJ, Homberger DG (1988) Symposium: Questions, explanations, models and tests in morphology: the interaction between hypotheses and empirical observations. Am Zool 28:183–288

Bock WJ, Wahlert G von (1963) Two evolutionary theories – a discussion. Br J Philos Sci 14:140–146

Bock WJ, Wahlert G von (1965) Adaptation and the form-function complex. Evolution 19:269–299

Böker H (1935–7) Vergleichende biologische Anatomie der Wirbeltiere. Fischer, Jena

Caplan AL, Bock WJ (1988) Haunt me no longer. Biol Philos 4:443–454; 472–474

Davis DD (1958) The proper goal of comparative anatomy. Proc Cent Bicent Congr Biol, Singapore, pp 44–50

Davis DD (1964) The giant panda. A morphological study of evolutionary mechanisms. Fieldiana Zool Mem 3:1–339

Dullemeijer P (1972) Explanation in morphology. Acta Biotheor 21:260–273

Dullemeijer P (1974) Concepts and approaches in animals morphology. Van Gorcum, Assen

Dullemeijer P (1980) Functional morphology and evolutionary biology. Acta Biotheor 29:151–250

Dullemeijer P (1985) Diversity of functional morphological explanation. Acta Biotheor 34:111–124

Dullemeijer P, Barel CDN (1977) Functional morphology and evolution. In: Hecht MK, Goody P, Hecht B (eds) Major patterns in vertebrate evolution. Plenum, New York, pp 83–117

Dyke C (1988) The evolutionary dynamics of complex systems. A study in biological complexity. Oxford Univ Press, New York

Gans C (1985) Vertebrate morphology: Tale of a phoenix. Am Zool 25:689–694

Gans C (1989) On phylogenetic constraints. Acta Morphol Neerl Scand 27:133–138

Gould SJ (1970) Evolutionary paleontology and the science of form. Earth Sci Rev 6:77–119

Gould SJ, Lewontin RC (1979) The spandrels of San Marco and the Panglossian paradigm: a critique of the adaptationist programme. Proc R Soc London Ser B 205:581–598

Gould SJ, Vrba ES (1982) Exaptation – a missing term in the science of form. Paleobiology 8:4–15

Gutmann WF (1966) Zu Bau und Leistung von Tierkonstruktionen, 4–6. Abh Senckenb Naturforsch Ges 150:1–106

Gutmann WF (1971) Der biomechanische Gehalt der Wurmtheorie. Z Wiss Zool 182:229–262

Gutmann WF (1972) Die Hydroskelett-Theorie. Aufs Reden Senckenb Naturforsch Ges 21:1–91

Gutmann WF (1978) Wie werden Muskeln beim Wachstum verlängert? Nat Mus 108:267–273

Gutmann WF (1987a) Organismus and Konstruktion. I. Das energetisch getriebene System und die Evolution. Nat Mus 117:165–172

Gutmann WF (1987b) Organismus und Konstruktion. II. Wie die Eigenheiten der Lebewesen von Philosophie, Biophysik und Morphologie verpasst werden. Nat Mus 117:288–298

Gutmann WF (1988) The hydraulic principle. Am Zool 28:257–266

Gutmann WF (1989) Die Evolution hydraulischer Konstruktion. Organismische Wandlung statt altdarwinistischer Anpassung. Kramer, Frankfurt/M

Gutmann WG, Bonik K (1981) Kritische Evolutionstheorie. Gerstenberg, Hildesheim

Hesse R, Doflein F (1910–14) Tierbau und Tierleben in ihrem Zusammenhang betrachtet, 2 vols. Teubner, Leipzig Berlin

Homberger DG (1986) The lingual apparatus of the African grey parrot, *Psittacus erithacus* Linné (Aves: Psittacidae): description and theoretical mechanical analysis. Ornithol Monogr 39:xi+233

Homberger DG (1988) Models and tests in functional morphology: The significance of description and integration. Am Zool 28:217–229

Kraemer L, Bock WJ (1989) Symposium: Is the organism necessary. Am Zool 29:1055–1197

Lenior T (1982) The strategy of life. Reidel, Dordrecht (reprinted: 1989, Univ Press, Chicago)

Liem KF, Wake DB (1985) Morphology: current approaches and concepts. In: Hilderbrand M, Bramble DM, Liem KF, Wake DB (eds) Functional vertebrate morphology. Harvard Univ Press, Cambridge, pp 366–377

Mayr E (1961) Cause and effect in biology. Science 134:1501–1506

Mayr E (1976) Evolution and the diversity of life. Harvard Univ Press, Cambridge

Mayr E (1982) The growth of biological thought. Diversity, evolution and inheritance. Harvard Univ Press, Cambridge

Mayr E (1985) Weismann and evolution. J Hist Biol 18:295–329 (reprinted: Mayr E (1988) Toward a new philosophy of biology. Harvard Univ Press, Cambridge)

Nagel E (1961) The structure of science. Harcourt, Brace, New York

Osse JWM, Barel CDN, Dubbeldam JL (eds) (1989) Explanations in functional morphology. Acta Morphol Neerl Scand 27:1–179

Pranger R (1990) Towards a pluralistic concept of function. Functional statements in biology. Acta Biotheor 38:63–71

Reif WE (1983) Functional morphology and evolutionary ecology. Paläontologie 57:255–266

Reif WE, Thomas RDK, Fischer MS (1985) Constructional morphology: The analysis of constraints in evolution. Acta Biotheor 34:233–243

Seilacher A (1970) Arbeitskonzept zur Konstruktions-Morphologie. Lethaia 3:393–396

Seilacher A (1972) Divaricate patterns in polecypod shells. Lethaia 5:325–343

Seilacher A (1979) Constructional morphology of sand dollars. Paleobiology 5:191–221

Seilacher A (1989) Vendozoa: organismic construction in the Proterozoic biosphere. Lethaia 22:229–239

Stümpke H (1961) Bau und Leben der Rhinogradentia. Fischer, Stuttgart (reprinted: 1967, The snouters. Form and life of the rhinograds. Univ Chicago Press)

Szalay FS, Bock WJ (1991) Evolutionary theory and systematics: Connections and differences between neontology and paleontology. Syst Evol Forsch (in press)

Thomas RDK (1979) Constructional morphology. In: Fairbridge RW, Jablonski D (eds) The encyclopedia of paleontology. Dowden, Hutchinson & Ross, London

Vogel K (1989) Constructional morphology and the reconstruction of phylogeny. Abb Naturwiss Ver Hamburg 28:255–264

Wake DB (1982) Functional and evolutionary morphology. Persp Biol Med 25:603–620

Wake DB, Roth G (eds) (1989) Complex organismal functions: integration and evolution in vertebrates. Dahlem Workshop Reports, Life Science Research Reports 45. John Wiley & Sons, Chichester New York

Weber H (1955) Stellung und Aufgaben der Morphologie in der Zoologie der Gegenwart. Zool Anz, Suppl 18:137–159

Weber H (1958) Konstruktionsmorphologie. Zool Jahrb Abt Allg Zool Phys Tiere 68:1–112

Functional Aspects of Morphology

WERNER NACHTIGALL *

Summary

The chapter deals with the terms morphology, functional morphology and constructional morphology. Their meaning is discussed in the light of older and newer statements expressed by experts from these fields of research. Especially the methods, aims, and significance of constructional morphology are formulated, including hydroskeletal aspects.

1 Introduction

The complex which will be discussed here includes descriptive morphology, functional morphology, and – mainly – constructional morphology. What are the specific characteristics of these fields of research? I shall try to formulate them in general terms and support the statements with citations.

2 Descriptive Morphology

The first task of a morphologist is to describe structures.[1] The ability to recognize a structure within the object under observation depends largely upon the method used. Structures are assorted according to aspects of similarity and their positions within a system.[2] This may be carried out at various levels, at subcellular and cellular level or that of tissues, organs, or the organism as a whole. Functional criteria have no special, or at least no decisive role to play.[3] They are, how-

---

* Zoologisches Institut der Universität Saarbrücken, Universität Bau 6, D-6600 Saarbrücken, FRG

[1] "Die Morphologie oder die Formenlehre... hat zunächst alles zu beschreiben, was äußerlich erkannt werden kann, wie Größe, Farbe, Proportion der Teile..." (Hertwig 1919).

[2] "Gleichermaßen wichtig ist es jedoch, die Bauprinzipien der Organismen zu erfassen, die Beziehung der einzelnen Teile zur Gesamtorganisation des Individuums zu studieren, die wechselseitige Abhängigkeit der Organe... zu analysieren.... Dies alles ist Aufgabe der modernen Vergleichenden Anatomie" (Frick 1965a).

[3] "The historical aspects of tracing a system to a similar one that occurred earlier may even be done on a prescientific level of historical narratives" (Mayr 1982).

N. Schmidt-Kittler and K. Vogel (Eds):
Constructional Morphology and Evolution

ever, never completely excluded, even though functional observations usually occur spontaneously and "according to common sense".[4]

The desire to obtain a complete recording of all observable structures at as many levels as possible has been more or less achieved in the field of descriptive morphology. This is apparent in the case of certain species such as human beings and domestic animals, which have been closely studied, and in the comparative observation of forms and their particular relationships (comparative anatomy, comparative morphology).[5] Morphology incorporates specific methods of research, its own laws of deduction, and its own results, for example, aspects of homology.[6] With its criteria of similarity, it was largely responsible for providing a foundation for fundamental biological questions – the retracing of evolutionary or ontogenetic development.[7] For a functional view of the organism morphology provides the necessary basis. For a complete understanding of evolution, morphology is necessary, but not sufficient, and therefore not adequate as a sole method.[8]

3 Functional Morphology

The functional morphologist tries to understand a structure from its functional aspects, or to describe and explain the functional correlation between individual structural elements.[9] To this end he will use a physical-technical approach rather than a physical-chemical or other approach, i.e., he will use relevant knowledge from physics, engineering, and other scientific disciplines.[10] The combination of

[4] "...neue zweckmäßige Formen entstehen können und wieder durch Verbesserungen der Konstruktionen herbeigeführt werden, ähnlich wie die anfänglich primitiven Konstruktionen unserer Maschinen durch den Gebrauchszweck und die Konkurrenz immer mehr vervollkommnet und hoch entwickelt werden" (Wurmbach 1970).

[5] "Anatomy is the science of the observation and description of structure. ... Morphology is the science of the interpretation of observed structure" (Hildebrand 1974).

[6] "The Reichert-Gaupp theory means, 'daß das (primäre) Kiefergelenk der Reptilien dem Hammer-Amboß-Gelenk im Mittelohr der Säugetiere homolog ist und die Säugetiere ein neues, ein sekundäres Kiefergelenk erworben haben'" (Frick 1965b). After Gaupp 1912.

[7] "Ontogenetic explanation tries to trace a certain ontogenetic state from a earlier one, using a system of definitions and typological boundary conditions" (Russell 1916).

[8] "Morphologie als Form- und Gestalt-Beschreibung ist beim Fehlen von Konstruktions- und Funktions-Einsichten ein wichtiges, legitimes und sehr schwieriges Geschäft, das zur Ordnung von Organismen sehr wichtig ist. Es reicht zur Erstellung phylogenetischer Zusammenhänge nicht aus" (Gutmann 1972).

[9] "It is sometimes differentiated between functional morphology I (question: 'What is an organ good for?') and functional morphology II (question: 'How does an organ function?'). Functional morphology II is taken to be synonymous to constructional morphology ('Konstruktionsmorphologie'). Functional morphology I in this sense means a comparison between morphological structures and the most important biological functions. Possibilities of functional explanations are better when more physiological and ecological aspects are taken into account" (Reif 1985a).

[10] Analyzing the transformation in the copulatory organs from the Araneus- to the Larinaria-type it is stated: "Die Form der Einzelteile allein vermag keinerlei Hinweise darauf zu liefern, welcher der Typen der plesiomorphe und welcher der apomorphe ist. Betrachten wir die genannten 'Merkmale' jedoch unter technischen Aspekten, so müssen wir feststellen, daß ihre Unterschiede das Gesamt-

these two aspects occurred very early in the course of research.[11] Since then it has proved to be very useful in innumerable research projects.

It is often a simple task to recognize the main function of a structure, i.e., to explain[12] this structure. This provides a guideline which enables one to choose the appropriate descriptive and explanatory vocabulary which leads to an explanation. To explain something means to recognize and classify it as something exceptional within a general complex, thus bringing the formulation of a question onto a higher, i.e., more general level.[13]

This approach is already enclosed[14,15] in the words functional morphology. Although the meaning of the term 'function' is not at all unequivocal it is commonly used as the counterpart to the term structure. Separated from its function, a structure may be classified as a basis for typological descriptions[16], but since the two build a unit and complement each other like an image and reflection, the essence of a structure can only be understood in unison with its function or in the light of its adaptative value.[17] Only then can it be incorporated into other structure-function-correlations at a higher level of understanding, leaving prescientific aspects of finality and reaching causal interpretations.[18,19]

---

gefüge so stark verändern, daß die beiden verglichenen Kopulationsapparate in mehreren Teilbereichen unterschiedlich funktionieren müssen. Es kann deshalb nur versucht werden, durch einen Vergleich der Mechanik der beiden Typen einen Weg zu finden, wie der eine Typ in den anderen transformiert werden kann" (Grasshoff 1973).

[11] One of the first formulation of modern science was the defining of criteria of similarity for animals similar in shape, but differing in size (Galileo Galilei; translation von Oettingen, 1890).

[12] "Ist es gelungen, einen Vorgang auf eine lückenlose Kausalkette zurückzuführen, so wird er als 'erklärt' bezeichnet. Auch der Biologe muß versuchen, ... Kausalanalyse zu betreiben" (Wurmbach 1970).

[13] "Erklären bedeutet das Phänomen in ein kausales System von Ursachen und Wirkungen einordnen, so daß einerseits seine Ursachen bekannt sind, andererseits seine Wirkungen auf andere, benachbarte Phänomene durchschaubar werden" (Nachtigall 1972).

[14] "Die Nominalform zu 'Fungieren' wie zu 'Funktionieren' ist 'Funktion'. Diese Gleichheit ist Ursache vieler Irrtümer. ... Deshalb ziehe ich den Vorschlag vor, bei Betrachtungen dieser Art von 'Konstruktionen' zu sprechen" (Mollenhauer 1970).

[15] The ambiguity of the term 'function' was demonstrated by using a water tap as example ('Zapfventil'): "Unsere etymologischen Betrachtungen legen nahe, das Wasserspenden als die *Funktion* des Ventils anzusehen. ... Die Einrichtung des Ventils für diese besonderen Aufgaben ist die *Konstruktion*, auf Grund derer es funktioniert. Diesen Unterschied zwischen dem 'Funktionieren' und dem 'Fungieren' wollten auch Bock und von Wahlert 1965 herausstellen – wenn auch nicht in letzter Konsequenz" (Mollenhauer 1970).

[16] "... zeigt sich, daß Morphologie als Studium der biologischen Form an sich ein Phantom ist. Ein solches Forschungsprogramm kann immer nur zur Typologie und zur metaphysischen Konzeption von Archetypen führen" (Reif 1982).

[17] "Structures are often described as adaptations. For instance, the very strong forelimbs of moles may be described as adaptation for burrowing. This means that...ancestral moles with stronger forelimbs were better able to burrow and were consequently favored by natural selection" (Alexander 1975).

[18] "Eine Untersuchung auf Funktionen hin ist Finalitätsbetrachtung. Die gestellte Frage lautet: Wofür? Wozu? .... Einsicht in den Ablauf einer Tätigkeit oder das Spiel von Kräften gewinnt man durch Kausalbetrachtungen. Die gestellte Frage lautet: Warum? ... Der Sinn, die Rolle eines Gliedes eines Organismus für dessen Leben kann sich in ganz verschiedenen Forschungsbereichen aufspüren lassen. Und daher hat der Funktionsbegriff so vielerlei Aspekte.... Das erschwert seine begriffliche Fassung so sehr" (Mollenhauer 1970).

[19] Basic aspects of interpretation in science are discussed (Nachtigall 1972).

The functional backing up of a structure is, of course, prone to possible misinterpretation, especially in the case where a correlation cannot be proved experimentally. The danger of this happening is particularly great in the functional interpretation of fossil structures and in the hypothetical reconstruction of evolutionary changes. Nevertheless, functional morphology is indispensable if we are to attain as complete an understanding of evolution as possible.

From the very beginning, the functional way of thinking and experimental results of physical sciences were incorporated in the scientific research of biological phenomena[11,20] and have been present throughout the history of science. As late as the 1960's, however, their significance was still drastically underestimated. They, too, are important even though they are not sufficient.

4 Constructional Morphology

Whereas the attention of a functional morphologist is drawn more to the function of individual structures and elements, that of the constructional morphologist is focused upon the functional analyses of morphological constructions. How are such constructions defined? When the term constructional morphology (Konstruktionsmorphologie) was introduced[21], and when it was used thereafter[22] and partly redefined[23], the definitions were never very precise, but pointed towards a technical construction. Let us now consider the consequences of assuming that constructional morphology is equivalent to a technical construction.

A construction, whether technical or biological, functions as a whole.[24] Its various elements are interlinked and entwined in such a way that one cannot understand a system as a construction, i.e., in the above-mentioned sense, if one were to omit just one single essential parameter.

Accepting this point of view, one is faced with the difficulty that, strictly speaking, a biological system can no longer be freely defined: at the start of a new line of research, one is never sure whether the elements lying outside a definable area will not finally prove to influence it (perhaps even essentially). This leaves

---

[20] The first analysis of animal locomotion according to modern definition of physiology of movement was made by Borelli (1685).

[21] "Morphologie geht aus von dem reinen Bau und seinem Werden, berücksichtigt dabei die Dynamik des Werdens, dehnt sich aus auf die Konstruktion samt deren Dynamik, auf das Werden der Konstruktion und die Dynamik dieses Werdens in der Ontogenese, im Generationengang und in der Evolution" (Weber 1950/51 after Reif 1985b).

[22] It was stated that the term Konstruktionsmorphologie is used in two senses: firstly that of the analysis of a functionally integrated system of organs and the onto- and phylogenetic morphogenesis of such a system, secondly in the analysis of nonhistorical ("bautechnische") limitations of organic forms due to geometrical laws of morphogenesis and regulation. The first aspect, going back to Weber (1958), was, for example, followed by Gutmann and Peters (1973) and Bonik and Gutmann (1978) the second aspect, for example, by Seilacher (1970); Reif (1975), and Thomas (1979) (Reif 1985b).

[23] In paleontology the term Konstruktionsmorphologie was used in a broader sense to combine constructional morphology according to Weber and functional analysis according to Reif (Funktionsmorphologie I) with morphogenetic analyses (Seilacher 1970).

[24] "Konstruktion ist die Einheit aus Bau und Leistung samt ihrer Dynamik, ihrem Werden und der Dynamik ihres Werdens" (Weber 1950/51).

us with only one sensible course of action, that of studying the whole organism as a construction; all separable groups of elements are therefore only subsystems of the system construction.

The laboratory biologist who wishes to understand a living system which is available for experimental research will have little difficulty, if he makes the right choice of subsystems, in recognizing its principal function and in choosing his subsequent functional morphological action accordingly. However, should he try to study the organism as it is, as a whole (i.e., as a biological construction), it will be almost impossible to find a guideline[25] since the latter will now have an aspect of indefinite complexity. In addition, the organism will appear in a great variety of time-space states during its life time. Thus, to define constructional morphology as the analyzation of a complicated biological construction in order to understand it completely (as an engineer is able to understand the complex construction of a car) is impracticable (and according to some scientists, even unnecessary[26]). In a strict sense, it would make scientific work impossible because biological constructions or biological machines are usually much more complicated than technical ones. This is also true for the hydraulic view of organisms: Organisms can be best treated as hydraulic machines[27]. It applies to the larvae and adults of lower invertebrates and at least to the earlier ontogenetic stages of all living systems.

The machine aspect of an organism[28] is indeed important for a detailed as well as for an holistic approach[29]; neither should the energetic aspect[30] be neglected. Sometimes the only appropriate way[31] to describe a biological system or subsystem is to treat it as a machine or as an apparatus[32,33], although there is a certain danger of sailing into pure physics and analyzing a biological construction

---

[25] "Dem technischen Denken fiele es leicht, die Organe eines Wirbeltieres im Inneren unserer Leibeshöhle anders anzuordnen, als wir sie in Wirklichkeit vorfinden, ohne daß sie deshalb schlechter funktionieren müßten" (Portmann 1959).

[26] "So hat Nachtigall (1978) völlig recht, daß es schwer ist, die Ökonomisierung komplizierterer Systeme zu erfassen. Er übersieht aber, daß diese Schwierigkeit überall besteht und die Erfassung der Komplexheit (wenn sie möglich wäre) die Frage nach der Ökonomie gar nicht mehr erforderlich machen würde" (Bonik und Gutmann 1978).

[27] "Mittels der Hydraulik-Theorie wird im komplexen Feld der Biologie eine klare morphologische Erklärungsebene definiert... ." (Gutmann und Bonik 1980).

[28] Gutmann states that organisms function as energy converting and energy transducing machine-like systems (Gutmann, this Vol.).

[29] "Hydraulik ist nicht ein addierender Verband, sondern ein geschlossenes Gesamtsystem. Alles wirkt mit allem zusammen. Das kann man nicht sehen oder unmittelbar erfassen. Es ist nur in einem geschlossenen mechanischen Erklärungsansatz verständlich zu machen... ." (Gutmann und Bonik 1980).

[30] Dullemeijer, introducing energetic aspects in constructional mechanics, asks how the various constructions cost in terms of amount of material in production, i.e., how much the animal has to pay for it in energy (Dullemeijer, this Vol.).

[31] Physical formulations which point in the right direction, but which are by far not sufficient for a satisfactory explanation have been made for example by Bonik (1978).

[32] A part of the wing mechanism of calliphorid flies can be treated as a gear-box construction (Wisser and Nachtigall 1990).

[33] To define hierarchy of complexity it has been suggested to distinguish between appartuses and machines. Schmidt-Kittler states that machines are simply more complex variants of apparatuses normally working on the basis of more than one principle (Schmidt-Kittler, this Vol.).

as a technical system per se[34] (which, by the way, does not matter as long as this point of view is not overestimated).

There is another aspect that makes biological constructions so unique. From the very beginning of evolution, organisms develop within the organic and physical (mechanical) boundaries[35], possibilities, and restraints of living matter that organizes itself in certain environments. Thus every organism is the result of its ontogeny on the one side and the whole phylogenetic process on the other side. Therefore organic constructions can be analyzed in two principally different ways. One may regard a fly sitting on a table as an existing construction and take for granted that it is as it is. That is what a biomechanic usually does (without necessarily integrating himself in the class of functional or constructional morphologists[36]). On the other hand, one may regard this fly as a construction which has its personal ontogenetic history[7] as well as its phylogenetic history as a species. That is what the functional evolutionist does (combining models and methods from paleontology and considering a wide field of aspects[37,38] up to evolutionary constraints on the genetic basis of morphogenesis[39]).

The first considerations always form a foundation on which the second ones are based, whilst intrinsic and extrinsic factors complement each other.[40] The biomechanic does not deny that ontogenetic and evolutionary aspects are of crucial importance for the true nature of the constructions he analyses,[41] but due to the practical difficulties coupled to the analysis of biological constructions he rarely reaches the desirable goal of including evolutionary aspects.

One can escape the dilemma of overwhelming complexity in biological constructions with the help of a practical, but, strictly speaking, methodically incorrect, restriction. One chooses the most complicated functional complex which involves as many subsystems as possible (or ideally all), and treats it as the principal

---

[34] "Regarding the physical adaptations of fish for swimming it has been stated: 'Weil die physikalisch ausgerichtete Hydrodynamik ... biologische Systeme wie eine Art denaturierter Flugzeuge oder Unterseeboote versteht ..., nimmt sie ihre wichtigen allgemeinen Einsichten nicht ernst, die biologisch unglaublich weitreichende, wenn auch simple Erklärungen liefern; da sie nicht auf die Vielfalt biologischer Anforderung achtet, bemerkt sie nicht, daß die Beachtung der Reibungsverminderung ... einseitig und damit unbiologisch ... ist'" (Bonik und Gutmann 1978).

[35] The aspect that physical boundary conditions are of special evolutionary importance has been summarized by Nachtigall (1978).

[36] Vogel states that obviously one cannot prove well-defined boundaries between constructional and functional morphology and biomechanics, especially since each field has a long tradition. The use of one of these terms only indicates the point of each case (Vogel, this Vol.).

[37] Being focused on functional as well as morphogenetic (and architectural) constraints, Seilacher states that constructional morphology forces a dynamic view of biological form (Seilacher, this Vol.).

[38] Vogel states that constructional morphology deals with shape generation, shape control, and mechanical coherence of the entire biological body and that this also includes its ontogeny and evolution. Neither functional morphology nor biomechanics encompass such extensic concepts (Vogel, this Vol.).

[39] Tautz states that it was possible to define a hierarchy of genetic interactions which determine the major outlines of the body pattern (Tautz, this Vol.).

[40] Liem states that a simultaneous analysis of both the intrinsic and extrinsic determinals seems to be the best approach to expense the emergations of variation in design (Liem, this Vol.).

[41] Duncker states that only by proceeding to a complementary comparative and experimental analysis of the integrative systemysations in the evolutionary development of organisms we can advance to a modern constructional morphology as an integrative part of evolutionary biology (Duncker, this Vol.).

function.[35] At the same time, one assumes that the constructive essence of an organism is optimally adapted to this principal function. This then enables one to describe – at least tentatively – the constructional morphology of the organism. It has been stated that looking at the "apparently most important" aspect of a construction does not make too much sense.[42] For practical reasons, however, this may be the only possible approach and the start of a causal chain of analysis. Furthermore, many biological constructions, systems, or overriding complexes have been analyzed to a satisfactory degree by using this method.

Accordingly, the process of movement may be considered as an overriding complex, a view which was taken very early in scientific research[20] and which lead to principal insights.[43] In this case, it is reasonable to assume that the evolutionary links of an organism are also valid for movement constructions. Therefore a constructional morphologist working on the theory of evolution is in a position to determine how something new developed from something old, by starting with the principles of physics already acquired and by keeping them in mind whilst altering the basic parameters. This is more than just morphologically based phylogeny.[44] The "constructive predicament" or the type of "logic" arising from the construction-dependent course of physics will act as a guideline. This has proved to be an heuristic principle and has pulled the physical-technical approach out of its slumber.

On the other hand, this approach has demanded evidential value.[45,46] It is often based on assumptions[47] and models[48] and either denies the principal importance of experimental verifiability[49] or uses insufficient calculations. Experimental verifications of the basic aspects of the hydroskeleton theory are still lacking;

---

[42] "Nachtigall schlägt vor, das 'mutmaßlich bedeutungsvollste' Element als entscheidend herauszunehmen und zu behandeln. Was aber ist einem System das wichtigste? .... Faktisch bleibt nichts anderes übrig, als die Optimierung und Ökonomisierung von Teilsystemen zu erfassen und zu zeigen, daß sie zur Optimierung des Gesamtsystems beigetragen haben müssen" (Bonik und Gutmann 1978).

[43] "Beispielsweise hat er (Borelli) die Newton'sche Formulierung 'Actio et reactio sunt equales' (Erstes Newton'sches Gesetz) vorweggenommen durch die am schwingenden Vogelflügel gewonnene Erkenntnis, daß ... die vom Flügel auf die Luft ausgeübte Kraft gleich ist der entgegengesetzten Widerstandskraft" (Nachtigall 1987).

[44] "Dennoch ist ... darauf zu bestehen, daß die morphologisch begründete Phylogenetik teleologisch und idealistisch ist. Die Distanzierung von der morphologisch verbrämten Phylogenetik ... kann nicht deutlich genug ausfallen, weil auf morphologischer Grundlage phylogenetische Rückschlüsse überhaupt nicht zu sichern sind" (Gutmann 1972).

[45] "Meine Modelle enthalten in den Darstellungen der heute vorliegenden bekannten Organismen die *faktischen* Konstruktionszusammenhänge..." (Gutmann 1972).

[46] "Die Vorfahren der Wirbeltiere müssen ... gestreckte Wurmkonstruktionen, hydraulische Apparate gewesen sein. Dieser Rückschluß beruht auf konstruktiven Zwängen des Bewegungsapparats ..." (Bonik und Gutmann 1978).

[47] "In unserem gemeinsamen Grundmodell zur Stammesgeschichte der tierischen Konstruktionen (Bonik, Grasshoff und Gutmann 1976a, b) haben wir gezeigt, daß die Vorfahren der heutigen Metazoen nur gallertgestützte Vielzeller mit gitterartiger Muskelverspannung gewesen sein konnten" (Bonik 1978).

[48] This is demonstrated in detail in the article *Modellvorstellungen als Hauptelement phylogenetischer Methodik* by Peters und Gutmann (1973).

[49] Duncker states that for the elaboration of the functional nature of the structural basis of an investigated function a complementary interaction of comparative investigation and causalanalytic experimentation is essential (Duncker, this Vol.).

a quasi-experimental approach[50] does not seem to be sufficient. A controversial discussion has developed around this approach; its comprehensive claim has not been accepted[51], although admittedly it is a particularly relevant and powerful tool.[52] In order to understand evolution as completely as possible it is necessary, although not sufficient.[53]

There is not a single competent theory or set of theories, and thus, basically, only insufficient criteria are available today according to which a scientist may work when dealing with one of the major basic problems in biology, i.e., the evolution of higher-order constructions from lower-order constructions. Strictly speaking, it is impossible to deal with the question in this form. The fact that scientists, despite everything, still occupy themselves with this problem is an asset shared by other basic phenomena in the field of natural science, e.g., cosmological problems. They incorporate the same principal difficulties. It takes a good deal of scientific ingenuity to overcome these, i.e., to analyze and measure what is analysable and measurable[54] and to fill the gaps with theoretical considerations in such a way that the whole approach does not appear too improbable.

5 Morphology: Subdivisions

What about the relations and subdivisions mentioned? It has been proposed that one should refrain from using terms like functional and constructional and either to ascribe everything to the general cover "morphology" or to ascribe the functional and dynamic aspects[37] to "morphodynamics". It is my opinion, however, that subdivision is necessary, as is the case in all fields of science and furthermore, that the latter term is equivocal (aspects of morphogenesis of systems[55] and functional aspects of existing systems) and therefore could well lead to a continual misunderstanding. The term engineering morphology, on the other hand, simply characterizes a methodical (mostly biomechanical) approach. I therefore propose that we continue to use the three terms discussed in this chapter according to our current understanding as characterized in the following.

---

[50] Essential aspects of this approach have been proposed by Gutmann (this Vol.)

[51] Standing out against the emotional discussions of the 1970's, a somewhat more convincing review was given by Osche (1978) *Das Homologisieren als eine grundlegende Methode der Phylogenetik.*

[52] Dullemeijer states that this line is a major one, but this does not imply that other constructions cannot therefore arise without hydraulics (Dullemeijer, this Vol.).

[53] Dullemeijer states that this hydraulic system is ontogenetically and phylogenetically a basic condition but it does not account for the enormous variety of animal forms and constructions (Dullemeijer, this Vol.)

[54] Duncker states that this analysis should always start with the typological analysis of structures and/or functions which constituted the basic structural and functional systems. In the next step a functional analysis should attempt to reveal the functional meaning of structural components. In the third and final step a system analysis must be undertaken. To a certain extent these three steps of analysis describe in their sequence the historical development of morphological research (Duncker, this Vol.).

[55] To demonstrate the importance of morphogenetic constraints – in addition to the phylogenetic and functional ones – as determinants of biological form a triangular schema has been proposed combining the three aspects of "biological function", "morphogenetic fabrication", and "phylogenetic tradition" under the roof of constructional morphology (Seilacher, this Vol.).

"Descriptive morphology" or just "morphology". Listing and/or classifying descriptions of morphologically identifiable parts of subsystems in a system following time-independent lines of geometrical classification and/or leading to time-dependent lines of ontogenetic or phylogenetic complications.

"Functional morphology". Establishing connections between structures and their functions at any level of complexity, from a single-structure to single-function correlation to those degrees of complexity that are still analyzable.

"Constructional morphology". Analyzing a biological substrate as a construction, i.e., as a system in which very different subunits are integrated to keep it working. The system must be investigated analytically in order to understand it as a complex of diverse structure-function interactions. (Retracing evolutionary lines including hydraulic aspects is certainly a typical construction-morphological approach).

As indicated, one cannot clearly differentiate between "functional" and "constructional" morphology. Nontheless, the latter requires a more integrated approach to a morphological system or complex, and is certainly the more difficult. For detailed information see Dullemeijer (p. 311).

Acknowledgments. I wish to thank Mrs. Winifred Pattullo for revising my English version, and Mrs. Ingrid Schwarz and Mrs. Irmtraud Stein for typing the manuscript.

References

Alexander R McNeill (1975) The chordates. Cambridge Univ Press

Bonik K (1978) Quantitative Betrachtungen zur Gallertoid-Hydroskelett-Theorie. Evoluierende Systeme III. Kramer, Frankfurt/M, pp 22–38

Bonik K, Gutmann WF (1978) Die wichtigsten Beziehungen zwischen Fischgestalt und Hydrodynamik. Evoluierende Systeme III. Kramer, Frankfurt/M, pp 58–86

Borelli JA (1685) De motu animalium. In: Glugd Batav (ed) Angeli Barnabi, 2. (Translation: Mengeringhausen 1927, Ostwald's Klassiker). Engelmann, Leipzig

Dullemeijer P (1974) Concepts and approaches in animal morphology. Van Gorcum, Assen

Frick H (1965a) Probleme und Ergebnisse der Vergleichenden Anatomie heute. Naturwiss Rundsch 18 (6):227–237

Frick H (1965b) after Gaupp E (1912) Die Reichert'sche Theorie (Hammer-, Amboß- und Kieferfrage). Arch Anat Entwicklungsgesch Suppl 1912

Grasshoff M (1973) Konstruktions- und Funktionsanalyse an Kopulationsorganen einiger Radnetzspinnen. Aufs Reden Senckenb Naturforsch Ges 24:129–151

Gutmann WF (1972) Die Hydroskelett-Theorie. Kramer, Frankfurt/M

Gutmann WF, Bonik K (1980) Borelli und die Folgen – kann man Mechanik in Lebewesen sehen? Nat Mus 110 (9): 263–274

Hertwig R (1919) Lehrbuch der Zoologie, 12th edn. Fischer, Jena

Hildebrand M (1974) Analysis of vertebrate structure. John Wiley & Sons, New York

Mayr E (1982) The cross of biological thought. Belknap Press, Cambridge, Mass

Mollenhauer T (1970) Betrachtungen über Bau und Leistungen der Organismen, Teil I. Kramer, Frankfurt/M

Nachtigall W (1972) Biologische Forschung. Aspekte, Argumente, Aussagen. Quelle & Meyer, Heidelberg

Nachtigall W (1978) Evolutionsaspekte und physikalische Grenzwerte. Zool Jahrb Anat 99:109–114

Nachtigall W (1987) Vogelflug und Vogelzug. Rasch & Röhring, Hamburg Zürich

Oettingen A von (1890) Unterredungen und mathematische Demonstration über zwei neue Wissenszweige, die Mechanik und die Fallgesetze betreffend. Ostwald's Klassiker der exakten Wissenschaft, 11. Engelmann, Leipzig

Osche G (1978) Das Homologisieren als eine grundlegende Methode der Phylogenetik. Aufs Reden Senckenb Naturforsch Ges 24:155–175

Peters DS, Gutmann WF (1973) Modellvorstellungen als Hauptelement phylogenetischer Methodik. Aufs Reden Senckenb Naturforsch Ges 24:26–38

Portmann A (1959) Einführung in die Vergleichende Morphologie der Wirbeltiere. Schwabe, Basel Stuttgart

Reif WE (1982) Functional morphology on the procrustean bed of the neutralism debate. Notes on the constructional morphology approach. N Jahrb Geol Palaeontol Abh 164:46–59

Reif WE (1985a) Konzepte und Geschichte der Funktionsmorphologie. Aufs Reden Senckenb Naturforsch Ges 35:107–142

Reif WE (1985b) Organismus und Selektion – Probleme der Evolutionsbiologie. Aufs Reden Senckenb Naturforsch Ges 35:107–142

Russell ES (1916) Form and function. A contribution to the history of animal morphology. Murray, London

Seilacher A (1970) Arbeitskonzepte zur Konstruktionsmorphologie. Lethaia 3:393–395

Weber H (1950/51) Skriptum zur Vorlesung "Einführung in die Terminologie und Methode der Konstruktionsmorphologie", winter term 1950/51. Univ Tübingen, FRG

Wurmbach H (1970) Lehrbuch der Zoologie, vol 1. Fischer, Stuttgart

An Outline of Morpho-Constructional Analysis

NORBERT SCHMIDT-KITTLER [1]

Summary

A concept of morpho-constructional analysis is outlined which describes functions of organic features in terms of transformation of energy input into effect output on the strength of physical principles. The dependence of effect outputs on constructional parameters, such as geometrical configurations and material properties, is illustrated in a vector model. It shows that performances of organisms, on principle, can be provided in equivalent manner by different constructional solutions. Options exist in regards to (1) how constructional parameters can be provided by an organism, (2) how particular effects or performances can be produced, and (3) how organizational units can be constrained by other features or properties of the organism. Together with the fact that selective forces are only placed on effect outputs or performances, this explains why evolving organisms, on principle, dispose of considerable degrees of freedom in developing constructional solutions to environmental demands. On the basis of the model it can further be demonstrated that optimality and economy of organic features have a fundamentally distinct meaning.

1 Introduction

Constructional morphology in its research strategy conforms to the general methodology of the reductionistic approach of biosciences which can be characterized as analytical and descriptive. Observation and analysis of organic structures and processes form the first step. Description of the analyzed phenomena as particular organic materializations of physical, chemical, and geometrical laws forms the second step. Once the basic hypothesis of reducibility of biological phenomena to inorganic laws is accepted, there are no other hypotheses needed within the approach.

In the analysis a distinction is made between laws and their exemplifications: while the laws express necessities concerning the relationships between theoretical magnitudes, the exemplifications of the laws become realized on the strength of contingent values which the magnitudes take on in the individual cases considered. In an artificial apparatus the laws due to appropriate choice of constructional parameters are employed in a way leading to the desired effect. The particular causal plexus responsible for this effect is called the principle of action of that

[1] Institut für Geowissenschaften, Paläontologie, Universität Mainz, Saarstr. 21, D-6500 Mainz, FRG

N. Schmidt-Kittler and K. Vogel (Eds):
Constructional Morphology and Evolution

apparatus. Machines are simply more complex versions of apparatuses normally working on the basis of more than one principle.

Reducing organisms to the language of inorganic laws, parameters, and effects seems to be incompatible with the tremendous diversity of life. In practice, even seemingly simple organic structures and processes prove to be causally interwoven with so many other parts of the living system that precision of statements in one respect is normally acquired at the price of neglect of other aspects of the organic system. In the analytical approach followed here no principal distinction is made between organic and machine examples. The fundamental difference between machines and organisms or organic systems, however, must be seen in the fact that in the latter the parameters determining the effects are, without exception, organically produced and organically maintained. The present approach deals only with the physical aspect of the organism-performed parameters. We must keep in mind that it cannot express the quality these have beyond this approach due to their ontogeny and evolutionary history.

2 Materialization of Principles

Consider the forearm of a tetrapod. It can materialize at least three different principles: the forearm as a whole, this means the ulna and radius, together with the elbow articulation as the pivotal axis, can act as a two-armed lever. The forearm as part of the anterior extremity can form a pillar carrying part of the load of the body; and the ulna together with the radius can work as a torsional system, turning the hand from a supinal to a pronal position and back again. Any of these performances can be analyzed mechanically, which means that the produced effects can be described in terms of the principles employed, the constructional properties of the forearm, and the acting forces. It would be possible to show experimentally that properties and effects are correlated on the strength of the employed principle.

Generally speaking, we can distinguish between (1) properties which are essential to the materialization of the principle, (2) properties which are not necessary to the materialization of the principle but cause specific modifications of the effect, and (3) properties which have no influence on the effect at all, hence are completely contingent in regards to the principle.

In the example of the two-armed lever the ulna as a rigid bony element and its articulation to the upper arm form the constructional elements which are essential to the materialization of the principle. A property of the second category is the shape of the articulation surfaces of the elbow joint determining the degree of freedom of movement of the forearm. In bovids, for instance, the forearm is constrained to movement within a sagittal plane. Properties of the third category are, for instance, the forearm's composition of two bones and the shape of these bones. Both of the latter, however, are essential to the materialization of the torsional principle which allows the forearm to be twisted.

The latter fact demonstrates that features or properties may change their significance depending upon the principle actually employed. In order to give them a precise meaning on the adopted approach, we must express them in a quantifiable manner as parameters involved in the materialization of one or another principle. This is why in the following, instead of features or properties, we speak of parameters.

The forearm as a bioconstruction, strictly speaking, does not materialize the principles but has the potential to materialize them through the action of the forces of muscles which form part of the construction. If the forearm works as a lever or a torsional system, the effect can only be produced through muscular forces. The materialization of the principle is a dynamical process passing through a sequence of different states. Contrary to this, if the forearm is employed as a pillar, muscular forces are needed only to bring the bones into the appropriate position but not for maintaining this position. The principle in this case is materialized by a constructional state under the equilibrium of statical forces. An example for the latter would be the position of rest of a horse.

In the example of the tetrapod forearm, mechanical and statical principles become materialized through properties or parameters which are exclusively provided from the side of the organism. Other examples can be found where principles are realized through combinations of organic and environmental parameters.

A turtle turned on its back is not able to re-establish its normal life position; but the moment it touches a rigid and heavy object with one of its legs, the object is used as an abutment for employing the leg as a composed lever. In this moment the rigidity and unmovability of the environmental object, e.g., a stone, form parameters as essential as the constructional parameters of the leg of the turtle. The muscular forces of the animal through their action establish an interface between environmental and organic parameters.

Another example would be the scissors principle materialized by the jaws of carnivorous turtles, the occlusal surface of the cheek teeth of many hypsodont rodents, or the carnassial teeth of Carnivora. Constructional parameters essential to the realization of the principle are two parallel-oriented edges meeting at a sliding contact due to the action of the jaw muscles. The muscular action of the construction as such does not produce a scissors effect. This can only be performed if food material is brought between the edges. The actual realization of the principle then depends upon the relationships between the sharpness and hardness of the edges and the toughness of the food material. This means that the employment of the principle leading to the effect of cutting food implies an interface between organic parameters and environmental parameters. The horny edges of the jaws of a turtle would not, for instance, be sharp and hard enough to produce the very high punctual stresses required for cutting the tendons of a vertebrate. The scissors principle (in this admittedly hypothetical case) would not operate because of a mismatch of organic and environmental parameters.

Consider the eye constructions of arthropods and vertebrates: although employing optical principles in different ways, their interfaces with the environment are established through the same parameters, which are the refractive index and the shape of the lenses.

The above discussed examples show that bioconstructions can realize quite different categories of principles. For a complete enumeration we should mention principles conforming to statical, mechanical, hydraulical, and optical laws respectively. With regards to the parameters involved, it is meaningful to distinguish between internal and peripheral materializations of principles. While in the former exclusively organic parameters can be found, the second type represents combinations of organic and environmental parameters.

3 Operational Units

As already shown earlier, the significance of parameters can only be specified in connection with the principles they are involved in, so that the latter form the basic elements of the analysis. This means that any constructional or operational complex can be described in terms of materializations of principles, which define its basic units. Such a unit would be composed of (1) the principle materialized (to be characterized through the indication of the inorganic law involved), (2) the energy input required for its materialization, (3) an enumeration of the parameters involved, and (4) the state or effect produced through the materialization of the principle. A general model is given in Fig. 1. For its application in practical analysis see Figs. 2 and 3.

The constructive elements on which the application of a principle is based need not appear in explicit form in the model. They are part of the characterization of the principle. Nor is it necessary to express in the model that, of course, energy is required for ontogeny and maintainance of any biological macrostructure. The actual energy input affecting the construction can appear in different forms, such as mechanical forces, statical forces, vibrations transmitted from a medium, or light falling in. The applicability of the model in the case of statical forces perhaps also needs some further explanation.

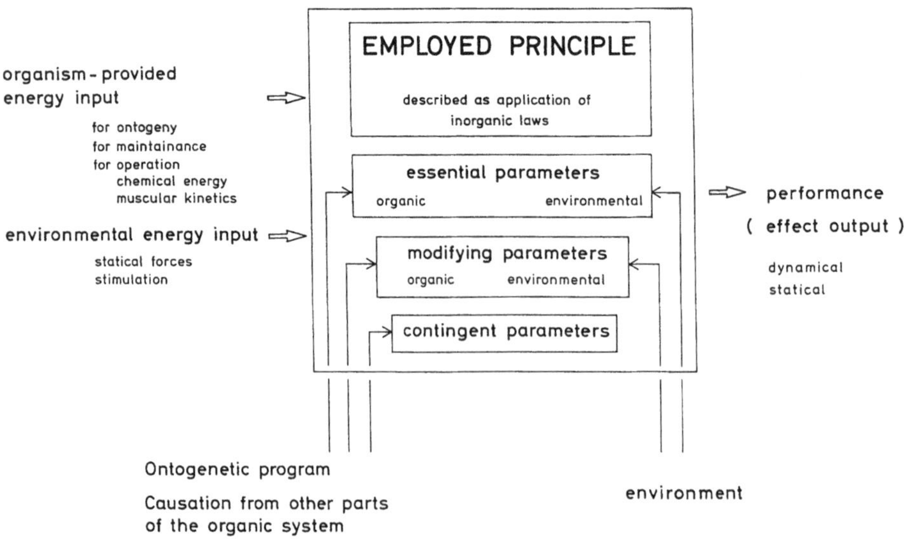

Fig. 1. Operations of bioconstructions can be described as transformations of energy input into effect output by virtue of morpho-constructional parameters in conformation with physical principles

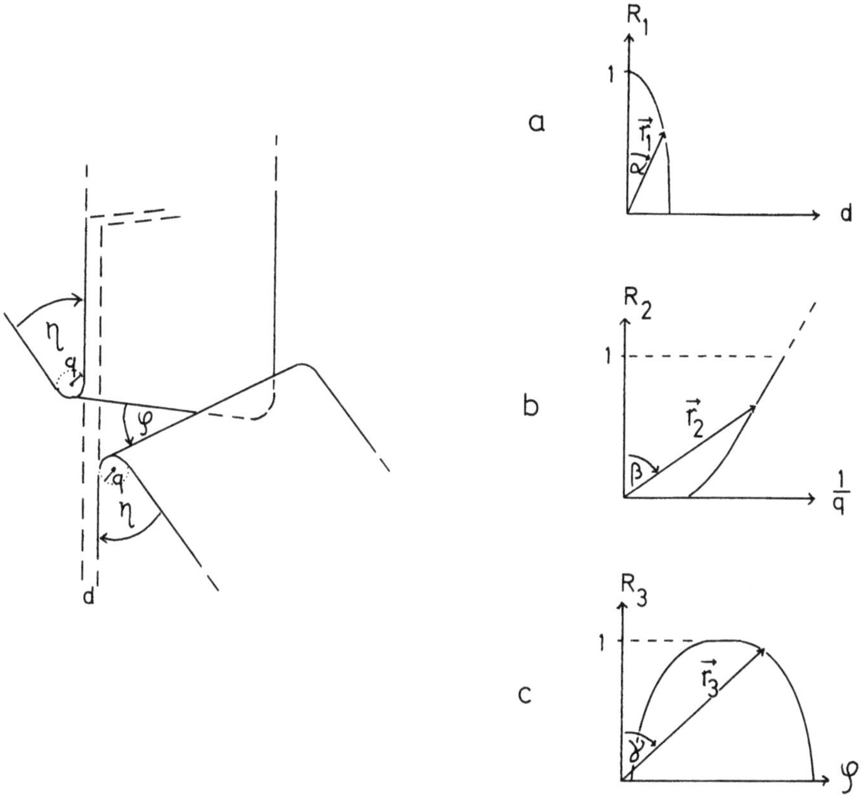

Fig. 2. *Left* Two cutting edges forming angle φ, passing one another at the distance d and possessing sharpness l/q; radius q depends upon angle η formed by the two surfaces of the cutting blades and on the degree of abrasion of the edges caused by their action. *Right* Nonquantitative characterization of the constructional parameters determining the efficiency of cutting of two edges; d, q, φ: constructional parameters; $\vec{r}_1$, $\vec{r}_2$, $\vec{r}_3$: contributions of the parameters d, q, and φ to the resultant effect output. **a** Dependence of the partial vector $\vec{r}_1$ on the distance of the cutting edges d; **b** dependence of the partial vector $\vec{r}_2$ on the sharpness of the edges expressed as the reciprocal of radius q; **c** dependence of the partial vector $\vec{r}_3$ on the cutting angle φ

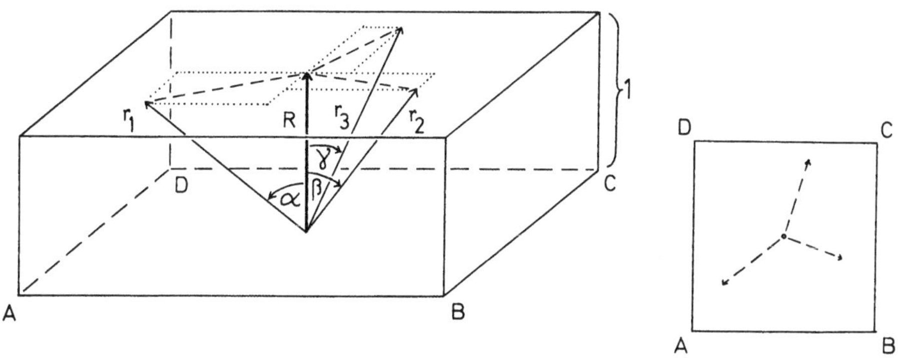

Fig. 3. Vector model expressing the contributions $\vec{r}_1$, $\vec{r}_2$, $\vec{r}_3$ of the parameter d, q, φ (see Fig. 2) to the resultant effect output R

While in a lever construction energy in the form of acting forces is evidently required for the employment of the principle, energy input is less obvious, e.g., in a statical construction because it normally does not lead to visible results. The energy input in this case is represented by forces placed upon the construction from the environment (e.g., gravitation, hydrostatic pressure) and also the organism (e.g., insertion points of muscles). The output is represented by the repartition of stresses caused by the forces and brought to an equilibrium with counteracting forces emerging from the constructions. The forces affecting the construction thus cause its transformation from a stress-free to a stressed state, which in the case of deficiencies in the construction can even be released in the form of fractures, signifying an irreversible transformation of states.

Bioconstructions materializing optical principles such as dispersing or converging lenses do not have dynamical effects either, but cause transformations of the spatial distribution of light energy from an input state to an output state.

It is important to see that principles can be materialized and also abandoned as a consequence of the change of parameter values. A constructional element, for instance, can only operate as a lever if it possesses a minimal required stiffness relative to the forces acting on it. Strictly speaking, the effect the applied forces cause divides into one portion leading to deformation of the lever element and one portion turning the lever element. If the share leading to deformation increases to become predominant, the lever effect gradually vanishes. In case no other changes occur and the magnitude of the acting forces remains constant, variation of the stiffness parameter within or beyond a certain range of values would be decisive to the materialization or abandonment of the principle. Similarly, in the above-discussed scissors example, the cutting effect vanishes very rapidly if the distance of the planes within which the shearing edges move against each other is increased (Fig. 3). The same could be observed if the shearing edges became increasingly rounded. From this and any number of other examples the general statement can be distilled that principles become materialized by virtue of essential parameters taking on values within critical limits.

4 A Model of Parameter Influences on Effects

The relative importance of the contribution of a parameter to the resultant effect output, in principle, can be expressed in a quantified manner within a Cartesian diagram as a curve possessing a more or less steep maximum. In the scissors example, for instance, the curve describing the correlation between the distance of the planes of movement of the shearing edges (d) and the performed cutting effect (when all other parameters remain constant) possesses a very marked maximum. The value of the cutting effect is maximal at the distance of the planes d taking on the value zero, and decreases very rapidly if the distance value becomes greater (Fig. 2a). The influence of the sharpness of the edges (q) on the effect output is characterized by a rather steep curve as well (Fig. 2b). The maximal effect contribution which can be acquired depends, however, upon the material of which the edges consist, since it determines the maximal forces which can be developed on the edges without destroying them. If we consider, instead, the influence of the cutting angle (φ) formed by the shearing edges, we find a curve with a comparably broad and flat effect maximum (Fig. 2c), which signifies that the resultant cutting effect is not very sensible to changes within a certain range of the cutting angle.

The parameters and their variable contributions to the resultant effect output can be illustrated by a geometrical model using vector representations (Fig. 3). The effect output performed by the operational unit may be named the resultant vector R (vertical in its orientation to the plane ABCD), than the parameters involved in the operational units may be expressed as partial vectors (r). The particular category and kind of parameter may be represented by the orientation of the vertical projection of the respective partial vector on ABCD.

Optimal efficiency of the operational unit can be expressed by the resultant vector taking on the value 1. In this case the projections of the partial vectors on the resultant vector are all equivalent to 1. This means that the parameter values are optimal in regards to their contributions to the effect output. Change of parameters to less optimal values corresponds to diminished length of the resultant vector $\vec{R}$, such that it takes on values between 1 and 0. In the scissors example of Fig. 2, the contribution of the parameters d, q, and φ to the resultant cutting effect of the edges $\vec{R}$ is expressed by the partial vectors $\vec{r}_1$, $\vec{r}_2$, $\vec{r}_3$ respectively. The model demonstrates that any suboptimal effect corresponding to a value of $\vec{R}$ below 1, can be caused by more than one, even many constellations of parameter values d, q and φ. In terms of organization of living beings and evolution this means that a considerable space of options exists for answering particular demands. As shown later (Sect. 5 and 9) this is of major importance to an understanding of organic evolution.

The model, in principle, can also be applied to complexes formed by causally interconnected operational units. In this case the resultant vector represents the terminal output of the complex or system. The partial vectors represent the effect resultants of the operational units, which in turn are determined by their parameter components. Generally, depending upon the systemal level we consider, a particular effect may have the meaning of a resultant or of a parameter.

The fact that the influence parameter changes can have on the effect output of an employed principle is determined exclusively by the latter, bears an important consequence: mutations of the genetic code, which are ultimately responsible for any parameter change, are controlled in their influence on the organic system by the employed principles and the systemal properties which derive from them. Whether genetic mutations can be significant to an organism is in the first place a matter of systemal organization and only in the second place a matter of the environmental context.

5 Optional Pathways of Organization

The model of parameters and effects discussed in the previous section and the kinds of causal interconnections of operational units open up a very interesting view on the properties of organic systems, and particularly on the fact that, in detail, most performances of organisms can be provided in an equivalent manner by different constructional solutions. We find optional freedom in (1) how con-

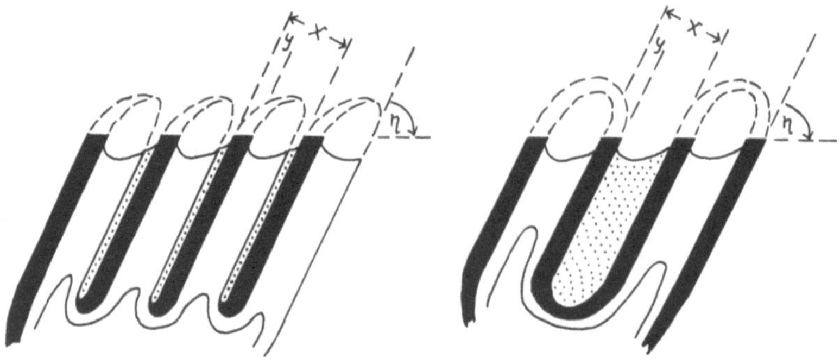

Fig. 4. Cross-sections of hypsodont rodent teeth vertical to the enamel crests. Parameters relevant to cutting efficiency, such as thickness of enamel bands y, distance of the enamel bands from each other x, and the angle of the cutting edges η exhibit the same values although they are provided by very different types of constructions. *Black* enamel; *white* dentine; *dotted* cementum

structional parameters can be provided by an organism, (2) how particular effects or performances can be produced, and (3) how operational units or complexes can be constrained by other constructional properties of the organism.

1. We can find many cases where identical parameter constellations are caused by quite different constructional configurations. This can be exemplified by the alternative ways in which the scissors principle is realized in different rodent species: although in Fig. 4 all parameters which are relevant to the effect output have the same values, we deal with quite distinct structures built by quite different ontogenies. This demonstrates that organisms can dispose of considerable degrees of freedom in providing the constructional parameters which are relevant to the materialization of a principle.

2. From the model exemplified by Figs. 2 and 3 and discussed in Section 3 it can be easily seen that only in the case of optimal efficiency are the parameters fixed at their optimal values. Suboptimal effect outputs corresponding to lengths of the resultant vector below 1 can be caused by many different constellations of parameter values. This shows once more that a considerable degree of optional freedom exists in how organisms can materialize principles.

3. If units of performance are interconnected, forming constructional or operational complexes, it is due to effect/parameter links. Following the direction of causation we can distingiush a prior unit which through its effect output influences the posterior unit in such a way that the terminal effect output of the latter is improved or diminished. We particularly find that it is irrelevant to the operation of the just posterior unit how the influencing parameter value or stimulation is produced by the prior unit. It is, for instance, irrelevant to the cutting effect of the carnassials of a carnivore whether the precision of movement of the shearing edges caused by the lever action of the mandible is due to the shape of the jaw articulation, or is controlled by the sliding contact of the canines, or (which is improbable) exclusively controlled by the neural stimulation pattern of the jaw muscles. It shows that a specific parameter value decisive to a particular cutting effect can be produced by quite different constructional options for constraining

the movement of the mandible. A great number of additional examples could be cited in order to illustrate the general fact that in an elementary chain of causation different constructional options of the prior unit may influence only one parameter of the posterior unit.

Options of providing operational parameters, of producing effects, or causing constraints occur not only on the most basic level of organization. In fact, the systemal particularities described in (1) to (3) apply to all organizational levels. Options are involved in any integration of operational units to higher complexity including the uppermost systemal levels of organic nature (e.g., strategies of survival, see Sect. 8). We should even expect that options of organization are the more numerous the more complex the considered constructions or operational systems are.

The aspects discussed in this section were not paid much attention in the existing literature, although they are of major importance to an understanding of organic evolution. One of the implications is that selectional forces placed on species have only an indirect influence on their organizational pathways.

6 Two Kinds of Optimality

The values taken on by the parameters of a unit of performance determine not only whether an effect conforming to a principle emerges or vanishes, but also the relative magnitude of the effect performed. We can – at least theoretically – consider outstanding cases where the operation energy is completely transformed into a particular effect output. We call them optimal materializations of a particular principle. The fact that such cases never actually occur in nature or in the technical world does not at all diminish its heuristic importance in the frame of the analysis followed here.

An optimal lever would quantitatively transform a force applied on the lever bar into lever movement, without loss of energy through friction in the lever articulation and without temporarily storing part of the applied force in the internal stress field produced within the lever bar. Energy loss due to friction between two surfaces and stress fields established within rigid bodies occurs in conformation with physical laws which have nothing to do with the operation of the lever as such. They signify instead factors disturbing the operation of the lever. Optimal materialization of a principle thus signifies transformation of energy from an input state to an output state solely on the strength of that principle, without draining away part of the energy through changes of state in conformation with other physical principles.

Degrees of optimality following this definition can be measured in principle by the inverse ratio of the total amount of energy (E) flowing into an operational unit and the amount of energy absorbed by the materialization of the principle considered (Em). Optimal materialization of a particular physical principle, $Em/E = 1$, means maximum correlation between energy input and effect output on the strength of that principle. Materialization of a principle vanishes if Em/E approaches zero. We note that optimality considerations of this kind do not have recourse to factors outside the operational unit considered. We may speak of optimality of effect.

It is important to see that optimality in this strictly physical sense is not identical with optimality in regards to a particular demand which a construction or apparatus is expected to fulfill. In bioconstructions demands are derived from internal systemal needs or from the environment, and in technical examples they are stipulated by the task facing an engineer.

The essential difference between the two kinds of optimality is that constructions or apparatuses, which are equivalent in optimality expressed through the Em/E ratio, may serve a particular stipulated demand to different degrees due to parameters which are independent of that E/Em ratio. Lever constructions of equivalent optimality may, for instance, be more or less adjusted to produce a certain effect, say a catapult effect, depending upon the relative lengths of the lever bars forming modifying parameters of the constructions. Grasshoppers are, for example, distinguished by great elongation of two particular elements of their legs as an adjustment to performing big leaps. Optimality in this sense can only be formulated in regards to factors outside the considered operational unit or complex. We may call it optimality of performance.

There are, hence, basically two possible ways of optimization of a construction or operational system. First: optimization can be the result of general improvement of material properties and coherence of parameters in regards to materialized principles. Second: optimization can be due to improved adjustment of parameters to demands imposed by conditions outside the considered constructional or operational complex.

We choose the letter scale as a mechanical model. It works on the principle of the two-armed lever compensating the weight of an object put on the scale by the increase of the length of the lever on which the reference weight is mounted. This increase is produced through the fact that the two lever arms balancing the two weights form an obtuse angle. Precision of measuring can be optimized by general constructional parameters, such as (1): the orientation of the lever arm on the side of the weight to be measured; it should be horizontal; (2) the subdivision of the measure; it should be in accordance with a particular nonlinear function; (3) the quality of articulation of the pivotal axis of the lever; it should work with as little friction as possible. Quite different from the improvement of accuracy through optimizing these parameters is the improvement of sensibility of the instrument we acquire, if we fold the reference weight upwards, and thus shorten the lever length of the reference weight. It has the effect that within a narrower range of measuring, the pointer deflection per weight difference to be measured increases markedly. Further, change of the magnitude of the reference weight would result in a shift of the range of increased accuracy of measuring towards another section of the measure.

A biological example would be the sense of hearing in higher tetrapods. General constructional improvement is acquired, e.g., by the integration of two additional bones (os articulare, os quadratum) to the vibration-transmitting system of the middle ear primarily formed of only one element (stapes), and the elongation and coiling of the trochlea of the inner ear. Both changes result in an extending of the frequency range of the audiogram. Independently of this, improved sensibility within a specific interval of the frequency range is attained by the change of only a few parameters, such as the surface ratio of the vestibular membrane and the basis plate of the stapes, the relative weight of the ear ossicles and their stiffness of coupling (Fleischer 1978).

To mention only one further example, we can cite the legs of tetrapods in which general optimization of the apparatus of muscles and tendons and also of the precision of articulations of the leg bones is superimposed by the development of specific length ratios of the bones in connection with particular environmental demands (e.g., hanging in trees, moving through swamps).

7 Conceptional Consistency with Experimental Biosciences

Up to now we have only dealt with the constructional aspects of organisms without taking into consideration the fact that bioconstructions are built of biomolecules and organically produced minerals. Also, in order to operate, they not only need energy supply provided by metabolism but also require neural control. It is claimed here that all domains of the organic system, although materializing quite different categories of inorganic laws, can basically be dealt with in conformation to the concept followed in the previous sections. This means that all biological phenomena can ultimately be analyzed in terms of the employed principles, parameters, and input/output relations. In neural processes the basic unit would be represented by the electrophysiological stimulus. In molecular biology and physiology – irrespective of the extreme variety and complexity of processes and products – only few principles can be found. These are the principles of chemical bonding. Diameters of atoms and ions represent essential parameters. Concentrations of ions and molecules in the chemical milieu and also, e.g., thermodynamical states, have the significance of modifying parameters. Interestingly enough, the nucleotide sequence of the DNA in this view has the significance of a contingent parameter.

The basic units of organization are in part directly interrelated through effect/parameter links or effect output/effect input links, thus forming systems. Through particular kinds of connections of the basic units, control and regulation principles become materialized. They can be found in the various organizational domains on all levels of complexity, and in most cases are based on transmission of neural signals or endocrine information. Well-known quite complex regulation systems are, for example, responsible for the maintainance of blood pressure or body temperature. One of the most simple systems materializing the negative feedback principle is, for instance, the stoma apparatus in the epidermis of land plants, regulating transpiration.

Instead of giving only rough outlines, it would be desirable to carry out the analysis in the various biological fields in detail in order to show that the concept suggested in the previous sections covers organic phenomena in general.

8 Economy

In a previous section the forearm of tetrapods was taken as a general example for demonstrating how principles of action are materialized in organisms. Together with the hand and the humerus, the forearm works as a composed lever system and should, in principle, be analyzable as such in any particular example. If we look, however, at the many different species-characteristic length ratios found in the leg bones of, e.g., the bovids, we understand that explanations in terms of principles parameters and effects cannot grasp the essential differences. These are related more to different modes of movement than to precisely identifiable physical parameters, and hence cannot be understood independently of the niches the species occupy. Considerations about niches, however, are already clearly beyond the approach of causal morphology.

In Section 5 we saw that the energy expenditure/effect balance (Em/E) describes the way in which a principle is materialized by the involved parameters. The question, however, as to what this balance means to an organism can only be considered within the ecological context. It depends upon the use the organism makes of it, so that it forms a part of its strategy of life in facing environmental conditions. Thus it may occur that a part of an organism, both from a point of view of systemal coherence and of adjustment to environmental parameters, fulfills a particular performance nearly optimally. This notwithstanding, it may become too expensive in regards to the total energy household of that organism due to changes in its life strategy. What counts in this case is not optimality but economy expressed as the relation between energy consumption and fitness contribution. The methodology dealing with these kinds of questions is based upon the observation of selective processes. Its central theoretical term is that of fitness, which can be defined in terms of a stochastical value function.

9 Organizational Options and Selection

The repartition of fitness contributions of the various performances of an individual or a species is a matter of its strategy of survival. It is partly provoked by the environment which "sets prices" on particular strategies. There are, however, no constraints from the side of the environmental challenges in regards to the specific kind of devices developed in order to cope with them. Rayner (this Vol.) gives very interesting examples of bats living within the same habitat on different feeding strategies. The strategy adopted influences the organizational system by determining the repartition of selectional pressures on the various subsystems, down to the performances of the most basic organizational units. As a consequence of our considerations on options involved in organizational pathways (Sect. 4), we found that selective pressures on the various levels of organization can frequently be answered in alternative ways. Beside this, decrease of relative fitness contribution of a subunit of an organic system may be compensated by improvement of relative fitness contribution of another subsystem, so that the performance output of the system as a whole remains practically unchanged. This means that selection discriminates in any particular case between variations of performances but not between the different ways in which one and the same performance is provided.

10 Conclusion

Organic evolution proceeds in conformation with the laws of the inorganic world but owes its essential properties to the fact that it is influenced by two independent sources of indeterministic causation: mutations of the genetic code and environmental changes. At the end of Section 3 we stated that mutations induce changes of organic parameters, which latter, according to the physical principles they materialize, generate effects. These may become useful or disturbing to the organism, depending upon the properties of the system they affect. It is, hence, the systemal organization which ultimately determines how mutations become expressed in the adult and in this way become exposed to selective forces of the environment. This is what Vogel (this Vol.) calls the "internal aspect" of selection. Further, genetic change, prior to exposure to selective constraints on the part of the environment, undergoes selective constraints resulting from inherent properties of the ontogenetic program. We can speak of the systemal and ontogenetic filter, through which mutations must pass prior to becoming relevant to the interaction of organisms with their environment.

Environmental conditions on the part of the organisms provoke strategies of survival. According to the strategies adopted, selective constraints of different strength are placed upon them, leading in part to high degrees of adjustment to specific environmental factors or patterns of factors. The kinds of influence that environmental demands exert on the organism were discussed in Section 8, where we stated that selective constraints are placed first of all upon effect resultants which on the next lower level of organization prove to be provided by highly structured systems, which in turn are composed of subsystems and so forth, down to the most basic units. In principle, any of the causal links between parts of systems or subsystems, down to the lowermost organizational level, comprises a certain degree of freedom in regards to the parameter combinations responsible for the caused effects (Sect. 4). This is the reason why organic systems, even where they are submitted to severe environmental constraints, are removed in their evolutionary options from the immediate external influences.

Systemal properties of the organisms determine the kind of how genetic change and environmental influences interfere. The interferences which actually occurred along the geological time axis make up the irreducible historical quality of organisms.

Reference

Fleischer G (1978) Evolutionary principles of the mammalian middle ear. Adv Anat Embryol 55

Concepts of Constructional Morphology

KLAUS VOGEL [1]

Summary

A definition of constructional morphology is offered and its relations to functional morphology and biomechanics are described. Starting from the introduction of this term and kind of morphological analysis by Weber, it is described as a compromise between a morphogenetic and a phylogenetic factor and the biological role (Seilacher). The importance of biomechanics in morphogenetic processes is emphasized. Arguments are offered that evolution is not in the first place the story of ecological adaptations but the history of constructional improvement. Natural selection has impact not only on protection and matter and energy acquisition but also on the economy and optimality of the matter and energy-consuming and -converting machinery of the organism.

1 Constructional Morphology, Functional Morphology, Biomechanics: a Research Field with Different Foci

The cover design of our Symposium Volume exhibits the greatest possible pretension of constructional morphology: a body surrounded by a flexible envelope, which is tightly filled with a fluid or with viscid material, automatically assumes a spherical shape; at least as long as the surrounding envelope maintains uniform firmness and no exterior forces act upon the body. This is true for protists as well as for multicellular organisms. Every deviation from the spherical shape must be actively generated by a bracing fiber or muscle system. This is also true for skeleton-bearing organisms. Skeletons are phylogenetically and ontogenetically secondary structures, their origin and development require preformed hydraulic bodies. In the figured trilobite the bracing muscles and connective tissue are responsible for maintaining the structural shape during the molting process, which subsequently becomes copied by the excretion of a new ectoskeleton.

It follows that constructional morphology, as I and some other participants understand it, deals with rules of shape generation, shape control, and mechanical coherence of the entire biological body. This also includes its ontogeny and evolution.

Schmidt-Kittler's and my definition of constructional morphology (this Vol.) is as follows. Constructional morphology is the study of organisms and their

[1] Geologisch-Paläontologisches Institut, Universität Frankfurt, Senckenberganlage 32–34, D-6000 Frankfurt/M., FRG

N. Schmidt-Kittler and K. Vogel (Eds):
Constructional Morphology and Evolution

parts as coherent systems and subsystems that obey statically and dynamically (at rest and in movement) physical principles and processes. The choice of this term also focuses attention on potentials and limitations set on these systems by ontogenetic growth and evolutionary history. It includes exploration of how mechanical and chemical forces and processes influence each other and how construction, behavior, and environment act and react on each other.

I will return to other definitions and concepts of this field later. Neither functional morphology nor biomechanics encompasses such extensive concepts. Functional morphology asks how an organ works (often in relation to its extra-organismic environment) and how function influences its form. Biomechanics deals with the mode of operation of muscle-skeleton systems and often neglects ontogeny and phylogeny. Typically, the subjects used for biomechanical research are arthropods, insects, and vertebrates. They possess organs such as arms, legs, and wings, the operation of which can be studied relatively independently of the construction as a whole. Moreover, the physical laws governing these organs are relatively simple (e.g., lever laws). This is an important difference compared to lower invertebrates. It renders possible significant experiments.

Obviously one cannot draw well-defined boundaries between constructional and functional morphology and biomechanics, especially since each field has a long tradition. The use of one of these terms only indicates the focal point of each case. For reasons just mentioned, constructional morphology is, in my opinion, the higher-ranking term because it includes functional morphology as well as biomechanics. I shall explain this in more detail.

In 1955 Weber introduced and defined constructional morphology. He deplored the state of morphology and reported (my translation): "...abstraction from function, neglect of construction and self-restriction to formal homologies resulted in the failure of descriptive morphology ("reine Morphologie") especially in the field of evolution" (p. 154). However: "... an evolutionary morphology cannot refrain from function, because no evolutionary change of form can be explained without consideration of how the investigated component parts and forms function. Every stage of evolutionary change must produce an operational constructive unity. This must be understood in line with every stage of ontogeny, from which evolution is composed" (p. 154). "I call this working method, which has been created as a combination of descriptive morphology and comparative functional analysis, constructional morphology. Construction should be understood as the incorporation of form and operation (function). This includes its dynamic, its development, and the dynamic of its development" (p. 155). Methodically, descriptive morphology ("reine Morphologie") and constructional morphology are equivalent; however, both are imperfect without its counterpart. Weber's ideas on units of form and function as well as the incorporation of ontogeny and evolution into a constructional morphological approach should be kept in mind during the course of this presentation. Due to his early death he could not complete his work. A more detailed analysis of ideas and aims behind Weber's morphological investigations was published by Reif in 1985 (1985a, b).

2 Constructional Morphology as a Compromise

Regarding Weber's studies, Seilacher (1970) gave the term "Konstruktionsmorphologie" a new connotation. Since that time many biologists and paleontologists relate this term to a specific context. R. Thomas (1979) introduced this concept into the American literature.

According to Seilacher, constructional morphology analyses the influence of three factors on organismic form, i.e., a historical-phylogenetic factor, an ecological-adaptive factor (later renamed biological role) and a so-called Bautechnik factor or morphogenetic-fabricational factor. Two corners of this triangle had been considered already by other authors as being essential or even the only significant determinants of constructional and functional morphology. They view organisms as members of historical transformation processes which can be explained in terms of their ancestry, but modified for adaptation to their respective environments and life habits.

2.1 Morphogenetic Aspect and Ontogenetic Limitations (Epigenetics, Biomechanics)

I will return to the latter-mentioned aspects later, and now focus attention on the new factor, architecture. Architecture recognizes laws of geometry and cybernetics, physical or chemical properties of the available materials, and limitations in possible growth programs. These determinants are considered to be ahistorical and do not result from adaptation. This is what Dullemeijer (this Vol.) calls "universal constraints". However, as Seilacher (1972) demonstrated on divaricate patterns in pelecypods, they can be used for secondary adaptational processes.

Since morphogenetic programs influence ontogenetic development, let me first refer to some authors who discuss the canalizing effect of developmental processes and the constraints they impose on realization of optimal technical solutions.

The concept of "canalization" was introduced in 1956 by Waddington. He wished to characterize his observations that phenotypes are apparently buffered against ecologically induced morphological or physiological alterations. Phenotypes are highly stabilized systems. Only a part of all possible character variations become realized due to some constraining and suppressing mechanism. Apparently, there are only a limited number of stabile developmental pathways. This suggests that regulatory interactions at the genetic and epigenetic level control the development and evolution of phenotypes. Waddington spoke of an "epigenetic landscape" to characterize the field of morphological and physiological variation. In this field "steady states" are distributed discontinuously. Their "force field" directs evolutionary change and determines which biological novelties will actualize. According to Alberch (1980, p. 657), the importance of epigenetic control is illustrated "by the fact that different genetic mutations can result in the same morphological outcome" (cf. Schmidt-Kittler and Vogel, this Vol.).

Liem and Wake (1985:373) described morphogenetic limitations and canalizations as follows: "The particular aspect of development that concerns us is epigenetics – the control of gene expression by the microenvironments encountered by cells during development, or, more generally, the mechanisms by which genes express their phenotypic effects". "Internal co-ordination" by genetic and developmental influences are considered as a "main directive agency" in organic evolution by Whyte (1967).

The concert of genetics, epigenetics, and development of structural and functional complexes and their relation to the respective environment was analyzed and exemplified by Wake and Larson (1987) with their investigations on the feeding system of lungless salamanders. The limited evolutionary licenses of component parts bound into patterns of functional complexes were recognized by Dullemeijer (1958, 1974, 1980). In his contribution to this Volume he calls them "local constraints".

Waddington seems to suggest that canalization appeared via natural selection (for discussion see Ho and Saunders 1979, p. 587). Other authors, in contrast, interpret epigenetic mechanisms as "selection ratchets" and thus as an argument against neo-Darwinism. Ho and Saunders (1979) express this view in an extreme way. They combine it with a rather Lamarckian idea that "genetic assimilation" of environmental influences is possible, and compare organismic structures with dissipative structures, as if organisms were dirigible at pleasure by external influences. In my opinion there is no reason to believe that the epigenetic system, whatever this really may be, has not been generated and is not being maintained by natural selection.

Before I undertake to describe an important deficiency of the epigenetic hypotheses, I would like to mention some other aspects of morphogenetic analysis. These are geometric growth gradients (for example D'Arcy Thompson 1942; Rudwick 1968; Seilacher et al. 1968) or "morphophysiological" gradients, as were assumed, for instance, by Urbanek (1960, 1963, 1966) to explain proterogenetic evolution of graptolite colonies. Or further, the attempts of Raup (1968) and others to reduce morphological complexity to sets of a few simple instructions. Fractional geometry can contribute to this latter point of view. Raup (1966) stimulated the constructional morphology. He presented a three-dimensional block diagram of theoretically possible shapes of coiled shells and showed that actually existing shapes occupy only relatively small regions of this block. Apparently, not all geometrically possible forms are biologically possible or functionally efficient. Plausible constructional (functional) reasons can be specified to explain growth limitations in coiled shells.

At the end of this morphogenetic chapter I must draw attention to the fact that most contributions to epigenetic processes ignore an important factor, which to me is as effective as the genetic-biochemical action. I am speaking of the biomechanical component, a factor which should be of special interest for constructional morphologists.

Chemical messenger substances, chemical reagents and reactions need space in which they will be transported, as well as trails on which they can be carefully directed to induce the desired reaction at the right place. How is this space generated, shaped, maintained and structured? This can only happen by mechanical forces. The space must be "blown up" and if need be prevented from becoming globular. A mechanical coherent system must be established for shape control and for efficient mechanical reactions to biochemical and other signals. It follows that genetic information and biochemistry are canalized or at least mutually influenced by laws of mechanics. I feel supported in propagating this idea by Bereiter-Hahn's contribution to this volume. An organism's successful origin and growth depends on those genetic and biochemical signals and reactions which set mechanical processes in motion, or that are at least, in accordance with mechanical properties and coherence of bioconstruction. "To think that heredity will build organic beings without mechanical means is an unscientific piece of mysticism" (His 1888, after Gould 1970, p. 89).

There are very few publications which draw attention to dynamic biomechanical forces in ontogenetic development. One is a paper on diatoms by Schmidt (1987). She concludes from her observations that the development of the skeleton is governed by strict mechanical control. Position, growth and shape of silica deposition vesicles is determined by molding forces which apply pressure and tension. Molding is executed by specific neighbour vesicles, by microfilaments and other elements of the cell interior. Similar conclusions regarding the involvement of microfilaments and microtubules in scale formation of chrysomonads have been published by Brugerolle and Bricheux (1984).

Looking at metazoans, Bandel (1982) showed that the morphological change in archeogastropods from a caplike to a trochospiral shell during early ontogeny results from retractor muscle traction. Further observations on biomechanical influences during the ontogeny of gastropods are in print (Edlinger). In echinoids, food groove and interlocking element patterns, internal pillars, lunule notches, keyholes, and other morphological traits of the skeleton are constrained by the special mode of plate mosaic growth (Seilacher 1979), which is based on the pneu character of echinoid construction.

In the embryonic vertebrate the neural tube and the brain must be stretched and expanded hydraulically (Vogel and Gutmann 1981). Muscles which are only capable of shortening cannot grow without being stretched. This needs mechanical, for instance hydraulic, forces (Gutmann and Bonik 1981).

All these authors seem to come to one conclusion: we cannot explain the quality of an organism, its organs, or organ systems without considering its ontogeny. Ontogeny is an essential aspect of constructional morphology.

2.2 *Ecologic-Adaptive Aspect*

I shall now return to Seilacher's triangle of constructional morphology. It was mediated to English-speaking colleagues by Thomas (1979). He limited constructional morphology to skeleton evolution. This version is not very consistent, because skeletons can only "copy" that which is ontogenetically and phylogenetically predetermined by the soft body.

Thomas (1979) added the term "natural selection" to the ecologic-adaptive corner of the triangle. This characteristic is attributed only to this corner. It reflects a view which has been prevailing since Darwin, that evolution occurs through adaptation of the organisms to their respective environment and is only kept going by this kind of selection. Since the concept of selection plays a major role for my own and some other participants' understanding of constructional morphology, I am going to briefly comment on it.

Many evolution theorists argue as if natural selection results only in external adaptation. They believe that environmental change is essential in order for evolution to continue. I will exemplify this by some typical quotations.

In 1965 Bock and von Wahlert defined selection as "the action of any factor of umwelt on an organism..." (p. 291) and evolutionary adaptation as "the long-term, hereditary adjustment of an organism to a particular set of environmental conditions..." (p. 296). This understanding of selection seems to be shared by Dobzhansky et al. (1977, p. 8), who described evolution as follows: "Organic evolution is a series of partial or complete and irreversible transformations of the genetic composition of populations, based principally upon altered interactions with their environment. It consists chiefly of adaptive radiations into new environments; adjustments to environmental changes that take place in a particular habitat, and the origin of new ways for exploiting existing habitats... ." Valentine (1975) tried to interpret the origin of "ground-plans", for example those appearing in the Lower Cambrian, as a response to environmental conditions and as "adaptive" strategies.

Vermej (1987) presented the most recent example demonstrating a purely environmental perception of selection. According to him the main selective agent is "escalation" by the predator-prey relation. Alberch (1982, p. 19) characterized the traditional position of neo-Darwinism as follows: "...evolution is viewed as the process of adaptation to changing environments under natural selection".

Lorenz (1978) followed the same line, asserting that morphology is always a mirror of the environment: "Adaptation of an organism or of one of its organs to an aspect of the external environment means copying this aspect in a certain sense. By form and function, the horse hoof is a copy of prairie ground and its physical qualities, in the same sense as the fins of a fish are a copy of the water, or the eye, one of the sun. Should you not believe in a pre-stabilized harmony between organism and environment, you must at least come to the conclusion that copying environmental information has somehow become introduced into the organismic system" (p. 46, my translation).

From my point of view it is wrong to think of selection as a force that can be explained exclusively by organism-environment relations although this point of view is held by most evolutionary scientists. It cannot explain the origin and change of organismic constructions. I will return to this problem.

2.3 Phylogenetic Aspect

Following the morphogenetic and the ecologic-adaptive aspects, I proceed to the third factor that is indispensable for the understanding of organismic constructions: the phylogenetic-historical aspect.

Some of those colleagues working on biomechanics, including participants of this symposium, ignored this point. Nachtigall (1971) emphasized that the biologist cannot principally answer the questions "how did it develop?" and "why this way and not another?"; "we can deal only with the forms as they exist now" (p. 14). This idea is consistent when experimentation is permitted to be the *only* approach to biological constructions.

On the contrary, those biologists and paleontologists who are especially interested in the phylogeny of respective organism groups or who are generally interested in elucidating evolution cannot exclude the historical point of view. Furthermore, many scientists are convinced that analyses of constructional morphology can contribute to the understanding of evolution in a most revealing way.

Morphology cannot be reduced to physical formulas. It would remain unexplained why water-living vertebrates are never equipped with screw propulsion, why flying animals use different materials to build their wings such as chitin, feathers, or skin; why land-living vertebrates walk on only four instead of six or more legs like other animals; or how it came to the functionally problematic coupling of air-tube and esophagus in vertebrates. The reason is that organisms are not created on the drawing-board. Jacob (1977) described this characteristic as "evolution and tinkering".

Which ancestral traits are successfully taken over by the following generations is also a result of selection. These features are not judged by their value for the adult organism alone, but also for their function during the ontogenetic growth process. The pure mechanical role of morphological elements in early ontogeny has not been explored as it should be within the frame of constructional morphology.

3 The Internal Machinery Aspect

I will now elaborate on the emphasis of an internal machine aspect. It relates to the mechanical forces that enforce the morphology of an organism and make its movements possible. It relates to the coherent mechanical structure which is the basic indispensable condition for every expedient physiological process in an organism. For example, how does a coelenterate maintain the form of a cylinder and which role do the mesenteries and septae play (Grasshoff 1984)? Which morphological elements exert shape control in a worm? Which ones permit peristaltic movements without quantitative change of cross section and which ones keep the animal's length constant (Gutmann 1977)? How can the respective biomechanics be changed to work more efficiently?

By which means can the ammonite animal detach itself from its last septum, during its shell growth and advance in its living chamber? Which activities and forces influence its backside, the "septal membrane", during this process, and which elements are responsible for its characteristic shape, which is copied by the subsequently secreted septum (Seilacher 1988)?

Which are the respective mechanical prerequisites for the formation of different skeletons, ontogenetically and phylogenetically, without destructive influence on mobility (Vogel and Gutmann 1981)? What are the internal advantages of different skeletons for the respective construction?

The last questions can help me clarify what I mean by "internal machinery aspect" and why I believe it is necessary to emphasize this aspect and to confront it with the environmental influence on evolution. Most authors discussing functions of exoskeletons consider their value for protection against predators and unfavorable environmental conditions. The same is true of the protist skeletons.

A huge number of publications on foraminifera, especially paleoecological investigations, place environmental relations as the sole role of these shells (ecologic-adaptive aspect). Before us, nobody had asked whether a shell could be a tool for shape control, especially for nonspherical cells. Most important, it serves as a very economic shape control, because otherwise energy consuming activity of the cytoskeleton would be necessary to maintain a nonspherical shape against the hydraulic pressure of the cell interior (Bereiter-Hahn 1985; Vogel and Gutmann 1988). In foraminifera, no one has asked whether the shell could enhance or make possible relatively independent mechanical activities of the interior and exterior cytoplasm in a very economic way. Alternatively, this could only be achieved by energy-consuming activity of cytoplasm, because cells react as units to every mechanical stress due to their hydraulic nature. Nobody had asked about the importance of a shell or its parts in keeping the nucleus or organelles in position in spite of a constantly streaming and moving cytoplasm. Bonik (1978, 1979) was the only person who asked such questions while investigating diatoms and presented respective observations.

These examples of mechanical restrictions and possibilities (which have been treated extensively in the publications mentioned) emphasize that internal morphological determinants exist, which cannot simply be explained as adaptations to special environmental conditions, but result from the biomechanic nature of morphology. They are the architectural base for the statics and dynamics of the respective Bauplan. They are the substructures upon which environment-related adaptations can be saddled. Compared to Valentine (1975), who attempted to explain Baupläne as results of adaptive strategies, we see the driving force of Bauplan evolution as processes of economizing and optimizing the internal machinery, the coherent construction; processes that work relatively independently of special environmental conditions. The qualities of the respective bioconstructions have determined which new environments could be conquered successfully. Further adaptation was of secondary importance.

I have purposely introduced an "internal machinery aspect" to characterize morphological and physiological traits that are necessary elements of a mechanical and energetical coherent construction and which cannot be explained without consideration of this unity. I would like to draw your attention to a problem that is generally neglected when discussing selection and evolutionary theory. There is no reason to assume that selection only controls the external relations of an organism, only improving the efficiency of matter and energy *intake* along with the aspect of protection. Inevitably the economy and optima of matter and energy *transfer within* the organism and of *energy conversion* is being tested as well (at least as long as the law of energy conservation is valid in biology as it is in physics and chemistry). The addition of a notochord into a metameric worm (Gutmann 1977) or the development of an opening muscle in brachiopods substituting a bulky indirectly opening system (Gutmann et al. 1978) achieves energetic advantages no matter what type of environment the animal is living in.

Bock and von Wahlert (1965) were the first to link selection consistently with the relative energy expenditure of the respective organisms. But apparently they only consider that part of energy expenditure relevant, which relates directly to the environment.

Our concept of a selection that improves the efficiency of innerorganismic energy conversion adds the necessary counterpart to a selection which operates on the energy expenditure in the environmental relations of an organism.

Some earlier authors have, it is true, realized that organisms are more than the sum of external adaptations and that they cannot be solely explained by ecologic-adaptive argumentations (plus ontogenetic and phylogenetic limitations). However, they seldom treat this topic with consistency. A very few examples should be presented. As early as 1951, Franz emphasized that every organism is equipped not only with special aptitudes but also with general qualifications (for example homeothermy). These give the organism all-round adaptability which makes it possible to cross environmental barriers and to change its environment. Essential evolutionary progress, according to Franz, does not take place by increasing adaptation but by increasing general qualification. Moreover, it was Franz as well, who linked selection with energy expenditure. According to him, biological progress is also when organisms can produce the same activity despite a reduction in energy. He argues about improvements of structure and function in organisms which can be measured by their efficiency: compare this with Urbanek (1988). Kier (1965) explains phylogenetic changes of Paleozoic echinoids as follows: "All the changes do not seem to require a change of habitat to explain their origin but resulted from the gradual improvement of the animal as a living mechanism" (p. 446).

Gould (1970) denies that the early diversifications of phyla can be understood as classical adaptive radiations. Instead they are "sets of competing experiments in basic design" or "the best mechanical design" (p. 78). "The evolution of most major groups is not the story of ecological variations... , but a history of mechanical improvement" (p. 111).

Gutmann propagates the autonomy of the organism, with very strong and consistent arguments. Morphology, activities, ontogeny, and evolution can only be explained on the basis of the internal machinery (a series of publications, e.g., Gutmann and Bonik 1981; Gutmann 1988). This machinery determines which environmental aspects will be used in which way. "Whether a certain environmental event is or is not followed by internal change, which change will occur and what intensity it will have strictly depends on the internal dynamics of the organism", and: "The environmental event only acts as a trigger for the internal change; it does not determine what will happen internally. This is true for all physiological and morphological changes an organism can undergo, including the domain of sensory physiology and behavior where this closure seems to be a paradox. But again: whether a certain environmental event can influence the behavior of an organism, and of what type and intensity the response will be, depends on the existence, general properties, and current states of specific sense organs and neuronal systems for the guidance of behavior" (Roth 1982: 42).

Those constructional morphologists who, like myself, emphasize the bioconstruction primacy prior to the environment, join Gould and Lewontin (1979) in their attack against the "adaptationist program". However, in contrast to their criticisms, we do not argue the role of selection. Their misconception of selection, which by the way, had already been presented by Schindewolf (1950, p. 427 and 1969, p. 109), follows from the traditional Darwinian view, that only special environmental conditions are capable of selection. However, if you concede that the matter and energy consuming and converting part of the organismic machinery contributes to fitness as well as matter and energy acquisition organs, it must necessarily be exposed to selection as strongly. More than that, it is the coherent unity of the organismic construction that determines how its subsystems can respond to selective forces. It sets the framework for possible morphological, physiological, and behavioral changes during the optimization and economization process (Peters 1983) driven by selection.

4 Methods and Pretension of Constructional Morphology

If this concept is realistic, it follows that there cannot be any other morphology than constructional morphology. One cannot fully understand an organism whose morphological description does not recognize or make use of the machinery aspect. This idea has been expressed cautiously by Bock and Homberger (1988, p. 186) and is expressed in a consequent manner by Gutmann (1977, 1988). Further, if this is true, then systematics must be built on constructive insight, as long as it also pretends to be "natural systematics" based on consanguinity (Vogel 1989).

Finally, if it is true that (1) biological form is enforced by mechanical power, (2) every motoric activity is mechanically controlled, (3) ontogeny cannot work without mechanical processes, and (4) phylogeny is a series of mechanically coherent constructions, then it follows that every phylogenetic reconstruction must be guided by biomechanical insight. Since constructional morphology "appears to be the only approach to phylogenetic reconstruction that is based on the principle of causality", one should "therefore expect it to have precedence over other methods" (Vogel 1989).

I concede this is a high pretension. The reason for criticism is the question of methodology. Compared to genetics, biochemistry, molecular biology, and all disciplines that claim to represent the "modern biology", constructional morphology as characterized above appears to have a serious fault.

Essential claims cannot be supported by experiments. Thus, "testability" of constructional morphology would seem to be a problem (Lauder 1981). Most experimental approaches to our discipline have investigated organs of insects, arthropods, and vertebrates. Their biomechanics can be studied relatively independently of the whole construction. They obey relatively simple sets of laws, for example lever laws. Further, the relationship between the organism's body or its surface and the surrounding medium, for instance during swimming or flying, can be and have been studied experimentally (e.g., Nachtigall 1977; Rayner 1981; Reif 1981; Preuschoft 1987).

However, problems do arise when we attempt to study invertebrates other than insects and arthropods, which are of special interest for every life history researcher. Their shape and motoric activity are governed by hydraulic forces. (This, by the way, is also true for the ontogenetic growth of the above-mentioned insects, arthropods, and vertebrates.) In a hydraulic system any local intervention effects the state of the whole system. Experimentation therefore is very difficult (Bereiter-Hahn 1985) or may even be inappropriate in order to obtain realistic results.

Hence, the method used in many constructional morphological studies is the comparison with paradigms. This is similar to Rudwick's (1961 and later) method for functional morphology of fossil animals. The paradigms are stimulated by physical and technical models, apparatuses, and machines. The study begins with a hypothesis on the character of a bioconstruction and its functions that leads to the formulation of a paradigm. If those elements which in the paradigm are responsible for the coherence of the mechanical architecture and its typical functions are also present in the biological object, then one concludes that it works according to the same principles. Consequently, these are comparative methods. In contrast to "classic" comparative morphology, those comparisons aim at technical and physical paradigms.

I do not think that this method is faulty. On the contrary, compared to experiments, in which the frame conditions must be kept as simple as possible ("reductionism") it has significant advantages. It permits an integrative view.

The organism is more than the sum of its biochemical-physiological reactions. Understanding these reactions does not explain the organism. The reaction must take place at the right time and place, and that within a construction network in which all parts are mechanically and functionally coupled. Chemical experiments investigating muscle activity can only explain the muscle contraction. The stretching movement necessary before the next contraction is generated by antagonistic muscles that must be appropriately placed. This three-dimensional order of statics and dynamics of bioconstruction just mentioned can only be approached by constructional morphology. Of course, an organism is not based on mechanics alone; but mechanics is the indispensable foundation, the scaffold, and the addressee of all other physiological qualities and processes. Even the order and placement of the neural system cannot be understood without the constructive frame. In fact, it gains meaning by being placed in local and functional relation to machinery.

The true value of the individual mechanisms within an organism can only be comprehended by using an integrative view. One of the most impressive arguments in favor of this statement is a study from Bereiter-Hahn and Strohmeier (1987). They showed that the osmotic pressure produced by the chemical activity of ion pumps is not sufficient to shape a cell, either as a tight-sphere or slacknonsphere. Instead it can use a contraction of the cytoskeleton to secrete water against osmotic pressure.

I do not wish to repeat or prolong the endless dispute between reductionistic and "holistic" views (also see Caplan 1988). I only wish to emphasize that arguments such as the Alpbach symposium *Beyond Reductionism – New Perspectives in the Life Sciences* are as valid as they were 20 years ago (Koestler and Smithies 1969). "...the present danger does not really lie in the loss of universality on the part of the scientist, but rather in his pretence and claim of totality" (Frankl 1969, p. 397). To quote Alberch 1980: 665: "...recent evolutionary theory has been plagued by a strong reductionist approach.... This reductionist philosophy is clearly inadequate, when we deal with features like morphology and behaviour". Duncker (this Vol.) also considers the reductionist approach to phenomena of nature as a destructive danger. Instead, he propagates the comparative method.

It will be exciting to observe the progress of computer simulation and to what extent it will be able to mediate between integrative and reductionist views.

5 Concluding Theses

1. Morphology is mechanics. This is true not only for deformation during movement; the general shape of an organism and all of its organs, the shape of a protist, a coelenterate, or a chordate, is enforced by mechanical power.
2. Ontogeny must also be understood as a mechanical process. An essential part of genetic information must be translated into mechanical work, otherwise growth would not occur. Mechanics is an essential part of "epigenetics".
3. Selection not only tests adaptation to the environment but also economy and optimality of the organismic machinery as a whole, i.e., also the quality of the matter and energy consuming and converting part of the organism.
4. Evolution is not in the first place a story of ecological adaptations, but a history of constructional improvements that optimize the relation between matter and energy input and output including offspring. The bioconstruction determines "which ecological niches" the organism can successfully conquer.
5. Assuming the correctness of these theses, constructional morphology should replace *traditional morphology* as far as possible in *morphological description,* and *phylogenetic reconstruction.*
6. The integrative approach of constructional morphology using physical and technical paradigms, should be considered a necessary counterpart to the experimental (and mostly "reductionistic") method of biosciences.

References

Alberch P (1980) Ontogenesis and morphological diversification. Am Zool 20:653–667
Alberch P (1982) The generative and regulatory roles of development in evolution. In: Mossakowski D, Roth G (eds) Environmental adaptation and evolution. Fischer, Stuttgart New York, pp 19–36
Bandel K (1982) Morphologie und Bildung der frühontogenetischen Gehäuse bei conchiferen Mollusken. Fazies 7:1–198
Bereiter-Hahn J (1985) Architecture of tissue cells. The structural basis which determines shape and locomotion of cells. Acta Biotheor 34:139–148
Bereiter-Hahn J, Strohmeier R (1987) Hydrostatic pressure in metazoan cells in culture: its involvement in locomotion and shape generation. In: Bereiter-Hahn J, Anderson O, Reif WE (eds) Cytomechanics. Springer, Berlin Heidelberg New York, pp 261–272
Bock W, Homberger DG (1988) Preface. Am Zool 28:185–187
Bock WJ, Wahlert G von (1965) Adaptation and the form-function complex. Evolution 19:269–299
Bonik K (1978) Die Entstehung der Kieselalgen – ein stammesgeschichtliches Modell. I. Die Entwicklung der Schale. Nat Mus 108:267–273
Bonik K (1979) Die Entstehung der Kieselalgen – ein stammesgeschichtliches Modell. II. Die Konsequenzen der Schalenbildung. Nat Mus 109:1–9
Brugerolle G, Bricheux G (1984) Actin microfilaments are involved in scale formation of the chrysomonad cell Synura. Protoplasma 123:203–212
Caplan AL (1988) Rehabilitating reductionism. Am Zool 28:193–203
D'Arcy Thompson W (1942) On growth and form. Cambridge Univ Press, Cambridge/Mass, 1116 pp
Dobzhansky T, Ayala FJ, Stebbins GL, Valentine JW (1977) Evolution. Freeman, San Francisco, 572 pp

Dullemeijer P (1958) The mutual structural influence of the elements in a pattern. Arch Neerl Zool 13 (Suppl):174–188
Dullemeijer P (1974) Concepts and approaches in animal morphology. Van Gorcum, Assen, 264 pp
Dullemeijer P (1980) Functional morphology and evolutionary biology. Acta Biotheor 29:151–259
Dullemeijer P, Barel GDN (1977) Functional morphology and evolution. In: Hecht MK, Goody PC, Hecht BM (eds) Major patterns in vertebrate evolution. Nato Adv Stud Inst Ser A. Life Sci. Plenum New York London, pp 83–117
Edlinger K (in press) Biomechanik als Entwicklungsfaktor bei Anneliden- und Conchiferenlarven. Cour Forsch Inst Senckenberg, Frankfurt
Frankl VE (1969) Reductionism and nihilism. In: Koestler A, Smithies JR (eds) Beyond reductionism – new perspective in life sciences. Macmillan, New York, pp 396–416
Franz V (1951) Zum jetzigen Stand der Theorie vom biotechnischen Fortschritt in der Pflanzen- und Tiergeschichte. Biol Gen 19:368–386
Gould SJ (1970) Evolutionary paleontology and the science of form. Earth Sci Rev 6:77–119
Gould SJ, Lewontin RC (1979) The spandrels of San Marco and the Panglossian paradigm: a critique of the adaptationist programme. Proc R Soc London Ser B 205:581–598
Grasshoff M (1984) Cnidarian phylogeny – a biomechanical approach. Palaeontogr Am 54:127–135
Gutmann WF (1977) Phylogenetic reconstruction: theory, methodology, and application to chordate evolution. In: Hecht MK, Goody PC, Hecht BM (eds) Major patterns in vertebrate evolution. Nato Adv Stud Inst Ser, Ser A: Life Sci 14:645–669. Plenum, New York London
Gutmann WF (1988) Die Evolution hydraulischer Konstruktionen. Kramer, Frankfurt/M, 201 pp
Gutmann WF, Bonik K (1981) Kritische Evolutionstheorie. Gerstenberg, Hildesheim, 225 pp
Gutmann WF, Vogel K, Zorn H (1978) Brachiopods: Biomechanical interdependences governing their origin and phylogeny. Science 199:890–893
Ho MW, Saunders PT (1979) Beyond neo-Darwinism – an epigenetic approach to evolution. J Theor Biol 78:573–591
Jacob F (1977) Evolution and tinkering. Science 196:1161–1166
Kier PM (1965) Evolutionary trends in Paleozoic echinoids. J Paleontol 39:436–465
Koestler A, Smithies JR (1969) Beyond reductionism – new perspectives in the life sciences. Macmillan, New York, 435 pp
Lauder GV (1981) Form and function: structural analysis in evolutionary morphology. Paleobiology 7:430–442
Liem KF, Wake DB (1985) Morphology: Current approaches and concepts. In: Hildebrand M, Bramble DM, Liem KF, Wake DB (eds) Functional vertebrate morphology. Harvard Univ Press, Cambridge/Mass, pp 366–414
Lorenz K (1978) Das Wirkungsgefüge der Natur und das Schicksal des Menschen. Piper, München Zürich, 367 pp
Nachtigall W (1971) Biotechnik. Quelle & Meyer, Heidelberg, 127 pp
Nachtigall W (1977) Biophysik des Fliegens. In: Hoppe W, Lohmann W, Markl H, Ziegler H (eds) Biophysik. Ein Lehrbuch. Springer, Berlin Heidelberg New York, pp 537–550
Peters DS (1983) Evolutionary theory and its consequences for the concept of adaptation. In: Grene M (ed) Dimensions in Darwinism. Cambridge Univ Press, Cambridge London New York, pp 315–327
Preuschoft H (1987) Funktionsweise der Sehnen, Beanspruchung und Form der Phalangen und Metacarpalia (Referat). Verh Dt Anat Ges 81:279–292
Raup DM (1966) Geometric analysis of shell coiling: general problems. J Paleontol 40:1178–1190
Raup DM (1968) Theoretical morphology of echinoid growth. In: Macurda DB (ed) Paleobiological aspects of growth and development, a symposium. J Paleontol 42 Mem 2:50–63
Rayner JMV (1981) Flight adaptations in vertebrates. Symp Zool Soc London 48:137–172
Reif W-E (1981) Oberflächenstrukturen und -skulpturen bei schnell schwimmenden Wirbeltieren. Paläontol Kursbuch 1:141–157
Reif W-E (1985a) Konzepte und Geschichte der Funktionsmorphologie. Aufs Reden Senckenb Naturforsch Ges 35:107–131
Reif W-E (1985b) Konstruktionsmorphologie als biologisches Arbeitskonzept bei Hermann Weber. Aufs Reden Senckenb Naturforsch Ges 35:133–142
Reif W-E, Thomas RDK, Fischer MS (1985) Constructional morphology: the analysis of constraints in evolution. Acta Biotheor 34:233–248

Roth G (1982) Conditions of evolution and adaptation in organisms as autopoietic systems. In: Mossakowski D, Roth G (eds) Environmental adaptation and evolution. Fischer, Stuttgart, pp 37–48
Rudwick MJS (1961) The feeding mechanism of the Permian brachiopod Prorichthofenia. Palaeontology 3:450–471
Rudwick MJS (1968) Some analytic methods in the study of ontogeny in fossils with accretionary skeletons. In: Macurda DB (ed) Paleobiological aspects of growth and development, a symposium. J Paleontol 42 Mem 2:35–49
Schindewolf OH (1950) Grundfragen der Paläontologie. Schweizerbart, Stuttgart, 505 pp, 32 pls
Schindewolf OH (1969) Über den „Typus" in morphologischer und phylogenetischer Biologie. Akad Wiss Lit Mainz Abh Math Naturwiss Kl 1969:57–131
Schmidt AM (1987) Morphogenetic forces in diatom cell wall formation. In: Bereiter-Hahn J, Anderson O, Reif W-E (eds) Cytomechanics. Springer, Berlin Heidelberg New York, pp 183–199
Seilacher A (1970) Arbeitskonzept zur Konstruktions-Morphologie. Lethaia 3:393–396
Seilacher A (1972) Divaricate patterns in pelecypod shell. Lethaia 5:325–343
Seilacher A (1973) Fabricational noise in adaptive morphology. Syst Zool 22:451–465
Seilacher A (1979) Constructional morphology of sand dollars. Paleobiology 5:191–221
Seilacher A (1988) Why are nautiloid and ammonite sutures so different? N Jahrb Geol Paläontol Abh 177:41–69
Seilacher A, Drozdzewski G, Haude R (1968) Form and function of the stem in a pseudoplanktonic crinoid (*Seirocrinus*). Paleontology 11:275–282
Thomas RDK (1979) Constructional morphology. In: Fairbridge RW, Jablonski D (eds) The encyclopedia of paleontology. Dowden, Hutchinson & Ross, London, pp 482–487
Urbanek A (1960) An attempt at biological interpretation of evolutionary changes in graptolite colonies. Acta Palaeontol Pol 5:127–210
Urbanek A (1963) On generation and regeneration of cladia in some Upper Silurian monograptids. Acta Palaeontol Pol 8:135–254
Urbanek A (1966) On the morphology and evolution of the Cucullograptinae (Monograptidae, Graptolithina). Acta Palaeontol Pol 11:291–544
Urbanek A (1988) Morpho-physiological progress. In: Nitecki MH (ed) Evolutionary progress. Univ Chicago Press, Chicago London, pp 195–216
Valentine JW (1975) Adaptive strategy and the origin of grades and ground-plans. Am Zool 15:391–404
Vermeij GJ (1987) Evolution and escalation: an ecologic history of life. Univ Press, Princeton/NJ, 527 pp
Vogel K (1989) Constructional morphology and the reconstruction of phylogeny. Abh Naturwiss Ver Hamburg (NF) 28:255–265
Vogel K, Gutmann WF (1981) Zur Entstehung von Metazoen-Skeletten an der Wende vom Präkambrium zum Kambrium. Festschr Wiss Ges Goethe-Univ Frankfurt/M. Steiner, Wiesbaden, pp 517–537
Vogel K, Gutmann WF (1988) Protist skeletons: biomechanical preconditions and constructional utilization (Teichert-Festschrift). Senckenb Leth 69:171–188
Waddington CH (1956) Genetic assimilation of the bithorax phenotype. Evolution 10:1–13
Wake DB, Larson A (1987) Multidimensional analysis of an evolving lineage. Science 238:42–48
Weber H (1955) Stellung und Aufgaben der Morphologie in der Zoologie der Gegenwart. Zool Anz Suppl 18:137–159
Weber H (1958) Konstruktionsmorphologie. Zool Jahrb Abt Allg Zool Phys Tiere 68:1–112
Whyte LL (1967) Directive agencies in organic evolution. J Theoret Biol 17:312–314

Part II Bioconstruction: Operational Interconnections, Behavior, Environment

Dynamic Similarity in the Analysis of Animal Movement

R. McN. Alexander[1]

Summary

Dynamic similarity is a concept from physical science, related to the more familiar concept of geometric similarity. Two motions are dynamically similar if one could be made identical to the other by uniform changes of the scales of length, time and force. This chapter asks whether different-sized animals move in dynamically similar fashion.

When gravitational forces are important (as in running), dynamic similarity is possible only if the motions being compared have equal Froude numbers (v^2/gl, where v is velocity, g is gravitational acceleration and l is a length – leg length in studies of running). Studies of running show that mammals ranging from small rodents to rhinoceros generally move in approximately dynamically similar fashion, when their speeds are such as to make their Froude numbers equal.

When elastic or viscous forces are important, dynamic similarity requires equal Strouhal numbers or Reynolds numbers, respectively.

The concept of dynamic similarity has proved valuable in studies of running but has not so far contributed much to studies of swimming or flight.

1 Geometric Similarity

This chapter makes comparisons between the movements of different-sized animals. Do the movements of elephants resemble those of mice, and do sharks move like minnows? I will try to demonstrate the value and limitations of a concept from physical science, the concept of dynamic similarity. However, it will be convenient to start by reminding readers of the related concept of geometric similarity, which is concerned not with movement but with form.

Two animals or other objects are geometrically similar if one could be made identical in form to the other by multiplying all its linear dimensions by the same factor. For example, a triangle of sides, 3, 4 and 5 cm is geometrically similar to one of sides 6, 8 and 10 cm. If geometrically similar objects are made of materials of the same density (as related animals generally are), their linear dimensions must be proportional to the cube roots of their masses.

In many cases, related animals of different sizes are quite close to being geometrically similar. For example, the snout-vent lengths of lizards ranging in mass

[1] Dept. of Pure and Applied Biology, University of Leeds, Leeds LS2 9JT, UK

N. Schmidt-Kittler and K. Vogel (Eds):
Constructional Morphology and Evolution

from 0.5 g to 7 kg are proportional to (body mass)$^{0.33}$ (Pough 1980). The lengths of the principal limb bones of mammals ranging from 3-g shrews to 3-t elephants are proportional to (body mass)$^{0.35}$ (Alexander et al. 1979). In other cases, related animals are much less close to geometric similarity. If we examine the family Bovidae alone, instead of the mammals as a whole, we find limb bone lengths proportional to (body mass)$^{0.26}$ (Alexander et al. 1979). The wings of hummingbirds have spans proportional to (body mass)$^{0.53}$ (Rayner 1988).

2 Dynamic Similarity

Two motions are dynamically similar if one could be made identical to the other by multiplying all lengths by one factor, all times by a second factor and all forces by a third one. Dynamic similarity is possible only in certain circumstances, which depend on the nature of the forces involved (Duncan 1953).

To discover some conditions that are often important, consider the simple models of running animals (masses mounted on springs) that are shown in Fig. 1. Imagine them bouncing along, like children on pogo-sticks. In each bounce, kinetic energy is converted to potential energy and back again. For dynamic similarity the models must have equal ratios of kinetic energy $\frac{1}{2}mv^2$ to potential energy mgl: thus, they must have equal values of v^2/gl. This is a dimensionless number, called the Froude number. More generally, motions in which gravity is important cannot be dynamically similar unless they have equal Froude numbers v^2/gl, where v and l are a speed and a length characteristic of the motion.

Each time the bouncing models land on the ground, their springs are compressed and recoil: kinetic energy is converted to elastic strain energy and back again. For dynamic similarity they must have equal ratios of kinetic energy to elastic strain energy. If the springs were squashed flat on the ground, reducing l to zero, the strain energy would be $\frac{1}{2}Kl^2$. The ratio of this to the kinetic energy $\frac{1}{2}mv^2$ is Kl^2/mv^2. K/m is $4\pi^2f^2$, where f is the natural frequency with which the mass would vibrate up and down on the spring (Alexander 1983). Thus dynamic similarity requires equal values of f^2l^2/v^2 and so of its square root, the Strouhal number fl/v. This is a general condition that applies to all motions in which elastic forces are important.

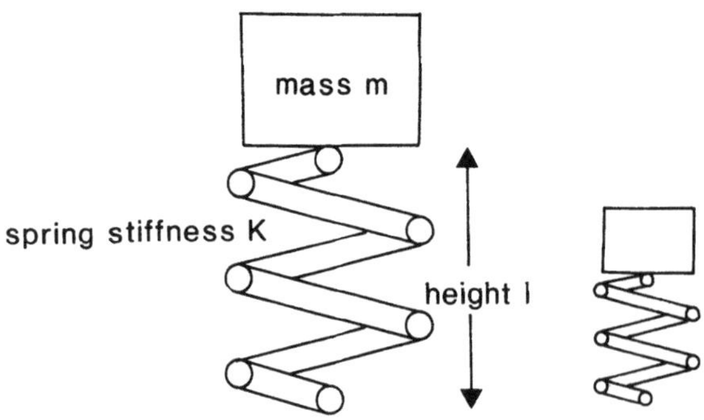

Fig. 1. Simple models of running animals of different sizes

If gravity and elastic forces are both important, dynamic similarity requires equal ratios of gravitational potential energy mgl to strain energy $\frac{1}{2}Kl^2$: it requires equal values of mg/Kl. When the models rest on their springs they compress them by mg/K, so this condition says that the models must deform in geometrically similar fashion under their own weight. This is the condition of elastic similarity, discussed by McMahon (1973).

The frequency used in defining the Strouhal number need not necessarily be a natural frequency of vibration: it may be the frequency of a periodic motion such as the stride frequency of a mammal or the tail beat frequency of a fish. In that case, f/v is the reciprocal of the stride length and equality of Strouhal numbers merely means that stride lengths are proportional to the linear dimensions of the body.

If viscous forces are important, dynamic similarity requires equal Reynolds numbers $\rho vl/\mu$, where ρ is the density and μ the viscosity of the fluid. If surface tension σ is important it requires equal Weber numbers $\rho v^2 l/\sigma$. Other conditions apply for other kinds of forces (Duncan 1953).

3 Running

Gravity is important for running animals, so equal Froude numbers are needed for dynamically similar movement. One of the characteristics of dynamically similar movement is that stride length should be proportional to leg length. Alexander (1976) predicted that mammals of different sizes should have equal relative stride lengths (stride length/leg length) when walking or running at equal Froude numbers (v^2/gl, where v is forward speed and l is leg length). He found that this was approximately true. Later, Alexander and Jayes (1983) formulated a more general hypothesis, that quadrupedal mammals of different sizes tend to move in dynamically similar fashion, when travelling with equal Froude numbers. We called this the dynamic similarity hypothesis.

Similar animals can walk or run in dynamically similar fashion only when their Froude numbers are equal. Even in that case there is no necessity for them to do so: for example, one might take relatively long strides and another relatively short ones, or one might trot and another gallop. However, there is a good reason to expect them to move in dynamically similar fashion. If one animal has adjusted its gait to minimize the mechanical power required of its muscles, a similar animal travelling at the same Froude number must move in dynamically similar fashion to minimize its power. There are some difficulties with this argument because the relationship between mechanical and metabolic power in running is not straightforward but I have argued (Alexander 1989) that gaits that minimize mechanical power will generally be advantageous.

Alexander and Jayes (1983) tested various predictions of the dynamic similarity hypothesis. The mammalian gaits that are generally recognized (walk, trot, gallop, etc.) are distinguished principally by the phase relationships between the movements of the feet. The hypothesis predicts that these phase relationships will be the same, for different mammals travelling at the same Froude number. Figure

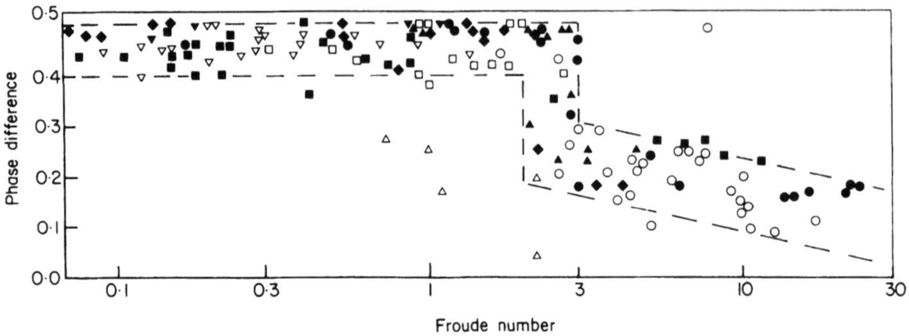

Fig. 2. A graph of the phase difference between the forefeet, against Froude number, for ●, dogs (*Canis*); ■, sheep (*Ovis*); ◆, camel (*Camelus*); ▲, ▼ rhinoceros (*Rhinoceros* and *Ceratotherium*); ○ ferret (*Putorius*); □, rat (*Rattus*); △, jird (*Meriones*); and ▽, coypu (*Myocastor*). (Alexander and Jaynes 1983)

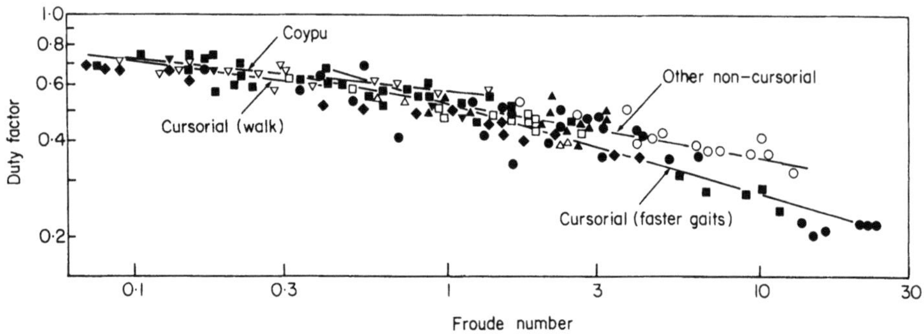

Fig. 3. A graph of the duty factor of the forefeet (the fraction of the stride for which each is on the ground) against Froude number. *Symbols* as in Fig. 2. (Alexander and Jayes 1983)

2 shows that this is approximately true, for the phase difference between the two fore feet. The sharp change that occurs (with very few exceptions) between Froude numbers of 2 and 3 is the change from trotting to galloping. The hypothesis also predicts that at the same Froude number, different animals will keep their feet on the ground for equal fractions of the stride. This proved to be approximately true (Fig. 3). Further, the patterns of force exerted on the ground should be similar: this also we found to be true. Finally (as already noted) relative stride lengths should be equal. This proved not to be true of mammals as a whole, but approximately true within each of three groups, the non-cursorial and cursorial mammals (Fig. 4) and the primates (Alexander and Maloiy 1984). Non-cursorial mammals are those that run on strongly bent legs, such as rats and ferrets. Cursorial mammals such as dogs and antelopes run with their legs much straighter. At any particular Froude number, primates take relatively longer strides than non-cursorial mammals, which in turn take longer strides than cursorial ones.

Pennycuick (1975) had previously measured the stride frequencies of a diverse group of cursorial mammals, walking, running or galloping in their natural habi-

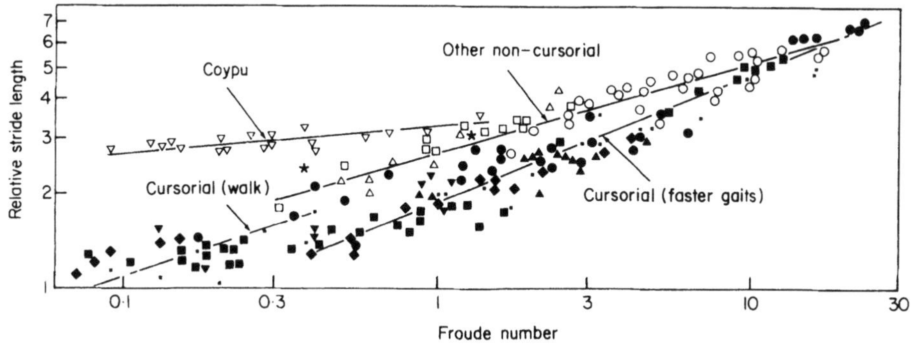

Fig. 4. A graph of relative stride length against Froude number. *Symbols* as in Fig. 2, with the addition of (*star*) cat and (*small dot*) horse; note that *filled symbols* refer to cursorial and *open symbols* to non-cursorial mammals. (Alexander and Jayes 1983)

tat in Kenya. He found that in each gait, stride frequency was *inversely* proportional to the square root of leg length. The dynamic similarity hypothesis suggests that the Froude number v^2/gl corresponding to the preferred speed within each gait should be the same for different mammals, and that at this Froude number they should all use the same relative stride length (L/l, where L is stride length). Since their speeds v would then be proportional to the square root of leg length, stride frequency (v/L) should be inversely proportional to the square root of leg length, as it is observed to be.

Strictly speaking, dynamic similarity is possible only between geometrically similar animals in which (for example) leg diameter is the same fraction of leg length. However, the diameters of legs seem less essential to the motion than their lengths, and it seems reasonable to look for something close to dynamic similarity between a stout-legged buffalo and a slender-legged gazelle.

We have seen that if gravity and elastic forces are both important, dynamic similarity requires elastic similarity. For animals made of the same materials, elastic similarity requires larger animals to have relatively thicker legs (McMahon 1973). Ungulate mammals including the antelopes (Bovidae) are especially well adapted for exploiting tendon elasticity in running (Alexander 1988). It may be no coincidence that the Bovidae differ from non-ungulate groups of mammals in being more nearly elastically than geometrically similar in the dimensions of their skeleton and tendons (Alexander et al. 1979, 1981).

The dynamic similarity hypothesis was originally proposed for quadrupedal mammals but it also holds quite well for bipedal hopping mammals and for birds (Hayes and Alexander 1983). It seems likely to hold also within major groups of reptiles and of arthropods. For example, cockroaches (*Periplaneta*) make an abrupt change of gait between slow and fast walking (Kozacik 1981). I predict that cockroaches of different sizes will be found to make the change at the same Froude number. (In testing this prediction, it will be convenient to use the overall length of a leg as the length dimension in the Froude number because the height of the hip joint from the ground, which has been used in studies of mammals, would be difficult to measure in cockroaches.)

4 Flight

Gravitational forces are important for flying animals, because the lift on their wings must support their weight. Therefore, dynamically similar movements of flying animals and of the air in their wakes are possible only if they move with equal Froude numbers. Viscous forces are also important because they act in the boundary layer of air close to the animal. Therefore, dynamic similarity requires equal Reynolds numbers. For animals of different sizes to have equal Froude numbers v^2/gl their speeds v must be proportional to the square roots of their linear dimensions l. For them to have equal Reynolds numbers $\rho vl/\mu$ their speeds must be inversely proportional to their linear dimensions. Therefore they cannot have their Froude and Reynolds numbers both equal: precise dynamic similarity is impossible for flying animals of different sizes.

We may nevertheless ask, whether animals of different sizes show anything approximating dynamic similarity in their flight. Like aircraft, they get the lift needed to support their weight by giving the air, through which they pass, a downward induced velocity. If different animals flew in dynamically similar fashion their induced velocities (w) would be proportional to their airspeeds (v). By a standard argument presented (for example) by Alexander (1977)

$$w/v = 4\,mg/\pi\rho\, b^2 v^2 \ , \tag{1}$$

where mg is the animal's weight, ρ is the density of the air and b is the wingspan. Dynamic similarity would require the animals to travel at speeds that gave them equal values of the expression on the right hand side of Eq. (1). For geometrically similar animals of equal density, m would be proportional to b^3 and this expression might be interpreted as a multiple of bg/v^2: that is, of the reciprocal of a Froude number. We will also use equation 1 in a different form, expressing it in terms of the wing area S and the aspect ratio A ($= b^2/S$)

$$w/v = 4\,mg/\pi\rho\, ASv^2 \ . \tag{1a}$$

A different line of argument that takes no account of similarity criteria suggests optimum speeds for flight (Ward-Smith 1984). An aircraft or flying animal has a minimum power speed (at which the mechanical power required to keep it airborne is minimal) and a maximum range speed (at which the work required to travel unit distance is minimal). Depending on the particular circumstances, animals should fly at one or other of these speeds, or at some multiple of one of them. This requires

$$(mg/\rho S v^2)(AC_{DO})^{-\frac{1}{2}} = K \ , \tag{2}$$

where C_{DO} is the zero lift drag coefficient and the constant K takes one value for the minimum power speed and another for the maximum range speed. Geometrically similar animals would have equal values of the aspect ratio A. They would also have approximately equal values of C_{DO}, since drag coefficients are affected rather little even by large differences of Reynolds number, in the range of Reynolds number that apply to birds, bats and all but very small insects (Alexander 1983). Geometrically similar animals that flew at equivalent optimum speeds and so had equal values of K [Eq. (2)] would also have very nearly equal ratios of induced velocity to airspeed [Eq. (1 a)].

The conditions represented by Eqs. (1 a) and (2) both require animals of different sizes to use equal lift coefficients $2\,mg/\varrho Sv^2$. Unfortunately, few reliable measurements have been made of the airspeeds of flying animals, from which lift coefficients can be calculated. However, Rayner's (1985) data on the speeds of (mainly terrestrial) birds and Pennycuick's (1987) data for seabirds both seem to show that small birds use lower lift coefficients than large ones. In any case, birds of different sizes are not generally geometrically similar; they tend to have wing spans proportional to (body mass)$^{0.39}$ (Rayner 1988).

Birds and bats use two flying gaits, a slow one in which lift is produced only in the downstroke and a fast one in which lift is produced throughout the wing beat cycle (Rayner et al. 1986). We might predict that the transition will be found to occur at equal Froude numbers in different species but so far the transition speed has not been determined precisely for any species.

Hummingbirds and many insects hover by beating their wings in a near-horizontal plane. We might predict that their hovering movements will be dynamically similar but, because the body is stationary in hovering, we cannot use its velocity in tests for dynamic similarity. It is more appropriate to use the induced velocity w, the velocity at which the animal's wing beats make air flow past it. The wing length r is an appropriate length. If animals of different sizes hover in dynamically similar fashion they will have equal Froude numbers w^2/gr and equal Strouhal numbers fr/w, where f is wing beat frequency. The theory of helicopters tells us that w is proportional to $m^{0.5}/r$, where m is body mass (Alexander 1983). Thus the Froude number is proportional to m/gr^3 and the Strouhal number to $fr^2/m^{0.5}$. For different animals to have equal Froude and Strouhal numbers their wing lengths must be proportional to $m^{0.33}$ and their wing beat frequencies to $m^{-0.17}$. Weis-Fogh (1973) studied a wide range of hovering animals from 1 mg mosquitoes to 20 g hummingbirds. His data (from his Table 4, excluding species marked with asterisks) show wing lengths proportional to $m^{0.37}$ and wing beat frequencies to $m^{-0.30}$. Thus Froude numbers were proportional to $m^{-0.11}$ and Strouhal numbers to $m^{0.07}$, and the requirements of dynamic similarity seem not be met. For example, the Strouhal number is predicted to be twice as large for a 20-g hoverer as for a 1-mg one. This implies that the ratio of vortex ring spacing in the wake, to ring diameter, will be half as large for the larger animal. A further requirement of dynamic similarity would be that the Reynolds numbers $\varrho wr/\mu$ should be equal but this is clearly impossible for animals of different mass if w is proportional to $m^{0.5}/r$ (if air density ϱ and viscosity μ are constant).

5 Swimming

Carangiform fish and sharks of different species and sizes make very similar movements when they swim (Bainbridge 1958; Webb and Keyes 1982). Tail beat amplitude, the wavelength of the body waves and the distance travelled per tail beat cycle are all about proportional to body length, but they change a little with speed. Gravity has no importance for a neutrally buoyant fish, so Froude numbers have no significance. Viscous forces are important and we might predict that fish would swim in dynamically similar fashion when their Reynolds numbers were equal (a large fish swimming slowly and a small one swimming fast might have equal Reynolds numbers). If there was dynamic similarity of fish and water movements, different-sized fish would have equal drag coefficients. However, Webb et al. (1984) calculated drag coefficients from the kinematics of swimming of trout (*Salmo*) and showed that they are different for trout of different sizes at the same Reynolds number.

6 Conclusion

In its original form, the dynamic similarity hypothesis of Alexander and Jayes (1983) predicted that quadrupedal mammals, walking or running at speeds that gave them equal Froude numbers, would move in dynamically similar fashion. This prediction has been found to be approximately correct for major gait parameters of a wide variety of mammals. However, with respect to stride length, quadrupedal mammals fall into three groups: cursorial and non-cursorial mammals and primates. The hypothesis has been extended successfully to bipedal mammals and birds and seems likely to be useful also in analysis of terrestrial gaits of arthropods. It has general value in relating size, speed and gait in legged locomotion.

The hypothesis might be extended further to include swimming and flight. We might predict that different animals using the same style of locomotion (flapping flight, undulatory swimming etc.) will move in dynamically similar fashion whenever their speeds are in the ratio that makes this possible. However, different-sized animals cannot fly in dynamically similar fashion because their Froude numbers and Reynolds numbers cannot simultaneously be equal. A more promising prediction is that in similar circumstances, flying animals of different sizes will choose speeds that are equal multiples of the minimum power speed.

The extended dynamic similarity hypothesis would predict that geometrically similar fish that were neutrally buoyant and swam well below the surface would swim in dynamically similar fashion, whenever their speeds were in the ratio that made their Reynolds numbers equal. (Gravitational forces would become important and Froude numbers also would have to be equal if the fish were denser than water or if their swimming generated surface waves). This prediction would not be very useful in comparing fish of grossly different sizes because they would be unlikely to use overlapping ranges of Reynolds numbers: large fish generally swim faster than small ones and so have much larger Reynolds numbers. In any case, calculated drag coefficients at the same Reynolds number are not the same for trout of different sizes (Webb et al. 1984).

Thus the usefulness of the dynamic similarity hypothesis seems at present to be confined largely to walking and running.

Acknowledgment. Dr. J.M.V. Rayner made useful comments on a draft of this paper.

References

Alexander R McN (1976) Estimates of speeds of dinosaurs. Nature (London) 261:129–130
Alexander R McN (1977) Swimming. In: Alexander R McN, Goldspink G (eds) Mechanics and energetics of animal locomotion. Chapman & Hall, London, pp 222–248
Alexander R McN (1983) Animal mechanics, 2nd edn. Blackwell, Oxford
Alexander R McN (1988) Elastic mechanisms in animal movement. Univ Press, Cambridge
Alexander R McN (1989) Optimization and gaits in the locomotion of vertebrates. Physiol Rev 69:1199–1227
Alexander R McN, Jayes AS (1983) A dynamic similarity hypothesis for the gaits of quadrupedal mammals. J Zool London 201:135–152
Alexander R McN, Maloiy GMO (1984) Stride lengths and stride frequencies of primates. J Zool London 202:577–582
Alexander R McN, Jayes AS, Maloiy GMO, Wathuta EM (1979) Allometry of the limb bones of mammals from shrews (*Sorex*) to elephant (*Loxodonta*). J Zool London 190:155–192
Alexander R McN, Jayes AS, Maloiy GMO, Wathuta EM (1981) Allometry of leg muscles of mammals. J Zool London 194:539–552
Bainbridge R (1958) The swimming speed of fish as related to size and to the frequency and amplitude of the tail beat. J Exp Biol 35:109–133
Duncan WJ (1953) Physical similarity and dimensional analysis. Arnold, London
Hayes G, Alexander R McN (1983) The hopping gaits of crows (Corvidae) and other bipeds. J Zool London 200:205–213
Kozacik JJ (1981) Stepping patterns of the cockroach *Periplaneta americana*. J Exp Biol 90:357–360
McMahon TA (1973) Size and shape in biology. Science 179:1201–1204
Pennycuick CJ (1975) On the running of the gnu (*Connochaetes taurinus*) and other animals. J Exp Biol 63:775–799
Pennycuick CJ (1987) Flight of auks (Alcidae) and other northern seabirds compared with southern Procellariiformes: ornithodolite observations. J Exp Biol 128:335–347
Pough FH (1980) The advantages of ectothermy for tetrapods. Am Nat 115:92–112
Rayner JMV (1985) Flight, speeds of. In: Campbell B, Lack E (eds) A dictionary of birds. Poyser, Calton, pp 224–226
Rayner JMV (1988) Form and function in avian flight. Curr Ornithol 5:1–66
Rayner JMV, Jones G, Thomas A (1986) Vortex flow visualizations reveal change in upstroke function with flight speed in bats. Nature (London) 321:162–164
Ward-Smith AJ (1984) Biophysical aerodynamics and the natural environment. John Wiley & Sons, Chichester
Webb PW, Keyes RS (1982) Swimming kinematics of sharks. Fish Bull 80:803–812
Webb PW, Kostecki PT, Stevens ED (1984) The effect of size and swimming speed on locomotor kinematics of rainbow trout. J Exp Biol 109:77–95
Weis-Fogh T (1973) Quick estimates of flight fitness in hovering animals including novel mechanisms for lift production. J Exp Biol 59:169–230

Cytomechanics and Biochemistry

JÜRGEN BEREITER-HAHN [1]

Summary

Cells are comprehended as self-organizing structures. The relationship between the quality of molecules as determined by the genome and mechanical forces developed by the structures into which these molecules are integrated is exemplified by discussion of generation of cell shape, interaction with the extracellular environment, and by microcompartmentation. Cytomechanics and biochemistry are shown to represent complementary modes of description of any living cell. As a consequence of the observation of "network causality", classical logics may be insufficient to describe the cause/effect relations established in self-organizing systems.

1 Different Laws = Different Entities?

Our analytical mode of research favors concepts reducing the highly complex organismic systems to certain dominating aspects, such as "information transducers" or "chemical machines". Concomittant with this way of thinking is the restriction to linear causal relationships (causa efficiens). The concept of hypercycle (Eigen and Schuster 1978) taught us the existence of complicated circular causal relationships, i.e., any effect is cause for the next reaction and by this, any effect finally is drawn back, causing its own cause. Thus self-organizing systems overcome the idea of a "circulus vitiosus" as being a forbidden construction in terms of classical logics. However, even this concept seems to be too simple to describe the branched and interdigitating cause/effect network realized in a living organism. A tiny section of such a network causality will be outlined at the end of this contribution (Fig. 2).

The development of any organism in the course of evolution and during ontogenesis is based on a continuity of

- information,
- structure,
- chemical reactions,
- interaction with its environment.

[1] Cinematic Cell Research Group, Johann Wolfgang Goethe Universität Frankfurt/M., Senckenberganlage 27, 6000 Frankfurt/M., FRG

N. Schmidt-Kittler and K. Vogel (Eds):
Constructional Morphology and Evolution

These features are mutually interdependent, their separate treatment results from our inability in multidimensional description and research. A few examples are thought to elucidate the interdependence of these features. Structure itself may provide the information for its own replication (i.e., biological membranes), or it may influence the informational flux between environment and organism or between subsystems in an organism, as was proposed by the model of "dynamic reciprocity" (Bissell and Aggeler 1987, see below). Because of the dynamic character of any biological structure, it results from the continuity of chemical reactions; biological structures are dissipative structures. Interaction with the environment occurs by exchange of matter, of energy, by acquisition of information about the environment, and by mechanical interaction, which can be considered to be a special case of energy and information exchange. Structural continuity assures the continuity of force transmission throughout ontogeny.

A similar description of organisms has been proposed by Gutmann and Bonik (1981) constructing a bionomial circle including the levels of metabolism, biomechanics, and hydromechanics. These three levels were considered to represent separate entities integrated into the bionomial circle. The reasons for level separation are derived from methodological requirements (each level has to be described by its specific laws). However, whether the non-reducibility of the laws characterizing each level is a reason sufficient to assume the levels being separate entities rather than comprising different aspects of an entire system (organism) remains a matter of discussion. With the present chapter I intend to provide some evidence which is in favor of the hypothesis that mechanics and biochemistry are *complementary* aspects of biological structure. Negligence of either – chemical and mechanical – properties prevents the understanding of the construction of an organism and its function.

I will not discuss synthesis and post-translational modifications of cytoskeletal proteins and of those endowing cytoskeletal structures with their specific mechanic abilities. It is trivial that biochemical reactions produce the machinery which generates the forces shaping a cell.

2 The Cell, a Mechanical System

The lowest level of living organisms is that of the cells. Cells have a defined shape, and many are motile, thus cells are structures subjected to the laws of mechanics and they are mechanically interacting with their environment.

The two main structures determining cytomechanics are:

- a flexible envelope (plasma membrane),
- a fluid-penetrated fibrillar meshwork.

Additional factors include cell walls, attachment to solid substrata, specific propagating organelles (flagella, cilia). The interaction of the envelope and the fluid-penetrated fibrillar meshwork makes a cell a hydraulic system (Bereiter-Hahn 1977): the flexible plasmamembrane encloses a liquid-filled space with intercalating polymers (Fig. 1). These fibrillar polymers reinforce the plasmamembrane and may connect various regions of a cell by traversing the cytoplasmic

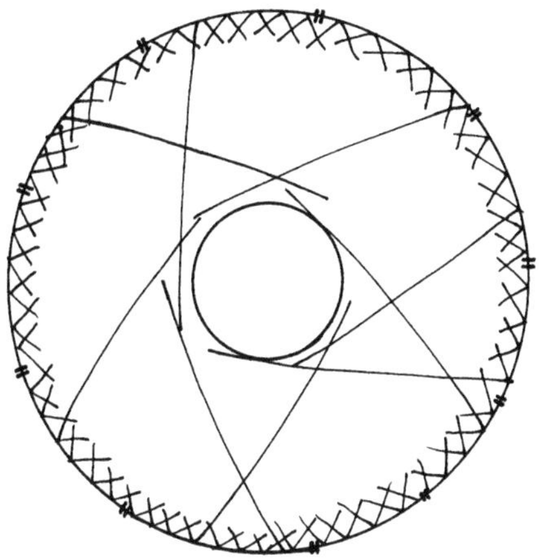

Fig. 1. Schematic representation of filament arrangements (actin fibrils) in a spherical cell. From the submembrane cortical fibrillar meshwork fibrils extend into a loose fibrillar system traversing throughout the cytoplasm. The *dark double spots* in the membrane represent Na/K-pumps, which also contribute to the internal pressure of this hydraulic system

space; the nucleus represents a second internal hydraulic system often enclosed in a fibrillar "basket" and its envelope being connected with some of the cytoplasmic fibrils (intermediate f.). Components of the fibrillar meshwork are endowed with the ability to contract actively (actomyosin), others (intermediate filaments) are assumed to contribute their elastic properties to the maintainance of cell shape, the third group (microtubules) exhibits some flexural rigidity forming a scaffold along which organelles may be displaced and which may serve as an abutment for the contractile cytoplasmic meshwork. The mechanical properties of these fibrillar elements are determined by the arrangement and intermolecular forces acting in between their monomer constituents. During differentiation and morphogenesis these fibrillar systems may develop separately; however, at a certain stage of development they interact with each other. A convincing example of such a sequence has been revealed for the erythrocyte differentiation (Lazarides 1987).

3 Forces in the Hydraulic System Result from Enzyme Activities

If the fibrils are distributed homogeneously, a free-living cell forms a sphere, any shape deviating from the sphere requires a nonuniform arrangement of the fibrillar system or nonuniform force generation. Assuming the presence of a semi-permeable membrane and steady-state conditions, such a spherical cell may either be in osmotic equilibrium with the surrounding or the cytoplasm may be hypertonic. In the latter case, the membrane is subjected to tension at the surface, the fibrillar meshwork has to counteract the difference in osmotic pressure. In the first case, only the flexural rigidity of the membrane and the associated cortical fibrillar meshwork contribute to the mechanical properties of the cortex; it remains highly flexible. If the gel is of considerable density, the gel itself may develop elasticity and thus stiffness (Oster and Odell 1984). Thus shape alone is not indicative of the physiological state of a cell (whether relaxed or contracted).

An additional consequence has to be drawn from these considerations, any contraction of the fibrillar meshwork if anchored to the membrane, gives rise to an osmotic pressure difference between the cell and its environment which (in the state of equilibrium) balances the contraction-generated pressure. This was stated first by Layrand and coworkers (1972). Thus any activities controlling the ion content of a cell interfere with the mechanical properties and activities of this cell. Osmotic pressure differences are generated by the Na^+/K^+-pump ($2\,K^+$ are transported into the cell and $3\,Na^+$ are pumped out at the same time) causing an electrical and an osmotic potential. The $Na^+/K^+/2\,Cl^-$ cotransport system and ion channels represent further factors involved in the regulation of cell volume by changing ionic content. In fact, the situation is even more complex because of Na^+/Ca^{2+} exchange (and other calcium and proton transporting systems) which may initiate contractions by increasing cytosolic calcium concentration. This example shows that activities of ionic transport systems and of enzymes not belonging to the group of "mechanoenzymes" are taking part in pressure (i.e., mechanical force) generation. Thus the numerous factors controlling cellular activities (i.e., differentiation, proliferation) by shifting ion fluxes (Durham and Walton 1982; Leister et al. 1985; Frantz 1985; Paris and Pouyssegur 1986; Rozengurt 1983; Rubin et al. 1979) disturb the mechanical equilibrium as well.

4 Dynamics of Structure Are Related to Mechanical Activities

Where the membrane loses its connection to the fibrillar cortex beneath, it becomes extended to the degree allowed by the lateral molecular interactions. If the extension exceeds these cohesive forces, the membrane ruptures. This happens only in cases of very fast swelling. Stretching of the membrane opens ion channels (Lansman et al. 1987) and thus induces the polymerization of actin at its inner surface, giving rise to a new cortical layer. Oriented fibrils can be formed from an isotropic meshwork if it is fixed at least at two sites (e.g., Opas 1987; Brix and Stockem 1987). This process normally requires the formation of adhesion sites to a solid substratum. Assembly of actin fibers at the edge of a membrane is a common phenomenon (Tilney 1983; Svitkina et al. 1986; Bershadsky et al. 1980), and in locomoting cells it yields the material for a continuous flow of material from the edge towards the cell center (for detailed discussion see, e.g., Bereiter-Hahn 1985; Bereiter-Hahn et al. 1981, 1990a). There the actin fibrils disassemble and the short pieces or the oligomeres percolate in the direction of movement, providing the material for polymerization at the leading edge. Polymerization and depolymerization could be a direct consequence of actomyosin contraction and lamella extension, resulting in a high concentration of F-actin in the first case and a low F-actin concentration in the latter, thus shifting the chemical equilibrium between F- and G-actin. This in turn could drive assembly at the membrane and disassembly close to the cell center (Alt 1990).

The influence of filament assembly and disassembly on cytoplasmic mechanical properties is obvious: viscosity changes; only F-actin can form fibrillar bundles and reinforce the hydraulic system, only F-actin promotes contraction together with myosin, bundles of F-actin can be used to build structures with considerable stiffness, as are microvilli (Mooseker 1974; Tilney and Mooseker 1971) or the acrosomal processes (Tilney et al. 1973).

5 The Role of F-Actin in Microcompartmentation

Actin is one of the most frequent proteins in a cell (10–20% of total protein mass), and it exhibits affinities to many molecules. In addition to the specific actin-binding proteins, the class of macromolecules best known for their ability to associate with F-actin are enzymes of the glycolytic pathway (Clarke and Masters 1976; Lynch and Paul 1988), which also bind to another group of cytoskeletal elements, the microtubules (Karkhoff-Schweizer and Knull 1987). These associations result in an enhancement of enzyme activity. One functional significance of this association is the supply with ATP by substrate chain phosphorylation to drive contractions (Tillmann and Bereiter-Hahn 1985), another one seems to be an involvement in the control of actin and tubulin polymerization (Bershadsky and Gelfand 1981, 1983). This can be shown for instance by polymerizing G-actin in the presence of glycolytic enzymes. Some of them enhance the initial rate of polymerization, a process which depends on the presence or absence of a suitable substrate (Blum and Bereiter-Hahn 1988, 1990).

After trypsinization, metazoan cells in culture are rounded and have to reorganize their cytoskeleton during the subsequent period of spreading. The early phases of spreading entirely depend on the supply of energy from respiration (Bereiter-Hahn et al. 1990b) while spread cells, having a well-developed actin system, cover about 50% of their energy demand from aerobic glycolysis (Tillmann and Bereiter-Hahn 1986; Bereiter-Hahn et al. 1987; Korohoda 1971, 1972).

Another point is the limitation in free mobility of adenine nucleotides. Under certain conditions only 20% of adenine nucleotides may leak out from cells (cell cultures) with membranes made permeable for small molecules (Woiteneck 1989, diploma thesis; Frankfurt/M.); NMR-studies of cardiomyocytes also point to a restricted mobility of ATP and ADP. However, the binding sites have not been identified yet.

The restriction of the free mobility of metabolites by more or less loose association with cytoplasmic matrix structures has been termed "microcompartmentation" (for overview see, e.g., Clegg 1988). This concept opened a new and challenging field of research. It can be extended to RNA and to small organelles as are the so-called "free" ribosomes which are bound to intermediate filaments (Traub and Nelson 1982; Traub and Vorgias 1984).

The mutual interaction of F-actin and microtubules with glycolytic enzymes and of intermediate filaments with RNA are further examples of structures primarily devoted to cell mechanics being involved in the control of metabolic activities which most probably are not directly related to biomechanics.

6 Controlling Cell Biochemistry by Mechanical Stresses

In the last Section some examples of the interaction of elements of the cytoskeleton with enzymes and metabolites were described. If these associations are of physiological significance, any factors influencing the organization of the cytoskeleton should alter the physiological condition of a cell, i.e., it should be involved in control of differentiation and metabolic adaptation to a changing environment.

Genetic information, transcription, and translation are the first determinants controlling cell structure by providing the "building materials". However, just as bricks do not build a house, protein molecules do not shape a cell. The ionic conditions, energy supply, and (as discussed above) the presence of enzymes and metabolites control the polymerization of the cytoskeletal proteins to form fibrillar structures (e.g., Bershadsky and Gelfand 1981, 1983; Svitkina et al. 1986; Tillmann and Bereiter-Hahn 1985). Mechanical stresses acting on the cells induce a certain pattern of the fibers during the process of their formation (e.g., Opas 1987). Mechanical stresses also influence the activity of ionic channels, thus being involved in the control of the cytoplasmic ionic conditions (Lansman et al. 1987). The control of stress fiber arrangement in endothelial cells by rheological stresses is one of the best-known examples (Franke et al. 1987). Mechanical stimulation of cells in culture was found to promote differentiation of muscle (Vanderburgh 1988) and keratinocytes (Görmar et al. 1990); volume changes by membrane-regulated ion permeability underlie lens fiber elongation (Parmelee and Beebe 1988).

Mechanical interactions of cells with their extracellular matrix are required for the organization of the cytoskeletal fibrillar systems, and thus represent a system for exchange of information between the surroundings of a cell and its internal. This interaction represents a crucial prerequisite to metazoan morphogenesis during ontogenesis (McClay and Ettensohn 1987). The hypothesis of "dynamic reciprocity" between the extracellular matrix and gene activities (Bissell and Aggeler 1987) emphasizes the significance of this information exchange. Interactions of intermediate filaments with the nuclear pore complexes and with RNA are thought to represent the structural basis of a control of gene activities by the cytoskeleton. Intermingling of intermediate filaments with microtubules, with actin, and with the plasmamembrane establishes the connection to the outside of the cell and to different metabolic processes in the cytoplasm (e.g., the energy metabolism, see above). Dynamic reciprocity is assumed to underly the relationship of cell shape, locomotion and proliferative activity described among others by Castor (1970, 1971), Folkman and Moscona (1978), Kajstura and Korohoda (1982, 1983), and Korohoda (1971, 1972).

Recently, considerable evidence on the control of gene activities by the chemical and mechanical properties of the extracellular matrix has accumulated (Emermann and Pitelka 1977; Ingber and Folkman 1987; Medina et al. 1987; Opas 1989; Thie et al. 1989). On the other hand, the extracellular matrix is a secretory product of the cells attached to it. The spatial organization of this matrix is a consequence of the orientation of cytoskeletal elements in the secreting cells (Zylberberg et al. 1988) and stresses developed in the tissue and organs as well. The principle that mechanical stresses are involved in the development of structures applies for the organization of cells and of tissues and organs (the next levels of integration) as well [e.g., in brain morphogenesis (Gordon and Brodland 1987) and in connective tissue development (Scott 1987; Tözeren and Skalak 1988)].

7 Conclusions

The starting hypothesis was that biochemistry and mechanics are complementary forms to describe the supramolecular organization of cytoplasm and generation of cell shape. Some mutual interactions of chemical processes and mechanical properties and activities were exemplified. Figure 2 summarizes the considerations and provides an example of the network of causes and effects:

The monomer/polymer equilibrium of elements of the cytoskeleton (arrow 1) is controlled by the ionic states (arrow 3), and by the relative concentration of monomers and polymers, which can be influenced for instance by contractions (arrow 4). Contractions are controlled by ions (arrow 7), they depend on the general arrangement of cytoskeletal elements (arrows 2 and 5) and this arrangement is also influenced by interactions with the extracellular matrix (arrow 6). Cell shape and locomotion result from the activity of the entire machinery, including hydraulic pressure generated by contractions and the volume-regulatory mechanisms via ion transport processes (arrows 7, 9, 10). A further important factor in generation of cell shape is the interaction with the environment (other cells and extracellular matrix) (arrow 11).

Contractions, ion transport, and assembly/disassembly of fibrils require a steady supply with metabolic energy (arrow 8), which in part is provided by enzymes associated with fibrous components of the cytoplasm (arrow 12). This association by itself shifts the monomer/polymer equilibrium. Cell shape and motility and mechanical stresses (affecting cell shape) are involved in the control of gene expression and proliferation (arrow 13), as are shifts in ion activities (arrow 15), and all cellular constituents result from gene activities (arrows 14–16), thus closing the network circle. Not to make it too complicated, in this scheme many reactions have been omitted, such as, for instance, the role of hormone receptors and protein kinases. A similar reaction network has been proposed for density-dependent proliferation control by Bereiter-Hahn (1988).

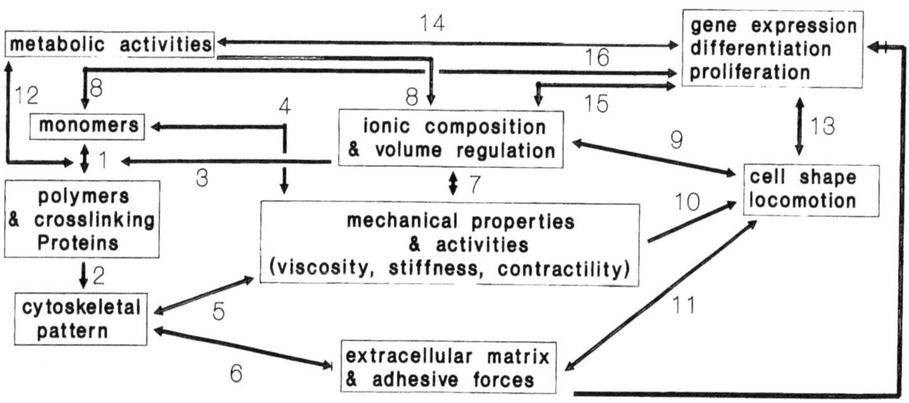

Fig. 2. Schematic representation of the network of mutual interactions of factors controlling biomechanics and the shape of cells. For detailed explanation see text

These considerations open the view for a concomittant increase in the complexity of chemical and mechanical interaction in organisms during evolution and ontogenesis. The current view of cells is dominated by its biochemical aspects; the mechanical aspects have not yet been regarded appropriately either in their influence on the metabolism of single cells or in their basic integratory function in multicellular organisms. Mechanical interaction with the environment is a basic achievement of cells, which in the course of evolution has changed from its significance in locomotion and feeding to a primitive form of signal transduction and reception, and has still remained an important factor in differentiation and morphogenesis.

Acknowledgments. The experimental work underlying this article has been supported by several grants from the Deutsche Forschungsgemeinschaft.

References

Alt W (1990) Mathematical models and analysing methods for the lamellipodial activity of leukocytes. Nato Adv Stud Conf Istanbul 1989, pp 403–422

Bereiter-Hahn J (1977) Die Zelle ein Pneu? Pneus Nat Tech IL9:152–156

Bereiter-Hahn J (1985) Architecture of tissue cells. The structural basis which determines shape and locomotion of cells. Acta Biotheor 34:139–148

Bereiter-Hahn J (1988) Involvement of microcompartmentation in the regulation of cell proliferation. In: Jones D (ed) Microcompartmentation. CRC, Boca Raton, pp 55–69

Bereiter-Hahn J, Strohmeier R, Kunzenbacher I, Beck K, Vöth M (1981) Locomotion of *Xenopus* epidermis cells in primary culture. J Cell Sci 52:289–311

Bereiter-Hahn J, Fava P, Fuhrmann C, Kajstura J, Münnich A (1987) What happens to culture cells when they become confluent? Eur J Cell Biol Suppl 17:6

Bereiter-Hahn J, Braun N, Vöth M (1990a) Continuity of movement and preservation of architecture during cell locomotion. In: Alt W, Hoffmann G (eds) Biological motion (in press)

Bereiter-Hahn J, Lück M, Miebach T, Stelzer KE, Vöth M (1990b) Spreading of trypsinized cells: cytoskeletal dynamics and energy requirements. J Cell Sci 96:171–188

Bershadsky AD, Gelfand VI (1981) ATP-dependent regulation of cytoplasmic microtubule disassembly. Proc Natl Acad Sci USA 78:3610–3613

Bershadsky AD, Gelfand VI (1983) Role of ATP in the regulation of stability of cytoskeletal structures. Cell Biol Int Rep 7:173–188

Bershadsky AD, Gelfand VI, Svitkina TM, Tint IS (1980) Destruction of microfilament bundles in mouse embryo fibroblasts treated with inhibitors of energy metabolisms. Exp Cell Res 127:421–429

Bissell MJ, Aggeler J (1987) Dynamic reciprocity: how do extracellular matrix and hormones direct gene expression? In: Cabot MC, McKeehan WL (eds) Mechanisms of signal transduction by hormones and growth factors. Liss, New York, pp 251–262

Blum S, Bereiter-Hahn J (1988) Lactate dehydrogenase (LDH) and phosphofructokinase (PFK) promote actin polymerization in vitro. Eur J Cell Biol 46 Suppl 22:58

Blum S, Bereiter-Hahn J (1990) Promotion of actin polymerization by LDH, PFK, GAPDH, aldolase and hexokinase. Fed Europ Biochem Soc 270:124 (Suppl)

Brix K, Stockem W (1987) Studies on microplasmodia of *Physarum polycephalum*. VII. Adhesion-dependent changes in the organization of the fibrillar actin system. Cell Biol Int Rep 11:529–536

Castor LN (1970) Flattening, movement and control of division of epithelial-like cells. J Cell Physiol 75:57–64

Castor LN (1971) Control of division by cell contact and serum concentration in cultures of 3T3 cells. Exp Cell Res 68:17–24

Clarke FM, Masters CJ (1976) Interactions between muscle proteins and glycolytic enzymes. Int J Biochem Bristol 7:359–365

Clegg JS (1988) On the internal environment of animal cells. In: Jones DP (ed) Microcompartmentation. CRC, Boca Raton, pp 1–16
Durham AC, Walton JM (1982) Calcium ions and the control of proliferation in normal and cancer cells. Biosci Rep 2:15–30
Eigen M, Schuster P (1978) The hypercycle. A principle of natural self organization. Part C: The realistic hypercycle. Naturwissenschaften 65:341–369
Emermann JT, Pitelka DR (1977) Maintenance and induction of morphological differentiation in dissociated mammary epithelium on floating collagen. In Vitro 13:316–328
Folkman J, Moscona A (1978) Role of cell shape in growth control. Nature (London) 273:345–349
Franke RP, Gräfe M, Dauer U, Schnittler H, Mittermayer C (1987) Stress fibres (SF) in human endothelial cells (HEC) under shear stress. In: Kiesewetter H, Jung F (eds) Blood fluidity and endothelial influences on microcirculation. Springer, Berlin Heidelberg New York, pp 989–992
Frantz CN (1985) Effects of platelet-derived growth factor on Ca^{2+} in 3T3 cells. Exp Cell Res 158:287–300
Gordon R, Brodland GW (1987) The cytoskeletal mechanics of brain morphogenesis. Cell state splitters cause primary neural induction. Cell Biophys 11:177–238
Görmar F, Bernd A, Bereiter-Hahn J, Holzmann H (1990) A new model of epidermal differentiation: induction by mechanical stimulation. Arch Derm Res 282:22–32
Gutmann WF, Bonik K (1981) Kritische Evolutionstheorie. Ein Beitrag zur Überwindung altdarwinistischer Dogmen. Gerstenberg, Hildesheim, 227 pp
Ingber DE, Folkman J (1987) Regulation of endothelial growth factor action: solid state control by extracellular matrix. In: Ingber DE (ed) Mechanism of signal transduction by hormones and growth factors. Alan R Liss, New York, pp 273–282
Kajstura J, Korohoda W (1982) Cellular metabolism and cell contact reactions. Stud Biophys 90:199–205
Kajstura J, Korohoda W (1983) Significance of energy metabolism pathways for stimulation of DNA synthesis by cell migration and serum. Eur J Cell Biol 31:9–14
Karkhoff-Schweizer R, Knull HR (1987) Demonstration of tubulin-glycolytic enzyme interactions using a novel electrophoretic approach. Biochem Biophys Res Commun 146:827–831
Korohoda W (1971) Interrelations of motile and metabolic activities in tissue culture cells. I. Respiration in chicken embryo fibroblasts actively. Folia Biol 19:41–51
Korohoda W (1972) Interrelations of motile and metabolic activities in tissue culture cells. III. Semisynchronous growth of chicken embryo cells preincubated in suspension. Folia Biol 20:333–338
Lansman JB, Hallam TJ, Rink TJ (1987) Single stretch-activated ion channels in vascular endothelial cells as mechanotransducers? Nature (London) 325:811–813
Layrand DB, Matveeva NB, Teplov VA, Beylina SI (1972) The role of elastoosmotic parameters in locomotion of myxomycete plasmodia. Acta Protozool 11:339–354
Lazarides E (1987) From genes to structural morphogenesis: the genesis and epigenesis of a red blood cell. Cell 51:345–356
Leister KJ, Wenner CE, Tomei LD (1985) Correlation of ouabain-sensitive ion movements with cell-cycle activation. Proc Natl Acad Sci USA 82:1599–1603
Lynch RM, Paul RJ (1988) Functional compartmentation of carbohydrate metabolism. In: Jones DP (ed) Microcompartmentation. CRC, Boca Raton, pp 17–36
McClay DR, Ettensohn CA (1987) Cell adhesion in morphogenesis. Annu Rev Cell Biol 3:319–345
Medina D, Li ML, Oborn CJ, Bissell MJ (1987) Casein gene expression in mouse mammary epithelial cell lines: dependence upon extracellular matrix. Exp Cell Res 172:192–203
Mooseker MS (1974) Brush border motility: microvillar contraction in isolated brush border models. J Cell Biol 63:231a
Opas M (1987) The transmission of forces between cells and their environment. In: Bereiter-Hahn J, Anderson OR, Reif W-R (eds) Cytomechanics. Springer, Berlin Heidelberg New York, pp 273–285
Opas M (1989) Expression of the differentiated phenotype by epithelial cells in vitro is regulated by both biochemistry and mechanics of the substratum. Dev Biol 131:281–293
Oster GF, Odell GM (1984) The mechanochemistry of cytogels. Physica 12D:333–350
Paris A, Pouyssegur J (1986) Growth factors activate the bumetanide-sensitive $Na^+/K^+/Cl^-$ cotransport in hamster fibroblasts. J Biol Chem 261:6177–6183
Parmelee JT, Beebe DC (1988) Decreased membrane permeability to potassium is responsible for the cell volume increase that drives lens fiber cell. J Cell Physiol 134:491–496

Rozengurt E (1983) Signalling mitogenesis in 3T3 cells: role of cycle nucleotides and ion fluxes. Cell Biol Int Rep 7:495–496
Rubin AH, Terasaki M, Sanui H (1979) Major intracellular cations and growth control: correspondence among magnesium content, protein synthesis, and the onset of DNA synthesis in BALB/c3T3 cells. Proc Natl Acad Sci USA 76:3917–3921
Scott JE (1987) Molecules for strength and shape: our fibre-reinforced composite bodies. Trends Biochem Sci 12:318–321
Svitkina T, Neyfakh AA jr, Bershadsky A (1986) Actin cytoskeleton of spread fibroblasts appears to assemble at the cell edges. J Cell Sci 82:235–248
Thie M, Schlumberger W, Rauterberg J, Robenek H (1989) Mechanical confinement inhibits collagen synthesis in gel-cultured fibroblasts. Eur J Cell Biol 48:294–302
Tillmann U, Bereiter-Hahn J (1985) Actin fibrillar system and energy-metabolism. Eur J Cell Biol 36 Suppl 7:66
Tillmann U, Bereiter-Hahn J (1986) Relation of actin fibrils to energy metabolism of endothelial cells. Cell Tissue Res 243:579–585
Tilney LG (1983) Interactions between actin filaments and membranes give spatial organization to cells. In: Satir BH (ed) Modern cell biology. Liss, New York, pp 163–199
Tilney LG, Mooseker M (1971) Actin in the brush-border of epithelial cells of the chicken intestine. Proc Natl Acad Sci USA 68:2611–2615
Tilney LG, Hatano S, Ishikawa H, Mooseker MS (1973) The polymerization of actin: its role in the generation of the acrosomal process of certain echinoderm sperm. J Cell Biol 59:109–126
Tözeren A, Skalak R (1988) Interaction of stress and growth in a fibrous tissue. J Theor Biol 130:337–350
Traub P, Nelson J (1982) Interaction of the intermediate filament protein vimentin with ribosomal subunits and ribosomal RNA in vitro. Mol Biol Respir 8:239–247
Traub P, Vorgias CE (1984) Differential effect of arginine modification with 1,2-cyclohexanedione on the capacity of vimentin and desmin to assemble. J Cell Sci 65:1–20
Vanderburgh H (1988) A computerized mechanical cell stimulator for tissue culture: effects on skeletal muscle organogenesis. In Vitro Cell Dev Biol 24:609–619
Woiteneck P (1989) Abhängigkeit der "Energy Charge" von der Dichte von Zellkulturen. Diploma Thesis, Univ Frankfurt/M.
Zylberberg L, Bereiter-Hahn J, Sire J-Y (1988) Cytoskeletal organization and collagen orientation in the fish scales. Cell Tissue Res 253:597–607

Constructional Principles and the Quasi-Experimental Approach to Organisms

WOLFGANG FRIEDRICH GUTMANN[1]

Summary

The basic theorems of constructional morphology as conceived by the hydraulic theory are presented. It is shown that physics and the law of conservation of energy prescribe the understanding of organisms as energy-converting, mechanically coherent systems, which are characterized by auto-formative and auto-mobile capabilities. The constructional properties have to be conceived as specific organismic principles. The constructional setup can be shown to determine the phylogenetic pathways over functionally viable intermediate stages. All restraints and irreversibility enforcing transformation steps are deducible from the constructional principles of the organisms.

The methodological applicability of the constructional principles is demonstrated in case studies for the hydraulic formative mechanisms of the vertebrate head and the determining influence of the biomechanical construction on the organization and the pattern generation of the nervous system in chordates and vertebrates.

1 Introduction

The maturity of a branch of science is indicated by the abstractness and the stricture of its theoretical tenets and the deductive character of its explanations. Biology can reach this stage in the field of morphology only if reliable basic principles of living organization can be shown to exist. These will be outlined in this chapter.

2 The Genesis of the Constructional Principles

By tracing the early phases of genesis of life on earth, irrefutable principles of living organization can be worked out. After decades of theorizing and experimental work there remains little doubt that the development of life started by abiotic processes on the surface of an early hot earth with large water-covered regions under an oxygen-free atmosphere. Strong radiation from the sun and the influence of heat on the mixture of gases resulted in the formation of a variety of organic compounds and their accumulation in water-filled basins. Some of

[1] Senckenberg-Museum, Frankfurt/M., Senckenberganlage 25, D-6000 Frankfurt/M., FRG

N. Schmidt-Kittler and K. Vogel (Eds):
Constructional Morphology and Evolution

these compounds began to interact and initiated molecular mechanisms typical and indispensible for life (Dose and Rauchfuss 1975; Day 1984).

After the accumulation of a diversity of organic components, closed structures in the form of vesicles could develop by mixing of a lipid scum with an underlying aqueous stratum containing the primitive and basic biochemical mechanisms of life. The driving power for the formation of eobiotic entities might have been convection or turbulence of some sort in a primeval soup (Bonik et al. 1977a).

These early closed entities consisted of a fluid filling in flexible membranes; the protocells were mechanically closed, capable of enlargement and of budding; by fusion larger units could emerge. As mechanical objects they show the characteristics of hydraulic units. We cannot be sure if the enclosed filling was really a fluid, it might already have contained a structural order determined by filamentous proteins which tended to form three-dimensional nets and gain anchoring contact in the lipid bilayer-membranes. Left to themselves, these units must have shown a tendency to round up and assume a spherical shape.

These primitive cellular units developed on diverging paths. The prokaryotic organisms, the bacteria, formed external cell walls to brace their tender bodies and acquire a determined shape. As mainly immobile and well-supported units, they were successful in reaching a high degree of sophistication in several synthetic and catabolic mechanisms. They invented the complicated biochemical machineries which can still be observed in extant living forms. In contrast to this, the transformation path to the eukaryotic organismic machines utilized the internal protein lattice to stabilize the cell. These tethering structures could actively determine specific shapes and suppress the tendency to form spherical shapes, thereby laying the foundation for auto-formation (Bonik et al. 1977a; Gutmann and Bonik 1981; Gutmann 1988a, b).

Organisms as Working Constructions.

Life achieved its organismic properties in a very early stage and remained firmly based on these first established constructional features. The more static stage of development was followed by the dynamization of the constructions by transition to motoric activity and automobile action, both of them based on energized mechanical machine properties.

The integration of myosin into the actin lattice allowed the energy-driven deformation of the hydraulic units. Mechanical activity became possible because the actin grid was integral in the coherence of the working whole. This means that the energy could be transduced from the molecular level to the working machine.

3 The Driving Energy

The motoric capabilities constitute the auto-formational and auto-mobile properties of organismic machines. All organisms are, at least in one short phase of their life, propulsive systems. The basic properties of animals can be only adequately worked out if these beings, in their wholeness, are conceived of as working machines and as propulsors.

The auto-formative and auto-motive properties of living entities are used for the acquisition of matter and energy, the performance of behavioral activity in the form of propulsion and reproduction. The differentiation and formation of the living machine is also effected by the work of coherent units. The formation of offspring is mechanical work and not due to the transmission of some imaginary genetical information; in the succession of generations the totality of organismal properties is handed over from the parents to their progeny.

There is no living activity which is not generated by mechanical work of the living machines. All chemical, molecular, and biophysical mechanisms are subservient in driving or coordinating mechanically based action of living machines.

The organismic constructions are able to work as long their organismal machines are operationally closed and possess an undisrupted connection from the molecular level, where energy conversion takes place, to the constructional frame, which, because of this coherence, is able to work if energy is appropriately transduced to it. How the driving mechanisms and the coordinating influence of molecular or physiological processes become effective can never be explained by simple continuation of the reductivistic principles because it is the coherence of the machine which determines how any kind of nonmechanical influence is actuated.

However, the raison d'être of organismic systems we investigate and explain as working systems cannot be elucidated by the intensification of morphological or even constructional studies.

Every functioning organism could be differently built. Every Bauplan represents a different mode of energy acquisition, energy transformation, and energy utilization in the environment. The basic constructional principles have to be used to reconstruct transformation paths leading to observable constructions. The explanations of organization have to be given in the form of reconstructions of the preceding constructional stages. Such models have a narrative touch and present the constructional preconditions and constraints which must have been realized in the preceding constructions in which they must have laid the foundation for the subsequent alterations, thus opening up the paths to new constructional morphoclines. In the course of such explanations, biomechanics in the newly defined sense provides the only useful, constraining, and determining principles.

4 The Specific Organismal Principles

From this level on, all invariant principles of organisms on all levels of complexity were firmly established. These are:

a) the hydraulic principle and with it the operational and mechanical closedness of the membrane around a fluid filling;

b) the mechanical coherence of membrane and all form-determining structures;

c) the auto-formative capabilities to use chemical compounds and energy from the metabolism to actively determine the shape;

d) the auto-mobile capabilities to deform the living machinery in a way to generate propulsion, to actively acquire matter and energy, and to produce offspring (Fig. 1);

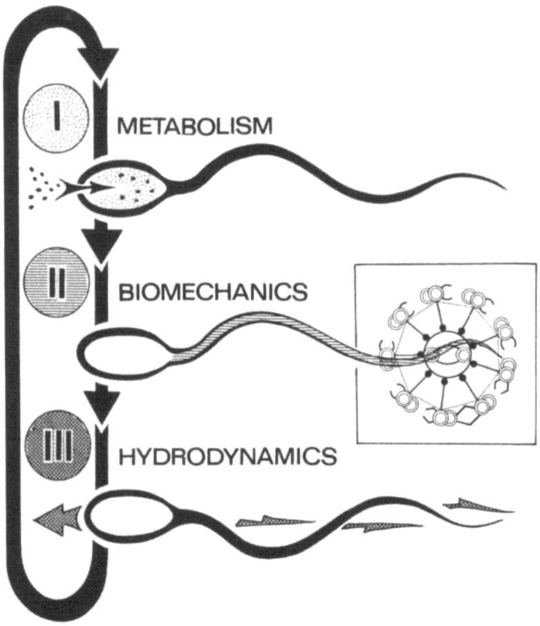

Fig. 1. Bionomic circle. Organisms are conceived as energy-converting systems which utilize chemical material and energy on the level of the metabolism (*I*), convert chemical energy in the morphological (mechanical) level (*II*) and inject the energy according to hydrodynamic laws into the medium (*III*). The activity must result in renewed input of matter and energy thereby closing the bionomic circle. Generation of progeny requires the reservation of energy for the generation of offspring

e) every living activity consists in the generation of mechanical work in the mechanical machinery and the utilization of this work of the living machine in the environment;
f) living activity can only be achieved because chemical energy is converted into mechanical energy and directed via the coherent machine to the working level;
g) the machine has to transfer energy in a directed form into the surrounding medium or onto the substrate to generate propulsive force according to Newtons principles;
h) as working machines, organisms are intrinsically active and endogeneously energized units and, as such, not bodies of the kind usually studied by physics. Coherence of the living machine is constituted from the very beginning of organismic life; the basic coherence of the living construction must not be thought of as composed of elements and must not be broken up into elementary structure (Fig. 2).

Every constructional aspect of the living machine can be tested and corroborated by morphological inspection. I would like to add here that there exists an absolutely reliable indicator for the non-understanding of the constructional principles just outlined: the postulate and the contention that manipulative experiments might be necessary or even useful approaches to corroborate them.

On the supposition of the validity of the outlined principles, morphology is constituted as a kind of engineering science, and has to be understood as the dis-

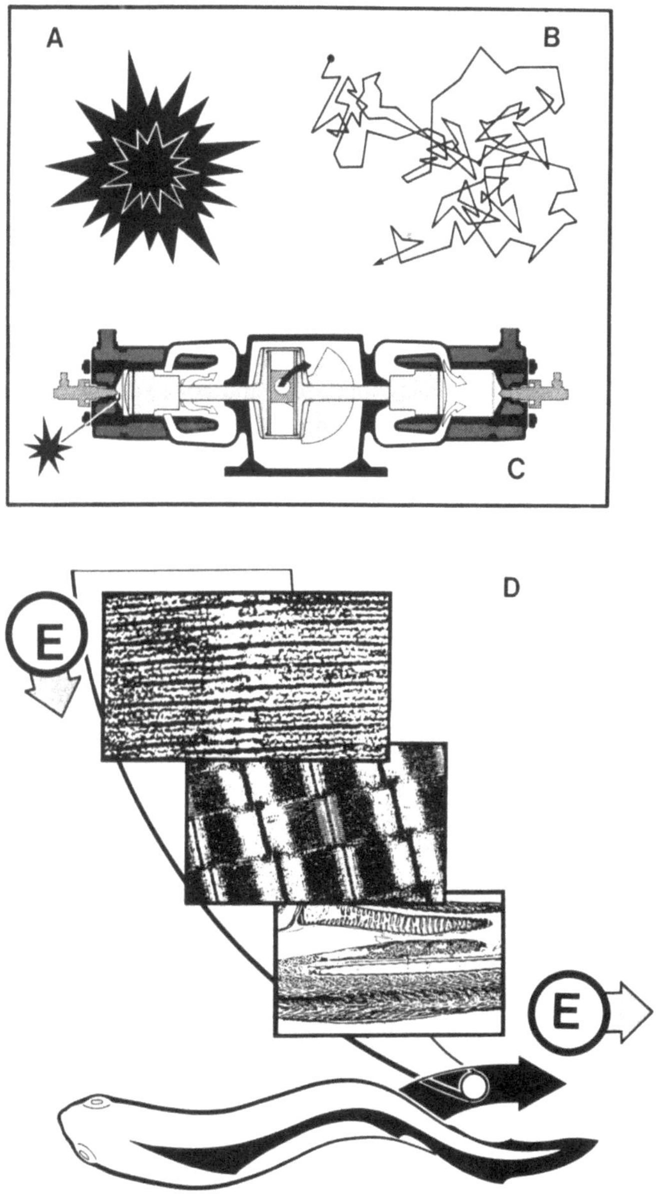

Fig. 2. Energy, machines, and organisms. **A** Energy has the tendency to dissipate freely. **B** Unrestrained Brownian movement of particles occurs when any kind of constraints are lacking. **C** Engine with its energy-converting sites and the constraining mechanisms which force the driving force on restrained paths and determine the useful work (Spektrum 1989). **D** Organismic machine in the form of a fish-like propulsor. The energy transformation occurs on the macromolecular level by the interaction of actin and myosin. The sliding fibers of the muscles connect to the connective tissue structures on the histological level. On this path energy is transmitted to the working construction. Biology and engineering are concerned with objects which become effective by the directing and constraining influence on energy which has a tendency to freely dissipate. From this follows that biology must never start from the abstractions of thermodynamics

cipline whose task it is to investigate mechanically coherent hydraulic systems. There can certainly be only one kind of morphology if the properties of the organismic machines are taken account of.

Constructional morphology must be based on very strict law-like principles; these are a set of abstract, absolutely reliable, and valid biophysical principles. As such they were almost entirely overlooked by form-oriented morphologists and by reductionistic physiologists and molecular biologists alike. Traditional biomechanics did not encourage the elucidation of mechanical coherence as preceding and superseding the formation of elements.

The validity and the structure of the organismic and constructional principles prescribe that morphology has to be based on a clear-cut methodology. How the organismic machine works has to be explained by starting from the description of the specific coherence of the system being analyzed, and continued with the elucidation of the energy conversion sites, the energy-canalizing system and the transmission over several structural steps to final transfer to the environment. As a result, behavioral activity and propulsion have to be explained by describing the mechanical work of the machine and the transfer of energy into the environment. Morphology requires the description of the organisms as working machines by the use of a mechanical terminology analogous to that in engineering.

Empirical investigation has to proceed according to the constructional principles. It has to elucidate the sites of energy conversion, the energy-directing structures, and the mechanical system with its constraints that enforce controlled activity. There is no use in adding up structures and elements to attain a reasonable understanding of the working construction. The mode of working of a living machine and the possible transformation sequences can only be deduced from a prior understanding of the constructional whole and the elucidation of the energy-converting and energy-directing structures within the coherent frame of the mechanically constrained working machine. In morphological studies, the coherent construction has to be worked out in the form of models. The form and the architecture of living beings have to be shown to be the result of form-enforcing structures in the coherent apparatus. It is the construction theory which guides every research undertaking.

Constructional morphology is not allowed to start from form, because the explanation of form can only be the result of explanatory endeavor and cannot be deduced from form as described by traditional morphology. Form does not show very much, because one cannot see the energy-conversion mechanisms, the energy-transducing devices, the mechanical coherence, and the form-enforcing mechanisms. A nonmorphological morphology has to be set up which will remain devoid of form. Form is a subjective side effect of the form-enforcing structures in the coherent frame of the living machine.

5 Consequences for Phylogenetic Research

Phylogeny is strongly determined by the principles just outlined. The phylogenetic models have to be set up by the strict application of the constructional tenets.

a) In the course of evolutionary change, the invariant constructional principles cannot be altered except for the sanction of self-annihilation. Organismic autonomy is inevitably complemented by self-destructivity.
b) Phylogenetic change can only proceed according to the internal constructional principles of living organization. Biomechanical properties have a guiding and directive influence on evolutionary transformation and determine the possible sequence of constructional alterations (Fig. 3).

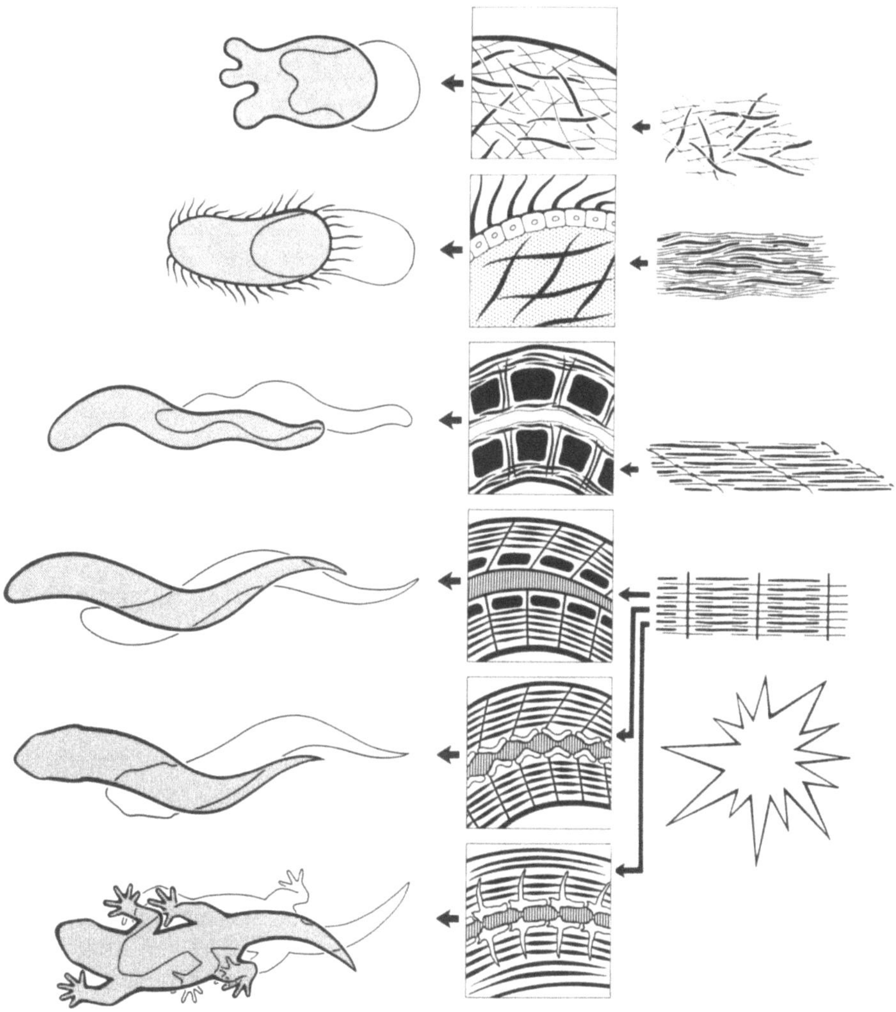

Fig. 3. Organisms function as energy-converting and energy-transducing machine-like systems. Energy conversion is generated by actomyosin fibers (*right column*); the energy is fed into the construction which is determined by the specific Bauplan properties (*middle column*). The energy is used for deformations of the body whereby energy is injected into the surrounding medium. The body form is indicative of the propulsor properties. The sequence is set up in a deductive way and describes the ordered transformation steps which have led to the existing metazoan constructions. None of the aspects outlined can be elucidated by the application of comparative morphology and homology research

c) Organisms can only utilize external conditions in a way allowed and licenced by the constructional set-up and its efficiency. Adequate energy and usable conditions for motoric and behavioral activity can only be determined from the prior knowledge of the working principles of the organismic machine.
d) Organismic change in the sense of increasing complexity can only consist of the formation of structures in the frame of the preexisting coherence. Any structure and element represents an internal differentiation of the constructional whole and can only contribute to activity and efficiency given in the stage prior to the formation of the structures. Phylogenetic transformation consists of the alteration of the coherence of the construction. In our model for the formation of skeletal structures, we have shown how the emergence of skeletal elements could enhance efficiency and improve the motoric performance of the living machines (Vogel and Gutmann 1981, 1988, in press a, ;, Grasshoff 1976, 1978, 1981, 1985, 1986).
e) In the course of constructional alteration, every major transformation step is dependent on the preceding constructional set-up. In many cases, modifications of the constructional machine will be accompanied by a loss of older ordering structures and patterns which cannot be developed again (Gutmann 1972; Bonik et al. 1977b; Bock 1988; Gutmann and Weingarten 1989). M. Eigen (1983) has recently called this directing mechanism strong temporality (Starke Zeitlichkeit). Evolutionary irreversibility can be firmly based on specific organismal principles.
f) The directing and constraining influence of (constructional) morphology is also effective in determining the change of the molecular and physiological levels. Here lies the real importance of morphology. Phylogenetic methodology has to follow strictly the methodological maxims which are deducible from the constructional explanation of organismic machines.

The application of the methods and maxims of constructional morphology puts strict restraint on the approach to morphological problems and enforces the obeyance of compeling principles. Morphology as a field of science and a teaching subject will have to be considerably revised.

All our attempts to explain morphology have to be based on clear-cut requirements: it is necessary to start with the invariant constructional properties and trace a trajectory from the fluid-filled vesicles to the organismic construction under investigation. It is also necessary to design these trajectories in a rational way and to insert those constructional stages which are conceived of as preconditions for the particular construction being explained. Every organismic construction and its phylogenetic morphocline have to be put on bifurcating trajectories starting from the fluid-filled and membraneously closed vesicles to the existing or fossil construction. Fossil and recent organisms have to be understood as evidence of model-like reconstructions (Gutmann and Bonik 1981; Gutmann 1988b).

The methodology advocated in our concept of constructional morphology can be equated with the methodological tenets proposed by W. Bock (this Vol.). The constructional principles conform certainly to nomological deductive explanation if the constitution of the organism as a coherent machine is taken account of in Bock's methodological frame.

The reconstruction of the sequence of organismic systems is also based on nomothetical deductive principles, although a narrative aspect comes in. However, there can be little doubt that the specific theory of the organism as a coherent hydraulic auto-formative and auto-mobile system deserves further theoretical elucidation because the understanding of coherent construction brings new problems to the field of philosophy of science.

6 Conclusions

If the preceding definition of the organism and the elucidation of its properties are correct, we can simplify theoretical reasoning. Most importantly, it is possible to invite physicists back to their work on abiotic aspects of nature according to the well-founded physical abstractions which have their own respected stance in science.

Organisms are, of course, not open systems in the sence of biophysics, they are operationally closed and the specific closedness has to be defined by the mechanical properties of the coherence of the hydraulic system.

Organismic autonomy becomes substantiated by the energized mechanical activity of the living machinery. Autopoiesis is just the empty word used by people who cannot understand the hydraulically generated form enforcement in ontogeny.

However, the great problem comes for morpholoy and traditional biomechanics. Biomechanics has firmly obeyed the principles of physical mechanics which have been elaborated since the 17th century. The all-embracing and overruling coherence and the superseding closedness of the organismic construction does not allow for a methodology which starts from elements and adds up complexity by integrating substructures in the form of mechanical components or molecular mechanisms. In my opinion, even integrative morphology has a strong smell of unlawful accretion of substructures or partial systems. Integrative morphology in the form advocated by some authors will never be able to achieve a reasonable insight into the coherence of the constructional whole whose prior existence provides the constructional preconditions for the development of any conceivable functional system or organ.

The paramount importance of coherence was well understood; Bereiter-Hahn et al. (1987b) refer to coherence as "mechanical continuum" and Walter Bock (1988) puts emphasis on it by calling it "connectivity". However, all theorems designed to cover the aspect of coherence should take account of the mechanically continuous machine with its energy-canalizing structures. It is essential to understand that the coherence of the living machinery is responsible for consuming the energy from the molecular level and for transducing the energy into the environment.

Morphology as we know it is based on description of form and architecture and on the comparison of configurational pattern according to the subjective prescriptions of homology research. Homology research in the German tradition, with all its circularity and reference to subjective impression of shape or gestalt in the scientifically irrelevant sense of Goethe, is no scientific methodology at all because it does not present the theory from which the methodological maxims can be deduced.

In order to gain free space for a new morphology which can only be based on an engineering access to the living machines, traditional morphology and the phylogeny based on the well-known traditional procedures will have to be profoundly revised and morphological knowledge has to be newly assembled according to strict principles (see the chapters by Vogel and Bock, this Vol.).

The concept outlined in the introductory passages of this paper claims that life and living activity are founded on specific vital principles which have nothing to do with vitalism. Vitalism is fended off by the emphasis placed on normal physical energy and nothing transcending this. The only transcendence necessary for new research strategies is the formulation of nonreductivistic methodological procedures. Only the liberation of biology from the domination of physiology and molecular biology will open up a breathtaking perspective for morphology as a science. The hydraulic principle and mechanical coherence are higher organismal principles of structure and these are specific nonreductionistic characteristics of living organization. The impact on morphology and physiology will be demonstrated by examples.

Theory and method of morphology must not be at variance with basic physical principles. Constructional morphology as a discipline which refers to the coherent morphological construction and the energy-converting capabilities of these systems can be understood on the basis of physical principles including the first and second laws of thermodynamics. Nothing in nature, in either the biotic or abiotic field, can occur and become manifest except by the influence of energy. Left to itself, all natural processes show an insuppressable tendency to dissipate.

It is the ability to suppress dissipation of energy, to direct and transform energy within the frame of the organismic machine and to constrain its mechanical activities in a suitable way that permits organic beings to exist. Organismic biology by its basis of constructional morphology insists that there are coherent and intrinsically active systems which are effective in suppressing the tendency of dissipation observable in molecular and energetic processes. Clearly constructional morphology demonstrates the existence of machine-like natural entities which are able to direct, canalize, and utilize energy by actions of the coherent living machine.

The process nature of life points to its reliance on nonequilibrium mechanisms and the requirements of an energy supply as the driving force. However, thermo-

Fig. 4. The relationship of thermodynamics and organismic biology. The *upper part* of the illustration presents a picture of the formation of dissipative structures from elementary disorder in a nonequilibrium condition. Coherence, except the one given in the containment, is lacking. The three stages shown are taken after Prigogine and Stengers 1988. The *lower part* signals organismic diversity from fluid-filled vesicles to the skeleton muscle system of tetrapod vertebrates. Organismic constructions are primarily determined by the mechanical coherence of the living entities which becomes effective in restraining or suppressing energy dissipation. The specific properties of organismic machines are listed on the *right hand side*. The body form contains the internal energy-converting and energy-directing structures, the body shape stands for the working apparatus which becomes effective as a propulsor. In the organisms mechanical coherence has predominance over the disintegrated molecules and all kinds of molecular and physiological mechanisms within the filling and the internal structures

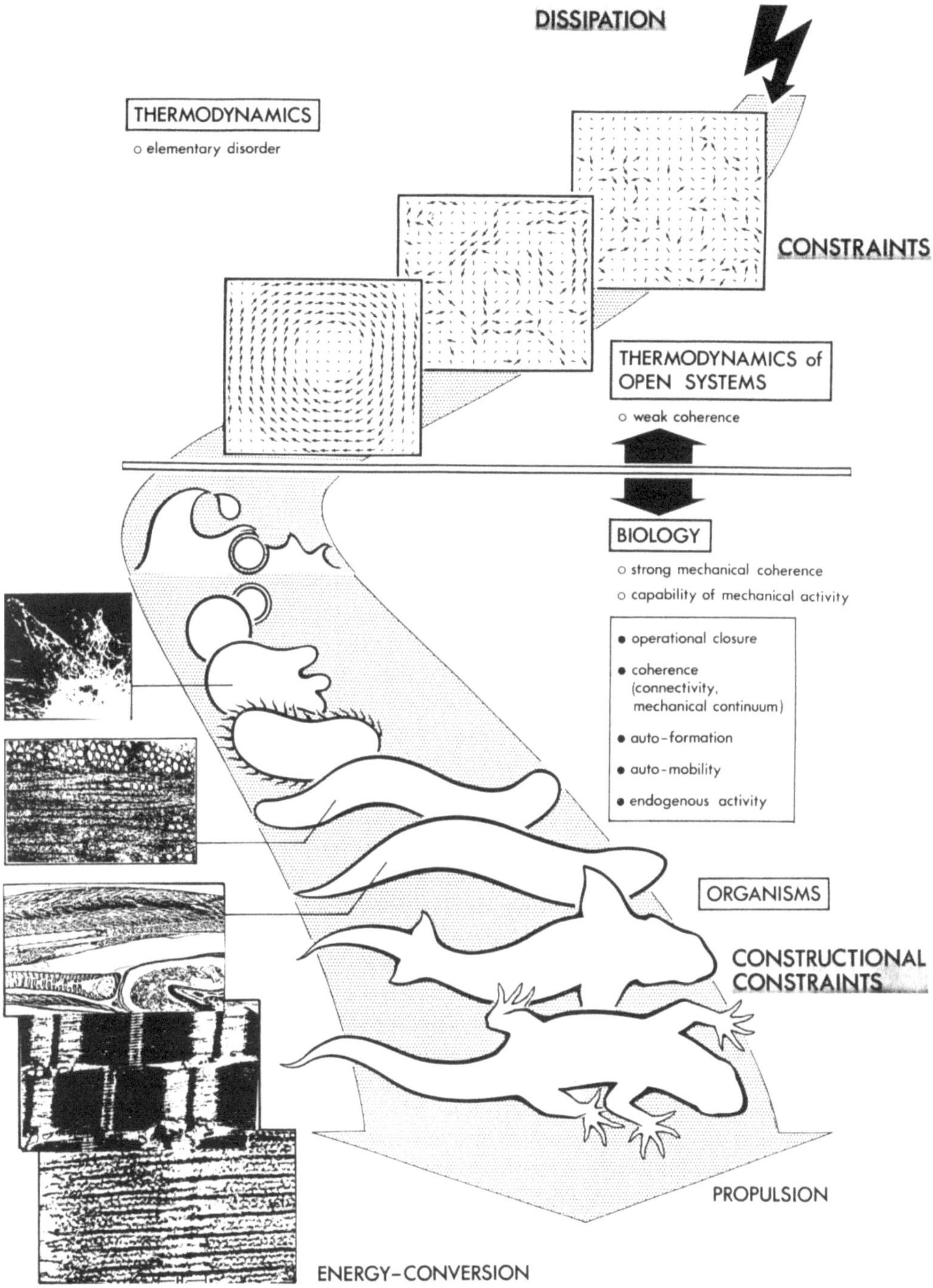
DISSIPATION
THERMODYNAMICS
o elementary disorder
CONSTRAINTS
THERMODYNAMICS of OPEN SYSTEMS
o weak coherence
BIOLOGY
o strong mechanical coherence
o capability of mechanical activity
• operational closure
• coherence (connectivity, mechanical continuum)
• auto-formation
• auto-mobility
• endogenous activity
ORGANISMS
CONSTRUCTIONAL CONSTRAINTS
PROPULSION
ENERGY-CONVERSION

dynamics of open systems which treats situations far from equilibrium conditions is also an insufficient and misleading approach to the understanding of organisms. It is misleading to equate dissipative structures with the constructional properties of living beings; thermodynamics of open systems cannot describe the energized activity and the mechanical work of organismic constructions (Fig. 4). It is the machine-like activity of organismic constructions which by the acquisition of matter and energy and the expulsion of waste products maintains the nonequilibrium situation of life in all organismal units (Gutmann and Weingarten 1988). The autonomy of organismic units is characterized by the capability to act by endogeneously generated activity; such capabilities have till now never been attributed to dissipative structures in the sense of thermodynamics of open systems.

Case Study 1. Nervous Coordination

It would be erroneous to believe that the revision of theoretical and methodological concepts could be restricted to the field of morphology; a strong impact and far-reaching consequences must become effective in physiology. Physiological functions have to be conceived as integrated and partly determined by the biochemical organization of the living beings in which they serve as driving and coordinating mechanisms. This has to be outlined in the following.

During the activity of propulsors, the nervous system serves as an essential structure. It sets the temporal pattern of the motoric activity, controls and coordinates the movement required for propulsion and behavior. Usually the pace-maker structures and the rhythmic and temporal pattern of activity of the nervous system are studied separately from morphology and without recognition of the determining influences and the demands constituted by the constructional frame (Grillner and Kashin 1975; Fetcho 1987). Apparently almost nobody seems to have noticed that the spatio-temporal pattern of nervous activity and its rhythmic spontaneity are dictated in its general properties by the requirements of the propulsors, i. e., the energized and mechanically working machine. In the frame of the hydraulic theory, it is obvious that pattern generation and order of activity which become manifest make sense only in relation to the working machinery. The working construction has its specific constraining aspects.

Evolution of the organismic constructions on the metazoan level was only possible if constructional changes were accompanied by appropriate modification of the spontaneous and highly ordered activity of the nervous system. As the nervous system is constituted from variable nervous elements with a high degree of freedom and modifiability in their interconnectedness, any conceivable pattern of activity and any kind of periodicity of pace-maker activity could become established. All variants showing maladjustments of biomechanical construction and neuronal activity will be superseded by better harmonized organisms. Therefore, the limited spectrum of patterns of activity which could be useful for propulsion and behavior can only be determined over selective mechanisms by the requirements of the whole working construction (Fig. 5).

Adjustment of construction and neuronal activity are only explicable on the basis of "internal" selective mechanisms; internal organismic integration of all

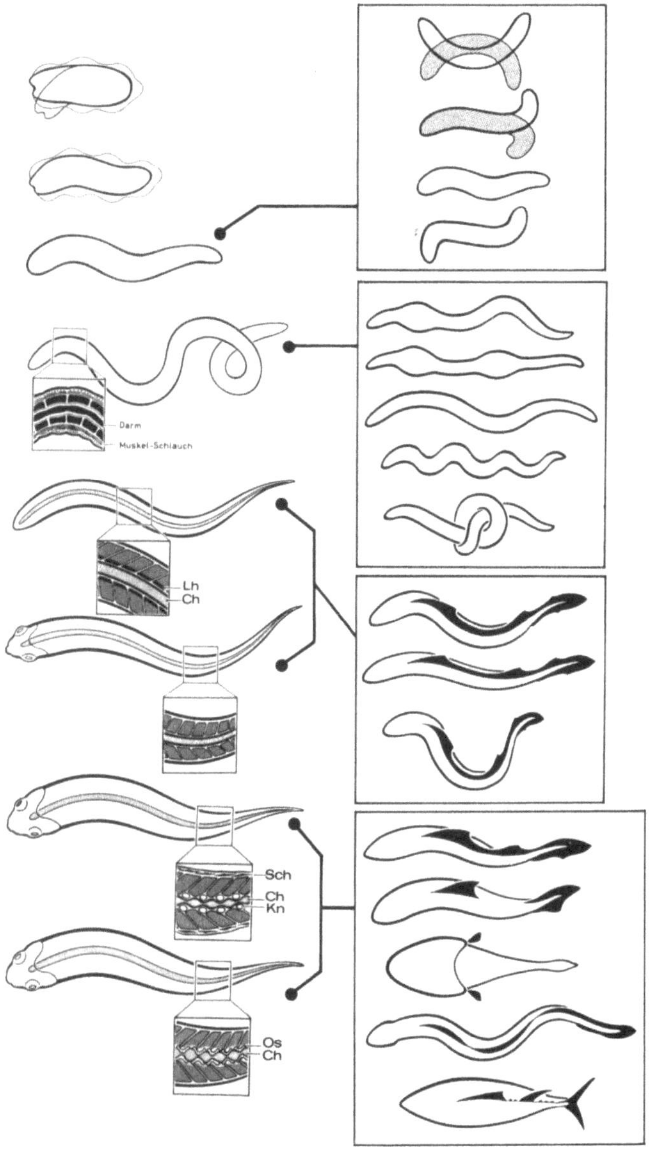

Fig. 5. Constructional transformation in propulsive systems is only possible if it is paralleled by the appropriate coordination pattern generated in the nervous system. The construction sets the demands and constraints which determine the useful pattern of an almost limitless spectrum of nervous activity. The correspondence of constructional stages (*left*), endogenously generated pattern and the possible, acceptable, and in the last resort most effective motoric performance (*right*) is indicated. Selection is a totally internal mechanism which determines the relationship and the integration of different structures of the energy-converting machine

physiological systems in all Bauplan constructions is established by these principles (Gutmann and Peters 1973; Bonik et al. 1977b; Vogel 1984).

In conjunction with the transformation of the worm construction and the formation of the chordate Bauplan, the coordinating pattern had to undergo considerable alterations. The motoric pattern of a muscle-braced worm was certainly different from the wriggling propulsion of the primitive chordates. In the length constant chordates nervous activity patterns required for peristalsis and for the control of the body shape became obsolete. The pertinent neural structures have disappeared and only those required for the remaining propulsive activity must have persisted.

One would have expected that the nervous system might have become simplified in the transition from worms to chordates. However, this is not the case. The reason for the formation of the voluminous neural tube and its complex neuronal network lies in the massivity and compactness of the longitudinal muscles. A great number of muscle cells had to be activated and the rate of contraction of the muscles controlled by extremely delicate mechanisms which required a voluminous neuronal apparatus.

In all stages, nervous activity must have been synthetic and spontaneous. A hydraulic system of the coelomic worm level and on all stages leading to the chordate Bauplan must have acted as constructional wholes. Therefore, the nervous system, which is primarily coextensive with the mechanical machine, has also to act as a whole in generating all-encompassing patterns of activity (Fig. 6).

Case Study 2. The Evolution of the Vertebrate Cephalon

The vertebrate cephalon is the rigid rostral region of the body of a coelomate construction. If one tries to give a phylogenetic explanation for the development of the head and its organization, it is necessary to start from the constructal conditions preceding the establishment of this organ (Gutmann 1969; Gutmann and Bonik 1981).

It is compeling to postulate that primitive coelomates, and therefore the predecessors of the chordates, must have used their whole bodies for propulsion. Even primitive chordates such as the acranians (but not the dwarfed tunicate larvae) must throw their whole body into sinusoidal movements, thereby generating thrust all over the length. The use of the whole body as a propulsor was absolutely necessary because in hydraulic constructions the motoric efficiency is so low and the body construction so easily deformable that the propulsive force must be consumed in the same region in which it is generated. Only in the course of a considerable rise of efficiency could the prevertebrates acquire a nonmotile anterior region. The increasing concentration of thrust generation in the hindpart of the body finds its expression in the formation of a caudal fin or a double system of caudal fins, as in typical fish.

The head region can only have formed in a step-by-step manner, starting with the formation of olfactory organs and the establishment of a primitive brain directly adjacent to the olfactory structures. In the next step the eyes, dorsal and lateral ones, developed as sense organs allowing orientation in the water and allowing perception of danger indicating photic changes. The establishment of the

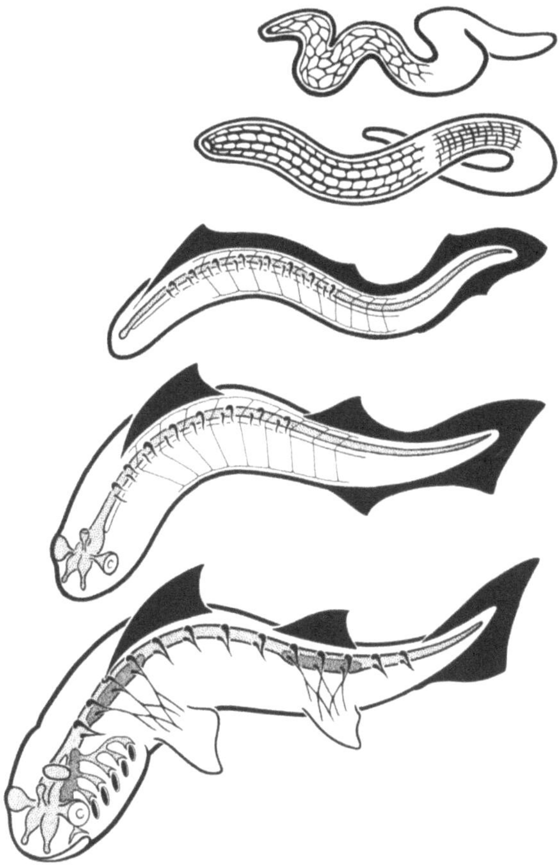

Fig. 6. In the course of chordate evolution the somites and the notochord developed on the dorsal side of the worm-like construction. Concentration of nervous structures followed the effector organs, the compacted muscles, thus generating the spinal chord with its metameric nerve connections to the muscles and the sensoric periphery. All other nervous configurations would require more building material and would be less economical. The nervous organization is dominated by the mechanical construction and strictly controlled by its demands. The spinal nerves are forced into the pattern of the myosepta which form tensile structures that connect the body axis with the compacted muscular structures. The brain could only develop in effective propulsors which allowed parts of the nervous organs to emancipate from the dominance of the propulsor. In the forerunners of the chordates, the nervous system must have been coextensive with the muscles in the whole body of the worm-like hydraulic system. Any kind of nervous concentration was precluded by the high deformability of the propulsive system

eyes, their improvement, and transformation into image-perceiving organs must have been accompanied by the formation of the next cerebral region, the mesencephalon. When the optical system came into existence by the modification of portions of the lateralis system and when in conjunction with it the hindbrain was formed, further somites had to disappear. As the position of the body in space could now be recorded by the new sensory structures, the dorsal eyes became obsolete and were reduced to much smaller structures.

The general selective advantage of the formation of large sense organs and the correlated brain structures is very obvious. Sense organs provide economization in the use of the propulsive system. Trial and error movements become the more expensive when the propulsor system gains complexity and efficiency. Chordates and primitive craniotes possess the best-energized propulsive machines of the animal kingdom. They could benefit considerably by establishment of the orientating and directing influence of the sense organs and the brain (Fig. 7).

Concomitantly with the formation and enlargement of the sense organs and the reorganization and restructuring of the anterior part of the spinal chord into a brain, a reduction of the somitic musculature had to occur. These muscles were gradually replaced in the course of cephalization. Some vestiges of the somitic muscles were, however, transformed into eye-moving elements and became integrated into the cephalon. This alteration of the muscles can be easily explained in a gradualistic way when the selective advantage in the enhancement of vision by saccadic movements is taken into account.

The whole transformation process and the establishment of the cephalon was only possible if two prerequisites were fulfilled. The reduction of the somites must have been paralleled by the establishment of stiff skeletal structures in the form of cartilage. This is the indispensible structural change of the mechanical frame in the phylogenetic sequence of the adult stages. Otherwise the cephalic region would have become instable and the sense organs and the brain could not have been held in their places.

In conjunction with skeleton formation, the ontogenetic requirements become effective. Skeletal structures cannot appear ad libitum, and in the first steps of their formation they have no auto-formative capabilities. They must borrow their basic form from preexisting soft structures. The prerequisite for their emergence lies in the formation of soft structures which achieve a stable and undeformable situation in the cephalic region of the early embryo. In the case of the vertebrate head, the constructional structures which generate the frame for the anlagen of the cartilaginous skeleton are clearly visible and have been described again and again since the middle of the 19th century. However, the biomechanical conditions seem never to have been understood.

In the early stages of ontogeny, the anterior part of the neural tube becomes considerably enlarged; it is transformed into its typical inflated shape. Usually, the anterior region of the central nervous system is called the brain, and the inner fluid-filled cavities the ventricles. The inflated state of the brain is strongly indicative of the employment of the hydraulic mechanism of enlargement by means of a fluid filling (Fig. 8).

For the process of phylogeny we can be quite sure that the inflation of the cerebrum replaced the disappearing and receding anterior somites which gave way to the forming brain and the large sense organs. The inflation mechanism must have appeared very early in the ontogeny of all stages of cephalogenenis. So in the course of the enlargement of the brain via an early inflation of the ventricles, the somites could become entirely reduced in the head region. The enlarged ventricles took their places. This inflation process could become more pronounced and extended further backwards pari passu with the perfection of the cephalon and the replacement of further somites. In this way, the inflated brain became

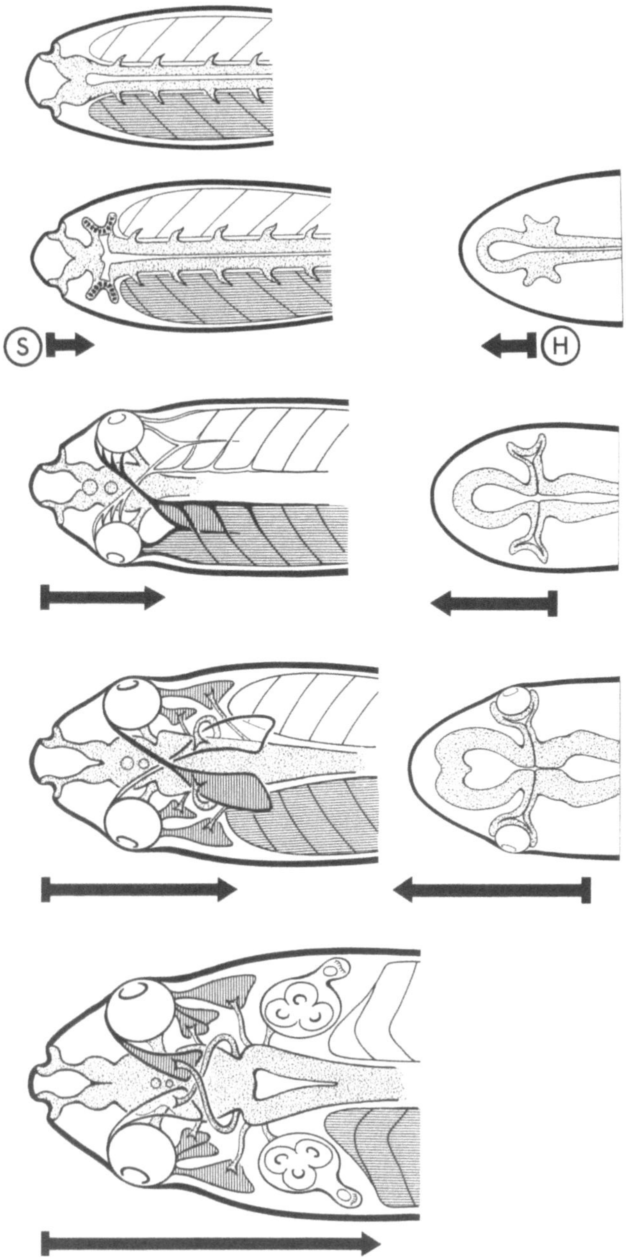

Fig. 7. The vertebrate head developed in the anterior region by replacing the somitic structures. Formation and enlargement of the nasal organs, the eyes, and the optic capsule required the concomitant formation of inflating hydraulic mechanisms of the ventricles during ontogeny shown in the *right column*. In the course of the suppression of the somitic structures some elements were transformed into the eye muscles. As a consequence, one eye muscle was shifted into the opposite side and pulled its nerve (VI) behind. This is easily understandable if it is taken into account that the antagonism of the muscles was primarily determined by the integral whole of the somites in the presumptive head region. Cephalogenesis consisted in the suppression of segmentation in the head region

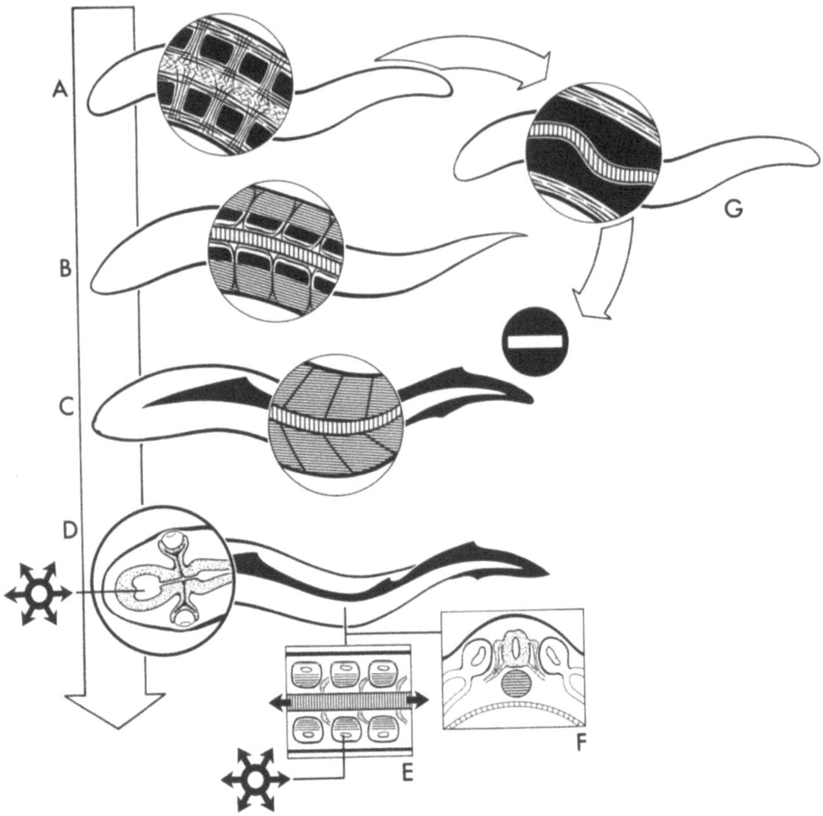

Fig. 8. The formative mechanism of the vertebrate trunk and the head region. **A** The evolution of the notochord required a densely spaced internal bracing as a constructional precondition. **B** The notochord cannot have developed in unsegmented or oligomeric forms because such constructions would have been instabile and the notochord could not have fulfilled any useful function. The predecessors of chordates can neither have been tubiculous nor oligomeric constructions. **C, D** In the predecessors of the craniotes the whole construction had to be homogenously organized. The somites provided the formative mechanisms of the whole body. The notochord serves as a stretching device. **E** The establishment of the head was achieved by suppression of the somites (and their formative influence on the rostral region) and the concomitant hydraulic inflation of the rostral region as a frame for the development of the skeletal stabilizing structures

helpful in ensuring the stability and in providing the frame for the differentiation of cartilaginous skeletal structures which served to stabilize the fully developed brain and the sense organs.

However, the hydraulic prefabrication of the morphological configuration of the head was not a question of the inflated brain alone. Fluid-filled cavities emerged also in the flanks of the cephalic region; they extended further ventral into the branchial bars. These head cavities have nothing to do with somites, as they are deficient in all metameric coelomatic structures. They generate the morphological frame in the flanks of the head and in the branchial apparatus. If we also take account of the hollow intestinal canal, we conceive the complicated hydraulically organized architecture consisting of the inflated brain, the foregut, and the head cavities. This constructional frame of soft, hollow, and hydraulically stabilized organs determines the sites for the formation of the skeletal structures. The head organization of hollow organs will certainly narrow down the passages through which the nerves from the brain can grow out and become organized in a suitable constructionally determined way. The dependence of nervous organization on biomechanical constraints is well established by studies which in their search for inductive mechanisms have entirely overlooked the mechanical (hydraulic) constraints. In the trunk region the nerves are forced into a metameric pattern by the densely spaced somites. Analogously, the head cavities and the gill slits are the guiding mechanisms constraining the formation of the branchial nerves.

The transformation of connective tissue into cartilage around the brain and the large sense organs was only possible after soft structures, the inflated brain, and the other head cavities generated the basic architecture and organization of the cephalon in early stages of ontogeny. The preformation of the skeletal elements by the soft structures remains valid in all stages and in every single ontogeny.

We must clearly understand that skeletal organization is never explicable if one starts from the stiff and mineralized elements. So the formation of the head in a gradualistic sequence could only occur in conjunction with the establishment of the appropriate ontogenetic mechanisms which utilized hydraulic principles in the form of enlarged ventricles, head cavities, and the anterior part of the intestinal canal.

The basic formation mechanism is only modified and faired over on all stages of vertebrate evolution according to the specific requirements arising on the different levels of the vertebrate sequence. However, in its basic processes, cephalogenesis was never really altered. All differentiations and specializations of the vertebrate cephalon require the prior formation of the basic hydraulic architecture.

There is one point that has to be emphasized: it is not possible to reconstruct the vertebrate predecessors and the organizational conditions prior to the formation of the head from the observable ontogenetic mechanisms in extant or fossil vertebrates. The process of generation and the production process of the structures vanish behind the products. Therefore, the usual attempts to explain the structures which one observes are futile. The explanations do not spring out of the observed objects. It is necessary to reconstruct the preceding constructional situations and the process of formation by using reliable abstract principles.

If this is taken into consideration, it becomes comprehensible that vertebrate morphologists may very well describe all details of structures in the vertebrate head but can never hope to infer the constructional preconditions by starting from the fully developed head organization (Starck 1978, 1979, 1982; Gans and Northcutt 1983; Gans 1988).

General Considerations

In the case of the formation of the vertebrate cephalon, it can be clearly demonstrated that the phylogenetic preconditions for the formation of the skeletal system lie in the preceding coelomate construction. The ontogenetic development of the skeletal structures is also determined by the hydraulic mechanisms of soft body organization in early ontogenetic stages. The skeletal structures develop as internal structures within the coherent frame of the hydraulic system. From this follows that the head region can never be explained in any sensible way by starting from the skeletal structures or other artificially separated elements (Gutmann 1969).

The hydraulic principle provides the one and only mechanism for the formation of morphological architecture in animals and the generation of shape in skeleton-free systems and in those organisms which are stabilized by skeletal elements. Skeletal structures develop as elements in a coherent system that is held turgescent by hydraulic mechanism. From this follows that the form of skeletal structures is enforced by the hydraulic system and its form-determining structures.

The formation of shape and the establishment of morphological organization are energy-driven processes that actuate mechanical work by means of hydraulic expansion and the restraining influence of other structures in a mechanically coherent construction. Shape and architecture are the result of the antagonistic activity of expansive hydraulic activity and the restricting interference of structures which suppress freedom of change, uncontrolled deformability, and simple spherical bulging of the coherent system. Thereby the required form and architecture are enforced. New mechanical approaches developed in recent years do not meet the necessary requirements as they were not based on the hydraulic principles and on mechanical coherence (Alberch and Kollar 1988).

When in the course of phylogeny the sequence of changes was strictly determined and the evolutionary pathway effectively constrained by the constructional setup of the living machine, the ontogenetic mechanisms and the favorable genetic changes had to follow suit. Genes and the genetically controlled formative mechanisms are always enslaved and forced into the constructional frame of the living organism.

Acknowledgments. In concluding, I would like to emphasize that this paper presents a highly concentrated overview of a long series of papers and several books which contain the theory and the reconstructions of animal Bauplans in a continuation of the prebiotic stages of evolution. The list of references has been kept short by referring to major review papers and books which should be consulted for additional citations.

The new methodology could only be developed step by step as a cooperative attempts of the Frankfurt school (K. Bonik, K. Edlinger, J. L. Franzen, M. Grasshoff, W. Maier, D. Mollenhauer, D. S. Peters, K. Vogel, M. Weingarten). Our work in Frankfurt began in the late 1960's and was strongly influenced by the papers of Bock and von Wahlert (1965), which we still consider a major break-through. There remains an obvious affinity with the later theoretical and methodological work of W. Bock and D. Homberger, and I hope that a high degree of agreement might be reached on the main issues. The illustrations were skillfully executed by Antje Siebel

References

Alberch P (1982) The generative and regulatory roles of development in evolution. In: Mossakowski D, Roth G (eds) Environmental adaptation and evolution. Fischer, Stuttgart New York, pp 19–36

Alberch P, Kollar E (1988) Strategies of head development: workshop report. Development 103 (Suppl):25–30

Bereiter-Hahn J (1987) Mechanical principles of architecture of eukaryotic cells. In: Bereiter-Hahn J, Anderson OR, Reif W-E (eds) Cytomechanics. Springer, Berlin Heidelberg New York, pp 3–30
Bereiter-Hahn J, Anderson OR, Reif W-E (eds) (1987a) Preface. In: Cytomechanics. Springer, Berlin Heidelberg New York, pp V–VI
Bereiter-Hahn J, Anderson OR, Reif W-E (eds) (1987b) Cytomechanics. Springer, Berlin Heidelberg New York, 294 pp
Bock WJ (1988) The nature of explanations in morphology. Am Zool 28:205–215
Bock WJ (1989) Organisms as functional machines: A connectivity explanation. Am Zool 29:1119–1132
Bock WJ, Wahlert G von (1965) Adaptation and the form-function complex. Evolution 19:269–299
Bonik K, Grasshoff M, Gutmann WF (1977a) Funktion bestimmt Evolution. Umschau 77(20):657–668
Bonik K, Gutmann WF, Peters DS (1977b) Optimierung und Ökonomisierung im Kontext der Evolutionstheorie und phylogenetischen Rekonstruktionen. Acta Biotheor 26:75–119
Day W (1984) Genesis on planet Earth. Yale Univ Press, New Haven London, 299 pp
Dose K, Rauchfuss H (1975) Chemische Evolution und der Ursprung lebender Systeme. Wissenschaftliche Verlagsgesellschaft, Stuttgart, 217 pp
Eigen M (1983) Evolution und Zeitlichkeit. In: Schriften C-F-v-Siemens-Stiftung (ed) Die Zeit. Oldenbourg, München Wien, pp 35–58
Fetcho JR (1987) A review of the organization and evolution of motoneurons innervating the axial musculature of vertebrates. Brain Res Rev 12:243–280
Gans C (1988) Craniofacial growth, evolutionary question. Development 103 (Suppl):3–16
Gans C, Northcutt RG (1983) Neural crest and the origin of vertebrates: a new head. Science 220:268–274
Grasshoff M (1976) Das 'Konstruktionsniveau' in der Phylogenetik. In: Schäfer W (ed) Evoluierende Systeme I und II. Aufs Reden Senckenb Naturforsch Ges 28:124–140
Grasshoff M (1978) A model of the evolution of the main chelicerate groups. Symp Zool Soc London 42:273–284
Grasshoff M (1981) Arthropodisierung als biomechanischer Prozeß und die Entstehung der Trilobiten-Konstruktion. Paläontol Z 55:219–235
Grasshoff M (1985) On the reconstruction of phylogenetic transformations. The origin of the arthropods. Acta Biotheor 34:149–156
Grasshoff M (1986) Cnidarian phylogeny – a biomechanical approach. Paleontog Am 54:127–135
Grillner S, Kashin S (1975) On the generation and performance of swimming in fish. In: Herman RM, Grillner S, Stein PSG, Stuart DG (eds) Neural control of locomotion. Plenum, New York London, pp 181–201
Gutmann WF (1969) Die Entstehung des Vertebrate-Kopfes, ein phylogenetisches Modell. Senckenbergiana Biol 50(5/6):433–471
Gutmann WF (1972) Die Hydroskelett-Theorie. Aufs Reden Senckenb Naturforsch Ges 21:pp 1–91
Gutmann WF (1988a) Die Selbstbewegerfähigkeit der Organismen als neues Naturprinzip? In: Universities Stuttgart and Tübingen (eds) Natürliche Konstruktionen. Mitt SFB Univ Stuttgart Tübingen 230(2):141–148
Gutmann WF (1988b) Die Evolution hydraulischer Konstruktionen. Organismische Wandlung statt altdarwinistischer Anpassung. Senckenberg Buch 65:200
Gutmann WF, Bonik K (1981) Kritische Evolutionstheorie – Ein Beitrag zur Überwindung altdarwinistischer Dogmen. Gerstenberg, Hildesheim, 227 pp
Gutmann WF, Peters DS (1973) Konstruktion und Selektion: Argumente gegen einen morphologisch verkürzten Selektionismus. Acta Biotheor 22(4):151–180
Gutmann WF, Weingarten M (1988) Organismus und Konstruktion III. Mechanische Arbeit von lebenden Konstruktionen sichert das thermodynamische Nichtgleichgewicht. Nat Mus 118(3):83–92
Gutmann WF, Weingarten M (1989) Studien zur Theoriengeschichte der Evolution. Nat Mus 119(2):55–62
Gutmann WF, Vogel K, Zorn H (1978) Brachiopods: biomechanical interdependences governing their origin and phylogeny. Science 199:890–893
Prigogine J, Stengers J (1988) Entre le temps et l'éternité. Fayard, Paris, pp 1–346

Starck D (1978, 1979, 1982) Vergleichende Anatomie der Wirbeltiere, vols 1–3. Springer, Berlin Heidelberg New York

Vogel K (1975) Funktionsmorphologie als Hilfsmittel paläontologischer Evolutionsforschung. Sitzungsber Wiss Ges JW Goethe Univ Frankfurt/M 13(1):5–19

Vogel K (1983) Zur gegenwärtigen Diskussion um die Makroevolution (punctuated equilibria). Paläontol Z 57(3/4):199–203

Vogel K (1984) Lebensweise und Umwelt fossiler Tiere. Biol Arbeitsber 39:171

Vogel K (1986) Origin and diversification of brachoipod shells: viewpoints of constructional morphology. In: Rachevoeuv PR, Emig C (eds) Les brachiopodes fossiles et actuels. Actes 1. Congr Int Brachiopodes, Brest 1985. Biostratigr Paleozoiquie Brest 4:399–408

Vogel K (1989) Constructional morphology and the reconstruction of phylogeny. Abh Naturwiss Ver Hamburg NF 28:255–264

Vogel K, Gutmann WF (1980) The derivation of pelecypods: role of biomechanics, physiology and environment. Lethaia 13:269–275

Vogel K, Gutmann WF (1981) Zur Entstehung von Metazoen-Skeletten an der Wende von Präkambrium zum Kambrium. Festschr Wiss Ges JW Goethe Univ Frankfurt/M. Steiner, Wiesbaden, pp 517–537

Vogel K, Gutmann WF (1988) Protist skeletons: biomechanical preconditions and constructional utilization (Teichert-Festschrift). Senckenbergiana Lethaea 69:171–188

Vogel K, Gutmann WF (in press a) Organismic autonomy in biomineralization processes. In: Crick R (ed) Origin, evolution and modern aspects of biomineralization in plants and animals. Plenum, New York, pp 45–56

Biomechanics in the Enamel of Mammalian Teeth

WIGHART VON KOENIGSWALD and HANS ULRICH PFRETZSCHNER[1]

Summary

The formation of mammalian enamel is strictly determined genetically but yields important functional adaptations on the levels of gross morphology and microstructure. Differentiation of enamel thickness is frequently found in hypsodont teeth of smaller and medium-sized mammals. It is mainly due to the position on the leading or trailing edge. In the enamel microstructure, the Schmelzmuster, defined as a specific combination of different enamel types, again shows a close relationship to the leading or trailing edge especially in small mammals. The various enamel types, defined by prism orientation, have different functional meanings. Some, like Hunter-Schreger bands, reinforce enamel to resist tensile forces and possess crack-stopping properties as well. In hypsodont molars of larger mammals, specific stresses occur near the enamel-dentine junction. The characteristic modified radial enamel found in this region is regarded as a special adaptation to these stresses. Modified radial enamel evolved convergently in several lineages. Because enamel types are mainly functionally determined, the taxonomical value of enamel structures is largely restricted to the level of Schmelzmuster.

1 Introduction

Mammalian teeth differ markedly from those of reptiles. Tooth replacement is reduced, forcing the mammalian tooth to serve for a much longer timespan as a cutting device compared to the teeth of reptiles. The quality of the mammalian tooth is improved by a more complex tooth morphology differentiated into various functional areas as well as the increasing complexity of the enamel structure during the mammalian evolution.

2 Levels of Enamel Investigation

Various levels of enamel investigation should be kept in mind when comparing the approaches and results of various authors. The lowest magnification shows the variation in the thickness of the enamel related to the masticatory movement. It is very helpful to study this character not in terms of traditional anatomical orientation but in functional terms such as leading and trailing edges (Greaves 1973).

[1] Institut für Paläontologie der Universität Bonn, Nussallee 8, D-5300 Bonn 1, FRG

N. Schmidt-Kittler and K. Vogel (Eds):
Constructional Morphology and Evolution

At higher magnifications, in the light microscope as well as in the SEM, the prisms become visible with a diameter of about 5 micron. Prisms are the mineralized paths of the migrating ameloblasts during enamel formation. Between the prisms interprismatic matrix (IPM) is often present. Its cristallites may differ in their orientation from the prisms. In many teeth the enamel is formed by layers of different types of enamel. Enamel types are defined by the orientation of the prisms. The three-dimensional arrangement of various enamel types within a tooth is defined as the Schmelzmuster (von Koenigswald 1977, 1980, 1982). Differences in the Schmelzmuster between leading and trailing edges obviously demonstrate its functional significance.

Normally, the enamel types do not discriminate taxonomic units; however, the Schmelzmuster may be used in several rodent groups to define taxa at the genus level.

Further magnification is needed to study the shape of the prisms themselves. Boyde (1964) and recently Martin et al. (1988) have defined various prism types according to their cross-section. To some degree prism types bear taxonomical information. The prism cross-section represents the shape of the secretory surface of the ameloblast, Tomes process, during enamel formation. To date no significant approach has been applied to the functional interpretation of the prism shape.

This chapter, concentrating on the biomechanical properties of the enamel, mainly considers tooth morphology and the Schmelzmuster, as visible at low or medium level of magnification. At this level of investigation the biomechanical approach is based on micromorphological observation in relation to the deduced stress pattern. An application of mathematical methods was recently initiated by Pfretzschner (in press).

3 Genetical Prerequisites

Mammalian bone is very different from enamel. In bone the functional properties are determined by the ability to rearrange the trabecula according to the directions of stresses. This optimal arrangement can only be achieved in that bone is rebuilt during the entire lifespan. Tooth enamel, however, is formed deep in the jaw and is shifted into its functional position after the final mineralization of the tooth. By the following placement into this functional position the tooth is exposed to the functional stress pattern. Since mineralized enamel cannot be rebuilt, all functional adaptations must be brought into the genome by way of natural selection.

4 Functional Importance of Enamel Thickness

The primitive mammalian tooth is covered by enamel of equal thickness. In tribosphenic molars, one side of some cusps bears primarily a thin enamel, which is abraded quickly and exposes the dentine core. This results in a sharp cutting edge formed by the cross-sectin of the thicker enamel on the remaining sides of the cusp. This cutting surface can be resharpened, as demonstrated impressively in the carnassials of carnivores.

Incisors of rodents and many other enlarged incisors of various mammalian groups use the difference in hardness of dentine and enamel in a similar way. The lack of enamel on the uncovered sides is the result of the ultimate reduction of enamel thickness.

In herbivorous mammals, using again the difference in hardness between dentine and enamel, the most efficient secondary occlusial surface was created in the molars (Fortelius 1985). In some larger, but especially in smaller mammals, a differentiation in thickness between leading and trailing edges can be observed. The marsupial genera *Vombatus* and *Lasiorhinus* reduced the enamel in the molars on the leading edge totally (Fig. 2). In *Otomys*, an African rodent, the leading edge of each cuspal ride is greatly reduced. The caviomorphs *Lagostomys* and *Chinchilla* completely lost the enamel on the leading edges (Fig. 1). In some arvicolid rodents, however, as in *Microtus*, the enamel of the trailing edge is reduced in thickness (Fig. 3).

Even when the mechanical effect of these reductions is not obvious, strict correspondence to leading and trailing edge indicates its functional significance. Reduction of the enamel generally results in equal spacing of the parallel cutting edges. A similar tooth construction is found in *Mammuthus* as well as in *Hydrochaeris*, achieved by an intercuspid cement filling (Fig. 4).

5 Functional Aspects of the Enamel Structure

5.1 Strengthening Against Abrasion

Since mammalian teeth must withstand abrasion by coarse food or attrition by the antagonist for a long time, a very resistant enamel is desirable. The ultrastructure of the enamel has some effect on the abrasion of the enamel. Rensberger and von Koenigswald (1980) demonstrated experimentally that the degree of abrasion is related to the orientation of the prisms. Areas in which the chewing forces and prism direction were almost parallel showed a higher resistance than those areas in which prisms and chewing forces formed wide angles.

The primitive radial enamel already present in many unspecialized dentitions seems to fulfill these basic demands, since the prisms rise apically in the desired direction, as in the Paleocene peryptichid *Carsioptychus* (von Koenigswald et al. 1987), in the marsupial *Didelphis* (Stern et al. 1989) or in the molars of *Cricetus* (von Koenigswald 1980). Young et al. 1987 demonstrated in *Phascolarctos* that the angle of the rising prisms differs in leading and trailing edges, which brings about an appropriate angle in both.

Abrasion, however, does not seem to be the basic stimulus for the development of new enamel structures. Nature found another much more effective way to compensate for high abrasion on the level of tooth morphology. Several mammalian lineages, feeding on highly abrasive food like grasses, developed hypsodont or even ever-growing teeth convergently. Therefore, the optimal solution in terms of abrasion is not the quality, i.e., an enamel of ultimate hardness, but a larger quantity of enamel available for abrasion.

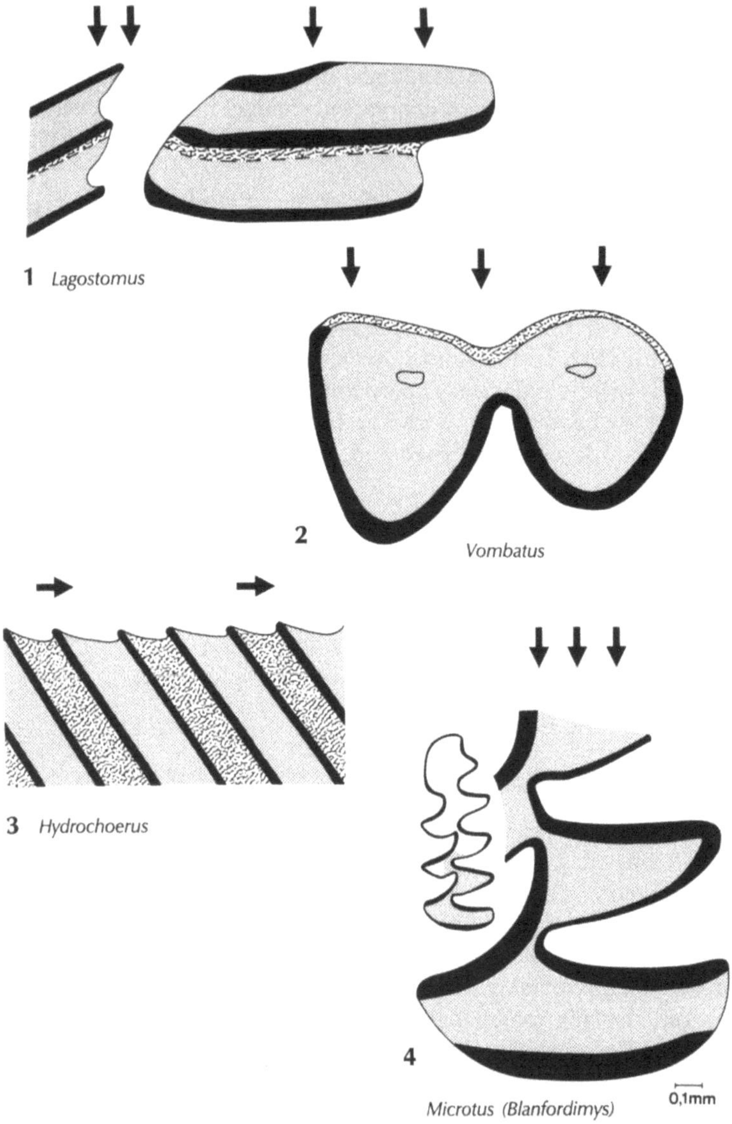

Fig. 1. Second lower molar of *Lagostomus* cross-section and longitudinal section. The enamel is totally reduced on the leading edge of the second element. The gap is filled solely by cement

Fig. 2. Lower molar of *Vombatus* in cross-section. Enamel surrounds only the folded trailing edge while a hard cement covers the leading edge

Fig. 3. Longitudinal section of lower third molar of *Hydrochaeris hydrochaeris* with almost equally cutting edges

Fig. 4. Lower first molar of *Microtus* (*Blanfordimys*) *roley* with a strong reduction of the enamel band in the trailing edge of each dentin triangle. (After von Koenigswald 1980)

In all figures *arrows* indicate the direction of chewing forces

5.2 Stress Distribution in Enamel

Pfretzschner (1988) formulated a hypothesis for enamel as a composite material (Fig. 5). The mechanical behavior of enamel is highly anisotropic, caused by the orientation of prisms and IPM. Enamel withstands compression in any direction fairly well, because the dense packing prevents the individual prisms from buckling. While tension parallel to the prism direction can also be sustained to a considerable degree, the prisms will split up easily if the tension forces are applied perpendicular to the prisms. This load application, tensile stress perpendicular to the prism direction, is the most dangerous one. Failure occurs through separation of adjacent prisms by interprismatic crack propagation.

A mathematical approach to stress patterns in teeth (Pfretzschner 1988) demonstrated the occurrence of tensile stress in enamel (Fig. 6). Orientation of the enamel prisms depends on the general shape of the tooth and on the load application. In low-crowned teeth tensile forces are circumferential and assume a maximum at the base of the crown. In the case of high-crowned molars of larger mammals, tensile stresses are oriented radially, possessing a maximum adjacent to the EDJ (Enamel – Dentine – Junction) and near the occlusal surface (Fig. 9).

5.3 Stress-Oriented Fibers in Enamel

In cancellous bone the linear elements are stout short beams, oriented in the direction of the trajectories of tensile as well as compressive stress. Enamel prisms are too slender to sustain considerable compression. Therefore, prism arrangements adaptive to the stress pattern are expected to be so that prisms take up tensile stress only. The ontogeny of the enamel allows prisms to start only at the EDJ and to end near the outer surface. No example was ever found where prisms are orientated exactly in the direction of the trajectories. Therefore the remaining component of the force vector vertical to prisms might be carried by decussated prisms or IPM crystallites.

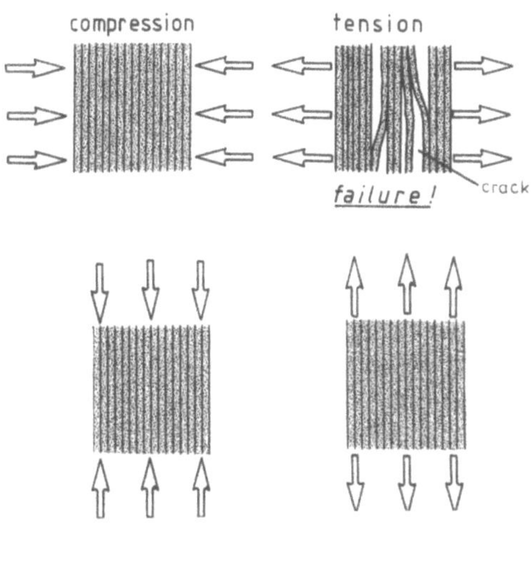

Fig. 5. Material behavior of enamel. Failure occurs most readily under tension perpendicular to the prisms. (Pfretzschner 1988)

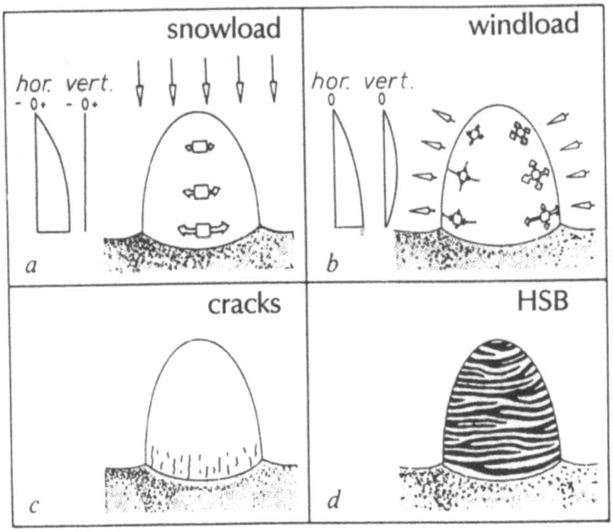

Fig. 6. Stress distribution on low-crowned unicuspid teeth. Various loads cause mainly horizontal tension forces (**a** and **b**). The Hunter-Schreger bands (HSB) lay horizontally (**d**), vertical to observed crack at the crown base (**c**), where the tension forces are maximal

In low-crowned teeth, horizontal layers with decussated prisms, the so-called Hunter-Schreger bands (HSB), are very common in most larger mammals. Indeed the horizontal prism decussation strengthens the enamel against major horizontal tension forces. As stated in the previous section, a different stress pattern appears in high-crowned molars, characterized by radial tension forces with a maximum near the EDJ. Indeed, all larger high-crowned mammals show a layer of a special enamel type close to the dentine, characterized by vertical decussation of linear elements. There are three convergent enamel types possible, the 3D enamel in proboscideans, vertical HSB in rhinocerotids and astrapotheres, and the advanced radial enamel common in artiodactyls, perissodactyls, notoungulates, and several other groups (Pfretzschner in press). These three enamel types occur only in high crowned molars and only in a thin layer (about 30% of the enamel thickness) adjacent to the EDJ.

The advanced radial enamel is of specific interest since here the thickened IPM arranged in "interrow sheets" (Boyde 1969) carries an important part of stresses instead of adjacent prisms. In this enamel type all prisms rise to the occlusal surface in radial rows. They are separated by thick plates of IPM with crystallites running downwards vertically to the prism direction (Figs. 7 and 8).

5.4 Relationship of Schmelzmuster and Stress Pattern

Only very basic functional qualities can be attributed to some of the various enamel types. Therefore, it is almost impossible to give a conclusive biomechanical interpretation for a complex Schmelzmuster. Genetical heritage dominates the Schmelzmuster and specific functional adaptations may be discrete.

Rodent incisors, for instance, are generally consisting of two layers, the inner one containing HSB and the outer one radial enamel. Any biomechanical interpretation must keep in mind that lagomorph incisors can fulfill similar functions

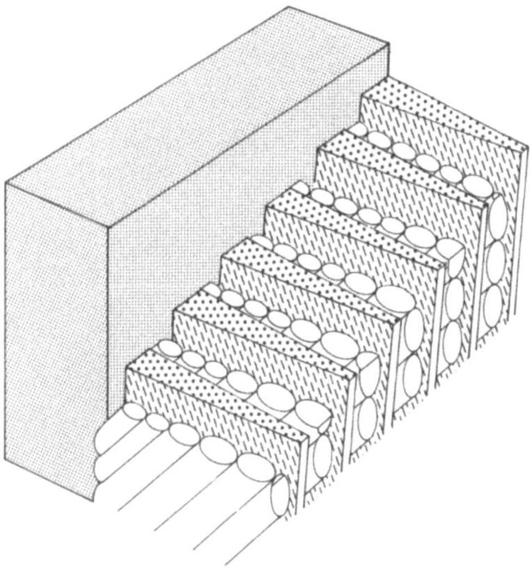

Fig. 7. In advanced radial enamel prisms stand in radial rows, separated by thick plates of interprismatic matrix (IPM). The crystallites of the IPM are orientated almost vertically to the prisms

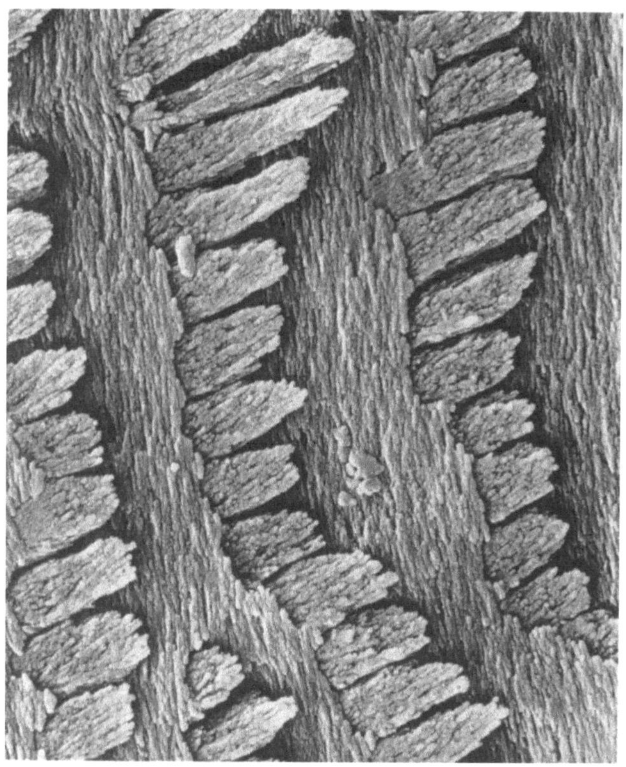

Fig. 8. Advanced radial enamel in a molar of *Phacochoerus*

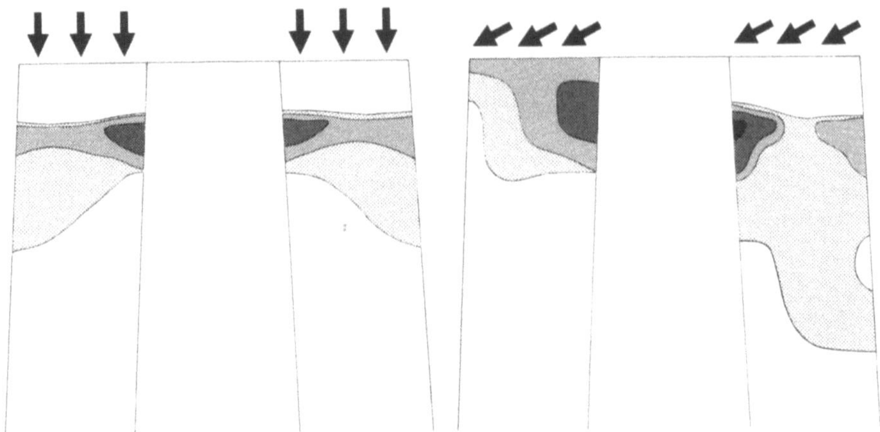

Fig. 9. Stress distribution in hypsodontid molars under various loading directions. The calculated maximum stress lies near the Enamel-Dentine Junction (EDJ). Tension stresses are directed radially

with a single layer. Even when this layer contains HSB as well, one must answer the question as to why other mammals with rodent-like incisors such as the Groeberidea (South American marsupials) (von Koenigswald and Pascual 1990) or Apatemyidae (Paleogene insectivores) (von Koenigswald 1990) have very different enamels.

On the other hand, the Schmelzmuster very often varies within a single dentition according to its functional units. The enamel of most enlarged incisors of small and larger mammals is distinctly different from that of their molars (von Koenigswald 1988). Some functional relationship seems to be plausible.

Therefore it is necessary to check the congruence of the stress pattern and the Schmelzmuster on a topographic level, even when the biomechanical properties of the enamel types are only insufficiently known.

In the enamel thickness a correlation between preferred cutting edges and the leading or trailing edges, respectively, of the tooth elements is obvious. Similar relationships can be described for the Schmelzmuster of various dentitions. For interpretation, however, each correspondence must be evaluated, since other factors might determine the Schmelzmuster, e.g., morphology or enamel thickness.

Hypsodont molars of herbivorous mammals are normally loaded in a single direction during mastication, so that leading and trailing edges are clearly defined (Greaves 1973). In large mammals like ruminants or elephants, no differentiation of the Schmelzmuster between the leading and the trailing edges was found. In small mammals like rodents, however, some groups show a very distinctive differentiation. The best-studied example are molars of arvicolids (von Koenigswald 1980, 1982) (Fig. 10). The complex Schmelzmuster of derived species distinguishes very clearly between leading and trailing edges. Even if the curvature of the enamel follows the same pattern with a concave enamel band on the leading edge and convex enamel on the trailing edge, respectively, it could be shown that only the stress pattern controls the Schmelzmuster (von Koenigswald 1980). The correspondence of Schmelzmuster and stress pattern to such a high degree is not

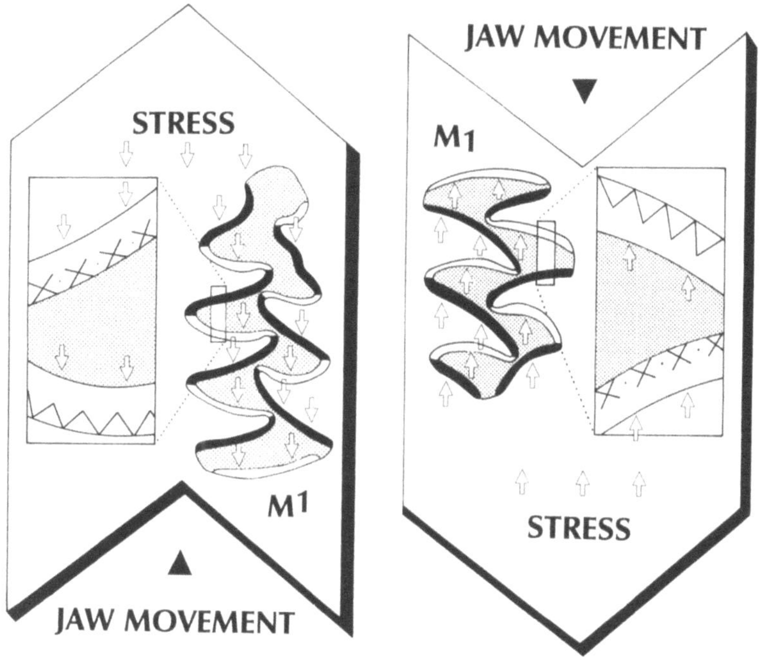

Fig. 10. Reversal of morphology and Schmelzmuster in lower and upper molars of the arvicolid *Mimomys*. (von Koenigswald 1982)

frequent and cannot be postulated for all cricetids. However, additional examples from other groups are known but not yet studied in detail.

5.5 Crack-Stopping Mechanisms

Even teeth of living animals show minute cracks in their enamel. Hairline cracks are particularly dangerous for any kind of homogenous material, since they concentrate stresses at the tip of the crack. Prismatic enamel is not homogenous and cracks prefer to follow the direction of prisms (Boyde 1976).

Different types of crack-stopping mechanisms have been detected. The first group is related to the direction of prisms (von Koenigswald 1988). Prisms penetrating through the enamel from the EDJ towards the outer surface in a straight line are highly endangered to be split by cracks. Even rather simple structures such as the unique turn of prisms as found in several enlarged incisors of marsupials (von Koenigswald and Pascual 1990) hinders a further prolongation of cracks, since the crack is deflected from its initial direction. Prism decussation, as found normally related to Hunter-Schreger bands, is highly effective if the bands are orientated parallel to the tension forces. Von Koenigswald et al. (1987), as well as Pfretzschner (1988), demonstrated how cracks entering the area of Hunter-Schreger bands were split and deviated several times so that the energy is reduced to such an extent that the crack cannot be propagated (Figs. 11 and 12). According to the fossil record, HSB were invented several times in parallel in various large

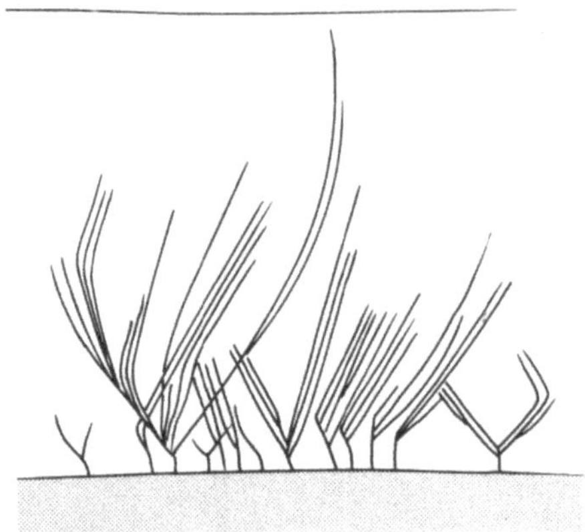

Fig. 11. Observed artificial cracks that split up in adjacent HSB and run into different directions. Thin section of a dog's canine

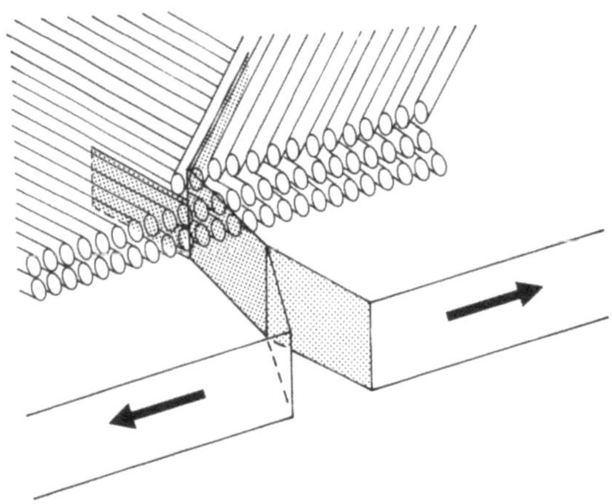

Fig. 12. Model of crack splitting in HSB. (von Koenigswald et al. 1987)

mammals, following the extinction of dinosaurs during the Paleocene (von Koenigswald et al. 1987). Even for these early arctocyonids it could be demonstrated that the horizontally arranged HSB stood parallel to the main tension forces.

In incisors and molars of some rodents, however, HSB were reduced in thickness during evolution until each band consisted of only a single layer of prisms, called uniserial HSB or lamellar enamel. In this type of HSB the prisms tend to stay within their band and a splitting parallel to these bands may be possible. In this case the orientation of the IPM is of increased importance (von Koenigswald 1990a). At a specific level of evolution the IPM turns from its primitive condition parallel to the prisms, which decussate at about 90°, to the third direction in space. In this case the enamel is strengthened by fibers in all three orthogonal directions, unbreakable for a hairline crack (Fig. 13).

Besides the decussation of prisms in HSB, bundles of prisms may cross over each other, like in the 3D enamel of elephants (Pfretzschner in press). In this enamel the Crook-Gordon mechanism (Fig. 14) may even act as a crack-stopping mechanism (Pfretzschner in press).

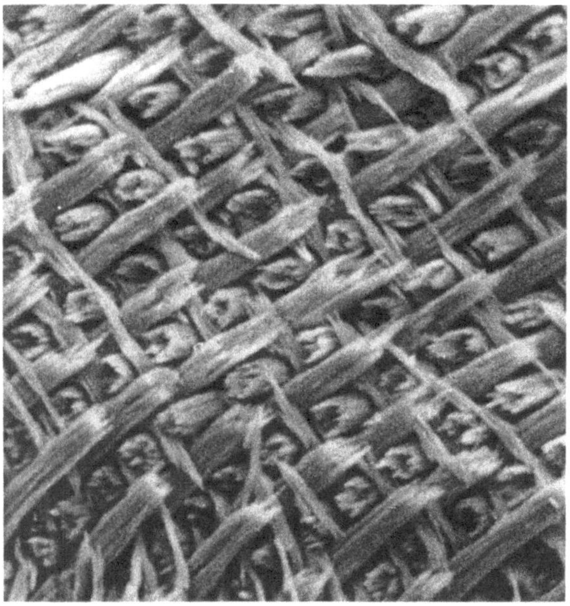

Fig. 13. Uniserial HSB in a rodent incisor (*Microtia*) with fibers orientated in all three directions

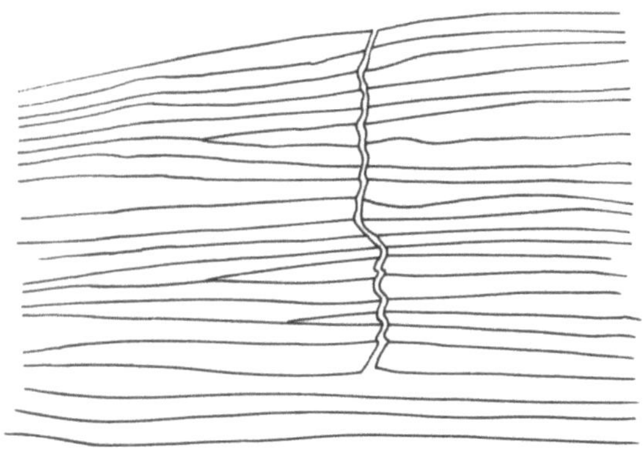

Fig. 14. Crack stopped by Cook Gordon mechanism in proboscidean enamel (*Elephas maximus*). After having crossed several prisms the crack stops in the interprismatic matrix (IPM)

6 Reduction and Remodeling of Enamel Structures During Phylogeny

If parts of the enamel structure are regarded as adaptative to some functional purpose, it might be theoretically possible that the Schmelzmuster is altered when functional requirements change. Several evolutionary lineages show a continuous development of the Schmelzmuster in the form of an increased differentiation (Bertrand 1988; von Koenigswald 1982; Pfretzschner in press). No examples have yet been found in which the rebuilding of the Schmelzmuster led to a more primitive evolutionary level.

The reduction of enamel thickness may cause the loss of the outermost layers of the Schmelzmuster, as seen in the trailing edges of molars of the arvicolid genus *Microtus* (von Koenigswald 1980). However, more rapid tooth eruption allows reduction of a part of the enamel band. This is definitely not a return to a more primitive situation.

Teeth of early cetaceans possess Hunter-Schreger bands, as is usual for large mammals. More modern whales reduced the enamel of their multified and morphologically simplified dentine teeth. In the various stages of the enamel reduction, the Schmelzmuster, as well as the prism type, have been changed and reduced (Ishiyama 1984, 1987; Sahni 1981). The loss of the Schmelzmuster is again related to a general reduction.

Lester et al. (1988) described a partial reduction of prisms in vampire bats in favor of an enamel free of prisms. They suggest "post-prismatic enamel" to distinguish this enamel from the primitive aprismatic enamel of reptiles or early mammals.

7 Conclusions

The general morphology of teeth shows many characters related to function. The thickness of the enamel, as well as the enamel as a material, can be interpreted in various aspects as being selected to fulfill its specific functions.

In contrast to bone, the structure of the enamel is determined genetically and cannot be rebuilt under functional stresses.

A single optimal type of enamel was not developed during mammalian evolution, but various different types can provide significant Schmelzmusters. Even when the quality of the various enamel types is not yet understood in detail, the Schmelzmuster often reflects the stress pattern during mastication. Therefore, the Schmelzmuster differs in the various functional areas even within the same dentition.

The genetical determination of the Schmelzmuster allows a continuous evolution of the Schmelzmuster, but a drastic change of construction seems to be rare. No examples for the return to a more primitive level of evolution are known. Therefore, in addition to the functional interpretation, phylogenetic information may be obtained by careful research of the enamel structure, as from any other morphological character.

Acknowledgments. Most of the results have arisen from various projects granted by the Deutsche Forschungsgemeinschaft. For valuable comments the authors thank Prof. Dr. Preuschoft, Bochum, and for the artistic work the authors are indebted to Mrs. D. Kranz, Bonn.

References

Bertrand P (1988) Evolution de la structure de l'email chez les Proboscidea primitifs: aspect phylogenetique et fonctionelle. In: Russell DE, Santoro JP, Sigogneau-Russell D (eds) Teeth revisited. Proc 7th Int Symp on Dental Morphology, Paris 1986. Mem Mus Nat Hist Nat, Paris (C) 53:109–124

Boyde A (1964) The structure and development of mammalian enamel. PhD Thesis, University of London, London

Boyde A (1969) Electron microscopic observations relating to the nature and development of prisms decussation in mammalian dental enamel. Bull Group Int Rech Sc Stomat 12:151–207
Boyde A (1976) Enamel structure and cavity margins. Operat Dent 1:13–28, 25 figs
Boyde A (1978) Development of the structure of the enamel of the incisor teeth in the three classical subordinal groups of the Rodentia. In: Butler PM, Joysey KA (eds) Development, function and evolution of teeth. Academic Press, London New York, pp 43–58
Fortelius M (1985) Ungulate cheek teeth: developmental functional and evolutionary interrelations. Acta Zool Fenn 180:1–76, 43 figs
Greaves WS (1973) The interference of jaw motion from tooth wear facets. J Paleontol 47:1000–1001, 1 fig
Ishiyama M (1984) Comparative histology of tooth enamel in several-toothed whales. In: Fearnhead RW, Suga S (eds) Tooth enamel IV. Elsevier, Amsterdam, pp 432–436, 8 figs
Ishiyama M (1987) Enamel structure in odontocete whales. Scann Microsc 1/3:1071–1079, 27 figs
Koenigswald W von (1977) *Mimomys* cf. *reidi* aus der villafranchischen Spaltenfüllung Schambach bei Treuchtlingen. Mitt Bayer Staatssammi Paläontol Hist Geol 17:197–212, 40 figs
Koenigswald W von (1980) Schmelzmuster und Morphologie in den Molaren der Arvicolidae (Rodentia). Abh Senckenb Naturforsch Ges 539:1–129, 136 figs
Koenigswald W von (1982) Enamel structure in molars of Arvicolidae (Rodentia, Mammalia), a key to functional morphology and phylogeny. In: Kurten B (ed) Teeth: form, function and evolution. Columbia Univ Press, New York, pp 109–122, 7 figs
Koenigswald W von (1988) Enamel modification in enlarged front teeth among mammals and the various possible reinforcement of the enamel. In: Russell DE, Santoro JP, Sigogneau-Russel D (eds) Teeth revisited. Proc 7th Int Symp Dental Morphology, Paris 1986. Mem Mus Nat Hist Nat (C) 53:148–165
Koenigswald W von (1990) Die Palökologie der Apatemyiden, Ausdeutung der Skelettfunde von *Heterohyus nanus* aus dem Mitteleozän bei Darmstadt. Palaeontolographica 210:41–77, 21 figs, 4 tabs
Koenigswald W von (1990a) Ein ungewöhnliches Schmelzmuster in den Schneidezähnen von *Marmota* (Rodentia, Mammalia). N Jahrb Geol Paläontol 180:53–73, 14 figs.
Koenigswald W von, Pascual R (1990) The Schmelzmuster of the Paleogene South American rodent-like marsupials *Groeberia* and *Patagonia* compared to rodents and other Marsupialia. Paläontol Z 64:345–358, 11 figs.
Koenigswald W von, Pfretzschner HU (1987) Hunter-Schreger-Bänder im Zahnschmelz von Säugetieren: Anordnung und Prismenverlauf. Zoomorphology 106:329–338, 17 figs
Koenigswald W von, Rensberger JM, Pfretzschner HU (1987) Changes in the tooth enamel of early Paleocene mammals allowing increased diet diversity. Nature (London) 328:150–152, 2, 4 figs
Lester KS, Hand SJ, Vincent F (1988) Adult phyllostomid (bat) enamel by scanning electron microscopy – with a note on dermopteran enamel. Scand Electron Microsc 2, 1:371–383, 38 figs
Martin L, Boyde A, Grine FE (1988) Enamel structure in primates: a review of SEM studies. Scann Microsc 2, 3:1503–1526
Pfretzschner HU (1988) Structural reinforcement and crack propagation in enamel. In: Russell DE, Santoro JP, Sigogneau-Russell DE (eds) Teeth revisited. Proc 7th Int Symp Dental Morphology, Paris 1986. Mem Mus Nat Hist Nat (C) 53:133–143
Pfretzschner HU (in press) Biomechanik der Schmelzmikrostruktur in den Backenzähnen von Großsäugern. Palaeontolographica
Rensberger JM, Koenigswald W von (1980) Functional and phylogenetic interpretation of enamel microstructure in rhinoceroses. Paleobiology 6/4:477–495, 17 figs
Sahni A (1981) Enamel ultrastructure of fossil mammalia: Eocene Archaeoceti from Kutch. J Palaeontol Soc India 25:33–37, 2 plates
Stern D, Crompton AW, Skobe Z (1989) Enamel ultrastructure and masticatory function in molars of the American opossum, *Didelphis virginiana*. Zool Linnean Soc 95:311–334, 11 figs
Young WG, McGowan M, Daley TJ (1987) Tooth enamel structure in the koala, *Phascolarctos cinereus*: some functional interpretations. Scann Miicrosc 1, 4:1925–1934, 4 tabs, 9 figs

The Functional Morphology of Gastropod Apertures

C. R. C. PAUL [1]

Summary

If snail shells are protective, the aperture forms a large gap in the protection afforded by the shell. Apertures function instantaneously in allowing emergence and withdrawal of the snail, and in growth as the generating curve of the shell. Opercula complete the protection afforded by the shell, but tight-fitting forms primarily prevent desiccation. The plane of the aperture reveals life orientation, except in high-spired shells and, together with other features of shell morphology, may reveal rates of locomotion and relationships to substrates. Unique terminal modifications to the aperture indicate maturity and involve one or more of the following: changes in (a) cross-section and (b) orientation, (c) reinforcement or (d) elaborations. All usually strengthen the outer lip and occur in snails which live for a long time after reaching adult size. Repeated apertural modifications indicate episodic growth and may also involve changes in coiling direction. Exceptions to these generalizations occur and specific cases need careful analysis.

1 Introduction

It is axiomatic that shells provide protection for gastropods by enclosing the soft tissues. Although it may seem self-evident, protection can be defined as any morphological, physiological or behavioral feature that tends to thwart the lethal or injurious effects of predators, parasites or the environment. So if the shell is basically protective the aperture represents a substantial gap. The paradigm (ideal design) for a protective envelope would be mechanically strong and completely enclose all soft tissue, leaving no gaps. However, such a structure would completely isolate the animal from the surrounding environment which contains all the food and oxygen necessary for life. The requirements of mechanical protection are diametrically opposed to those of all other vital functions and all protective skeletons are compromise structures. In gastropods this compromise is achieved by a temporal shift in behavior. Snails withdraw into their shells when threatened, but emerge to feed, breathe, mate etc., and the aperture enables them to do so. This paper reviews the morphology of gastropod apertures and associated structures against the generalized function of protection. However, each case needs individual scrutiny as exceptions occur commonly, but are often just as enlightening. Oth-

[1] Dept. of Earth Sciences, The University of Liverpool, P.O. Box 147, Liverpool L69 3BX, UK

N. Schmidt-Kittler and K. Vogel (Eds):
Constructional Morphology and Evolution

er aspects of gastropod morphology have been considered elsewhere (e.g., Linsley 1977), especially coiling (see Raup 1966 and references therein). Note, however, that the outline of the aperture represents Raup's generating curve. Thus apertures have both an instantaneous function involved with emergence and withdrawal of the body, and a growth function involved in the construction of the shell.

2 Opercula

Many gastropods have an operculum which seals the aperture when the snail withdraws, but this may involve detachment from the substrate. To complete the protection afforded by the shell, the operculum should be strong and fit tightly into the aperture. While many opercula match these requirements, an equal number are very loose-fitting with gaps all round when in place (Fig. 1). All terrestrial and freshwater and most intertidal marine operculates have tight-fitting opercula, although there are also wholely subtidal groups which do. Nevertheless, this suggests that one general function of tight-fitting opercula is to reduce water loss when exposed to the atmosphere. Rees (1964) figured modifications to the apertures of several terrestrial prosobranchs in which small pores or channels communicate with the exterior when the operculum is in place. Examples included cyclophorids and annulariids from Central America and S. E. Asia. Rees suggested that the common factor was a strongly seasonal climate which forced the snails to aestivate for prolonged periods, and that the breathing pores allowed gas exchange while reducing water loss. They also prevent penetration by all but the smallest organisms. Hermit crabs provide independent evidence of the basic protective effect of gastropod opercula. Solem (1974, Fig. 5, p. 97) illustrated an example of *Coenobita* with one claw closing the aperture of a *Nerita polita* shell. Loose-fitting opercula allow snails to withdraw further up the spire. Opercula are as morphologically varied as shells. Many snails use their opercula for other functions which may constrain the shape. Strombids have a blade-like operculum attached to the foot at one end and use it to probe sediment, aid locomotion and to right the shell when overturned (Solem 1974, Fig. 8, p. 102 and personal observations).

Fig. 1. Loose-fitting operculum of *Buccinum undatum* L. (Scales in mm unless otherwise indicated)

The vast majority of pulmonates are terrestrial and thus inevitably exposed to desiccation, yet pulmonates lack an operculum. Almost all prosobranchs have no modification to the mantle edge which reduces the gap between the body and the shell. When active they can potentially lose moisture from the entire lining of the mantle cavity as well as the emergent body surface. Pulmonates have a flap of the mantle which seals up the gap between the body and the shell edge completely. The pneumostome is a muscular pore in this seal which opens only when the snails breathe. Pulmonates lose water from the lung (= mantle cavity) only when the pneumostome is open and then at a very much reduced rate. Rates of water loss are lower when snails withdraw into their shells, but the difference is still significant and terrestrial operculates would desiccate very rapidly without an operculum. Surprisingly, Block (1971) found that epiphragms (temporary seals to the aperture produced by many pulmonates) do not affect rates of water loss. *Rumina decollata* even leaves a breathing pore in its epiphragm (Paul 1982, Fig. 8M–O). However, Block investigated snails which hibernate in winter. Epiphragms of aestivating snails are often thicker and almost certainly reduce water loss.

3 Orientation

Considered in isolation, limpet shells have the aperture at the widest part and seem very poor protection for the limpet. However, in life position the underlying substrate completes the protection afforded by the shell. *Patella* returns to the same spot after each tide and rotates the shell against the substrate before settling (Yonge and Thompson 1976, p. 59). This produces a very tight fit whether the shell or the rock or both are abraded, and again is probably primarily to reduce water loss. Limpets have exploited to the extreme the first protective response of any snail, namely to clamp the shell against the substrate. In doing so they have modified their shell shape so that the aperture can be held flush against the substrate. Snails with shells coiled through more than one whorl can only hold the aperture closely against the substrate if it is tangential to the preceding whorl (Linsley 1977). Tangential apertures can be produced by accelerating growth of the upper and inner parts of the whorl relative to the lower and outer parts, to produce the angles α and β (Fig. 2), or by changing the direction of growth before forming the final, mature aperture (see below).

Stereonets can be used to illustrate the orientations of aperture planes, or poles to these planes. In the conventional orientation of a dextral shell with the spire up and the aperture facing the observer, most poles to aperture planes plot in the SE quadrant (Fig. 3). The angles α and β can be used to plot these poles. Stereonets have two uses. Firstly, the poles to the actual aperture plane and to an exactly tangential aperture can be plotted. The angular distance between them (Fig. 4) is an objective measure of how closely the real aperture approaches the ideal position. Note, however, that the selective advantage of tangential apertures may relate to the absolute size of the gap between the aperture and the substrate, not the angular discrepancy. Small shells may have smaller gaps than large shells despite having a larger angular discrepancy. The second use of stereonets is to plot the normal life orientation of snails. By a quirk of the geometry of stereographic

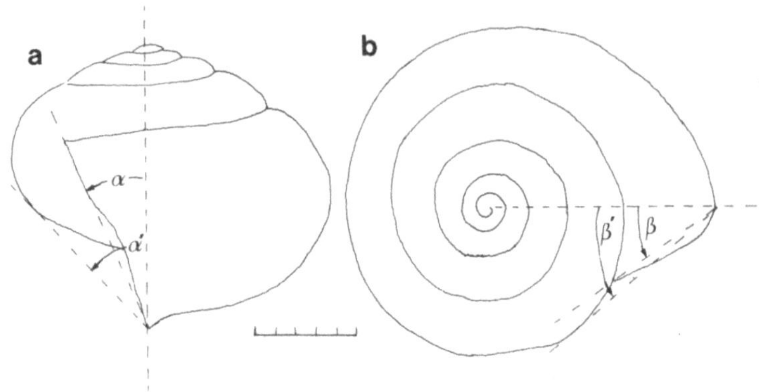

Fig. 2. a The angle α, and **b** the angle β for real and ideal (α', β') apertures of *Theba pisana* (Müller)

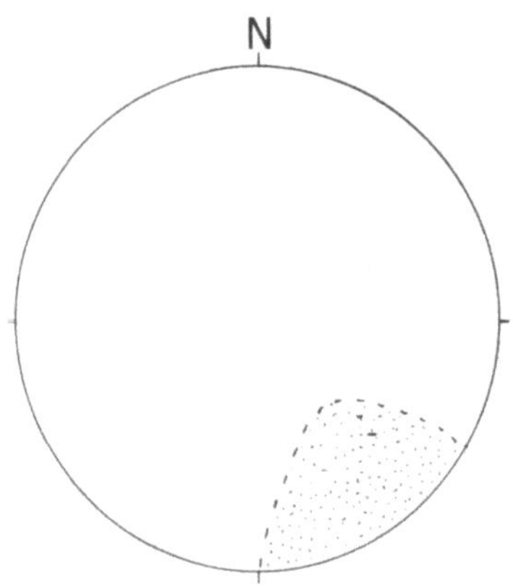

Fig. 3

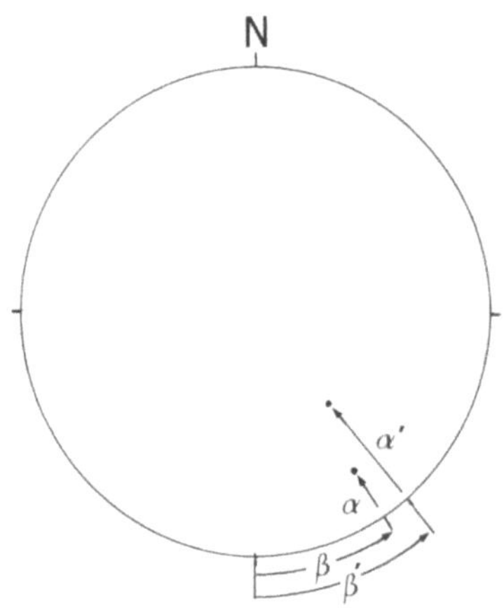

Fig. 4

Fig. 3. Area of stereonet within which most poles to aperture planes of dextral shells plot

Fig. 4. Angular distance between real and ideal aperture planes of *Theba pisana* (Müller) shown in Fig. 2

projections, if the pole to the aperture plane in the conventional orientation is plotted in the lower hemisphere, it is exactly coincident with the pole to the axis of coiling plotted in the upper hemisphere when the aperture is horizontal and the snail crawls due north (Fig. 5). Thus in practice once the pole to the plane of the aperture is plotted the life orientation is revealed.

Linsley (1977) showed that when snails are active the shells are usually balanced both fore and aft and left to right. This may well explain the long anterior canal of some shells such as muricids (Paul 1981). Linsley (1978) also showed that fast-moving snails have streamlined shells which lack elaborate ornament. The spire is held at a low angle to the substrate and often rotated to lie over the posterior foot. Conversely, slow-moving snails had highly sculptured shells often with the spire raised. The slowest group of all, Linsley's "shell draggers", have

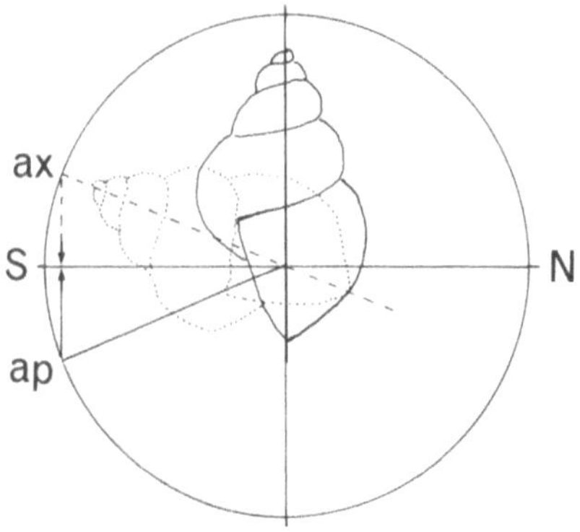

Fig. 5. Poles to aperture plane (*ap*) when axis is vertical, and to axis (*ax*) when aperture is horizontal, coincide if plotted in opposite hemispheres

shells which lie recumbent on the substrate. There is a maximum size, and in high-spired shells a maximum length, above which it becomes impossible to maintain the shell aloft especially when moving over loose substrates like sand, because the foot cannot exert enough grip on the substrate. Thus, although the pole to the plane of the aperture reveals life orientation in most shells, careful interpretation of individual cases is needed. High-spired shells are highly likely to have been recumbent in life. There is both a scale and a shape factor in this phenomenon, so that it is impossible to state a limiting size or shape (i.e., height to diameter ratio, H : D). A small shell with H : D = 3, for example, may well be held aloft, whereas a large shell with the same ratio and identical shape may lie recumbent.

Cain (1978 and references therein) has shown that in almost all snail faunas, both fossil and Recent, there is a bimodal distribution of the H : D ratio. Shells are either high-spired or discoidal. Furthermore, at least in terrestrial snails, Cain and Cowie (1978) demonstrated that high-spired snails were active predominantly on steeply inclined surfaces whereas discoidal snails apparently preferred more horizontal surfaces. It is possible that climbing vertical surfaces reduces frictional drag on high-spired shells. Certainly European clausiliids and Jamaican urocoptids, for example, are frequently found on steep surfaces with their shells hanging vertically under the influence of gravity (Fig. 7). Savazzi (1989) has recently described other features of the shell which can be used to infer an infaunal, burrowing mode of life. Although not stated explicitly, aperture planes of most infaunal snails are orientated at low angles to the spire. This enhances streamlining and facilitates movement through the substrate. In a dextral snail the angle α is measured anticlockwise (Fig. 2). Occasional snails have the plane of the aperture orientated in the reverse direction (i.e., with a negative angle α). In the extinct, Tertiary, freshwater snail, *Potamaclis* (Fig. 6), the plane of the aperture parallels the

Fig. 6. Life orientation of *Potamaclis turritissima* Forbes

substrate when the spire is recumbent, the normal living position. Paul (1989) interpreted this as an adaptation to living in rivers rather than lakes. With this orientation to the aperture, shell dragging would not raise the spire and thereby increase the chances of dislodgement in the current.

4 Apertures and Growth

Snail shells grow by accretion, small or large increments being added to the existing shell, which consequently retains a complete growth record. Growth lines reveal the positions of former apertures and therefore the orientation of the aperture plane can be reconstructed throughout growth, or from damaged fossil shells in which the last aperture is missing. Although a few snails apparently grow throughout their lifetimes (e.g., the freshwater lymnaeids), most reach a mature adult shell size and cease growth completely. The aperture is the weakest part of the shell, so that those snails which do cease growth invariably modify the final aperture. These modifications, provided they are unique, offer the best evidence that a fossil snail was fully grown when it died. It is also usually an indication that the snails lived for a considerable time after reaching maturity. In British land snails annuals (e.g. *Vitrina pellucida*, Paul 1975, p. 325) very rarely modify their apertures, but those that live for two or more years (e.g., *Helix*) invariably do. Terminal modifications to apertures fall into one or more of four main groups; changes in (a) cross-section and (b) orientation, (c) simple reinforcement, or (d) elaborations.

4.1 Changes in Cross-Section

One of the simplest terminal modifications of apertures is flaring, i.e., expanding rapidly like the bell of trumpet (Figs. 7, 16c). Flaring increases the surface area of

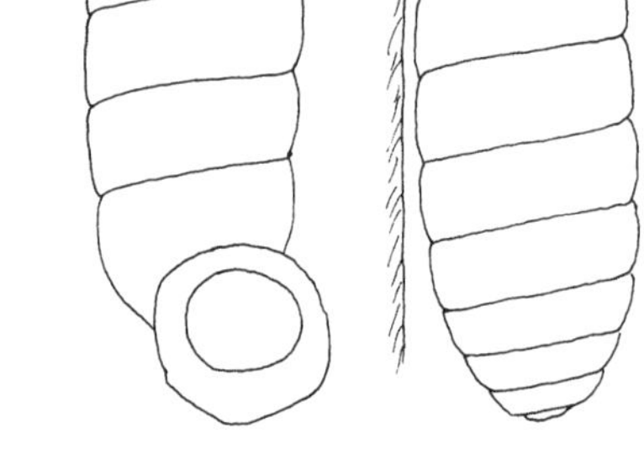

Fig. 7 a, b. *Urocoptis brevis* (Pfeiffer) showing flared, uncoiled aperture which allows snail to cling to vertical rock faces

the foot used to grip the substrate, but conversely once the snail is dislodged it provides larger access for predators. Simple constrictions of the final aperture are rarer, but both expansion and constriction achieve a stiffer shell by dint of curvature in more than one direction. Special modifications which tend to occlude the aperture are considered below.

4.2 Changes in Orientation

Many snails change the direction of coiling just prior to the final aperture to achieve a more nearly tangential orientation (Fig. 8). In European helicids a rough cline exists from northern Europe south to North Africa, which suggests it may be related to water loss during summer aestivation. Moroccan helicids have more nearly tangential apertures often achieved by substantial changes in the direction of coiling (Fig. 8). This modification is more effective with ground-dwelling habits and interestingly even Moroccan species of *Theba* and helicellines, which aestivate on plants well above ground, do not have tangential apertures. Terminal changes in coiling direction are not confined to land snails. The freshwater planorbid, "*Taphius*" *albicans*, turns the aperture slightly to the left when mature, while the marine snail *Aporrhais* has a negative angle α which allows the spire of the adult to be recumbent (Fig. 9). *A. pespelecani* is an infaunal detritus-feeder (Yonge and Thompson 1976, p. 90) and adults use the expanded outer lip as a shelter for a respiratory chamber (Fig. 9). Uncoiling is a special case of the tendency to reorientate the aperture at the end of growth. In Jamaican urocoptids,

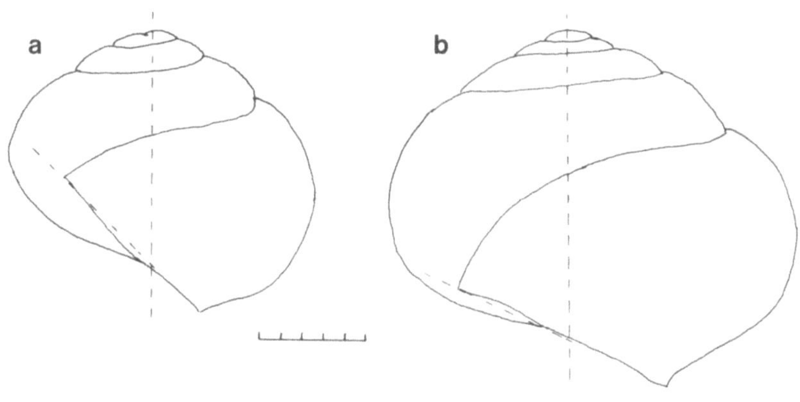

Fig. 8. Terminal reorientation of the aperture in **a** *Cepaea hortensis* (Müller) from England and **b** "*Archelix*" *abrolena* (Bourguignat) from Spanish Morocco

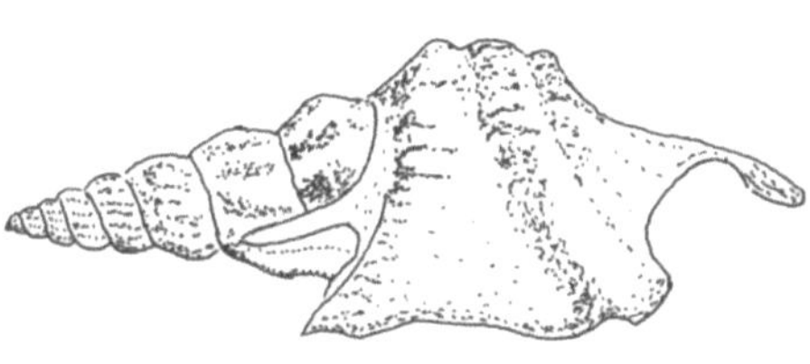

Fig. 9. *Aporrhais pespelecani* (L.) in living orientation

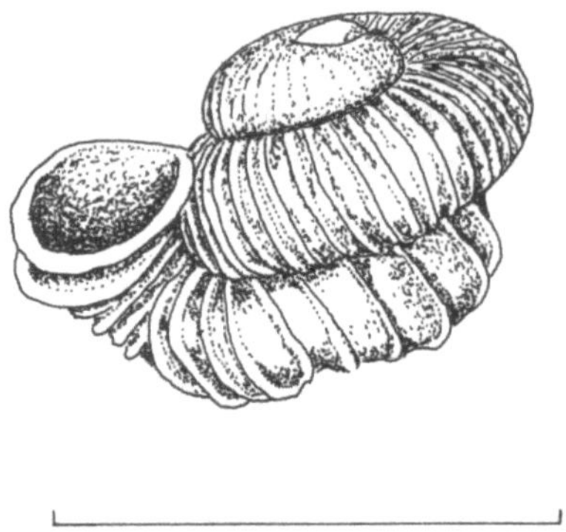

Fig. 10. Reorientation of the aperture in *Opisthostoma*

for example (Fig. 7), it often amounts to no more than the development of a completely free aperture allowing the snail to cling to vertical rock surfaces, but in some Cuban forms the uncoiled portion is straight and as long as the entire coiled shell. Uncoiling occurs in both land operculates and pulmonates. In some of the former it provides space for the breathing pore as well as producing a tangential aperture. *Opisthostoma* (Fig. 10) is another special case in which the adult aperture points upwards in the same direction as the spire. I can offer no suggested function for this bizarre, but widespread, modification.

4.3 Reinforcement

The growing edge of the shell is thin and fragile. Damaged growth lines are eloquent testimony to this. One of the commonest terminal modifications is thickening of the lip (Fig. 11) usually by an internal rib which may or may not be accompanied by an external constriction. Such modification is common in marine and terrestrial snails, but less so in freshwater forms. Vermeij (1978) has related thickening of the lip in marine snails to prevalence of durophagous predators (shell crushers). That such thickenings do offer effective protection is seen by the common occurrence of rodent-damaged land snail shells in Jamaica. Almost without exception these shells have the aperture entire but have been bitten through about half a whorl back (Fig. 12). External thickenings to the outer lip (Fig. 11b, c) are less common. Many examples are operculates and the presence of the operculum may well inhibit internal thickening.

4.4 Elaborations

These are often extensions or modifications of reinforcement to the outer lip and may occur internally or externally. Internally the rib may be produced locally into teeth or lamellae. However, in *Azeca* (Paul 1974) or the freshwater snail *Planorbula*, some or all of the teeth lie well back from the aperture. Solem (1974, p. 188) has interpreted these modifications as anti-predatory devices and it is easy to quantify the degree of mechanical protection afforded by apertural denticles. Paul (1974, p. 166) showed that the denticles of *Azeca* reduce the radius of the maxi-

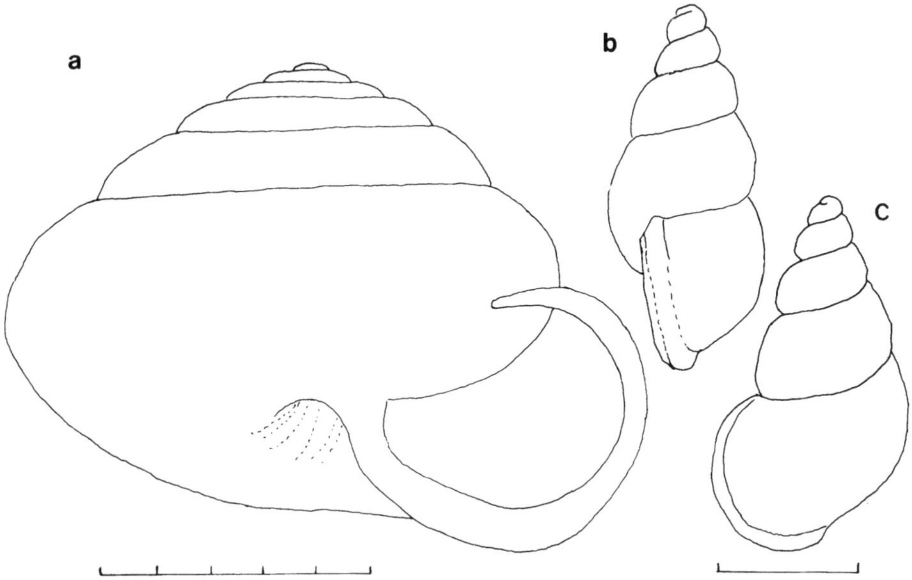

Fig. 11. Reinforcement of the outer lip by **a** an internal rib in *Trichia hispida* (L.) and **b, c** an external rib in an undescribed Pleistocene hydrobiid from Wallywash Great Pond, Jamaica

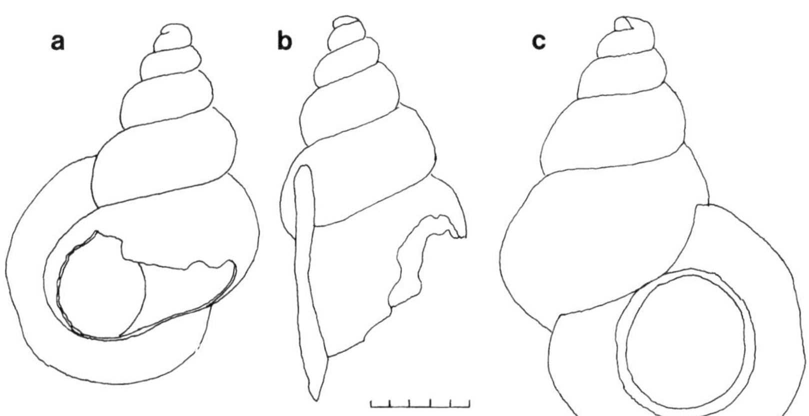

Fig. 12. a, b Rodent damage in, and **c** an undamaged shell of, *Tudora (Colobostylus) humphreysianum* (Pfeiffer), Glasgow, Jamaica

mum particle that could pass through the aperture by 38%. Clearly the denticles improve the protection of the aperture. Such mechanical analysis does not, however, demonstrate that this is the selective advantage conferred by denticles. In Britain and western Europe the genera *Azeca* and *Cochlicopa* have very similar habitats and overall shell morphology. However, *Cochlicopa* is by far the more

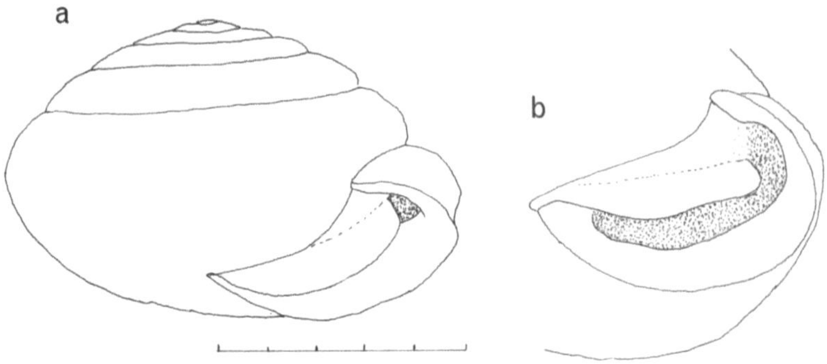

Fig. 13. **a** Occluded aperture of *Stenotrema monodon alicia* Pilsbry, and **b** oblique view to show slit-like opening. Town Bluff, Texas

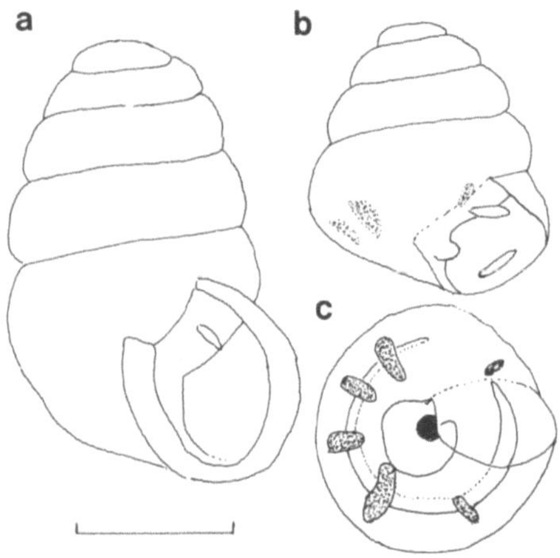

Fig. 14. **a** Adult and **b, c** juvenile shells of *Lauria cylindracea* (da Costa), Ibiza

widespread and abundant genus (Kerney and Cameron 1979). Whatever the benefits conferred on *Azeca* by the possession of lamellae, *Cochlicopa* is thriving without them. Solem (1974, p. 188) has shown that apertural lamellae may themselves be armed with tiny spines, or platelets that imbricate like the tiles on a roof. He argued that these were additional defences against insect predators, suggesting that the hairs on insect limbs would catch on the spikes or edges of the platelets making entry more difficult. Platelets are present on the North American land snail *Stenotrema*, some species of which have almost totally occluded apertures (Fig. 13). I suspect the imbricate platelets may also be an aid to the emerging snails, making it easier to extend the body through the constricted aperture. Retraction does not pose the same problem being effected by the columellar muscles, not by increasing blood pressure. Juveniles of the land snail *Lauria* possess several lamellae which adults lack (Fig. 14). *Lauria* is the only snail known to me that apparently has a mechanically better protected aperture as a juvenile than as an adult.

External varices or spines can also be regarded as elaborations of an external rib, but here I include extensive modifications such as occur in *Aporrhais* (Fig. 8)

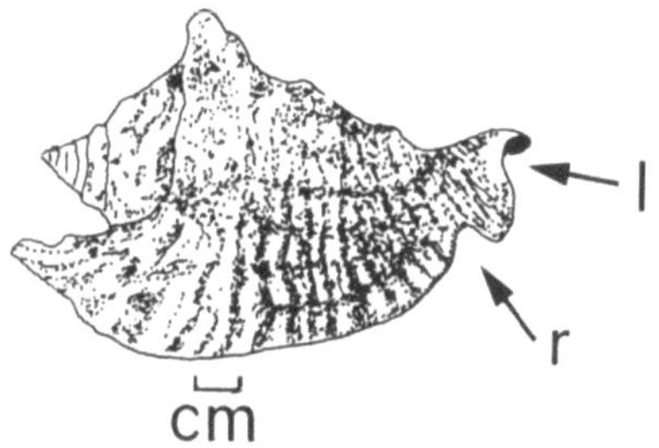

Fig. 15. Lateral view of *Strombus* (*Tricornis*) *raninus* Gmelin to show notches for left (*l*) and right (*r*) eyes

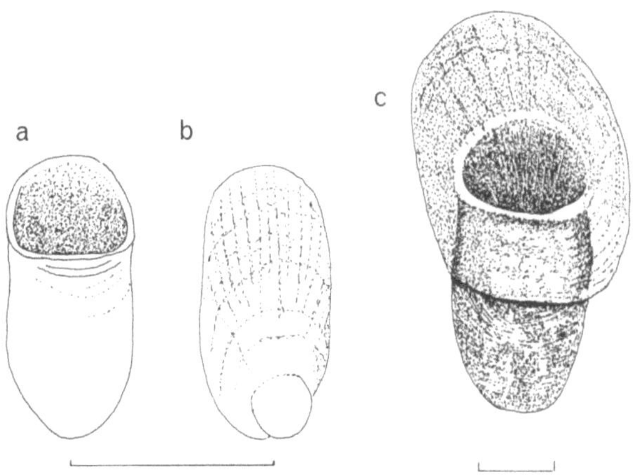

Fig. 16. a, b Septate aperture of *Gundlachia radiata* (Guilding), Pleistocene, Wallywash Great Pond, Jamaica, and **c** post-septate aperture of *G. meekiana* Stimpson, Recent, Ann Arbor, Michigan (**c** after Basch 1959, Fig. 1d)

or the spider shell, *Lambis*. While these obviously enlarge the shell, rendering it more difficult to swallow whole or to crush, the precise reasons for the patterns seen in individual species need detailed investigation. *Lambis* is said to include a rare case of sexual dimorphism among snails (Lindner 1975, p. 51). Elaborations may also include the development of terminal slits or notches. The most common and well known are slits or canals for the inhalent and exhalent currents in advanced prosobranchs. Species of *Rhinoclavis* develop both on the final aperture. All strombids develop special notches in the margin of the final aperture anteriorly and some posteriorly as well (Fig. 15). The left eye emerges along the anterior canal; an adjacent notch accommodates the right eye. In some species the posterior notch is used in righting the shell when overturned. The snail inserts the blade-like operculum in the notch, contracts foot muscles until the operculum snaps out of the notch and its blade-like form "skulls" the shell over to an upright position (personal observation).

Freshwater limpets originally attributed to *Gundlachia* develop septate apertures (Fig. 16a, b). Subsequently other species have been assigned to *Gundlachia* solely on the presence of septate forms, whereas some authors regard the occurrence of septate forms as a phenotypic response to drying out (see review in Basch 1959). The septa almost completely seal off the aperture. In many cases there is then a post-septate growth form with a new flared aperture (Fig. 16c). Whether genetically or environmentally induced, this is a rather special case of apertural modifications and since post-septate growth is known, cannot be regarded as a terminal modification.

5 Repeated Modification

The above examples all deal with terminal or near-terminal modifications to the aperture which are unique and occur only at maturity. Muricids and cymatiids typically produce repeated apertural modifications which indicate episodic growth. Both groups can add substantial increments to the shell, modify the aperture, and remain at that size for extended periods before adding another increment. Increments are as much as 2/3 of a whorl in cymatiids, 1/3 or less in muricids. Growth has only been described in detail for one muricid, *Eupleura caudata* (see McKenzie 1961). The addition of an increment took 3 weeks, but increments were only added once or twice a year. Regrettably one cannot age shells by counting increments. Vermeij (1979) has outlined the features which strengthen shells and undoubtedly such varices, especially those of cymatiids, strengthen not only the aperture but earlier whorls as well. Paul (1981) pointed out that the common addition of spiny varices three times per whorl found in some muricids ensures that there is always a series of spines pointing upwards (and backwards) in these shells (Fig. 17b). He also demonstrated that the lower pair of varices in

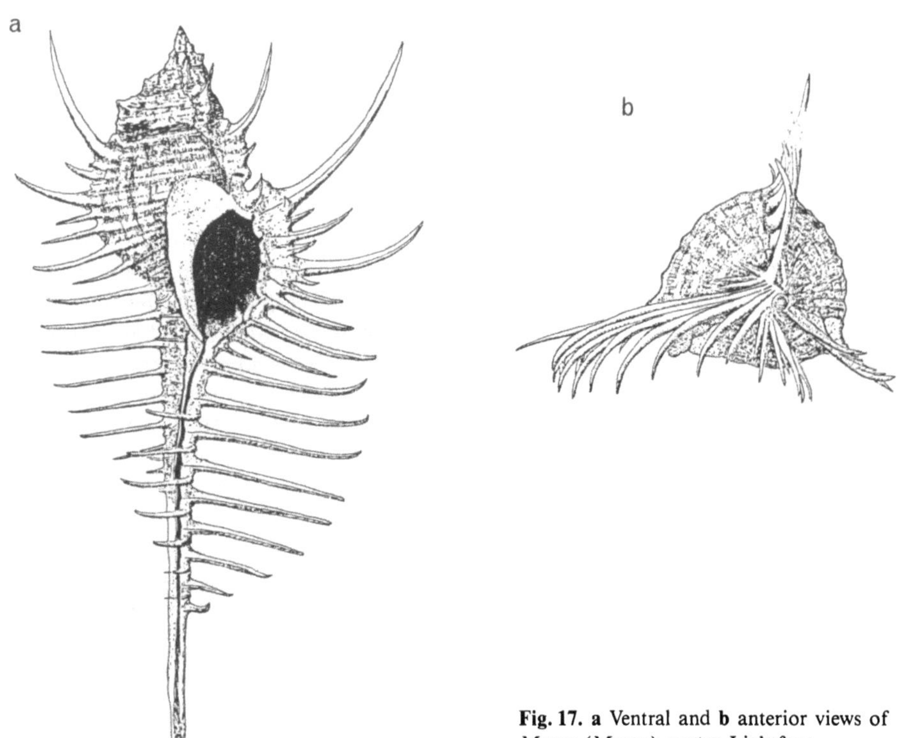

Fig. 17. a Ventral and **b** anterior views of *Murex* (*Murex*) *pecten* Lightfoot

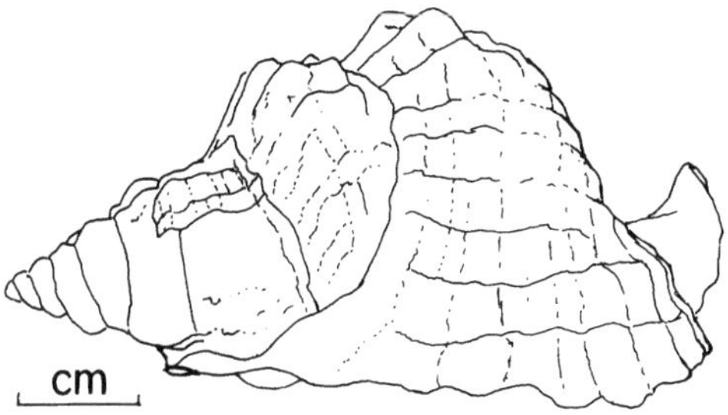

Fig. 18. Lateral view of *Distorsio (Rhysema) clathrata* (Lamarck) to show change of coiling direction at aperture

Murex pecten form a complete anterior cage within which the snail could feed unmolested (Fig. 17a). Repeated modifications may involve coiling direction as well as elaborations. The cymatiid *Distorsio* does both and this ensures that the spire maintains a low profile practically touching the substrate (Fig. 18). In *Biplex bubo* increments are half a whorl and the blade-like flanges of each modified aperture form a continuous snowshoe-like marginal rim around the shell. At the same time part of the spiral curve of each increment is omitted producing a flattened cross-section (Linsley and Javidpour 1980, Fig. 5).

6 Conclusions

1. Snail shell apertures have an instantaneous function associated with emergence and retraction of the body and a growth function during construction of the shell.
2. Tight-fitting opercula are primarily an adaptation to reduce water loss.
3. The plane of the aperture reveals normal life orientation except in very high-spired shells and, in conjunction with other features of shell morphology, may also suggest likely rates of locomotion and relationships to substrates.
4. Unique terminal modifications to the aperture indicate maturity and involve changes in cross-section and orientation, reinforcement or elaborations. These usually strengthen the outer lip.
5. Repeated apertural modifications indicate episodic growth and may include changes in coiling.
6. Exceptions to generalizations (2)–(5) above occur widely so individual cases need specific investigation.

References

Basch PF (1959) Status of the genus *Gundlachia* (Pulmonata, Ancylidae). Occ Pap Mus Zool Univ Michigan 602:9 pp

Block MR (1971) Epiphragms: some observations. J Conch London 26:388–409

Cain AJ (1978) Variation of terrestrial gastropods in the Philippines in relation to shell shape and size. J Conch London 29:239–245

Cain AJ, Cowie RH (1978) Activity of different species of land-snail on surfaces of different inclinations. J Conch London 29:267–272

Kerney MP, Cameron RAD (1979) A field guide to the land snails of Britain and North-west Europe. Collins, London, 288 pp

Lindner G (1975) Seashells of the world. Blandford, Poole, 271 pp, 64 plates
Linsley RM (1977) Some "laws" of gastropod shell form. Paleobiology 3:196–206
Linsley RM (1978) Locomotion rates and shell form in the Gastropoda. Malacologia 17:193–206
Linsley RM, Javidpour M (1980) Episodic growth in Gastropoda. Malacologia 20:153–160
McKenzie CR Jr (1961) Growth and reproduction of the oyster drill *Eupleura caudata* in the York River, Virginia. Ecology 42:317–338
Paul CRC (1974) *Azeca* in Britain. J Conch London 28:155–172
Paul CRC (1975) The ecology of Mollusca in ancient woodland. I. The fauna of Hayley Wood, Cambridgeshire. J Conch London 28:301–327
Paul CRC (1981) The function of the spines in *Murex (Murex) pecten* Lightfoot and related species (Prosobranchia: Muricidae). J Conch London 30:285–294
Paul CRC (1982) Pleistocene non-marine molluscs from Cala Salada, Ibiza. Geol J 17:161–184
Paul CRC (1989) The molluscan faunal succession in the Hatherwood Limestone Member (Upper Eocene), Isle of Wight, England. Tert Res 19:147–162
Raup DM (1966) Geometric analysis of coiling: general problems. J Paleontol 40:1178–1190
Rees WJ (1964) A review of breathing devices in land operculate snails. Proc Malac Soc 36:55–67
Savazzi E (1989) Burrowing mechanisms and sculpture in Recent gastropods. Lethaia 22:31–48
Solem GA (1974) The shell makers. Introducing molluscs. John Wiley & Sons, New York, 289 + ix pp
Vermeij GJ (1978) Biogeography and adaptation. Belknap, Cambridge, Mass, 332 pp
Vermeij GJ (1979) Shell architecture and causes of death of Micronesian reef snails. Evolution 33:686–696
Yonge CM, Thompson TE (1976) Living marine molluscs. Collins, London, 288 pp

Behavior plus "Pathology" – the Origin of Adaptations?

DIETER STEFAN PETERS[1]

Summary

The relationship of construction and behavior in the evolutionary process is discussed on the basis of comparison of malformed bills of avian species with the normal, but specialized bills of other species. The following insights emerged:
1. Construction and behavior can be evaluated only on the foundation of their interactions.
2. Construction and behavior can vary independently from each other within certain limits.
3. As a rule, behavior seems to act as a selective force on the variants of the construction. This means that the organism produces an own canalizing (selective) factor in addition to, and reaching beyond the canalizing limitations of the mere constructional constraints of its coherent system.
4. Through its behavior the organism involves parts of the environment in its activities. The concept of adaptation is meaningful only in this context.
5. The environment cannot bring about organismic variations; the environment is not creative, it can only eradicate.
6. The insight that behavior channels construction can be helpful in attempts to reconstruct the evolutionary process, even if it sometimes helps only to understand why a certain reconstruction cannot be achieved.

1 The Problem

From the multitude of bill shapes in birds, Figs. 1–6 show three very peculiar constructions. Each of them seems to be represented by two species. However, the examples are not of equal standard. Experienced ornithologists will see at a glance that only the birds on the right side have their "normal" bills, whereas the bills on the left are heavily misshaped. This may be a surprise for the uninitiated, since no obvious difference between the left and right specimens seems to be detectable. Why, then, should the bills on the left be malformations while the right ones are considered as highly specialized and "adapted"? (By the way, nothing is known about the genetic basis of these malformations. Nevertheless, it may be justifiable to use them in the following as if they were examples of hereditary variations).

[1] Forschungsinstitut Senckenberg, Senckenberg-Anlage 25, D-6000 Frankfurt/M., FRG

N. Schmidt-Kittler and K. Vogel (Eds):
Constructional Morphology and Evolution

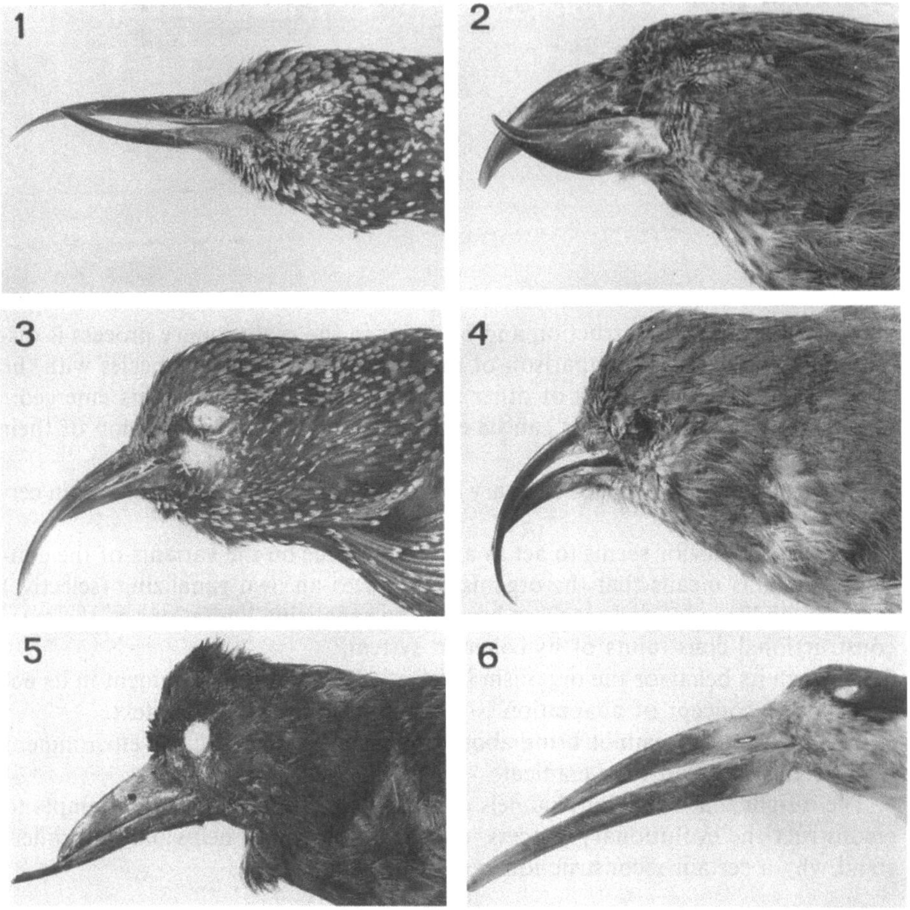

Fig. 1. Starling (*Sturnus vulgaris*) with deformed bill

Fig. 2. Crossbill (*Loxia curvirostra*)

Fig. 3. Starling (*Sturnus vulgaris*) with deformed bill

Fig. 4. Nukupuu (*Hemignathus lucidus*)

Fig. 5. Rook (*Corvus frugilegus*) with deformed bill

Fig. 6. Black skimmer (*Rhynchops nigra*). (All photographs: E. Haupt, Forschungsinstitut Senckenberg)

2 The Answer

By anatomical comparison alone the distinction cannot be understood: the bills from the left side seem to be equivalent to their counterparts on the right. In both cases the bills were integral parts of the coherent organismic apparatuses. Nor did the supposed malformation on the left result in complete dysfunction. This is demonstrated by the fact that the figured specimens wear an adult plumage. The exact age of the starlings is not known (they were collected in September and November respectively) but even if they died during their first year they must have survived at least several months on their own cost and gained enough food to afford not only mere survival but also a complete moult from the juvenile to the adult plumage. The rook survived at least 2 years.

The distinction between "normal" and teratological conditions in these very similar bills becomes meaningful only if we take the activities of the birds into account, particularly their feeding behaviors. Crossbills (*Loxia curvirostra*), it is true, feed on various plants and insects but they are especially effective in harvesting seeds from conifers, mainly from spruce. The divergent tips of their bill are extremely useful in the exploitation of the cones. Accordingly, spruce seeds make up the bulk of a crossbill's diet.

The bill of the nukupuu (*Hemignathus lucidus*) from Hawaii is an even more sophisticated instrument. It helps the bird picking small vertebrates and their larvae hidden in or behind the bark of trees. The upper and the lower jaw can be used in different ways. With the lower one the bird chisels openings in the bark in order to work its way near to concealed victims which are then removed with the long curved upper beak.

No such activities are part of the starling's feeding behavior. Instead, the starling frequently practises "prying" or "open-bill probing" while it forages. Some authorities claim that the effectiveness of this behavior accounts for the evolutionary and ecological success of starlings. Prying is best executed with a straight and pointed bill. Thus the two starlings from Figs. 1 and 3 must have been considerably disabled. As omnivorous birds, however, they managed somehow to survive for some time, but they certainly were less effective feeders than conspecific specimens with a normal bill.

The skimmer's (*Rynchops*) bill is very extraordinary, too, its much longer lower jaw being shaped like a knife blade. While foraging, the birds skim over water with their bills held open and the lower jaws immersed and functioning as organs of touch. The rook, of course, did not use its bill like that.

Now we can answer the question posed in the first paragraph. The distinction between "malformation" and "specialization" in the considered case does not refer to the mere anatomical construction but to the consistency of the construction with behavior.

3 Some Theoretical Considerations

The case of the bill shapes may be regarded as very simple. It can, however, promote the understanding of some aspects of evolution.

The evolutionary transformation or organismic patterns can be characterized as a process of optimization within the framework of given conditions (Bock and von Wahlert 1965; Peters and Gutmann 1971; Bonik et al. 1977). The starting point of any such transformation, logically as well as substantially, is the organism itself. The bills in our example were produced, of course, during ontogeny as parts of the avian organisms. The selection forces they were exposed to during this process were generated by the mechanical coherence and the behavior of the organism. Environment is involved only in so far as the activity of the organism interacts with environmental circumstances.

On earlier occasions two aspects of the energy-powered organismic machine have been distinguished (Peters 1983, 1984, 1985): (1) the organism as a "gatherer" of matter and energy and (2) the organism as a converter of matter and energy.

The first aspect is mainly "extrovert", as it refers to the biological role of parts of the construction in a behavioral context. From the viewpoint of evolution, the ultimate purposes of behavior are: to maximize the input of energy and matter, to minimize the loss of energy and matter, and to optimize the ratio between reproductive success and reproductive activity.

The second aspect is more "introverted" as it refers mainly to the construction and the working ("Funktionieren") of the coherent machinery. In this case improvement can be gained by optimizing the relation between efficiency and costs of the machinery and by optimizing the discharge of metabolic waste.

This distinction may well be illustrated by the bills from Figs. 1, 3, 5. Very probably the costs for the development of these bills were not very different from the costs for "normal" bills of starlings and rooks. The same may be true for the operating expenses of the jaw apparatus. Perhaps the lower jaw of the specimen from Fig. 3 could even be moved with greater ease than a normal one because of its relative shortness. Thus, concerning the aspect of the energy converter, there seems to be no major difference between the misshaped and the normal starlings or rooks respectively.

The opposite is true if we take into account the role of the bills in the feeding behavior, as was pointed out above. From this viewpoint the normal construction seems to be much more efficient than the modified one.

These considerations lead to the question of how construction and behavior are interconnected in the process of selection. From our example it seems obvious that behavior determines construction. On the other hand, the dominance of the construction seems to be self-evident because behavior is produced by the construction at any rate.

Thus the situation is rather confusing. To stay alive, an organism must maintain its bionomic circle, i.e., the organism has to be active. Only while it works is the efficiency of an organismic construction being tested. This applies particularly to structures involved in behavior, i.e., in extrovert activity. Usually the structures generating behavior (e.g., nervous system, muscles) are not identical with the effective structure (e.g., bill). Thus there is no definite firmly fixed relation between this structure and behavior. Behavior and construction are partially independent of each other. This laxity of relations is one of the principal preconditions of further evolution. Indeed, strictly perfect and exclusive adjustments or adaptations of any kind would thwart evolution, because any change then would be worse than maintaining the status quo. The alteration would result inevitably in a pathological condition of the organism or in a detrimental effect of the environment; the fitness would be deteriorated in both cases.

Let us return once more to Figs. 1–6. We learned that the three constructions may appear as pathological or adapted bills respectively, depending on how the birds make use of them. The conclusion seems obvious that in the species on the right side the very special construction of the bill evolved under the guidance and limitation of behavior. But, of course, behavior could not evoke the morphological transition from a simple straight bill to a more specialized one. It could only influence the maintenance or suppression of variants already existing. Morphological variation, as far as it is genetically fixed, is the product of nonidentical reproduction of the parental organisms, and it is independent of behavioral demands. If this is true, the situation becomes something of a picture puzzle. Does the construction evolve under the selective pressure of behavior or does the construction constrain the evolution of behavior?

No unequivocal answer can be expected in this case. The mentioned sloppiness of the link between construction and behavior does tolerate initial changes on either side of the interconnection. Most, if not all, organismic structures can be used in more than one activity. On the other hand, most activities can be realized with different structures. However, options are not unlimited; the more specialized a system already is, the narrower the limitations of further development seem to be. Intuitively, I suspect that in such cases behavior is the main selection force guiding the morphological transformation (Fig. 7). Considering, for instance, the well-known Darwin's finches (Geospizinae) or the even more impressive Hawaiian honeycreepers (Drepanididae), it is difficult to imagine that their radiation was pushed ahead primarily by morphological variation. Variation of behavior must have preceded morphological changes. The divergence of feeding behavior must have canalized the pathways of bill variation. There are good reasons to believe that the precedence of behavior is the rule in the evolution of structures involved immediately in the activities of the "gathering" machinery.

However, there may be exceptions to the rule, at least in theory. Suppose a small population of starlings, which by peculiar chance, all possessed a bill like in Fig. 3, and happened to invade an island previously free of starlings. As we know from our specimen, a starling with a nukupuu-like bill is able to feed, although not in the way most typical for normal starlings.

Supposing that the starling pioneers on the island can forage a sufficient amount of food to raise enough young to maintain the population, two pathways of further development are conceivable. (1) While foraging, the birds use different methods, among them also "prying", which is not very efficient in their situation. For this reason, and provided that genetically based variation of bill shape occurs, individuals with a less curved upper jaw and a smaller difference in length of the jaws will gain an advantage. Thus the process will lead to the evolutionary reconstruction of the straight "normal" bill. (2) There is no or a very low variation of bill construction. "Prying" continues to be disadvantageous and will be eliminated sooner or later, whereas a feeding behavior more consistent with the construction of the bill will be favored by selection. In this case a comparatively pathological feature would have become the starting point of an adaptive process.

The second alternative may happen in our fictitious population just because the initial competition is so different from the condition in a "normal" population. Insofar this example does not contradict our earlier made statement concerning pathological deviations. The evolutionary success of the malformed starlings is not likely to occur in a natural population, where the "wrong" bill will

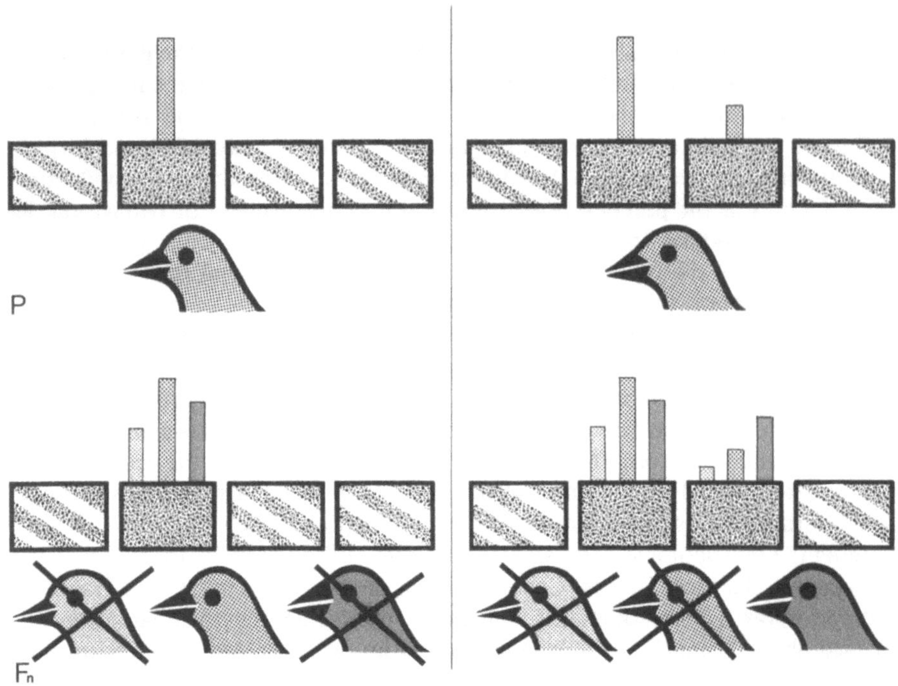

Fig. 7. The *quadrangles* represent various kinds of food potentially available. The birds, however, exploit only the food from the *densely dotted quadrangles*. The *columns* symbolize the amount of "fitness" derived from utilizing the particular food.

On the *left side* the parental generation (*P*) exploits only one resource to which it is optimally adapted. Provided that behavior does not change, descendants (*Fn*) deviating from the optimal construction will be handicapped (normalizing or stabilizing selection). On the *right side* the parental generation extended its feeding behavior and utilizes an additional source of food, albeit with only moderate success owing to the comparatively weak bill. Deviations in the direction of a stronger bill may now become advantageous (directional selection) (R. Klein-Rödder)

be outcompeted by the normal ones. In all probability the bills of the crossbill, the nukupuu, and the skimmer could not come into existence by sudden emergence amidst a population of straight-billed birds. Such a kind of evolutionary take-over would presuppose the pre-existence of a complicated behavior without consistent organs. Why should such a behavior have evolved? Slow continuous shifting of behavior followed by small changes of the constructions appears to be much more probable, as was postulated already by Mayr (1958, 1959). In this process it makes no difference whether behavior is genetically fixed or established by tradition.

What is the significance of the concept of adaptation in this context? What does it mean, for instance, that the bill of the crossbill is adapted to feeding on spruce cones? Some other birds also feed on cones, e.g., the great spotted woodpecker (*Dendrocopos major*). This species uses "anvils" for extricating the seeds from the cone. Crevices in bark or wood that may be further prepared and shaped by the bird serve as anvils. "... bird alighting at anvil holds new cone between breast and trunk while removing and dropping old one. ... New cone wedged into anvil upright or at ca. 45° with several blows of bill, and scales hammered off by downward blows enabling removal of seeds with tongue ..." (Cramp 1985). This description illustrates the elaborated feeding habit.

Obviously both species, the crossbill and the woodpecker, are efficient in feeding on spruce seeds despite having quite different bills. This example of "multiple evolutionary pathways" (Bock 1959) demonstrates very clearly that adaptation is not the morphological representation of certain environmental properties, as was sometimes claimed (Lorenz 1975). Adaptation is not simple adjustment of "form and function" channeled by a predominant umwelt; it rather comprises consistency of construction and behavior with regard to a property or to some properties of the umwelt, and it is brought about by, and only by the organism. Characteristically enough, even the classical instance of adaptive radiation, the already mentioned Darwin's finches, when examined closely, appears to exemplify the primacy of the organism in the adaptive process. Logically, Lack (1947) in his famed treatise maintained: "The evidence ... shows the fundamental importance of isolation in the differentiation of bird populations, and suggests that many of the differences between the forms are not correlated with differences in their environments. ... the point is that such new characters need not be related to possible differences between the environments ..., and they are evolved even if the ... environments are identical".

Hence Gutmann's (1989) suggestion to abandon the conception of adaptation seems to be justified. However, deliberation may be advisable. The primacy of organisms does not condemn the environment to insignificance. The "gathering" machinery of the organism interacts directly with particular parts of the environment. These parts, of course, are picked out, charged with specific significance, and thus organized as its umwelt (sensu von Uexküll 1909) by the organism. Nevertheless, the mere physical existence of these parts of the environment is independent of the organism at any rate. In contrast to the "converter" aspect of the organism, the "gatherer" aspect can be fully comprehended only if one knows something about the environmental circumstances involved in the activity of the organism. Suppose that the bills from Figs. 2, 4 and 6 were known only from fossil birds. I dare to maintain that we would be unable to reconstruct the habits of those birds, and this means that we would not be able to detect the polarity of the process of optimization leading to their particular constructions. Therefore, despite the laxity of relations mentioned earlier, a certain fit between the organismic machinery and the environment cannot be denied. If this kind of connection should be called adaptation, I would have no objections to it. Of course, there is no "creativity" of the environment implied. The environment is necessary but it cannot give rise to the slightest organismic novelty; it can at most eradicate.

4 Applied Conclusions

The preceding considerations are not purely academic. The awareness that behavior channels the adaptive process may be helpful when we attempt to reconstruct this process, as can be shown in the following example.

The phylogenetic relationships of the flamingos have been controversial for a long time. At least three different groups, namely Ciconiiformes (storks and allies), Anseriformes (waterfowl), and the charadriiform family Recurvirostridae (stilts and avocets), were alternatively regarded as their closest relatives. Flamingos are a highly specialized group of birds characterized by many diagnostic features including a unique construction of the bill. This bill is excellently suited to sieve small particles from a fluid. Indeed, flamingos feed mainly on small algae and vertebrates living in brackish water. A bill like this could have evolved only in birds that used to forage in water already before having been equipped with such a specialized feeding apparatus. Otherwise the transformation of the construction towards the specialized version of the flamingo bill would not have been an improvement of the construction. Hence one can predict that among the ancestors of flamingos must have been birds with a locomotory apparatus suited for swimming or – more probably – for wading, but without a flamingo-like bill. Whereas the contrary, namely a nonaquatic, terrestrial bird with a flamingo-like bill can be regarded as highly improbable.

From the fossil record the genus *Juncitarsus* Olson & Feduccia 1980 was taken for the earliest known representative of flamingos. It was not until 1987 that the skull of *Juncitarsus* became known (Peters 1987). It belongs to a complete skeleton recovered from the Eocene oil shales of "Grube Messel" that was described as *Juncitarsus merkeli*. The postcranial parts of this skeleton are the bones of a wading bird and resemble the bones of modern flamingos, but the bill is very different, and much more simply constructed than that of flamingos. As a matter of fact, this bird fits exactly the theoretical predictions just outlined. Concerning the relationships of the flamingos it was crucial to learn that the bill of the Messel bird displays a construction incompatible with those of storks and of waterfowl but recalling bills of charadriiform birds. Thus *Juncitarsus merkeli* corroborated very strongly the hypothesis that flamingos should be included in the Charadriiformes and that they are not closely related to the Ciconiiformes or the Anseriformes.

The hypothesis that behavior is the leading factor in the optimizing process reveals its resolving power also in other problems, for instance in the problem of parallel evolution of certain faculties in groups of closely related taxa. Well-known examples are the multiple development of eusociality in Hymenoptera, the multiple development of brood parasitism in the cuckoos, the multiple development of asymmetrical ears in owls, the multiple development of oil-foraging devices in bees, etc. Because parallel evolution can obscure cladistic relationships to a much greater extent than convergent evolution of distantly related groups, it appears very important to realize that parallel evolution becomes highly probably if behavior canalizes the process. I would like to exemplify the case, considering the oil-foraging bees.

As everybody knows, all bees (there are more than 20000 extant species known) feed mainly or even exclusively on nectar and pollen from flowers. This holds not only for the imagines but for the larvae, too.

Some plants specialized in producing oil in their flowers instead of nectar, and sundry species of bees utilize this source of nourishing food to provision their larvae (Vogel 1974). This was an exciting discovery, all the more since this habit evolved several times in quite different groups of bees and in different parts of the world. All oil-foraging species, however, evolved special constructions of their setae on the fore- and middle-legs or on the metasoma. There are specific differences in the shape of the setae and in the oil-gathering methods, but the collecting apparatus as a whole is so characteristic that, once it was known, the habit of oil foraging could be predicted on the basis of inspecting specimens in museum collections. The prediction was subsequently confirmed by field studies (Vogel 1981).

The multiple development of oil gathering in bees is less surprising if we consider that foraging in flowers is a habit shared by all bees, making the development of this new foraging technique easy.

Again, this example supports the view that behavior channels the maintenance and/or the elimination of competitive variations of the construction. Moreover, this conclusion leads to a second conclusion which can be most serious for phylogenetic studies. Indeed, it can negate some fundamental assumptions used in particular approaches to phylogenetic analysis. Let us assume that all bees became extinct long ago with the exception of the oil-foraging species. In this case, without any risk, one could bet one's bottom dollar that all taxonomists would be convinced that the oil-foraging apparatus is a synapomorphic feature indicating monophyly of the left-over bees. Only because the other bees did not become extinct do we know that this assumption would be wrong. This fictive example, however, arouses the embarrassing suspicion that the monophyly of all bees might also be an erroneous assumption. The habit and the apparatus of foraging nectar and pollen may have evolved several times, in analogy to the case of the oil bees.

Acknowledgments. I am very grateful to Walter Bock, New York, Wolfgang Gutmann, Frankfurt/M., and Adolf Seilacher, Tübingen, for their critical review of the manuscript.

References

Bock WJ (1959) Preadaptation and multiple evolutionary pathways. Evolution 13:194–211

Bock WJ, Wahlert G von (1965) Adaptation and the form-function complex. Evolution 19: 269–299

Bonik K, Gutmann WF, Peters DS (1977) Optimierung und Ökonomisierung im Kontext von Evolutionstheorie und phylogenetischer Rekonstruktion. Acta Biotheor 26:75–119

Cramp S (ed) (1985) Handbook of the birds of Europe, the Middle East, and North Africa, vol 4. Oxford Univ Press, Oxford NY

Gutmann WF (1989) Die Evolution hydraulischer Konstruktionen: Organismische Wandlung statt altdarwinistischer Anpassung. Kramer, Frankfurt/M

Lack D (1947) Darwin's finches. Cambridge Univ Press, Cambridge London New York

Lorenz K (1975) Die Rückseite des Spiegels. Piper, München

Mayr E (1958) Behavior and systematics. In: Roe A, Simpson GG (eds) Behavior and evolution. Yale Univ Press, New Haven, Conn, pp 341–362

Mayr E (1959) The emergence of evolutionary novelties. In: Tax S (ed) The evolution of life. Univ Press, Chicago, pp 349–380

Peters DS (1983) Evolutionary theory and its consequences for the concept of adaptation. In: Grene M (ed) Dimensions of Darwinism. Cambridge Univ Press, Cambridge London New York; and Editions de la Maison des Sciences de l'Homme, Paris, pp 315–327

Peters DS (1984) Evolutionstheorie – Zwangsläufigkeit und Grenzen. In: Kaiser P, Peters DS (eds) Evolutionstheorie und Schöpfungsverständnis. Pustet, Regensburg, pp 193–218

Peters DS (1985) Erneut vorgestellt: Das „Ökonomieprinzip“. Aufs Reden Senckenb Naturforsch Ges 35:143–153

Peters DS (1987) *Juncitarsus merkeli* n.sp. stützt die Ableitung der Flamingos von Regenpfeifervögeln (Aves: Charadriiformes: Phoenicopteridae). Courier Forschungsinst Senckenb 97:141–155

Peters DS, Gutmann WF (1971) Über die Lesrichtung von Merkmals- und Konstruktionsreihen. Z Zool Syst Evolutionsforsch 9(4):237–263

Uexküll J von (1909) Umwelt und Innenwelt der Tiere. Springer, Berlin

Vogel S (1974) Ölblumen und ölsammelnde Bienen. Steiner, Wiesbaden

Vogel S (1981) Abdominal oil-mopping – a new type of foraging in bees. Naturwissenschaften 67:627–628

The Function of Labyrinthodont Teeth: Big Teeth in Small Jaws

H. PREUSCHOFT [1], W.-E. REIF [2], C. LOITSCH [1], and E. TEPE [3]

Summary

Labyrinthodont teeth occur in crossopterygians, *Lepisosteus*, Devonian terrestrial amphibians, ichthyosaurs, and varanoids. That means they have evolved several times and seemingly independently. The question is why.

In biting, the tooth is exposed to compressive forces and to bending moments, which are resisted by the tooth.

For several cross-sections of two labyrinthine teeth we calculated the resistance properties: areas and moments of inertia in two planes (namely mesio-distal and bucco-lingual). The results do not deviate much, and show the ± linear increase of compressive strength, combined with a parabolic increase of bending strength.

From this we can conclude that the investigated teeth belong to predators which captured soft prey, not prey with hard shells.

Immature teeth are weaker than mature teeth, although they possess the same strength patterns. If we cut off the folds in a "young" tooth, the strength drops by 20–40%: a tooth with unfolded, ring-shaped cross-section would be much weaker. If one half of the folds is added to the remaining, peripheral part of the wall, strength increases by 8%. An old tooth, in which the plicidentine is thick, and the folds are in contact with each other, shows nearly no difference to a tooth with a smooth, unfolded wall of the same thickness. A tooth without folds would fulfill exactly the same task.

Labyrinthine teeth seem to occur in predators with large fangs, that are higher than the shallow jaw. Therefore, the high tooth cannot be fully developed in the shelter of a jaw. It has to erupt before it has reached its full strength. The arrangement of small amounts of dentine in folds provides higher resistance at an early developmental stage than would be available without folds. In a mature tooth, the folding offers neither advantages nor marked disadvantages over smooth dentine walls.

[1] Institut für Anatomie, Abt. Funktionelle Morphologie, Ruhr-Universität Bochum, Universitätsstr. 150, D-4630 Bochum 1, FRG
[2] Institut für Geologie und Paläontologie, Universität Tübingen, Sigwartstr. 10, D-7400 Tübingen, FRG
[3] Reinickendorfer Str. 120, D-1000 Berlin 65, FRG

N. Schmidt-Kittler and K. Vogel (Eds):
Constructional Morphology and Evolution

1 Introduction

Our approach – as a practical procedure, not as a philosophy – can be characterized by the following: We start off from shape (or behavior) and ask why it exists. We do not attempt to understand "shape" (of what: a species? or an element?) in a global sense, but instead in a rather confined way. In many cases (e.g., teeth), we are just interested in that particular character which distinguishes one form from another. In other cases (locomotion), however, we have to take into account the full complexity of an animal's body. We are convinced that we use our time best when we try to find explanations of only particular characters of a given shape, because this character distinguishes the form investigated from others. These explanations, we trust, can be generalized over many, often quite diverse, forms.

In the sense of Bock (this Vol.), we look for "nomological" answers. Since the best-formulated natural laws are physical laws, and since body shape in our preferred subjects – the mammals – is largely determined by the locomotor apparatus, and locomotion is something mechanical, we try to go back to mechanics. The relation between the "biological role" of a morphological trait of a given element and its shape is given and determined by its "mechanical function". The type of question we ask is: how does the organism profit from the possession of the trait under consideration? Sometimes knowledge of its mechanic function also informs us about the way in which the trait was acquired: either through the process of causal morphogenesis (= adaptive growth) or (through selection during evolution) as a means for saving energy, which gave the possessors of the trait improved chances of survival.

This approach will be exemplified with the paradigm of labyrinthine teeth.

2 Question

From Devonian times on, labyrinthine teeth are found primarily in crossopterygians and terrestrial amphibians, and also in ichthyosaurs. The characteristic feature of all "labyrinthodonts" is the construction of the teeth. In their upper parts the teeth have smooth surfaces and round cross-sections with a concentric dentine layer. Towards their bases the dentine wall is increasingly folded inwards. The folds project more or less into the pulp cavity. The whole tooth is covered by a thin layer of enamel (Fig. 1 a).

This morphologically conspicuous feature has evolved several times convergently. However, the phylogenetic relationships of all groups with labyrinthine teeth are not yet completely settled. The form of these teeth is noteworthy enough to seek an explanation of its mechanical function and its biological role. Within the last 6 years, these questions have been raised several times in the course of Congresses on Vertebrate Morphology.

Here, we present an attempt to analyze the functional significance of the complicated structure of labyrinthine teeth, or to find the functional and biological advantages of having labyrinthine teeth.

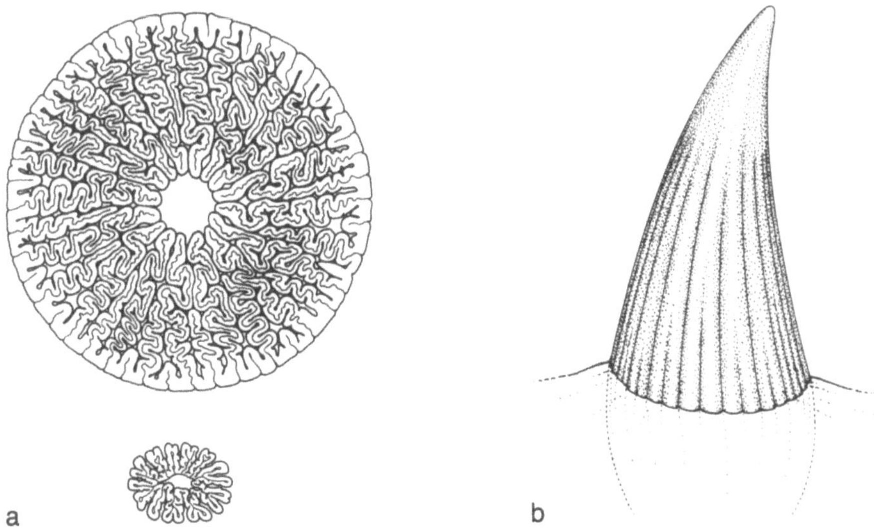

Fig. 1. Labyrinthine tooth. **a** Cross-sections. **b** Labial view

2.1 Taxonomic Distribution of Labyrinthine Teeth

Labyrinthine teeth are found in the following groups (data based on Conybeare 1824; Bystrow 1938; Besmer 1947; Chase 1963; Peyer 1968; Schultze 1969, 1970; Mazin 1983):

1. Lepisosteiformes, Neopterygii, Actinopterygii (Recent and fossil species of *Lepisosteus*, but not in *Paralepisosteus* from the Cretaceous; Fig. 10).
2. Osteolepiformes and Porolepiformes, Crossopterygii.
3. Labyrinthodontia, Anthracosauria, Seymouriamorpha, Microsauria (i.e. some amphibian groups; Fig. 11).
4. Ichthyosauria, Varanoidea, *Limnoscelis* (Captorhinomorpha) and *Champsosaurus* (Choristodera), i.e., some reptilian groups.

Labyrinthodonty has evolved at least four times independently: in the crossopterygians (the probable ancestors of the amphibians and reptilians), in the ichthyosaurians (whose ancestors are unknown, see Besmer 1947), in the Lepisosteiformes, and in the varanoids (see Schultze 1970). In the stratigraphically oldest ichthyosaurians and varanoids, labyrinthodonty is least pronounced or even not developed (Besmer 1947; Schultze 1970).

Earlier work is confined to descriptions of the structure of labyrinthine teeth. Owen (1840–1845) was the first to describe the general structure of labyrinthine teeth by means of a cross-section through a tooth of *Mastodonsaurus giganteus* from the Triassic. A detailed presentation of the microscopic structure of labyrinthine teeth of *Benthosuchus sushkini* was published by Bystrow (1938). He emphasized the growth of these teeth which starts from the already folded periphery and proceeds inward, adding layers of dentine gradually at the inner side of the tooth wall, thus filling up the pulp cavity. Bystrow also discussed the different ontogenetical ages of the teeth from different jaw fragments and also the replacement of the teeth.

The material of the labyrinthine teeth is called plicidentine, the enamel layer is very thin. According to the definition given by Schultze (1969), it consists of folded orthodentine, where the axes of the folds are in upright position and the intensity of the folding (number and relative depth of the folds) increases from the apex toward the base of a tooth. The relation between the vertical extension of the folded basal part and the smooth apical part of a tooth can differ considerably. However, we are not going into further detail on this. Schultze (1969, 1970) distinguished four different types of labyrinthine teeth. The criteria he used are geometry and intensity of the folding of the orthodentine, whether there is osteodentine in the pulp cavity or not, and the distribution of osseous tissue in the folds and at the base of the teeth. A functional differentiation and explanation of these four types has not been achieved yet, but they have been shown to be important taxonomical characters.

Before turning to the biomechanical analysis, it should be emphasized that labyrinthodonty exists only in some groups of predators which are equipped with large fangs: among actinopterygians only in *Lepisosteus*; among crossopterygians only in Porolepiformes and Osteolepiformes with large fangs, in amphibians, and some reptilians. Groups with large fangs but without labyrinthine differentiation are the struniiform crossopterygians (see below, Fig. 12), barracudas (Fig. 10d, e), crocodilians, many lacertilians, plesiosaurians, thecodontians, dinosaurs, many mammal-like reptiles, and mammals.

2.2 Biomechanical Considerations

As can be seen in Fig. 1b, the teeth are cone-shaped. Such a tooth easily penetrates the surface of a prey's body. It is also easily withdrawn while opening the jaws after biting because increasingly small cross-sections have to fit into the opening cut by the tooth. Elastic tissue slides into an apical direction because of its tendency to contract around the wound.

The teeth are exposed to different stresses according to the type of prey (see also in Preuschoft et al. 1974): When biting on hard surfaces (e.g., shells of molluscs, or skulls), the load is concentrated on the tip (Fig. 2 top left). The bite force has a component in longitudinal direction of the tooth (B_r) which creates compressive stresses that remain constant along the height of the tooth (Fig. 2 top second). (Stress distribution in teeth is discussed more extensively by von Koenigswald and Pfretschner, this Vol.). If the prey tries to escape, or if the predator moves faster than its prey, or if only the prey is squeezed between the jaws, an additional force component (B_h) occurs at right angles to the compression. It may be directed along the jaw in a rostro-occipital direction or perpendicular to the jaw (bucco-lingual direction). This force component causes shearing, and bending moments (Fig. 2 top third, see Preuschoft et al. 1985). These increase linearly from the tip towards the basis (Fig. 2 top third). Figure 2 top right represents a tooth shape optimized for sustaining such stressing. It is easy to understand that a tooth stressed in this way has to have a round cross-section, since the stresses act not only in one plane, as is demonstrated in Fig. 2, but also in other directions.

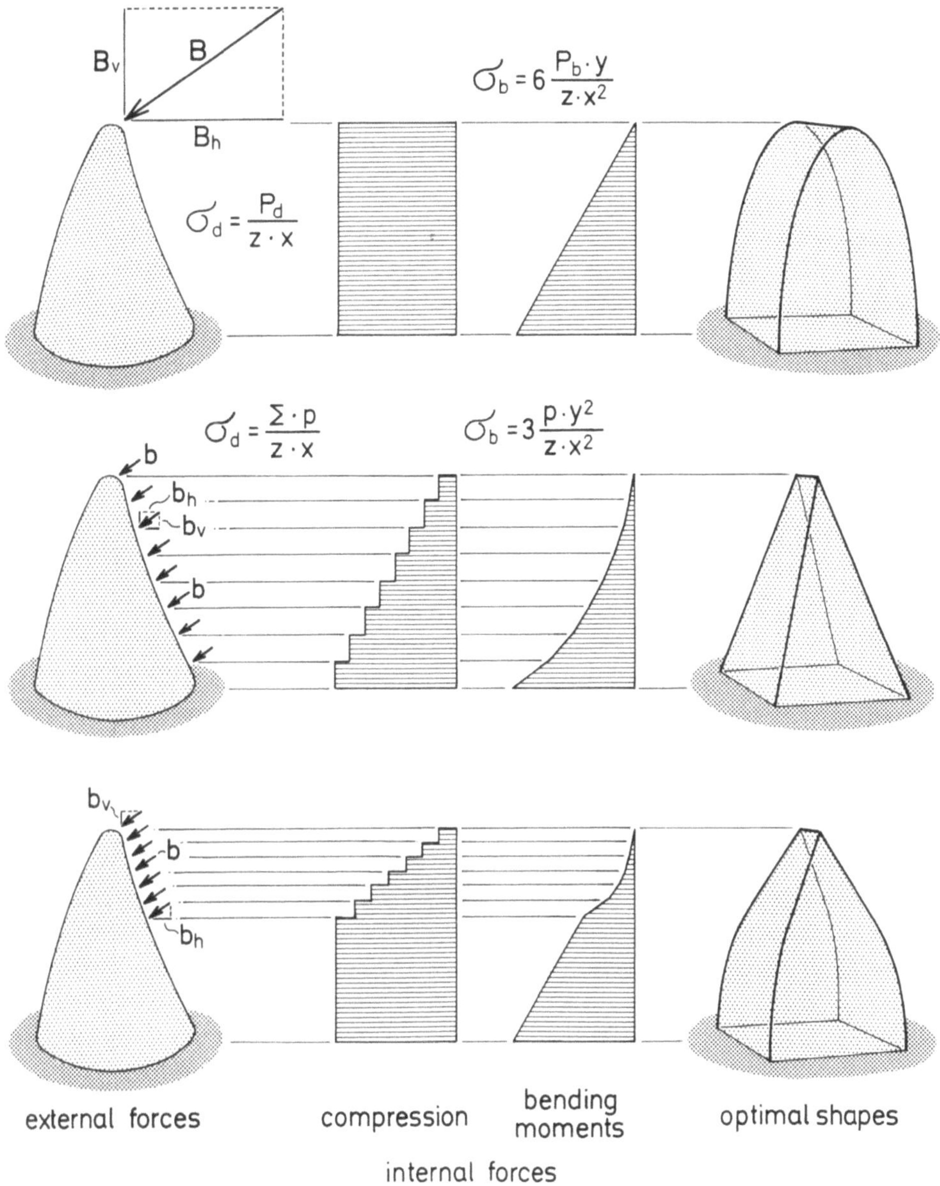

Fig. 2. Stressing of teeth. *Top row* biting on hard surfaces. *Middle* while penetrating into soft material or prey. *Bottom* while partly inserted into a soft body. *Left column* force application; *second column* compressive stresses caused by the vertical components; *third column* bending stresses caused by the (horizontal) shearing components of the bite force; *right column* "bodies of equal strength", i.e., optimal shapes against the assumed bite forces. By the formulas the internal stress σ can be determined. In a body of equal strength, σ has the same value at each point

However, if the tooth penetrates the prey's body surface, as in biting on the soft parts of vertebrates or cephalopods, the bite force is distributed along the surfaces of the tooth. Each partial force consists of a longitudinal and a transverse component (Fig. 2, middle). The longitudinal components of the partial forces (b) sum up from the tip to the basis; the transverse components cause bending mo-

Fig. 3. Photoelastic experiment to show the rapid increase of stresses in a sharp edge of a tooth (*top*) while biting on a hard surface (*bottom*)

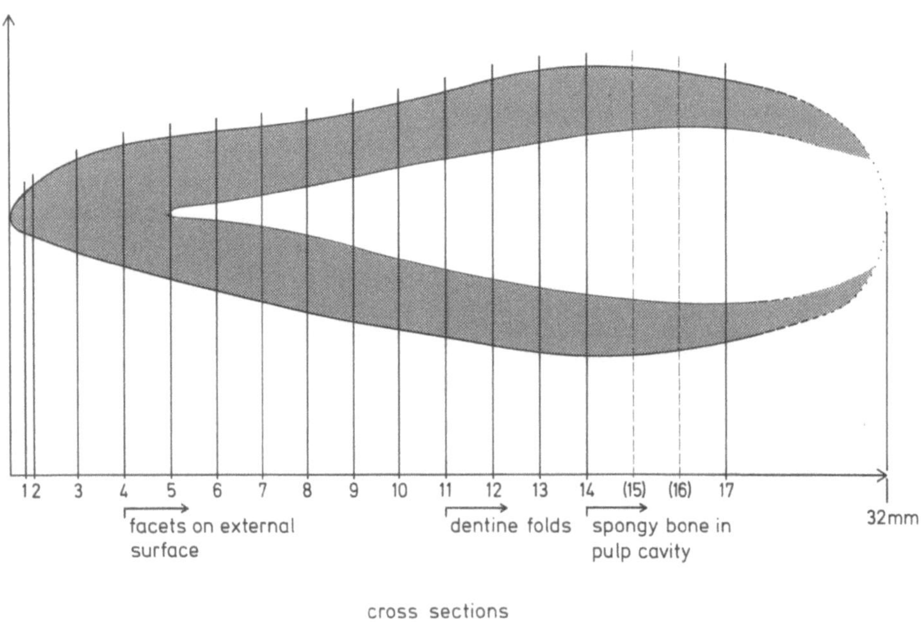

Fig. 4. Positions of the cross-sections through the investigated labyrinthine tooth of *Ichthyosaurus postumus*

ments which increase parabolically. In this case the bending also does not need to occur in the plane assumed in the figures. The transverse components can act in any direction, thus a round cross-section, that can be stressed from all sides, is optimal. If one direction of the bending components prevails, an elongation of the diameter in the direction of the highest or most frequent transverse forces adapts the tooth to this sort of stressing (see Valkenburgh and Ruff 1987). Figure 2, middle right shows a tooth shape optimal for resisting this sort of stressing.

Figure 2, bottom shows the stress pattern of a tooth which is only partly forced into a soft body. The compressive components sum up, and bending moments increase from the tip, as is shown in Fig. 2, middle. The not-inserted part of the tooth is stressed in the way demonstrated in Fig. 2, top.

As shown in earlier studies (Preuschoft et al. 1974), and as can be seen by a photo-elastic experiment (Fig. 3) the stresses in the tip increase in a bell-shaped way very quickly, if a load is applied to the tip. As a consequence, the tip is especially endangered when biting on hard objects. For this reason cone-shaped fangs usually have a dome-shaped tip (see Fig. 2, top).

The compressive forces, as well as the bending moments, must be resisted by a tooth. If the teeth are "adapted" to their function, they should possess strength properties which are well suited to sustain the above demonstrated stress patterns. Therefore we have estimated the resistance properties of teeth.

3 Materials and Methods

To determine the resistance or the tooth's stability against external forces, the shape-dictated resistance properties were calculated in two examples. From these results the function was deduced. Compressive and shearing components are distributed over the whole cross-sectional area (stress = force/area); bending moments, on the other hand, lead to stresses that increase towards the margins. These stresses are inversely proportional to moments of inertia ($I = \Sigma dF \cdot y^2$, that is the 4th power of the diameter).

One tooth of *Ichthyosaurus postumus* from the Jurassic was used. As could be seen from the cross-section (thickness of the wall, degree of folding) the tooth was mature. It measured 32 mm in height, and had a maximum diameter of 11.2 mm, which was 24 mm below the tip. The external surface of the tooth was equipped with longitudinal facets at the dome-shaped tip, shortly below it showed grooves running towards the basis. The ± round cross-sections show the facets in the upper part, towards the bases the cross-sections acquire ellipsoid outlines. The tooth was cut into slices of 1.7 mm thickness (Fig. 4), which were embedded and mounted on slides. The specimens were magnified photographically and the outlines of the tooth as well as of the pulp cavity were traced on transparent foil. Insufficient or damaged parts of the cross-sections were restored. This was impossible in the sections 15 and 16. Therefore the corresponding values are missing in all figures. In the cross-sections close to the base the typical folded structure was no longer discernible, because dentine and cancellous bone in the pulp cavity could not be distinguished reliably.

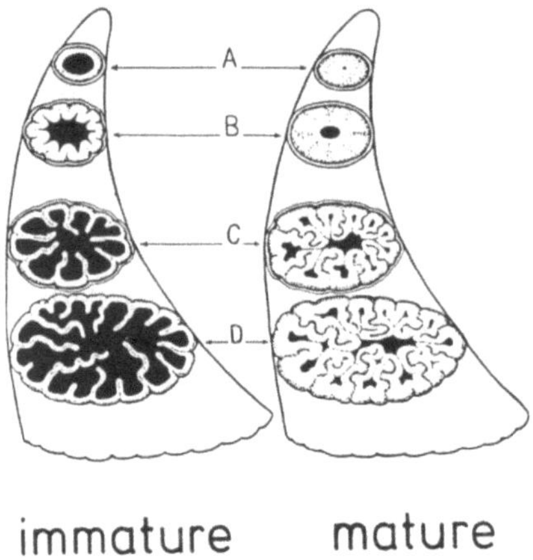

Fig. 5. Teeth of *Benthosuchus sushkini* with cross-sections (Bystrow 1938)

The resistance properties, namely cross-sectional areas and moments of inertia in two planes of the 14 cross-sections were calculated by means of a computer and a graphic tablet for:

1. the first discernible layer from the outside (enamel): this is formed together with the first thin dentine layer in the growing tooth and corresponds therefore to the minimum wall thickness of a "young" or immature tooth.
2. the whole dental wall from the external layer to the pulp cavity, i.e., the complete area of hard tooth material in the "old" or mature tooth.

The same procedure was applied to the cross-sections shown in Bystrow's (1938) Fig. 7 (Fig. 5). The positions of these cross-sections along the height of the tooth were taken for granted.

3. In the 13th cross-section we calculated the cross-sectional area and moments of inertia under modified conditions: in the immature state of the tooth, the folds were cut off.
4. In the same immature state, the folds were cut at about one half of their depths, the removed material of the folds added to the wall, thus creating a smooth pulp cavity with the same amount, but different distribution of material.
5. In the mature state, the external outline of the tooth and the internal outline of the pulp cavity were smoothened, so that a ring-shaped profile is produced. Its wall thickness corresponds to the distance from the external wall to that part of the folds which protrudes farthest inward.

Fig. 6. Cross-sectional areas at different levels of a tooth:
– mature tooth with thick dentine wall;
– immature tooth with thin dentine wall. **a** *Ichthyosaurus postumus*. Values decrease again beyond the alveolar margin. **b** *Benthosuchus sushkini*, according to Bystrow (1938). Since the tooth is attached directly to the jaw bone, no reduction of strength can be seen

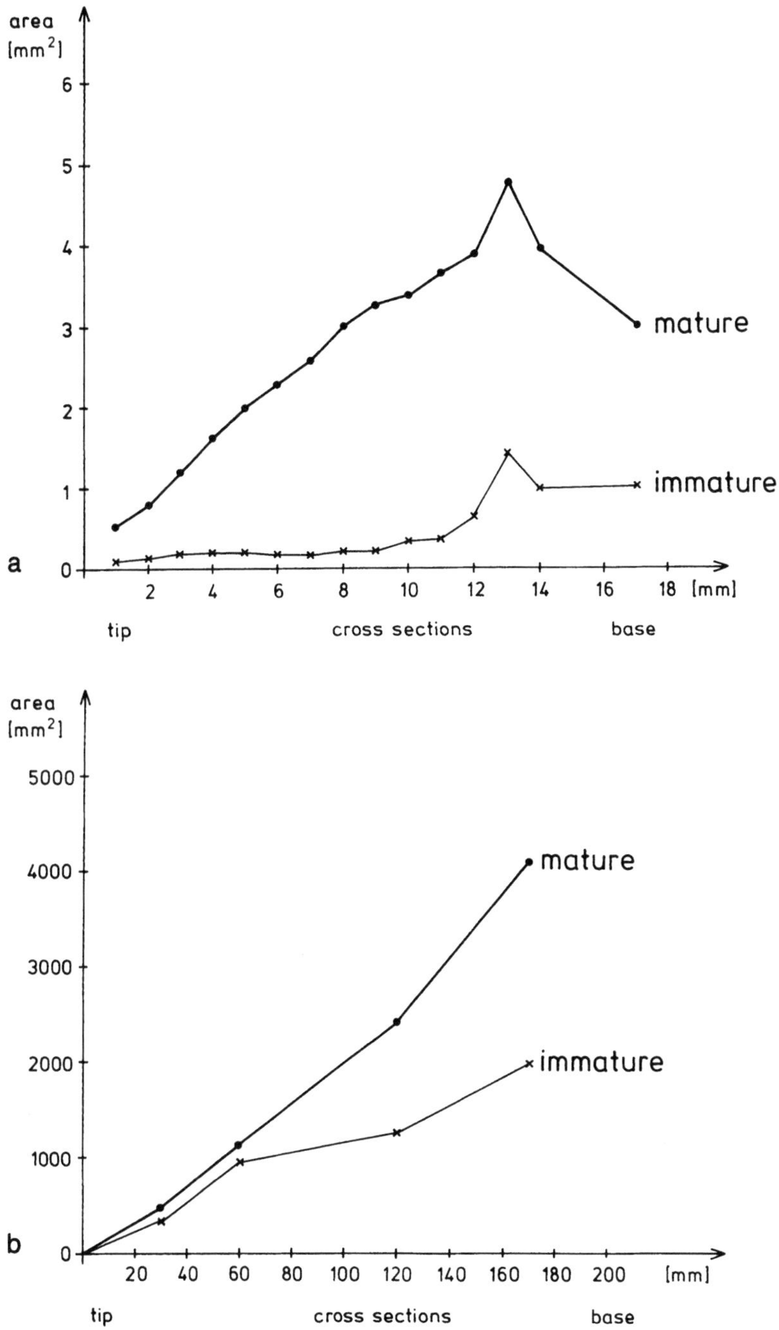
area
[mm²]
6
5
4
3
2
1
0
a
mature
immature
2
4
6
8
10
12
14
16
18 [mm]
tip
cross sections
base
area
[mm²]
5000
4000
3000
2000
1000
0
b
mature
immature
20
40
60
80
100
120
140
160
180
200
[mm]
tip
cross sections
base

4 Results

Figure 6a and b represents the cross-sectional areas of the respective immature and mature teeth of *Ichthyosaurus* and *Benthosuchus*. The areas are proportional to resistance against compression or shearing. These increase more or less linearly from the tip towards the base. The maximum is reached in cross-section 13, which represents roughly the level of the alveolar margin. From here on, the bite force acting on the tooth is transmitted gradually to the desmodont. The cross-sectional areas of the immature tooth clearly lag behind those of the mature one. The mature ichthyosaur tooth in the 13th cross-section is about four times stronger than the immature one. The teeth are suited to sustain shearing and compressive forces that increase from the tip towards the base. In immature teeth of ichthyosaurs the cross-sectional areas seem to remain constant over most of their height. This is to be expected if the load is concentrated on the tip (see Fig. 2). Near the base the cross-sectional areas increase quickly. Taken together, the immature teeth show a pattern of resistance similar to that of mature teeth, but are weaker.

Since only four cross-sections were available for *Benthosuchus* it was impossible to establish reliably the changes of its resistance properties. The findings, however, do not contradict the results obtained for *Ichthyosaurus*.

In Fig. 7a and b the moments of inertia are plotted against the height of the tooth. Heavy lines represent the resistance against flexion in a bucco-lingual direction, thin lines represent the bending resistance in a medio-distal direction. The bending resistance of the entire tooth wall (mature tooth) in *Ichthyosaurus* is clearly higher than the resistance of the external layer alone (immature tooth). For the entire dentine wall the maximum is $I = 3.3$ units, for the external layer it is $I = 1.3$ units. In both teeth the bending resistance increases parabolically from the apex towards the base. From the 13th cross-section on the values decrease. This is probably due to the fact that the mooring of the tooth takes over the moments in the alveola.

In Fig. 7b the moments of inertia are plotted for a mature and an immature tooth of *Benthosuchus*. Although only four sections could be considered, the graphs closely resemble those in Fig. 7a. Once again it is obvious that the highest bending resistance occurs in the bucco-lingual direction, and that it is higher in the "old" tooth as compared to the "young" tooth.

This means: the teeth are suited to sustain loads distributed over the height of a tooth and evoking either compressing or shearing, or bending stresses. Such distributed loads occur whenever the tooth penetrates into the (soft) body of a prey which moves relative to the tooth.

What mechanical consequences arise from the folding? Figure 8 allows us to compare the bending resistance of a folded and a smoothened immature tooth of *Ichthyosaurus*, in bucco-lingual or mesio-distal direction respectively. Between the tip and the 8th cross-section folding was not very pronounced. Therefore, no differences of strength can be found. Again the increase from apex to base is parabolic. The smoothened cross-section offers a bending resistance 20–40% *below* the values of the cross-sections with folds.

When one half of the folds is added to the external dental wall, the bending resistance is 8% *above* that of the true tooth. In a mature tooth, in which the folds

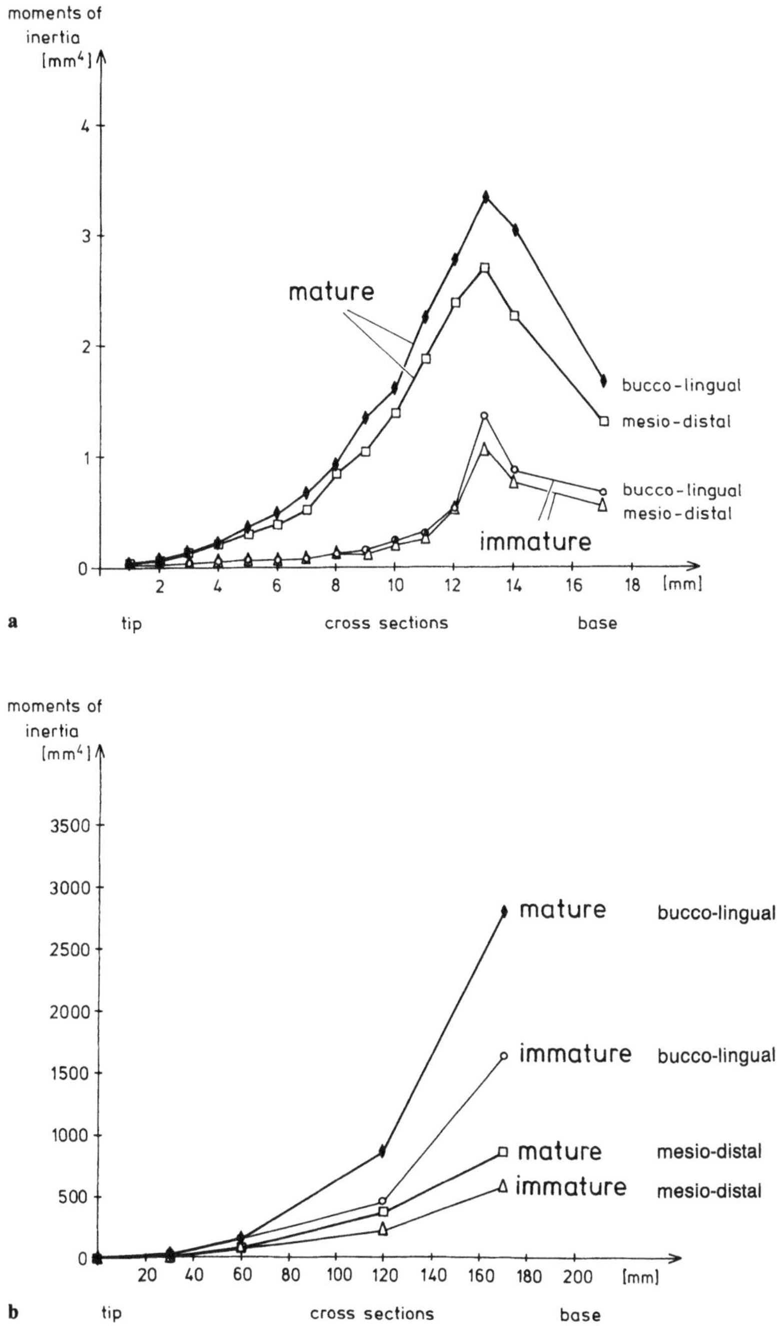

Fig. 7. Moments of inertia, as measure of bending resistance, on different levels of a labyrinthine tooth. Represented are values in bucco-lingual and mesio-distal directions, respectively:
- mature tooth with thick dentine wall;
- immature tooth with thin dentine wall. **a** *Ichthyosaurus postumus.* **b** *Benthosuchus sushkini*, according to Bystrow (1938). See also Fig. 7

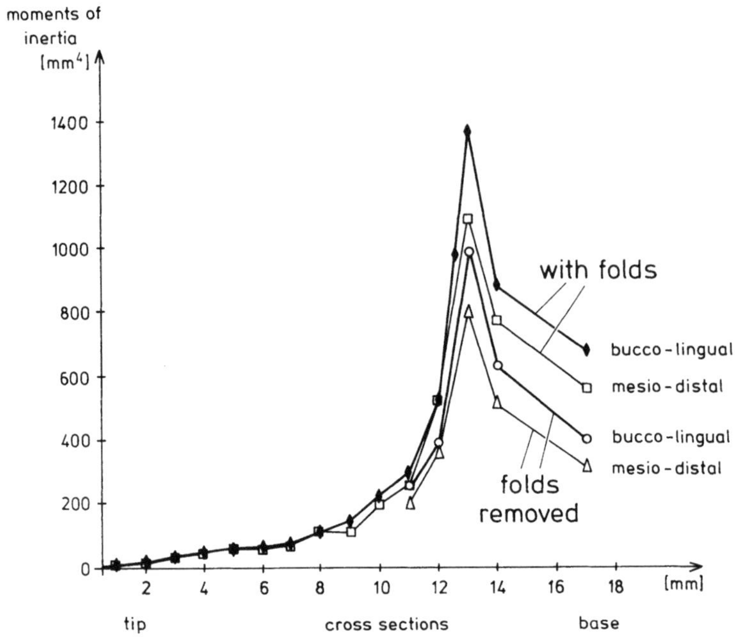

Fig. 8. Moments of inertia on different levels of a labyrinthine tooth (ichthyosaur) in bucco-lingual and in mesio-distal direction. The two graphs at the *top* refer to the external layer of dentine (that is an immature tooth). The two bottom graphs are based on values measured after removing the folds. Thus, they show the strength of a tooth with circular and smooth cross-section, without folding of dentine

touch each other, the difference is only 0.8% in favor of the tooth without folds. This difference is negligible. Thus, without folding, but with the same amount of material, a higher, or at least the same resistance could be achieved. Folding evidently is not a means to reach higher resistance while saving material.

The upper graph in Fig. 9 shows the bucco-lingual bending resistance of a mature tooth with smooth outlines and pulp cavity, that is with a massive, smooth, and circular cross-section, wall thickness being the same as the maximal depth of the folds. The values lie up to 10% *above* those of a true labyrinthine tooth.

5 Discussion

5.1 Earlier Approaches

Several authors have suggested explanations for labyrinthodonty. Peyer (1929) assumed that compartimentalization ("branching" or splitting of the pulp cavity formed by the dentine folds) helps to save material, and also that (in the case of

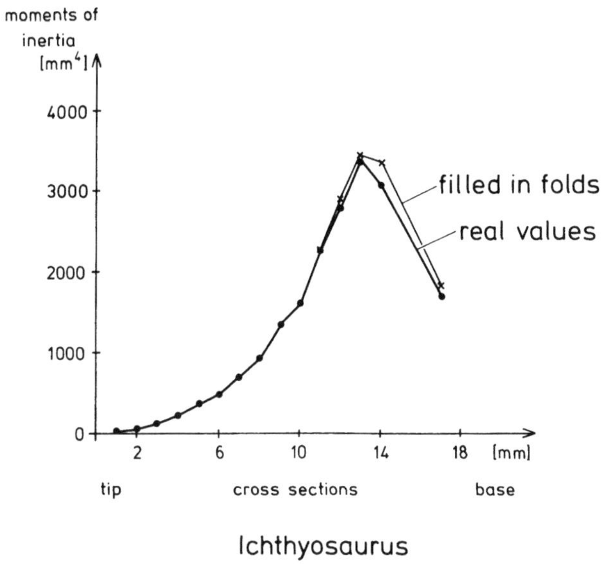

Fig. 9. Moments of inertia at different levels of a labyrinthine tooth:
- real values of a mature tooth with thick walls;
- values measured after filling in the folds to create a smooth circular profile with a wall as thick as the height of the folds

pleurodonty in *Varanus*) the disadvantageous oblique position of the plane to which the tooth is attached has to be compensated for by an enlarging of its surface area: this increase in size of the dental base leads to a "branching" of the pulp cavity and thus to folding of the teeth, i.e., labyrinthodonty. Odermatt (1940), Besmer (1947), and Peyer (1968) repeated the argument of saving material. Schultze (1969) did not support this idea. He objected by pointing out that the pulp cavity is filled with hard tissue in at least three of the four types of labyrinthine teeth he distinguished. He also found arguments against the assumption that folds of dentine make the dental wall rigid. First, in *Varanus niloticus* the blunt teeth (which are subject to higher (concentrated) stresses than the pointed teeth) show less folding than the latter. Second, plicidentine has been lost in the course of evolution in most of the reptilians, and never occurs in mammals. On the other hand, Schultze emphasized that the anchoring of a labyrinthine tooth depends on its external morphology, not on the internal intensity of folding. Deep ridges on the external surface of the tooth (not the existence of plicidentine inside) indeed improve the possibilities to anchor the tooth on the alveolar bone, and the deeper the alveola, the larger its surface which can be used by anchoring fibrous tissue.

Further, Schultze argued that plicidentine in the varanoids *Varanus* and *Heloderma* has no mechanical function since the teeth are immediately attached to the jaw bone, without osseous anchoring tissue in between. In the most basal part of the teeth the folds fuse, thus forming a honeycomb pattern. The aberrant forms of dentine and bone below the honeycomb structure of the dentine support

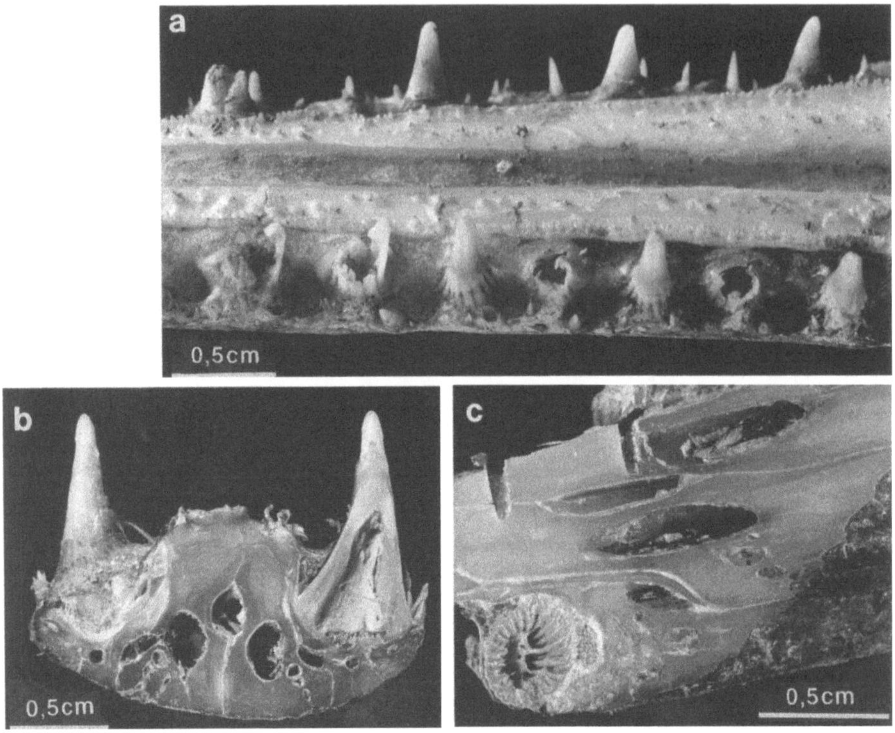

Fig. 10 a–c. *Lepisosteus* sp., Recent, upper jaw. **a** Oblique view. **b** Cross-section. **c** Oblique section, labyrinthine basal part of a tooth in the lower left. **d, e** *Sphyraena barracuda*, Recent. **d** Right lower jaw seen from the labial side. **e** Cross-section (in anatomical orientation) through the upper jaw (note the very deep attachment of the tooth that is cut longitudinally). **a–e** Collection W.-E. Reif

the tooth. Schultze agreed with Bullet (1942, p. 189) that the honeycomb pattern "has to be regarded as a compensation of the disadvantages of the pleurodont mode of attachment"; however, he said that this does not yet imply a functional explanation of plicidentine. On the other hand, he tried, however, to find a historical explanation by reviewing the taxonomic distribution of labyrinthodonty. "A functional interpretation of plicidentine ... is rejected" (Schultze 1969, abstract). In contrast to Schultze's methodological conception, the historical (evolutionary) explanation is no substitute for a functional explanation, instead, both are supplementary (Reif 1985). In cases of marked convergence, as in labyrinthodonty, the search for a functional explanation is a big challenge.

In the light of the odontological facts as summarized above, a number of explanatory attempts can be ruled out, since labyrinthodonty exists in numerous and diverse tooth-jaw relations and modes of attachment.

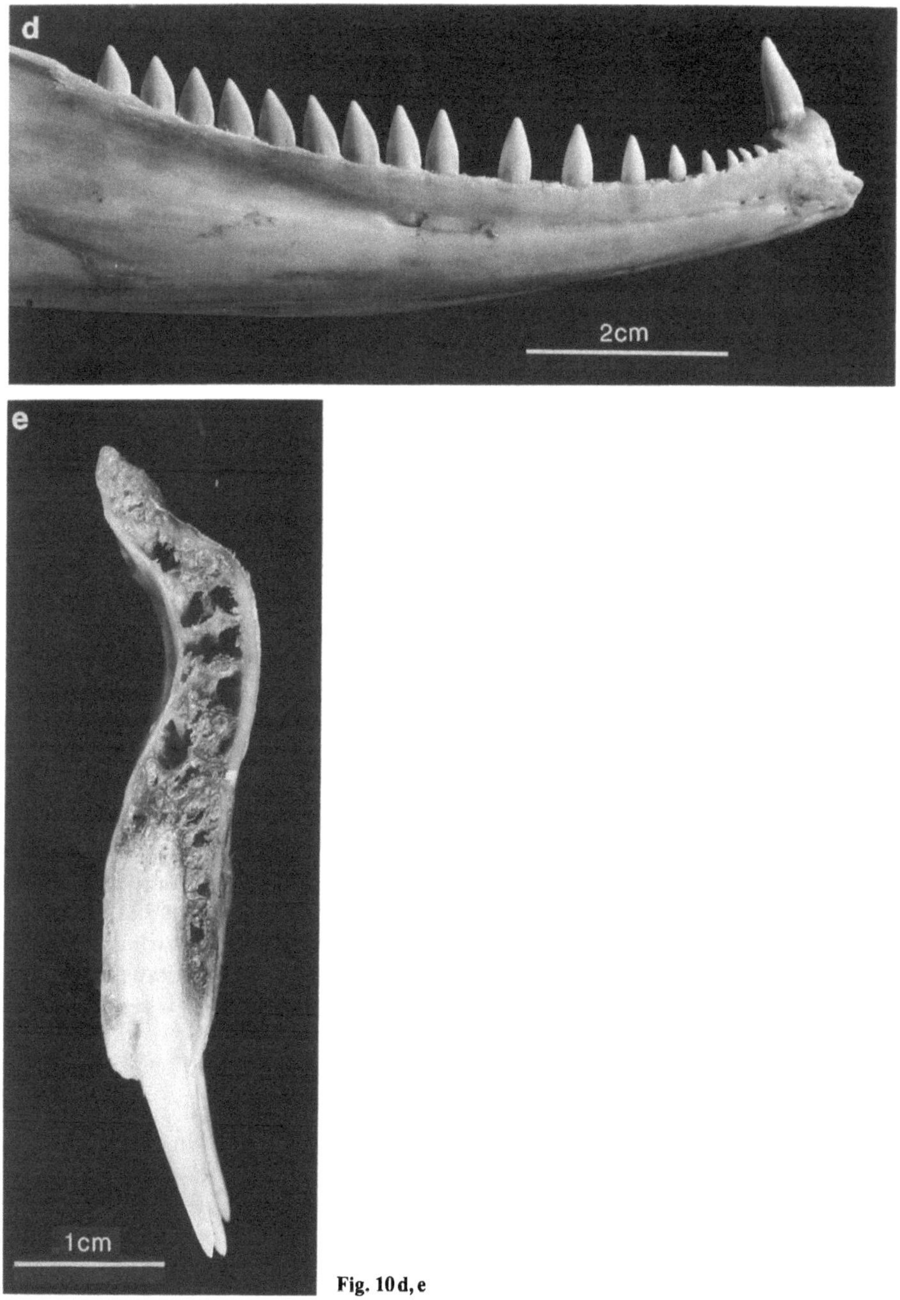

Fig. 10d, e

5.2 Anchoring and Embedding of the Teeth

Our explanation of labyrinthodonty is based on the observation that labyrinthine teeth occur in those forms which possess fangs of large size in relation to the height of the jaw. It should be noted in this context that the early ichthyosaurian *Mixosaurus cornalianus* (Triassic) possessed very small teeth without folding; that all the coelacanthiform crossopterygians possess small teeth without folding, and that plicidentine does not occur in those Osteolepiformes which have small-sized teeth (Besmer 1947; Schultze 1969). Notable are the large mesial fangs of the Struniiformes, which do not have plicidentine, and are attached to tooth whorls. In Porolepiformes and Labyrinthodontia very small teeth occur together with very large teeth within the same dentition. Here, both types seem to possess plicidentine.

Of importance for a functional explanation of labyrinthodonty is the fact that labyrinthine teeth occur in all kinds of tooth-jaw relations.

1. Pleurodont fixation in varanoids (the teeth are attached to the lingual side of the jaw).
2. Thecodont fixation in *Lepisosteus*, crossopterygians, labyrinthodonts, Triassic ichthyosaurians (the teeth are implanted into alveolae).
3. Aulacodont fixation (sensu Mazin 1983) in most of the Jurassic and Cretaceous ichthyosaurs (the teeth are implanted in a deep longitudinal groove in the jaw mandibular bones).
4. Acrodonty and prothecodonty in some crossopterygians and labyrinthodonts (the teeth are tied either to the flat mandibular bone or to shallow pits).

Likewise, labyrinthine teeth occur in all known modes of anchoring. In varanoids the dentine is connected directly to the jaw bone. In all other dentitions with labyrinthine teeth, trabecular, cancellous bone is built in late phases of tooth development and resorbed at shedding. In *Lepisosteus*, crossopterygians, labyrinthodonts, and in three primitive Triassic ichthyosaurs (*Svalbardosaurus*, *Grippia*, and *Utatsusaurus*; Mazin 1983), the tooth is attached by osseous tissue directly to the jaw bone (ancylosis). In the remaining ichthyosaurs, anchoring bone and jaw bone are connected by unmineralized fibers of connective tissue.

5.3 Reasons for the Cone Shape

What are the mechanical arguments and what are the biological roles that favor the general shape of the tooth? What conclusions can be drawn with respect to the food acquired by the possessors of labyrinthine teeth?

Several advantages for cone shape instead of cylindric shape are evident. The tooth penetrates the prey's surface more easily. It is easier to withdraw the tooth after biting, since the elastic tissue of the prey slides towards the tip of the tooth.

When a cone-shaped tooth is forced into a soft body, the compressive forces increase linearly along the tooth. A cross-section near the apex is stressed less than a cross-section close to the base. Only to sustain this increase of forces, the compressive strength towards the base has to increase linearly (Fig. 6). The same holds true for the shearing forces: they add up towards the basis and thus demand cross-sectional areas which increase linearly.

Labyrinthine teeth show the bending resistance necessary to sustain the moments that result when the tooth is driven in the body of a prey not armored by a hard shell. When built with minimal expenditure of material, the tooth's bending resistance should increase parabolically to sustain these stresses, with a maximum at the alveolar margin, where the tooth is attached to the bone. That this is indeed the case can be seen from Fig. 7.

If the tooth were constructed to sustain a force concentrated on its tip, the compression resistance should be on the same level throughout its height. A dome-shaped apex is less likely to break off when stress is applied obliquely than is a pointed tip (Preuschoft et al. 1974; Preuschoft 1989). It can be seen from the resistance properties that labyrinthine teeth – at least those of the species investigated – are adapted to penetration into soft prey, but not to perforate or crush hard shells. This would require a different construction.

5.4 Advantages of Folds Versus Smooth Walls

In Fig. 8 it can be seen that folds raise the resistance in immature teeth. Without folds, bending resistance is decreased by 20–40% in either direction.

In contrast, in mature teeth folds do not make a notable difference. If, however, the folds are shortened, and the material added to the dental wall in such a way that a pulp cavity with thick walls is produced, this tooth has a ca. 8% higher bending resistance than the tooth with the existing folds. With the same amount of material more resistance would be achieved without folding. A mature tooth shows slightly less bending resistance than a tooth with a smooth circular cross-section (Fig. 9): thus the folding represents no advantage! Hence, we have to ask why folding occurs at all.

The key for the solution is derived from the observations of Bystrow (1938, 1939) and the fact that labyrinthine teeth are found in predators with large fangs:

In all predators which possess labyrinthine teeth, the height of their fangs reaches or even exceeds the height of the respective jaw bone. This can be demonstrated for *Lepisosteus* (see Fig. 10), *Eusthenopteron* (Jarvik 1980, Vol. 1, Figs. 82 and 125; Vol. 2, Fig. 53, Osteolepiformes), *Rhizodus* (Owen 1840–1845, Plate 35; Osteolepiformes), *Holoptychius* (Müller 1985a, Fig. 500; Porolepiformes), *Mastodonsaurus* (Fig. 11; Labyrinthodontia), ichthyosaurs (Müller 1985b, Fig. 134; Owen 1840–1845, Plate 73/7), *Eogyrinus* (Panchen 1970, Fig. 4; Anthracosauria), *Seymouria* (Lehman 1955, Fig. 22; Seymouriamorpha), and *Euryodus* (Dechaseaux 1955, Fig. 31; Microsauria).

As a consequence of the periodic change of teeth, the jaws of the lower vertebrates contain teeth in different developmental stages side by side: from early developmental stages over fully grown to teeth which are just about to be shed. Very young tooth tips are missing in museum specimens since they were not yet tightly attached, and thus got lost after death. After eruption the teeth must have gained height very quickly. Since the immature tooth is placed within the row of functioning teeth, it is exposed to loads in the same way as the mature ones. This requires sufficient strength at an early developmental stage. The fact that the jaw bone is less high than the tooth itself makes it necessary that the tooth erupts very early in its morphogenesis, long before it has reached its definite size. At this

Fig. 11. *Mastodonsaurus* sp., Unterkeuper, Kupferzell, cross-section through a lower jaw with a large, worn labyrinthine fang. (Staatliches Museum für Naturkunde, Stuttgart, by courtesy of Dr. R. Wild)

stage, however, the dental wall is still very thin. This is exactly what we can see from Bystrow's drawings (1938, 1939). The youngest tooth which was tightly attached showed at its basis a wall which was about 1/15 of the wall of a mature tooth (*Benthosuchus*, Bystrow 1938, Figs. 8 and 13). At this stage the labyrinthine folding is already well developed. The relation between thickness of the wall and the shortest radius of the elliptic cross-section of the tooth is 1:30, thickness of the wall relative to the longest radius about 1:65. As demonstrated above, an immature tooth equipped with folds has a clear mechanical advantage as compared to a young tooth with smooth walls.

Labyrinthodonty therefore is very significant if the teeth have to erupt early, before they are completely formed, because of a lack of space in the jaws.

It is no problem for our explanation that mature labyrinthine teeth are not mechanically superior to teeth without folding. That labyrinthodonty occurs not only in large fangs but also in small teeth in the same jaw also is no objection against our hypothesis.

As a test to our hypothesis, the following observations can be used:

1. All the above-mentioned reptiles, mammals, and the barracuda (Fig. 10d, e) which are equipped with large fangs, but not with labyrinthine teeth, possess large jaw bones with deep alveolae. In these alveolae the teeth can mineralize to a large extent before they erupt.

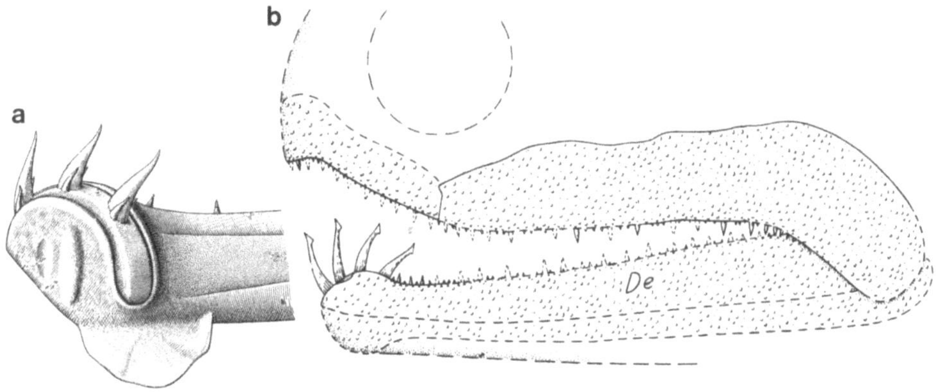

Fig. 12a, b. *Strunius rolandi* Upper Devonian, Lettland; **a** anterior part of right mandible in mesiodistal view with tooth whorl, and two small teeth; **b** *Onchodus jaekeli*, Upper Devonian, Wildungen. Left upper and lower jaw, with tooth whorl of the left mandible and small individual teeth. (**a** and **b** after Jessen 1966)

2. Among the four orders of crossopterygians, two have large fangs (Osteolepiformes, Porolepiformes); labyrinthine teeth occur in these two orders. The Coelacanthiformes (Actinistia) possess only very small teeth without folds. In the Struniiformes large fangs occur only in the mesial area of the lower jaws (Fig. 12; Jessen 1966). These fangs are not known to possess labyrinthine structure. However, these teeth are parts of tooth whorls; they are formed on the lingual side and transported on a "conveyor belt" (like the teeth of sharks, see Nobiling 1974) in labial direction until they reach their functional position on the edge of the jaw (Fig. 12). Further movement in labial direction transports the teeth into an area where their anchoring tissue is resorbed, with the consequence that they are shed. It is because of these tooth whorls that the problem of available space as detailed above for *Benthosuchus* and other labyrinthine teeth does not arise. It is not surprising therefore that labyrinthine teeth have not been found in the Struniiformes. The porolepiform *Holoptychius* has large normal fangs with plicidentine on the rami of the lower jaw. In addition, it has two tooth whorls in the mesial plane (Jessen 1966). It is unknown to us if the teeth on the tooth whorls have folds or not. (In all species of all groups investigated so far all teeth of any size are either labyrinthine or not labyrinthine. Hence it is quite possible that labyrinthodonty occurs in the tooth whorls of *Holoptychius*).

In comparison to a tooth with smooth and thin walls (see Bystrow 1938), the enlargement of the cross-sectional surface (as resulting from the folding) doubtless improves the transmission of the biting force which acts in the direction of the tooth's axis to the jaw bone.

Besides the increased moments of inertia and the increased resistance to compression in young labyrinthine teeth, it should be taken into account that the folds could also have importance for the eruption of the young tooth. They provide a large area of insertion for unmineralized fibers which anchor the tooth, and presumably also serve to transport the immature tooth in an apical direction. (It is still unknown where the forces come from that move an erupting vertebrate's tooth). Folds increase the elasticity of a tooth because they allow for movements within the dental material. However, there is no information about the true significance of such elastic deformations in teeth (in contrast to bones).

Finally, it is possible that the tooth material can not exceed a certain thickness, because the animal's organism is unable to built a thicker dentine wall. Reasons could be:

- The distances of diffusion in the growing tooth are limited.
- The folds provide the tooth with an additional inner surface, which could raise the speed of growth in the young tooth.

The resistance required then is achieved by folding the necessarily thin wall.

W. Maier (oral commun.) observed in *Varanus* that the smooth-walled tips of the teeth are quite long, and that the folds are confined to a section near the basis. This would fit our hypothesis, that the teeth develop smooth walls up to the size which can be housed in the jaw before eruption. The last phase of tooth growth is accomplished rapidly with the aid of the bending-resistant plicidentine.

6 Conclusion

Plicidentine and labyrinthodonty turn out to be an adaptive answer to a constructional or morphogenetic constraint in evolution (see Seilacher 1970; Reif et al. 1985). In general, the building of large fangs requires deep alveolae in which the teeth are fully mineralized before they erupt. Deep alveolae require high jaw bones. If these are not available, the large fangs have to erupt in very immature state. The adaptive answer to this constraint is the plicidentine, which gives the immature tooth mechanical properties optimal under the given conditions. The mechanical behavior of the mature labyrinthine tooth does not differ significantly from that of a normal tooth. Hence plicidentine can only be understood if morphogenesis and replacement of teeth are taken into consideration. This situation is reminiscent of the functional interpretation of the strongly folded septae of ammonites. Many attempts had been made to explain the shape of the fully mineralized septum until Bayer (1977a, b) demonstrated that the explanation must come from a morphogenetic point of view. Septae are mineralized membranes. They are subject to forces already before mineralization. If these forces would tear holes into the soft membrane, the mineralized product (if there were any) could never function.

References

Bayer U (1977a) Cephalopoden-Septen Teil 1: Konstruktionsmorphologie des Ammoniten-Septums. N Jahrb Geol Palaeontol Abh 154:290–366

Bayer U (1977b) Cephalopoden-Septen 2: Regelmechanismen im Gehäuse- und Septenbau der Ammoniten. N Jahrb Geol Palaeontol Abh 155:162–215

Besmer A (1947) Beiträge zur Kenntnis des Ichthyosauriergebisses. Schweiz Palaeontol Abh 65:21 pp

Bystrow AP (1938) Zahnstruktur der Labyrinthodonten. Acta Zool 19:387–425

Bystrow AP (1939) Zahnstruktur der Crossopterygier. Acta Zool 20:283–338

Bullet Ph (1942) Beiträge zur Kenntnis des Gebisses von *Varanus salvator* Laur. Vierteljahresschr Naturforsch Ges Zürich 87 (1/2):139–192
Chase JN (1963) The labyrinthodont dentition. Breviora Mus Comp Zool Cambridge Mass 187:13 pp
Conybeare WD (1824) Additional notices on the fossil genera *Ichthyosaurus* and *Plesiosaurus*. Trans Geol Soc London 2 Ser 1:103–123
Dechaseaux C (1955) Lepospondyli. In: Piveteau J (ed) Traité de paléontologie, vol 5. Masson, Paris, pp 275–305
Jarvik E (1980) Basic structure and evolution of vertebrates, 2 vols. Academic Press, New York London
Jessen H (1966) Die Crossopterygier des Oberen Plattenkalkes (Devon) der Bergisch-Gladbach-Paffrather Mulde (Rheinisches Schiefergebirge) unter Berücksichtigung von amerikanischem und europäischem *Onychodus*-Material. Ark Zool (Stockhol) Ser 2, 18 (14):305–389
Lehman J-P (1955) Anthracosauria. In: Piveteau J (ed) Traité de paléontologie, vol 5. Masson, Paris, pp 173–224
Mazin J-M (1983) L'implantation dentaire chez les Ichthyopterygia (Reptilia). N Jahrb Geol Palaeontol Mh 1983 (7):406–418
Moy-Thomas JA, Miles RS (1971) Palaeozoic fishes. Saunders, Philadelphia, 259 pp
Müller AH (1985a) Lehrbuch der Paläozoologie, vol III/1. Fischer, Jena, 665 pp
Müller AH (1985b) Lehrbuch der Paläozoologie, vol III/2: Reptilien und Vögel. Fischer, Jena, 665 pp
Nobiling G (1977) Die Biomechanik des Kieferapparates beim Stierkopfhai (*Heterodontus portusjacksoni = Heterodontus philippi*). Advances in Anatomy, Embryology and Cell Biology, vol 52, Fasc 6. Springer, Berlin Heidelberg New York
Odermatt C (1940) Beiträge zur Kenntnis des Gebisses von *Heloderma*. Vierteljahresschr Naturforsch Ges Zürich 85:98–141
Owen R (1840–1845) Odontography, 2 vols, 168 tabs. Bailliere, London
Panchen AL (1970) Anthracosauria. In: Kuhn O (ed) Handbuch der Paläoherpetologie, pt 5A. Fischer, Stuttgart, 84 pp
Peyer B (1929) Das Gebiß von *Varanus niloticus* L. und von *Dracaena guinensis* Daud. Rev Suisse Zool 36:71–102
Peyer B (1968) Comparative odontology. Univ Press, Chicago
Preuschoft H (1989) Biomechanical approach to the evolution of the facial skeleton of hominoid primates. In: Splechtna H, Hilgers H (eds) Trends in vertebrate morphology. Fortschritte der Zoologie, vol 35 (Spec Publ). Fischer, Stuttgart New York
Preuschoft H, Reif W-E, Müller WH (1974) Funktionsanpassungen in Form und Struktur an Haifischzähnen. Z Anat Entwicklunsgesch 143:315–344
Preuschoft H, Demes B, Meier M, Bär HF (1985) Die biomechanischen Prinzipien im Oberkiefer von langschnauzigen Wirbeltieren. Z Morphol Anthrop 76:1–24
Preuschoft H, Demes B, Meier M, Bär HF (1987) The biomechanic principles realised in the upper jaw of long snouted vertebrates. In: Sakka M (ed) Définition et origines de l'homme (morphogenèse du crâne et anthropogenèse). 3. Table Ronde Int CRNS Paris 5–8 juillet 1983. Éditions du centre national de la recherche scientifique, Paris
Reif W-E (1985) Konzepte und Geschichte der Funktionsmorphologie. Aufs Reden Senckenb Naturforsch Ges Frankfurt/M 35:107–131
Reif W-E, Thomas RDK, Fischer MS (1985) Constructional morphology: the analysis of constraints in evolution. Acta Biotheor 34:233–248
Schultze H-P (1969) Die Faltenzähne der rhipidistiiden Crossopterygier, der Tetrapoden und der Actinopterygier-Gattung Lepisosteus; nebst einer Beschreibung der Zahnstruktur von *Onychodus* (struniiformer Crossopterygier). Palaeontog It LXV:63–137
Schultze H-P (1970) Folded teeth and the monophyletic origin of tetrapods. Am Mus Nov 2408:10 pp
Seilacher A (1970) Arbeitskonzept zur Konstruktionsmorphologie. Lethaia 3:393–395
Valkenburgh B van, Ruff CB (1987) Canine tooth strength and killing behaviour in large carnivores. J Zool Lond 212:379–397

Complexity and a Coupled System: Flight, Echolocation and Evolution in Bats

J. M. V. RAYNER [1]

Summary

Flight places extreme pressures on mechanical and physiological systems capable of supplying energy for flight and structural strength to accommodate aerodynamic and inertial loads. Severe demands of echolocation are manifest mostly in neurology and behaviour. Echolocation with ultrasound is characteristic of microchiropteran bats, being used for orientation, prey detection and target ranging. Microbats are highly aerial animals relying almost entirely on flight for locating food; they represent an important model for studying the evolutionary interaction of echolocation and flight, two coupled adaptive domains.

This chapter reviews coupling between echolocation and flight in bats. The linked optimization of locomotory and sensory performance has been a major feature of the adaptation and radiation of bats.

1. *Mechanical Linkage Between Echolocation and Flight.* Flying bats omit echolocation calls during the upstroke when the pectoralis generates its peak force and the thorax is compressed. Microbats may not emit sound pulses while gliding.

2. *Call Design, Flight Patterns and Community Structure.* Bat communities are structured by the flight and echolocation in enabling a bat to move around its habitat and to find and catch prey. Species show a range of feeding patterns, which correlate with morphological adaptations favouring appropriate flight patterns and with echolocation call design.

3. *Echolocation and Prey Capture.* Bats hawking insects search for prey by echolocation. Prey location involves a decision on viability (catching the item must be mechanically possible and energetically profitable), and then the flight path must be controlled to intersect the prey. Sensory information, motor control and flapping flight force generation must be integrated, and echolocation and flight design must impose constraints on predation.

4. *Evolution of Bats, Flight and Echolocation.* Adaptive paradigms should be studied within a phylogenetic framework. Bats are traditionally considered a monophyletic group within the Mammalia but brain, eye-brain pathways and echolocation characters link megabats with primates. Further evidence is provided by the wings, which show significant anatomical and osteological differences, and the fossil record, which in non-echolocating megabats is markedly more recent than microbats. Bats do not form a natural taxonomic group but are diphyletic, and the membranous wing supported by elongated digits is homoplasic; megabats are probably a sister group of primates. Differences in size and the ab-

[1] Department of Zoology, University of Bristol, Woodland Road, Bristol BS8 1UG, UK

N. Schmidt-Kittler and K. Vogel (Eds):
Constructional Morphology and Evolution

sence of echolocation as a primitive character in megabats suggest possible differences in the evolution of flight in the two groups.

1 Introduction

Echolocation and flight are often quoted in biology textbooks as examples of extreme evolutionary and adaptational specialization. Flight is perhaps the most demanding of all locomotor adaptations; it places extreme pressures on the successful evolution of mechanical and physiological systems capable of supplying sufficient energy for flight and sufficient structural strength to accommodate the aerodynamic and mechanical loads experienced during flapping flight (Rayner 1987, 1988a, b). The demands of echolocation are manifest most obviously in neurology and behaviour, that is in the organization of the neural mechanisms by which calls are generated and echoes perceived and analysed (Pollak and Casseday 1989), and in the way in which an animal makes use of calls and their echoes (Griffin 1958). But because – like flight – echolocation is the means by which an animal orientates itself and finds its prey, there are also mechanical correlates of echolocation. Together with flight, echolocation has an important behavioural and ecological role in mediating the interaction between an animal and its environment.

Echolocation has evolved in various diverse and unrelated groups of aquatic or nocturnal mammals, and in a few nocturnal or caverniculous birds. Echolocation in air based on ultrasonic sounds is best typified by microchiropteran bats, which use it for orientation (detection of obstacles etc.), for prey detection, and – when prey has been located – for target ranging; ultrasonic signals are also used for communication with conspecifics and kin (Fenton 1985). Since they are highly aerial animals, almost all of which rely entirely on flight for locating and/or obtaining their food, microbats form an important model for studying the evolution and interaction of flight and echolocation, two coupled adaptive domains, which both demand extreme biophysical and biomechanical specialization.

In this chapter I review different aspects of the coupling between echolocation and flight in bats, and then show how this information is crucial in interpreting bat phylogeny and evolution. Each of the topics I consider indicates that the linked optimization of locomotory and sensory performance has been a major feature of the evolution, adaptation and radiation of bats.

The approach of considering coupled adaptive domains in a phylogenetic and evolutionary context exemplifies the important place occupied in biology by the methods of Konstruktionsmorphologie, which I see as study of the complex of structural, historical and physical/mechanical adaptations which permit an animal to survive and reproduce. Virtually every facet of the adaptation of microchiropteran bats is touched by their use of flight and echolocation, and study of these adaptations therefore implies consideration of a very wide range of aspects of bat biology. The investigator must always be aware of possible hidden implications of the system or systems being studied, and should beware of simplification: reduction of a system to just one easily defined physical, anatomical or developmental problem can readily be misleading (although granted to understand exactly how, as opposed to why, such a system functions it is of enormous benefit to consider its biomechanics in isolation). In this chapter I emphasize behavioural and ecological aspects of flight and echolocation; it is, however, inevitable that correlates of locomotor and sensory Konstruktion will be traced at every level of the animal's organization, and therefore studies of fine-scale adaptations associated with flight and echolocation are no less relevant to bat Konstruktionsmorphologie.

2 Mechanical Linkage Between Echolocation and Flight

Flying microchiropteran bats invariably omit search-phase echolocation calls synchronized with the wingbeat; calls are initiated during the final phase of the upstroke or the very start of the downstroke, and the sustained constant frequency calls of bats such as horseshoes (Rhinolophidae) continue through much of the downstroke (Schnitzler 1971, Heblich 1986; Fig. 1 a). [In the sole megachiropteran genus yet known to use echolocation, the cavern-roosting *Rousettus*, sounds are omitted during the downstroke, but consist of brief broad-band pulses generated by tongue-clicking, which are not comparable with microchiropteran calls (Herbert 1986). One megachiropteran species, *Eonycteris spelaea*, may use the noise of its wingbeats as a crude form of echolocation (Gould 1988).]

During the final phases of the upstroke the pectoralis muscle is lengthening, but must generate a large and increasing force against wing inertia to slow the upwards movement of the wing. The moment at the wing root from the pectoralis (depressor) is greatest at the start of the downstroke, remains large during the downstroke, and then reduces to zero in the early part of the upstroke; at slow speeds an elevating moment may be required to raise the wing, but at higher speeds the wing is raised by aerodynamic forces (Rayner 1986, 1987, 1988 a; Fig. 1 b). Electrophysiological experiments confirm that in birds (Dial et al. 1988; Jenkins et al. 1988; Goslow et al. 1989, 1990) and bats (Altenbach and Hermanson 1987; Hermanson and Altenbach 1981, 1983, 1985; Fig. 1 c) pectoralis activity is initiated during the final phase of the upstroke, correlating with onset of the largest mechanical loads. Observations on birds show that pectoralis activity corresponds to increasing compression of the thorax (Goslow et al. 1989, 1990; Jenkins et al. 1988), and if – as seems likely – the same occurs in bats, this would correspond with expiration, and hence with the opportunity to emit a call carrying maximum sound energy. Circumstantial support for the idea that synchronization of call emission is driven by mechanical forces from the flapping wings comes from the observation that occasional broad inter-pulse intervals in foraging pipistrelle bats (Schnitzler et al. 1987) may be associated with the brief glides which are characteristic of this genus (Thomas et al. 1990); during gliding, the thorax would not be deforming, and there would be no external mechanical force on the thorax driving synchronization of respiration.

Energy consumption associated with the production of echolocation calls in resting bats has been measured, and has been shown to be appreciable, approaching some ten times basal metabolic rate, thus reaching levels comparable to the energy demands of flight (Speakman et al. 1989); this surprisingly large value

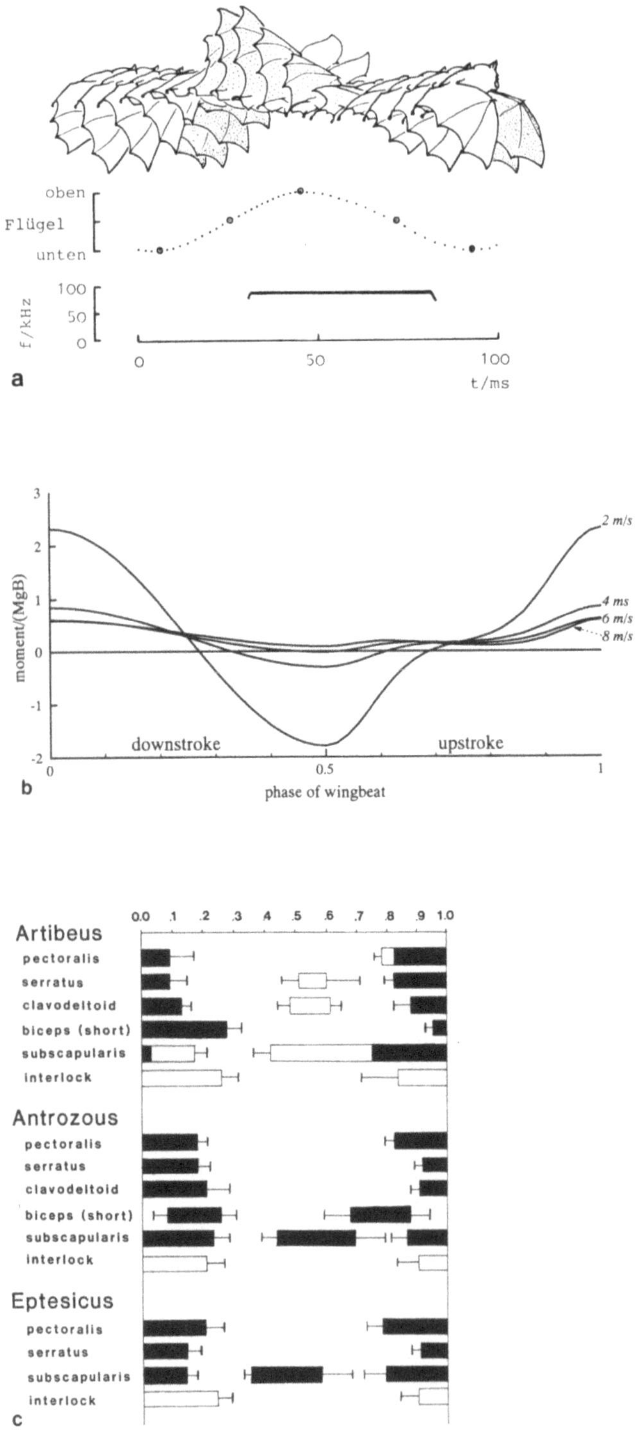
oben
Flügel
unten
f/kHz
100
50
0
0
50
100
t/ms
a
3
2
1
0
-1
-2
moment/(MgB)
2 m/s
4 ms
6 m/s
8 m/s
downstroke
upstroke
0
0.5
1
phase of wingbeat
b
0.0 .1 .2 .3 .4 .5 .6 .7 .8 .9 1.0
Artibeus
pectoralis
serratus
clavodeltoid
biceps (short)
subscapularis
interlock
Antrozous
pectoralis
serratus
clavodeltoid
biceps (short)
subscapularis
interlock
Eptesicus
pectoralis
serratus
subscapularis
interlock
c

probably represents the cost of creating the enhanced airflow needed for emission of calls with sufficient intensity. Synchronization of call emission with the wingbeat and expiration will therefore also serve to reduce additional energy demands due to echolocation while in flight (Speakman et al. 1989), and because of the quantity of air required for a useful call may indeed be essential. This may also explain why echolocation is rare in terrestrial vertebrates, in which appropriate airflows cannot be generated in association with locomotion, and also suggests that the evolution of flight and echolocation were linked: a nocturnal flier requires some means of sensing its surroundings, while a sufficiently loud call appears to be possible only when the animal is flapping its wings.

3 Echolocation Call Design, Flight Patterns and Community Structure

The pattern of bat communities is mediated by the roles of flight and echolocation, since both are closely tied to the ability of a bat to move around its habitat and to find and catch its prey. Eco-morphological correlations between wing morphology, flight muscle physiology, echolocation call design and ecological niche (habitat use, diet, foraging patterns, etc.) have been demonstrated within bats as a whole (Norberg and Rayner 1987; Norberg 1986, 1990a, b; Fig. 2) and within a range of different bat communities (= guilds) (e.g., Aldridge 1985, 1986; Aldridge and Rautenbach 1987; Baagøe 1987; Findley and Black 1983; Habersetzer 1986; Heller and von Helversen 1989; Jones and Rayner 1988, 1989; McKenzie and Rolfe 1986; Neuweiler 1989; O'Neill and Taylor 1986; Weid and von Helversen 1987; Fig. 3). In the course of the development of the communities to which each bat belongs, flight behaviour, constructional morphology associated with flight (wing design, skeletal structures, muscle performance etc.) and echolocation have simultaneously been refined and specialized.

These correlations represent the current state of a process of evolutionary optimization of form (in its widest sense) and behaviour, in which not only are a wide variety of adaptive domains linked within any individual bat species, but also interactions between bat species, prey insects, potential competitors and host plants all play an essential role. Although it is natural to consider this as an optimization process, it would be wrong to model it by a tendency to maximize the output of any one independent mechanical system or to optimize any one measure of performance. Rather, the tendency under natural selection for *inclusive*

Fig. 1. Synchronization of echolocation pulse emission and the wingbeat in flying bats. **a** Long FM/CF/FM call of *Rhinolophus ferrumequinum* showing how onset of the call coincides with the final phase of the upstroke (Heblich 1986). **b** Roll moments at the wing root for a noctule *Nyctalus noctula* flying at different speeds. The moments represent the forces to be applied by muscles of the pectoral girdle to flap the wings in the dorso-ventral plane, and represent the greatest components of aerodynamic and mechanical forces acting on the wing. They are calculated by aerodynamic theory, based on observed wingbeat kinematics and wake vortex structures, but do not take account of elastic or other deformation of the pectoral girdle (see Rayner 1986, 1987 for details of computation). **c** Electrical activity in muscles of three bat species, showing onset of depressor activity during the final phase of the upstroke, as predicted from **b** (Altenbach and Hermanson 1987)

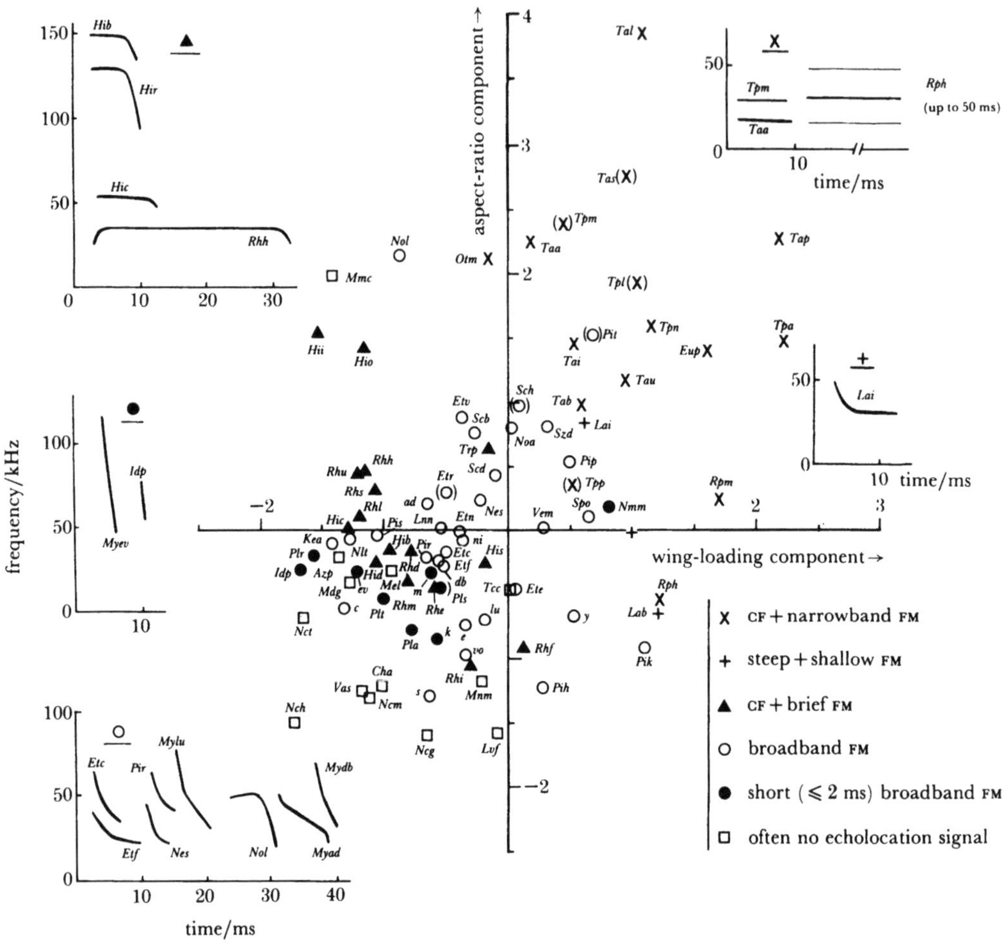

Fig. 2. Correlation between echolocation call design and wing morphology. The *axes* represent size-independent principal component measures of wing morphology (loading component is comparable to wing loading, or relative wing size, and aspect component to aspect ratio, or wing shape). Wing design measured by these components correlates with foraging behaviour, and shows significant taxonomic consistency. This figure shows that echolocation call structure is also consistent with morphology, which is evidence of strong ecological selective pressures for optimization of call structure, wing design and foraging mode (Norberg and Rayner 1987)

fitness to be maintained or increased will be manifest as optimization of a range of quantities [in the present context these might include power, speed, cost-of-transport, agility, manoeuvrability, sound intensity, prey detection performance, etc. (Norberg and Rayner 1987)]. Moreover, the adaptation of any individual species must not only allow it to do its job well (see below), but must also allow it to survive unpredictable and extreme events, such as fluctuations in weather or food supply; in animals as sensitive to environmental conditions as bats, it is probable that this somewhat intangible requirement for flexibility plays

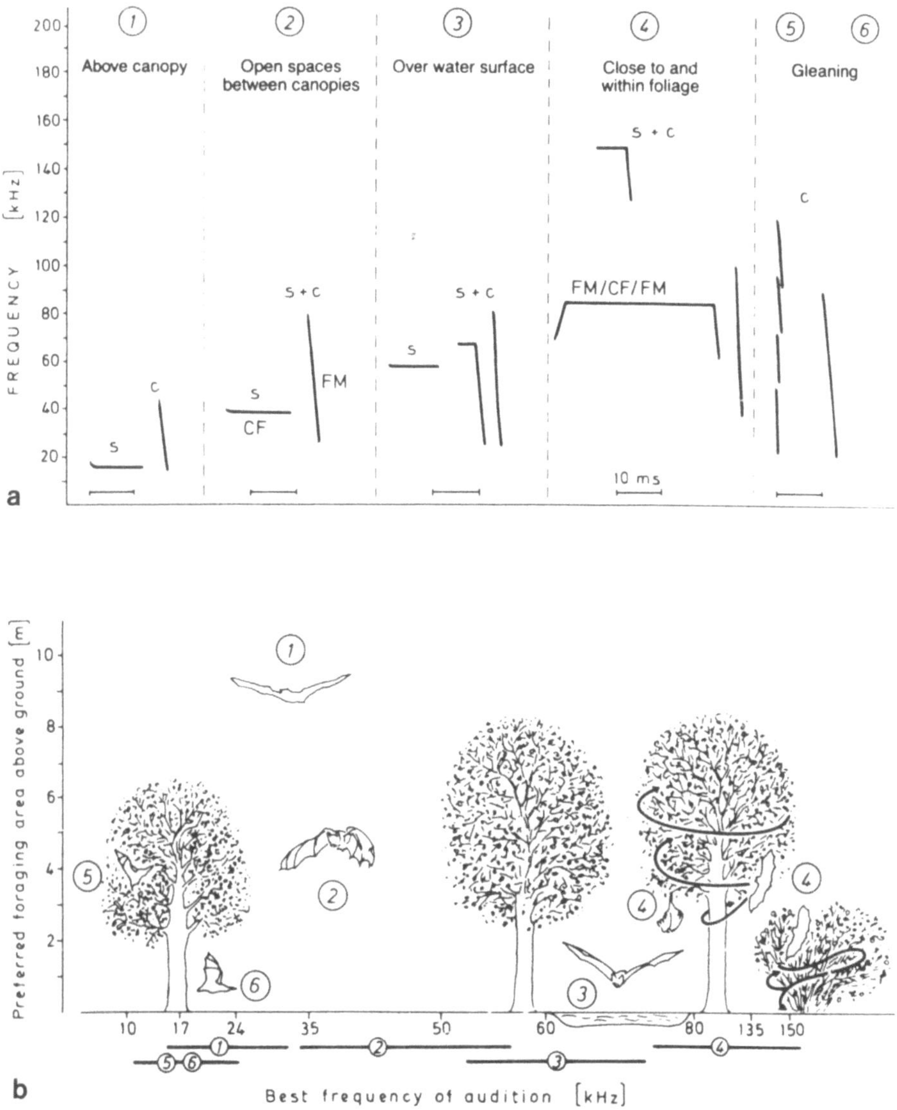

Fig. 3. The foraging niches and sonagrams of echolocation calls of insectivorous bats. **a** Types of echolocation signal emitted in different foraging areas, showing schematized sonagrams emitted while bats are searching for (*s*) and catching (*c*) prey. The *encircled numbers* refer to the foraging areas depicted in **b**. *FM* frequency modulated call; *CF* constant frequency call; brief FM calls are most common. **b** Foraging areas and their correlation to the frequency range of best audition in bats. *Encircled numbers* indicate foraging areas and modes: *1* hawking above canopy; *2* hawking in open spaces between canopy; *3* hawking and/or trawling over water surfaces; *4* hawking close to and within foliage (including flycatching); *5* foliage gleaning; *6* ground gleaning. The *thick lines below the abscissa* indicate the approximate frequency range of best audition in various bat species foraging in the different areas (Neuweiler 1989)

as important a role in evolution, adaptation and speciation as does specialization for specific niches.

The process of evolution and adaptation takes place within constraints; any animal species may be defined as the sum of the responses of all of its precursors to the mosaic of constraints they experienced, bearing in mind that the constraints will have changed through the course of geological time as environment has varied and Bauplan has developed. The constraints involved vary widely, and can range from ecological (the quantity of biomass or the types of prey a predator can extract from its environment) to mechanical (limitations of physical processes: force generation from the wings; stress limits on bones, muscles, tendons; information content of brief echolocation calls and echoes) and morphogenetic (Bauplan; adaptations forming an essential part of the phylogenetic heritage) (see Rayner 1988a for discussion of these constraints on bird flight). An important aspect of morphogenetic constraints is that an animal can become "locked in" to an adaptation: thus, a nocturnal aerial insectivore might perform sufficiently well by using vision rather than sound (cf. nightjars), but in the course of evolution of microbats the visual system has not been developed to the same extent as has echolocation and aural senses. It is difficult to conceive that a highly specialized microbat could successfully give up sound in favour of vision.

It is noteworthy that communities in different parts of the globe show remarkable uniformity in patterns of species differences and ecological niche occupation, even though the taxonomic groups involved may differ widely (e.g., Findley 1976; Norberg and Rayner 1987). (This phenomenon is by no means unique to bats.) It is not known whether the bats involved experienced similar patterns in phylogeny and development, or whether it is the result of some form of self-ordering at the ecological (rather than the species) level; the mechanism may be comparable to pattern generation and self-similarity in morphological structures (albeit with different underlying mechanisms), reflecting some of the constraints outlined above.

Temperate and sub-tropical insectivorous bat guilds include species with a range of feeding patterns, which include aerial hawking, flycatching, trawling for insects on or just over water, and gleaning from foliage and (rarely) from the ground (Fig. 3). Typically, each community contains one or a few species specializing in each foraging mode, and in contiguous biogeographical regions there is considerable taxonomic consistency (Norberg and Rayner 1987). More unusual feeding modes – including piscivory, carnivory, sanguivory and herbivory (fruit, pollen, nectar) – are more frequently encountered in tropical communities. Specific morphological adaptations [for instance in the shape and size of the wings (Norberg and Rayner 1987; Fig. 2)] have evolved to favour the flight movements appropriate for each mode of feeding: as an example, bats flycatching within woodland have broad, short wings conferring manoeuvrability in small areas, while aerial hawking bats have longer, narrow wings giving good agility; features of echolocation call design are correlated with flight morphology (Figs. 2, 3) because the calls are also specialized for feeding mode. Different acoustic structures in the calls permit determination of distance, relative motion or target texture, and these patterns can be traced taxonomically in inter-family comparisons (e.g., Simmons and Stein 1980; Schnitzler and Henson 1980; Norberg and Rayner 1987);

however, the mechanisms underlying small-scale interspecific differences in call design in this context are relatively poorly understood, and are influenced to a considerable extent by diet and habitat structure [see Aldridge (1986) and Jones and Rayner (1988) on myotids, Habersetzer (1986) on hipposiderids, and Jones and Rayner (1989) on horseshoe bats].

4 Echolocation and Prey Capture

The "classic" paradigm of echolocation and flight is represented by bats hawking insects (Griffin 1958; Griffin et al. 1960; Simmons 1987; Simmons et al. 1979; Simmons and Kick 1983; Webster and Griffin 1962). Hawking is defined as the mode of hunting in which a flying predator actively seeks and catches flying insect prey. Here again we see an evolutionary coupling between the two adaptive domains, although the coupling now also involves the bat's prey, which in some cases are sensitive to, and can interfere with, the hunter's echolocation. This is particularly significant for an insect which may not have the agility to control its flight path to evade a bat.

While searching for prey, a bat emits regular calls, usually at a rate of one per wingbeat (above). When it perceives an echo, the bat must decide whether the target is a food item, whether the food item is worthwhile or viable (in the senses that the energy gain to the bat must outweigh the cost of capture (and the chance of failure), that the bat may not be sufficiently agile or manoeuvrable to catch the prey, that the prey is of appropriate size or type, and that the prey may be sensitive to echolocation and may evade the bat or "jam" its signals), and then must control the flight path to intercept the prey (Fig. 4). As the prey is approached call structure changes and call repetition increases rapidly (the "feeding buzz") enabling precise control of flight to locate the prey (references above, also e.g., Neuweiler et al. 1987; Schnitzler et al. 1985, 1987; Jones and Rayner 1988, 1989; Kalko and Schnitzler 1989). Finally, the prey must be captured, in wings, tail membrane or mouth (Webster and Griffin 1961), and if the prey is large the bat may have to land on a perch to consume it. The hawking process involves the integration of sensory information, motor control, flapping flight force generation and behavioural decision making, often with very short response times. For this reason echolocation and flight design, so far as they can reasonably be considered separately, both impose constraints on predation; quite apart from the ecological pressures for specialization mentioned above, it is not surprising in view of the complexity of the hunting manoeuvre that morphology and call design are correlated at the community level; thanks to their echolocation and the specialization of their flight design microchiropteran bats are the only animals able to hawk flying insects at night.

In fact, hawking is a relatively unusual feeding mode in bats (Norberg and Rayner 1987), and is normal in perhaps only one third of insectivorous species. For bats with inappropriate flight design it is probably much harder to catch flying insects than is usually imagined, especially owing to the ability of many potential prey insects to detect ultrasound and to avoid the bats. However, most microchiropteran insectivores resort to hawking occasionally, particularly when

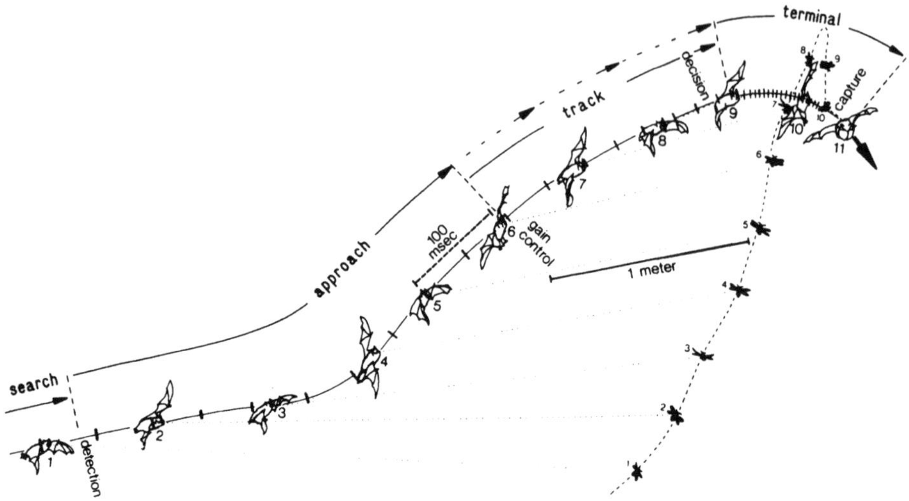

Fig. 4. Diagram of an interception manoeuvre during an aerial hawk by an echolocating bat. The bat and the target insect are both shown as *numbered images* based on stroboscopic photographs taken at intervals of 100 ms. A *thin dotted line* connects corresponding images of the bat and the target to show target range. The flight path of the bat produces a "sound track" of sonar signals, with the occurrence of each omission indicated by a *short bar* perpendicular to the flight path at the location of the bat when it produced the sound. Note the dramatic increase in call emission rate after image 8 during the "feeding buzz" (Simmons 1987)

aerial insects are relatively abundant; and variations on the hawking paradigm in which comparable mechanical and acoustic specializations can be identified are evident in other feeding modes. In each case, specializations in flight morphology and behaviour, in choice of prey type and in echolocation are closely interlinked. We do not know which of the various feeding modes of insectivores is primitive: it is most likely to be some form of surface gleaning, in which the bat does not need to integrate information concerning the movements of flying prey (Rayner 1986), but modern gleaners often show considerable specialization in echolocation strategies.

5 Bats Which Do Not Use Echolocation

Not all bats echolocate. As far as is known, all Microchiroptera are able to echolocate, although it is facultative in some species; frugivorous phyllostomids, for instance, locate food by smell, vampire bats find targets by heat (infra-red radiation), and some insectivores and carnivores eschew echolocation while hunting so as to avoid alerting their prey. In the Old World fruit bats (Megachiroptera) ultrasonic echolocation occurs only in the frugivorous genus *Rousettus*, in which sounds are generated relatively crudely by tongue-clicking, and echoes are used purely for orientation. With very few exceptions, megabats are entirely frugivorous or nectarivorous, but nonetheless they are predominantly nocturnal, and all megabats would find echolocation be of undoubted value for orientation, as in

Rousettus; its absence as a synapomorphy of megabats is somewhat surprising, although it is compensated to some extent by the relatively large eyes and the well developed visual system.

Some microbats such as the Megadermatidae and the vespertilionid genus *Plecotus* are termed whispering echolocators, using calls of very low intensity, or on occasion ceasing sound emission altogether (e.g., Fiedler 1979; Fenton 1984; Anderson 1990): these species glean for stationary (or non-flying) prey which might detect loud echolocation calls; megadermatids hunt small vertebrates in the manner of owls, by listening for prey-generated sounds, and *Plecotus* is believed to listen for the wingbeat noise of months warming up prior to taking flight. These bats cannot use echolocation for flight-path control during prey capture; it is essential that they hunt in habitats with which they are familiar. All such species have relatively large wings permitting slow, and therefore safe and controlled, flight; similar wing adaptations are found in ecologically-comparable nocturnal birds (Rayner 1988a).

6 Ontogeny of Flight and Echolocation

During development, juvenile bats use ultrasound for mother-young communication. Sounds emitted by juveniles tend to be complex, with many harmonics, but in their acoustic structure, and in particular their frequency/time pattern, some are reminiscent of adult echolocation calls. At the time of first flight the wings are not fully developed (P.M. Hughes, pers. commun.), and juveniles are beginning to use echolocation for orientation, if not for prey capture (e.g., Buchler 1980; Brown et al. 1983; Habersetzer and Marimuthu 1986; Konstantinov and Makarov 1987; Rübsamen 1987).

Synchronization of the onset of flight and echo-based orientation indicates endogenous control of behavioural ontogeny, but refinement of both processes requires learning and experience by the newly volant bats as foraging proficiency improves. Considerably more research is required to clarify the relative importance of learning and development in the ontogeny of bats.

7 The Evolution of Bats, Flight and Echolocation

In the preceding parts of this chapters I have explored some of the ways by which – in contemporary bats – flight and echolocation are interlinked, but I have said little about the evolution, separately or together, of the two adaptive domains, or about their relation to the evolution of bats. All adaptive and evolutionary paradigms should be studied within a phylogenetic framework (Padian 1982, 1987; Rayner 1988b, 1989a, b); this is particularly critical in relation to the problem of bat evolution since bats demonstrate such extreme morphological and behavioural specializations. Moreover, the problem is presently topical owing to recent reassessments – which have proved surprisingly contentious – of the origins and relationships of bats.

In the past it has been assumed by most workers that bats form a monophyletic group within the Mammalia, being related most closely to insectivores or other Archonta (e.g., Winge 1941; Van Valen 1979; Padian 1987). The membranous wing supported by elongated digits is unique to bats (although it appears in a modified form in the "flying lemur" *Cynocephalus* (Dermoptera) of South-East Asia, a possible sister group of bats; Leche 1886; Miller 1907; Pettigrew et al. 1989), and the complex of characters representing the wings and flight has traditionally been taken as synapomorphic of the Chiroptera (Padian 1987; Wible and Novacek 1988). The precise relationships of the Chiroptera with other mammalian groups have been obscure, partly because the first microbats appeared as far back as the early Tertiary, and partly because different lines of evidence have pointed to different sister groups; however, the apparent difficulty of evolving flight would to most authors appear to preclude a multiple origin of what we know as bats.

For mechanical reasons it seems inescapable that flight evolved through a gliding ancestor similar to (or perhaps belonging to) the Dermoptera, with the main evolutionary trend being the lengthening of the digits and the long bones of the arms and legs to spread the wing membrane and increase the surface are generating aerodynamic force. It is reasonable to imagine that ultrasonic communication, or perhaps echo-orientation, was already present in the proto-bats (as it is in insectivores), and became progressively refined to permit prey location as early flying bats improved flight performance and radiated to occupy more specialized ecological niches (Smith 1977; Simmons and Stein 1980; Fenton 1984; Kovtun 1985; Rayner 1986, 1988b, 1989a, b; Scholey 1986). It is unrealistic to imagine the first flying bats as hawking aerial insectivores, for this mode of feeding demands the highest refinement in both flight and echolocation (see above): gleaning of surface insects, perhaps while crawling rather than flying, is a more likely feeding mode.

There is, however, a major problem implied by this model which has never been successfully resolved: why do megachiropterans (with the minor exception of *Rousettus*) not use ultrasonic echolocation? The first bats in the fossil record appear in the Eocene (ca. 55 ma B.P.), and are very similar to modern microbats (Jepsen 1966, 1970), with brains and ears of a size indicating the capacity to echolocate (Novacek 1985); ecomorphological relationships of bats from the Messel deposit (49 ma B.P.) imply sophisticated community structures and behaviour similar to those of modern bat feeding guilds (Richter and Storch 1980; Habersetzer and Storch 1987; Norberg 1990b). The first megabats only appear in the fossil record much later (Oligocene, 34 ma B.P.), and, taken with assumed chiropteran monophyly, this would imply that megabats form a derived group within the Microchiroptera; yet there is no evidence that echolocation was primitive in Megachiroptera. Some authors have argued – despite the chronology of the fossil evidence – that megabats are the least derived (Winge 1941), while others claim that insectivory is primitive in bats, and that echolocation was redundant in a derived frugivorous megabat lineage and became lost; neither of these arguments is particularly convincing. Echolocation is such an invaluable adaptation for any nocturnal flier that its disappearance is hard to accept [echolocation can even develop in blind humans (Griffin 1958)], particularly since for nocturnal orientation vision – however well developed in megabats – is a poor substitute for echolocation. Echolocation has been retained in frugivorous phyllostomids of the New World, which in many other ways are comparable with megabats (Norberg and Rayner 1987), and moreover megabats show no obvious primitive (as opposed to unspecialized) characters in comparison to microbats.

Additional information is now available following the recent discovery of major differences between mega- and microchiropteran bats in the brain, in the retinotectal pathways linking eye and brain, and in the spinal chord (Pettigrew 1986; Pettigrew et al. 1989). The eye and brain structures in microbats are typical of those of generalized mammals, but the structures in megabats are remarkably similar to those in primates, and imply visual and somatosensory specialization in the absence of echolocation; there is no evidence for obligative echolocation in primates, and ancestral primates, like contemporary prosimians, were nocturnal. On the other hand, the microbat spinal chord is highly specialized, containing regions of grey matter to which some functions associated with the wings are devolved, while the megabat spinal cord is simple, similar to that of primates, and typical of generalized mammals. So numerous and so close are the similarities between megabats and primates that a close evolutionary relationship between the two groups seems inescapable. Finally, differences in bone proportions in the wing digits cast doubt on the validity of the use of wings as a synapomorphy of bats as a monophyletic group (Fig. 5; Pettigrew et al. 1989, but for contrary views see Wible and Novacek 1988). The assumption of bat monophyly must therefore now be questioned.

The set of characters drawn from flight and the wings and from sensory adaptations therefore provides a complex of conflicting information. The central question is whether bats form a monophyletic group, that is whether megabats and microbats share a common ancestor which possessed the main features of "bats", that is with flight (or some precursor of flight), and possibly echolocation. If not, "Chiroptera" is not a natural group, megabats and microbats would have different non-flying sister groups, and it becomes necessary to establish which of the main characters involved (flight, echolocation, spinal chord, eye and brain) are homoplasies and which are synapomorphies. Pettigrew et al. (1989) have given four alternative phylogenies for bats based on flight and the eye-brain characters, and I have modified these to include echolocation (Fig. 6). Either the features of the eye and brain are convergent between megabats and primates, or the presence of wings is homoplastic between mega- and microbats. Several of the phylogenies also imply the loss of a major adaptation: flight in primates in (a) (although it cannot seriously be argued that contemporary primates are flightless bats!), echolocation in primates in (b) and (c), or the derived inverted retinotectal pathway in Microchiroptera in (b), with reversion to the primitive state. It is undoubtedly possible for major adaptations to be lost in the course of macroevolution, but for a nocturnal lineage to make major alterations to a sensory domain without parallel evolution of an alternative, and while experiencing a complex and broadly similar sensory environment, seems most unlikely.

Accordingly, it seems on the presently available evidence that phylogeny (d) is most appropriate, and that flapping flight has evolved twice in the mammals. This is perhaps not greatly surprising, since in vertebrates gliding flight has

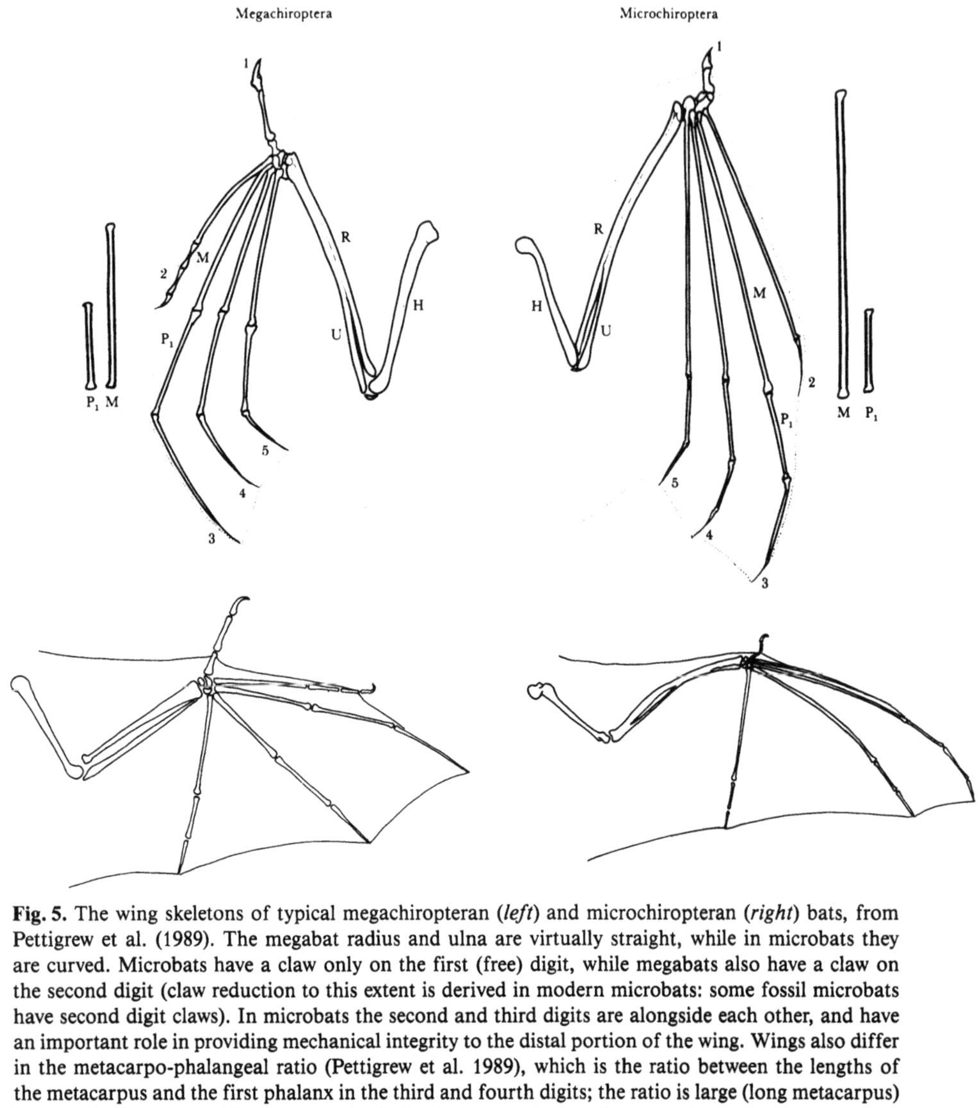

Fig. 5. The wing skeletons of typical megachiropteran (*left*) and microchiropteran (*right*) bats, from Pettigrew et al. (1989). The megabat radius and ulna are virtually straight, while in microbats they are curved. Microbats have a claw only on the first (free) digit, while megabats also have a claw on the second digit (claw reduction to this extent is derived in modern microbats: some fossil microbats have second digit claws). In microbats the second and third digits are alongside each other, and have an important role in providing mechanical integrity to the distal portion of the wing. Wings also differ in the metacarpo-phalangeal ratio (Pettigrew et al. 1989), which is the ratio between the lengths of the metacarpus and the first phalanx in the third and fourth digits; the ratio is large (long metacarpus) in microbats, and short in megabats (and also in Dermoptera and primates)

evolved at least 18 times, and flapping flight has evolved four (or five!) times. This phylogeny is also consistent with the shorter fossil record of megabats, with their primitive spinal cord and with the absence of echolocation as a primitive character. Megabats should be classified as "flying primates", probably with Dermoptera as sister group (Fig. 7; Smith 1977, 1980; Pettigrew et al. 1989); tarsiers and galagos are probably the closest living primates to the Dermoptera/Megachiroptera lineage. Relationships this distant in mammal evolution are hard to determine, and presently the precise relationships of microbats must remain undefined. This model is also supported by a range of other characters, including differences

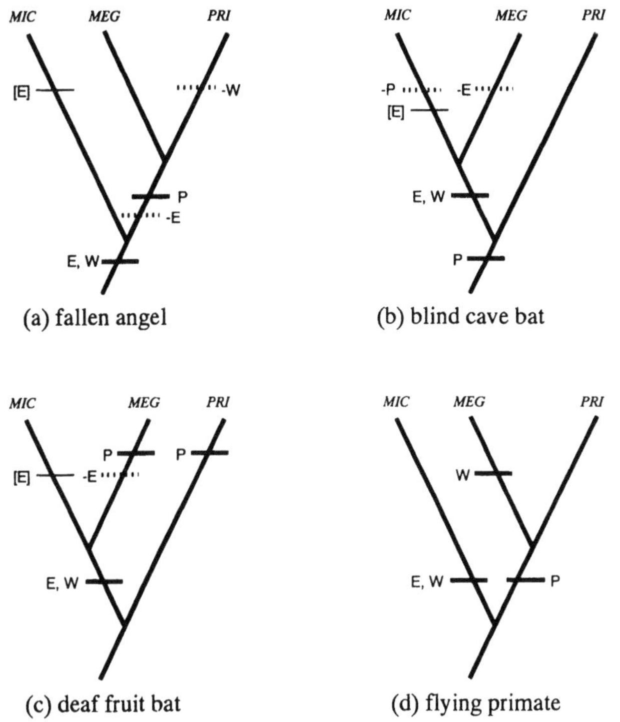

Fig. 6. Four alternative phylogenies for the Chiroptera proposed by Pettigrew et al. (1989), showing relationships of Microchiroptera (*MIC*), Megachiroptera (*MEG*) and Primates (*PRI*), together with three critical sets of characters: *W* wings and flight; *E* echolocation; *P* a complex of primate brain and eye characters. Loss of a character is indicated by a *dotted line.* Pettigrew et al. did not include echolocation in their phylogenies: interpretation is complicated when echolocation is introduced, and this character is indicated on the phylogenies with the assumption that it evolved in the microbat lineage simultaneously with flight (although this hypothesis has not been tested); alternative stages for the later evolution of echolocation are also indicated as [E]. **a** The *fallen angel* model, in which primates are flightless bats: this is not proposed seriously, and is included purely for completeness. **b** The *blind cave bat* model, in which Microchiroptera have lost primate-specific brain and eye features, and Megachiroptera have lost echolocation. **c** The *deaf fruit bat* model in which primate-specific characters are convergent in Megachiroptera and Primates, and again echolocation is lost in Megachiroptera. **d** The *flying primate* model, in which wings and flight have evolved twice, and all three character complexes form synapomorphies defining monophyletic clades. Only model **d** does not require the apparently anomalous loss of echolocation, and this model is also the most parsimonious way to account for all available data (from echolocation, brain, skeleton, penial morphology, haemoglobin sequences, Dermoptera, the fossil record and biogeography, etc.) (After Pettigrew et al. 1989)

in size, penis morphology, behaviour and biogeography between the two groups, and by some of the limited suitable chemical classification data (Pettigrew et al. 1989). If valid, it will demand major rethinking of the biology of bats and of the history and relationships of major mammal groups. Flight and echolocation have a major role to play in the discussion of bat evolution, and as the argument above indicates, point towards a diphyletic origin of bats. In the evolutionary context these characters can only be assessed by detailed study of their ecological, biophysical and biomechanical significance in modern animals. Interpretation of bat

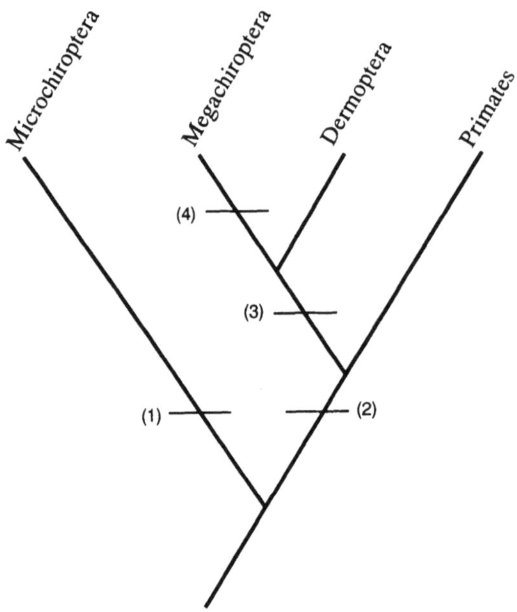

Fig. 7. Proposed phylogeny of the paraphyletic group "bats", showing relationships of the order Microchiroptera and Primates and the families Pteropodidae (= Megachiroptera) and Cynocephalidae (= Dermoptera) within the Primates. Character sets include the following: *1* Microchiropteran characters: elongated forelimbs and digits; patagium, attached between digits of manus and foot, and tail; laryngeal echolocation; enlarged cochlea; enlarged hind brain; grey matter in spinal cord; large or increased metacarpo-phalangeal ratio (Fig. 5); middle-temporal visual cortex. *2* Primate characters: enlarged neocortex; primate-like eye and brain; pendulous penis; single pair of pectoral mammae; low or reduced metacarpo-phalangeal ratio (Fig. 5); etc. *3* "Flying Primate" characters: relatively large initial size; brain specialized for aerial activity; elongated forelimbs; flattened vertebral ends of ribs; patagium, attached between digits of manus and foot, and tail; elongated middle phalanges of digits. *4* Megachiropteran characters: elongated digits spreading patagium; flapping flight. See Padian (1987), Wible and Novacek (1988), Pettigrew et al. (1989) and Beard (1990) for discussion and full definition of these and other characters. No sister groups for Microchiroptera within the Mammalia, and for Cynocephalidae/Pteropodidae within the Primates are proposed at present; tooth and claw architecture link Dermoptera and plesiadapiform primates (Rose and Simons 1977). Note that the phylogeny is supported both by structural and osteological characters, and also by features related to flight and echolocation, which are the main adaptive domains of "bats"

relationships based solely on structural characters and giving minimal consideration to their biological relevance would risk indicating a completely different phylogeny from that which from a broader view appears most robust. Phylogenetic inferences from Konstruktionsmorphologie should not be made in isolation from the biology of the Konstruktions being studied.

Acknowledgments. I am grateful to Neill Alexander, Gareth Jones and Jack Pettigrew for discussion and many forms of assistance, and for their comments on this manuscript. My work on animal flight and on bat echolocation and evolution is funded by the Royal Society and by grants from the Natural Environment Research Council, the University of Queensland and the University of Bristol Research Committee.

References

Aldridge HDJN (1985) Manoeuvrability and ecology in British bats. Myotis 23–24:157–160

Aldridge HDJN (1986) Manoeuvrability and ecological segregation in the little brown (*Myotis lucifugus*) and Yuma (*M. yumanensis*) bats (Chiroptera: Vespertilionidae). Can J Zool 64: 1878–1882

Aldridge HDJN, Rautenbach IL (1987) Morphology, echolocation and resource partitioning in insectivorous bats. J Anim Ecol 56:763–778

Altenbach JS, Hermanson J (1987) Bat flight muscle function and kinematics at the scapulo-humeral lock. In: Fenton MB, Racey PA, Rayner JMV (eds) Recent advances in the study of bats. Cambridge Univ Press, Cambridge, Mass, pp 100–118

Anderson MA (1990) Feeding behaviour and echolocation in the brown-long-eared bat, *Plecotus auritus*. In: Hanak V, Horacek I, Gaisler J (eds) European bat research 1987. Proc 4th Eur Bat Res Symp, Praha 1987. Charles Univ Press, Praha (in press)

Baagøe HJ (1987) The Scandinavian bat fauna: adaptive wing morphology and free flight in the field. In: Fenton MB, Racey PA, Rayner JMV (eds) Recent advances in the study of bats. Cambridge Univ Press, Cambridge, Mass, pp 57–74

Beard K (1990) Gliding behaviour and palaeoecology of the alleged primate family Paromyidae (Mammalia, Dermoptera). Nature (London) 345:340–341

Brown PL, Brown TW, Grinnell AD (1983) Echolocation, development, and vocal communication in the lesser bulldog bat, *Noctilio albiventris*. Behav Ecol Sociobiol 13:287–298

Buchler ER (1980) The development of flight, foraging and echolocation in the little brown bat (*Myotis lucifugus*). Behav Ecol Sociobiol 6:211–218

Dial KP, Kaplan SR, Goslow GE, Jenkins FA (1988) A functional analysis of the primary upstroke and downstroke muscles in the domestic pigeon (*Columba livia*). J Exp Biol 134:1–16

Fenton MB (1984) Echolocation: implications for ecology and evolution of bats. Q Rev Biol 59:33–53

Fenton MB (1985) Communication in the Chiroptera. Indiana Univ Press, Bloomington

Fiedler J (1979) Prey catching with and without echolocation in the Indian false vampire bat *Megaderma lyra*. Behav Ecol Sociobiol 6:155–160

Findley JS (1976) The structure of bat communities. Am Nat 110:129–139

Findley JS, Black H (1983) Morphological and dietary structuring of a Zambian insectivorous bat community. Ecology 64:625–630

Goslow GE, Dial KP, Jenkins FA (1989) The avian shoulder: an experimental approach. Am Zool 29:287–301

Goslow GE, Dial KP, Jenkins FA (1990) Bird flight: insights and complications. BioScience 40:108–115

Gould E (1988) Wing-clapping sounds of *Eonycteris spelaea* (Pteropodidae) in Malaysia. J Mammal 69:378–379

Griffin DR (1958) Listening in the dark. Yale Univ Press, New Haven

Griffin DR, Webster FA, Michael CR (1960) The echolocation of flying insects by bats. Anim Behav 8:141–154

Habersetzer J (1986) Vergleichende flügelmorphologische Untersuchungen an einer Fledermausgesellschaft in Madurai. In: Nachtigall W (ed) Biona Report 5, Bat flight – Fledermausflug. Fischer, Stuttgart, pp 75–104

Habersetzer J, Marimuthu G (1986) Ontogeny of sounds in the echolocating bat *Hipposideros speoris*. J Comp Physiol A 158:247–257

Habersetzer J, Storch G (1987) Klassifikation und funktionelle Flügelmorphologie paläogener Fledermäuse (Mammalia, Chiroptera). Cour Forsch Inst Senckenb 91:117–150

Heblich K (1986) Flügelschlag und Lautaussendung bei fliegenden und landenden Fledermäusen. In: Nachtigall W (ed) Biona Report 5, Bat flight – Fledermausflug. Fischer, Stuttgart, pp 139–156

Heller K-G, Helversen O von (1989) Resource partitioning of sonar frequency bands in rhinolophid bats. Oecologia 80:178–186

Herbert H (1986) Korrelation zwischen Flügelschlag und Ortungslautaussendung bei fliegenden und landenden Flughunden *Rousettus aegyptiacus*. In: Nachtigall W (ed) Biona Report 5, Bat flight – Fledermausflug. Fischer, Stuttgart, pp 157–168

Hermanson JW, Altenbach JS (1981) Functional anatomy of the primary downstroke muscles in the pallid bat, *Antrozous pallidus*. J Mammal 62:795–800

Hermanson JW, Altenbach JS (1983) The functional anatomy of the shoulder of the pallid bat, *Antrazous pallidus*. J Mammal 64:62–75
Hermanson JW, Altenbach JS (1985) Functional anatomy of the shoulder and arm of the fruit-eating bat *Artibeus jamaicensis*. J Zool London Ser A 205:157–177
Jenkins FA, Dial KP, Goslow GE (1988) A cineradiographic analysis of bird flight: the wishbone in starlings is a spring. Science 241:1495–1498
Jepsen GL (1966) Early Eocene bat from Wyoming. Science 154:1333–1339
Jepsen GL (1970) Bat origins and evolution. In: Wimsatt WA (ed) Biology of bats, vol 1. Academic Press, New York London, pp 1–64
Jones G, Rayner JMV (1988) Flight performance, foraging tactics and echolocation in free-living Daubenton's bats *Myotis daubentoni* (Chiroptera: Vespertilionidae). J Zool, London 215:113–132
Jones G, Rayner JMV (1989) Echolocation and foraging behavior of wild horseshoe bats *Rhinolophus ferrumequinum* and *R. hipposideros* (Chiroptera, Rhinolophidae). Behav Ecol Sociobiol 25:183–191
Kalko EKV, Schnitzler H-U (1989) The echolocation and hunting behavior of Daubenton's bat, *Myotis daubentoni*. Behav Ecol Sociobiol 24:225–238
Konstantinov AI, Makarov AK (1987) [Formation of echolocation in ontogenesis of the bat *Rhinolophus ferrumequinum*.] Zh Evol Biokhim Fiziol 23:98–109
Kovtun MF (1985) The evolutionary morphology of locomotion organs system in bats (Mammalia: Chiroptera). In: Mlikovsky J, Novak VJA (eds) Evolution and morphology. Acadecmica, Praha, pp 589–596
Leche W (1886) Über die Säugethiergattung *Goeleopithecus*. Kgl Svensk Vet-Akad Handl 21(11):1–92
McKenzie NL, Rolfe JK (1986) Structure of bat guilds in the Kimberley mangroves, Australia. J Anim Ecol 55:401–420
Miller GS (1907) The families and genera of bats. Bull US Nat Mus 57
Neuweiler G (1989) Foraging ecology and audition in echolocating bats. TREE 4:160–166
Neuweiler G, Metzner W, Heilmann U, Rübsamen R, Eckrich M, Costa HH (1987) Foraging behaviour and echolocation in the rufous horseshoe bat (*Rhinolophus rouxi*) of Sri Lanka. Behav Ecol Sociobiol 20:53–67
Norberg UM (1986) Evolutionary convergence in foraging niche and flight morphology in insectivorous aerial-hawking birds and bats. Ornis Scand 17:253–260
Norberg UM (1990a) Vertebrate flight: mechanics, physiology, morphology, ecology and evolution. Springer, Berlin Heidelberg New York Tokyo
Norberg UM (1990b) Ecological determinants of bat wing shape and echolocation call structure with implications for some fossil bats. In: Hanak V, Horacek I, Gaisler J (eds) European bat research 1987. Proc 4th Eur Bat Res Symp, Praha 1987. Charles Univ Press, Praha (in press)
Norberg UM, Rayner JMV (1987) Ecological morphology and flight in bats (Mammalia, Chiroptera): wing adaptations, flight performance, foraging strategies and echolocation. Philos Trans R Soc London Ser B 316:335–427
Novacek MJ (1985) Evidence for echolocation in the oldest known bats. Nature (London) 315:140–141
O'Neill MGO, Taylor RJ (1986) Observations on the flight patterns and foraging behaviour of Tasmanian bats. Aust Wildl Res 13:427–432
Padian K (1982) Macroevolution and the origin of major adaptations: vertebrate flight as a paradigm for the analysis of patterns. Proc 3rd N Am Paleontol Convention, vol 2, pp 381–392
Padian K (1987) A comparative phylogenetic and functional approach to the origin of vertebrate flight. In: Fenton MB, Racey PA, Rayner JMV (eds) Recent advances in the study of bats. CambridgeUniv Press, Cambridge, Mass, pp 3–19
Pettigrew JD (1986) Flying primates? Megabats have the advanced pathway from eye to midbrain. Science 231:1304–1306
Pettigrew JD, Jamieson BGM, Robson SK, Hall LS, McNally KI, Cooper HM (1989) Phylogenetic relations between microbats, megabats and primates (Mammalia: Chiroptera and Primates). Philos Trans R Soc London Ser B 325:489–559
Pollak GD, Casseday JH (1989) The neural basis of echolocation in bats. Springer, Berlin Heidelberg New York Tokyo

Rayner JMV (1986) Vertebrate flapping flight mechanics and aerodynamics, and the evolution of flight in bats. In: Nachtigall W (ed) Biona Report 5, Bat flight – Fledermausflug. Fischer, Stuttgart, pp 27–74
Rayner JMV (1987) The mechanics of flapping flight in bats. In: Fenton MB, Racey PA, Rayner JMV (eds) Recent advances in the study of bats. Cambridge Univ Press, Cambridge, Mass, pp 23–42
Rayner JMV (1988a) Form and function in avian flight. Curr Ornithol 5:1–77
Rayner JMV (1988b) The evolution of vertebrate flight. Biol J Linn Soc 34:269–287
Rayner JMV (1989a) Mechanics and physiology of flight in fossil and recent vertebrates. Trans R Soc Edinburgh Earth Sci 80:311–320
Rayner JMV (1989b) Vertebrate flight and the origins of flying vertebrates. In: Allen KC, Briggs DEG (eds) Palaeoecology and palaeoenvironments. Belhaven, London, pp 188–217
Richter G, Storch G (1980) Beiträge zur Ernährungsbiologie eozäner Fledermäuse aus der "Grube Messel". Nat Mus 110:353–367
Rose KD, Simons EL (1977) Dental function in the Plagiomenidae: origin and relationships of the mammalian order Dermoptera. Contrib Mus Paleontol Univ Mich 24:221–236
Rübsamen R (1987) Ontogenesis of the echolocation system in the rufous horseshoe bat, *Rhinolophus rouxi* (Audition and vocalization in early postnatal development). J Comp Physiol A 161:899–913
Schnitzler H-U (1971) Fledermäuse im Windkanal. Z Vergl Physiol 73:209–221
Schnitzler H-U, Henson OW (1980) Performance of airborne animal sonar systems, I Microchiroptera. In: Busnel R-G, Fish JF (eds) Animal sonar systems. Plenum New York, pp 109–181
Schnitzler H-U, Hackbarth H, Heilmann U, Herbert H (1985) Echolocation behavior of rufous horseshoe bats hunting for insects in the flycatcher-style. J Comp Physiol A 157:39–46
Schnitzler H-U, Kalko E, Miller L, Surlykke A (1987) The echolocation and hunting behavior of the bat, *Pipistrellus kuhli.* J Comp Physiol A 161:267–274
Scholey KD (1986) The evolution of flight in bats. In: Nachtigall W (ed) Biona Report 5, Bat flight – Fledermausflug. Fischer, Stuttgart, pp 1–12
Simmons JA (1987) Acoustic images of target range in the sonar of bats. Naval Res Rev 39:11–26
Simmons JA, Kick SA (1983) Interception of flying insects by bats. In: Huber F, Markl D (eds) Neuroethology and behavioural physiology. Springer, Berlin Heidelberg New York Tokyo, pp 267–279
Simmons JA, Stein RA (1980) Acoustic imaging in bat sonar: echolocation signals and the evolution of echolocation. J Comp Physiol 135 A:61–84
Simmons JA, Fenton MB, O'Farrell MJ (1979) Echolocation and pursuit of prey by bats. Science 203:16–20
Smith JD (1977) Comments on flight and the evolution of bats. In: Hecht MK, Goody PC, Hecht BM (eds) Major problems in vertebrate evolution. Plenum, New York, pp 427–437
Smith JD (1980) Chiropteran phylogenetics: introduction. In: Wilson DE, Gardner AL (eds) Proc 5th Int Bat Res Conf. Texas Tech Press, Lubbock, pp 233–244
Speakman JR, Anderson ME, Racey PA (1989) The energy cast of echolocation in pipistrelle bats (*Pipistrellus pipistrellus*). J Comp Physiol A 165:679–685
Thomas ALR, Jones G, Rayner JMV, Hughes PM (1990) Intermittent gliding flight in pipistrelle bats (*Pipistrellus pipistrellus*) (Chiroptera: Vespertilionidae). J Exp Biol 149:407–416
Van Valen L (1979) The evolution of bats. Evol Theor 4:103–121
Webster F, Griffin D (1962) The role of the flight membrane in insect capture by bats. Anim Behav 10:322–340
Weid R, Helversen O von (1987) Ortungsrufe europäischer Fledermäuse beim Jagdflug im Freiland. Myotis 25:5–27
Wible JR, Novacek MJ (1988) Cranial evidence for the monophyletic origin of bats. Am Mus Nov 2911
Winge H (1941) The interrelationships of the mammalian genera, vol 1: Monotremata, Marsupialia, Insectivora, Chiroptera, Edentata. Reitzles, Copenhagen

Biomechanical Aspects of the Wing Joints in Flies, Especially in *Calliphora erythrocephala*

A. WISSER[1] and W. NACHTIGALL[2]

Summary

In flies, the wings are driven indirectly by the dorso-longitudinal and dorso-ventral muscles. These indirect muscles transfer their power to the thorax capsule, which is capable of oscillating. Two clefts in the pleural wall of each side allow this capsule, a light weight construction with stiff ridges and thin plates of cuticle within, to oscillate. These oscillations are transduced by the scutellum and its two posterior tergal levers to the wing joints. Over a turning axis, running through two turning points on the underside of each wing joint, the wings move down- and upwards. First there is always a downstroke, often induced by a "starter" muscle. The pleural tension muscles contract the pleural wall of the capsule, so that the click mechanism during wing-beat may occur. During wing beat direct steering muscles change additionally the geometrical angle of attack of the wing, generating forces and moments for flight manoeuvers, and they steer the pleural gearbox to use it as a three-step switch – increasing unilaterally the beat amplitudes – or as a three-step stop – decreasing unilaterally the beat amplitudes.

1 Introduction

In flying insects, two principle types of wing drive are found: (1) wing drive and wing steering are controlled by direct flight muscles, or (2) these functions are separated in two functional muscle types. Indirect dorso-longitudinal (dlm) and dorso-ventral muscles (dvm) in the thorax capsule drive the wing-beat and direct muscles, ending on the elements of the wing joint, steer the fine movements of the wing joint during flight. Several authors have done investigations on the morphology and physiology of the thorax and the flight muscles of flies. We want to show the interactions of wing drive and -steering corresponding to functional models which simulate the movements of the thorax capsule and the wing base elements during flight very closely.

[1] Fachrichtung Zoologie der Universität des Saarlandes, Bau 6, D-6600 Saarbrücken, FRG
[2] Zoologisches Institut der Universität Saarbrücken, Universität Bau 6, D-6600 Saarbrücken, FRG

N. Schmidt-Kittler and K. Vogel (Eds):
Constructional Morphology and Evolution

2 Results

2.1 Wing Drive: the Thorax and the Indirect Flight Muscles

During flight (Fig. 1 a) the dorso-longitudinal muscles (Fig. 1 b, dlm) contract the anterior part of the thorax capsule, the meso-praescutum, and its posterior part, the meso-phragma. By this contraction the notopleural cleft (npc) is opened (Figs. 1 a, 2 a) and the pleural cleft (plc) closed (Figs. 1 a, 2 b) and, as indicated by the inserts, the wing moves downwards. The dorso-ventral muscles (Fig. 1 c, dvm) contract antagonistically the dorsal and ventral parts of the capsule, the scutum and the ventral episternum and epimerum. Now by this contraction the notopleural cleft (npc) is closed (Fig. 2 a) and the pleural cleft (plc) is opened (Fig. 2 b)

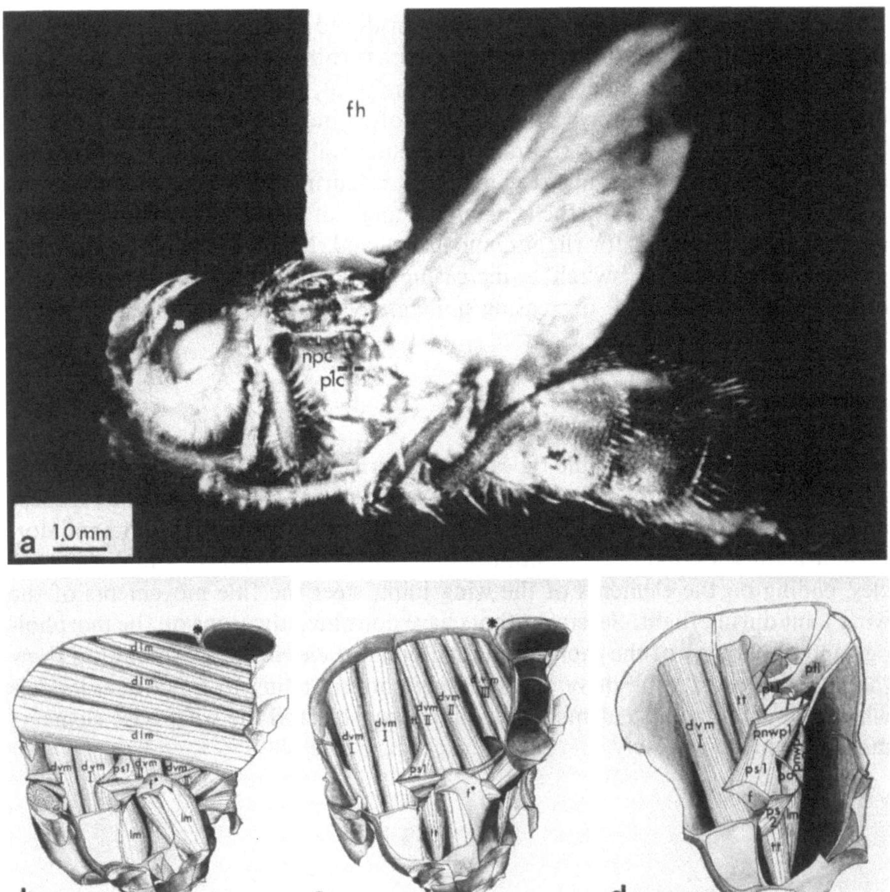

Fig. 1 a – d. Morphology of the wing drive. **a** Fixed flying fly (in a wind tunnel), notopleural and pleural cleft (*npc, plc*). **b** Dorso-longitudinal muscles (*dlm*). **c** Dorso-ventral muscles (*dvm*). **d** Tergo-trochanter muscle (*tt*). **b – d** 3-D drawings; *asterisk* scuto-scutellar suture; for further abbreviations see text

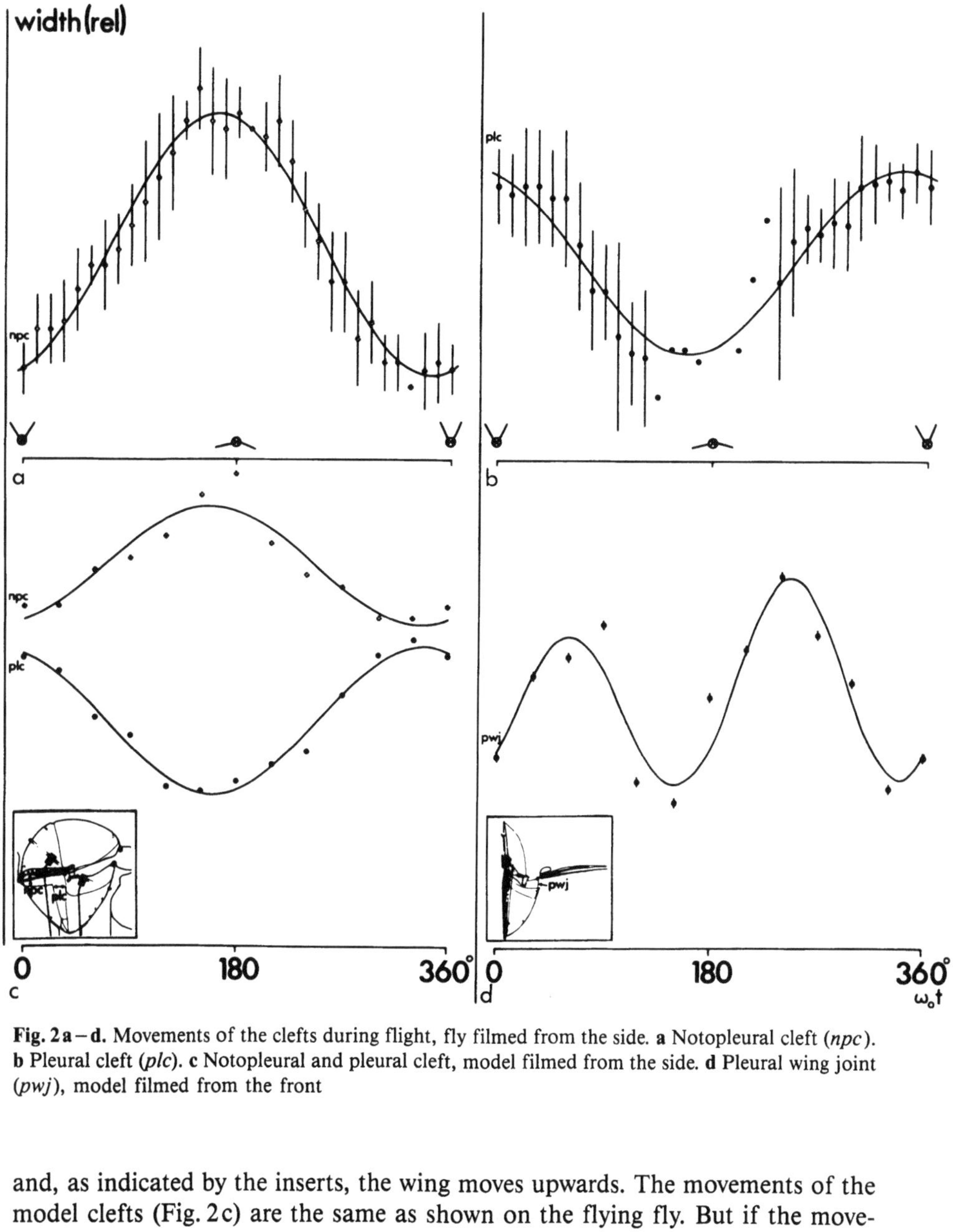

Fig. 2a–d. Movements of the clefts during flight, fly filmed from the side. **a** Notopleural cleft (*npc*). **b** Pleural cleft (*plc*). **c** Notopleural and pleural cleft, model filmed from the side. **d** Pleural wing joint (*pwj*), model filmed from the front

and, as indicated by the inserts, the wing moves upwards. The movements of the model clefts (Fig. 2c) are the same as shown on the flying fly. But if the movements of the pleural wing joint (pwj) of the model, filmed from the front, are analyzed, the plot shows two maxima during one wing beat cycle, corresponding to the click mechanism described by Boettiger and Furshpan (1952).

In flies, the right and the left wing are coupled mechanically by the posterior part of the scutum, the scutellum. On each side of the scutellum the posterior tergal lever extends towards the wing joint. If the dorso-longitudinal muscles (dlm) are active the rear edge of the scutellum moves downwards, turning around the scuto-scutellar suture (Fig. 1 b, c, asterisk), and the distal tip of each tergal lever pushes the pterale I (pt I), inserting between the anterior and posterior tergal lever (atl, ptl), upwards, driving each wing over the pleural wing joint (pwj) downwards. If the dorso-ventral muscles (dvm) are active, the movements are reversed and the wings beat synchronously upwards. The start of a flight is always accompanied with a downstroke, mainly induced by the tergo-trochanter muscle (Fig. 1 d, tt), the "starter" muscle (Nachtigall and Wilson 1967), which is a component of the start jump (Nachtigall 1978). The tension muscles, the pleuro-sternal muscles (Fig. 1 d, ps 1,2), and the tergo-pleural muscle (Fig. 3 c, tp) are used to change the tension of the thorax capsule.

2.2 *Wing Steering: the Wing, its Joint and the Direct Flight Muscles*

Regulation of the Geometrical Angle of Attack. In *Calliphora*, the leading edge of its wing is coupled to the anterior upper branch of pterale I (pt I ant), the main wing drive sclerite, by the head of costal (hco) and radial vein (hr, Figs. 3 a, b and 4 a, c). The wing spread is divided functionally into the anterior and posterior part of the wing (Fig. 3, aw, pw). Between them a bending zone (Fig. 3 a, b, bz) runs from the dorsal wing joint – between the rear edge of the head of the radial vein (hr) and pterale II (pt II, Fig. 3 b) – through the anterior cross vein (Fig. 4 b, ac) to the distal tip of the wing. This vein is formed like the flexible tube of a vacuum cleaner, to be bent easily.

When the dorso-longitudinal muscles during downstroke dominate (Fig. 3 b, A), the anterior part of the wing (aw) pronates along this bending zone (bz) at the upper turning point of the wing beat. In the main phase of the downstroke the anterior upper branch of pterale I (pt I ant) moves the backward bent end of the head of costal and radial vein upwards (hco, hr). Now the anterior ventral support of the wing, the pleural "klöppel" (k) and its radial fit (Figs. 3 b, 6 c), works and the anterior part of the wing (aw) still pronates (Fig. 3 a, b, *1*). In this phase of the wing beat the "klöppel" works also as a wing synchronous receptor for the flight muscles of flies (Miyan and Ewing 1984). At the lower turning point of the wing beat the anterior part of the wing (aw) supinates (Fig. 3 b, *2*) when the dorso-ventral muscles dominate (Fig. 3 b, B). During the upstroke phase the joint remains in contact with this anterior support. Now the attacking aerodynamic forces supinate this part further until the morphological upside gets the physiological downside (Fig. 3 c, *2*). This is called the "*basic*" automatic regulation of the geometrical angle of attack.

Fig. 3 a – c. Morphology of the regulation of the geometrical angle of attack. **a** SEM micrograph of the left wing (upside), bending zone (*bz*), two turning axes of pterale III (*ha*, *va*), (*x*) contact region: semi-circular part (*sp*), underside of the distal plate of pterale III. **b** Scheme showing the two types of regulation: *1*, *2* "basic" automatic regulation; *1* pronation and *2* supination of the anterior part of the wing (*aw*); *3 – 5* "*wing-drive-independent*" additional regulation; *3 + 4* pronated inclination of the posterior part of the wing (*pw*); *5* supinated inclination of this part. *c* Scheme of "*basic*" automatic (*1*, *2*) and "*wing-drive-independent*" additional regulation (*3 + 4/5*) during downstroke and during upstroke; *0* starting point

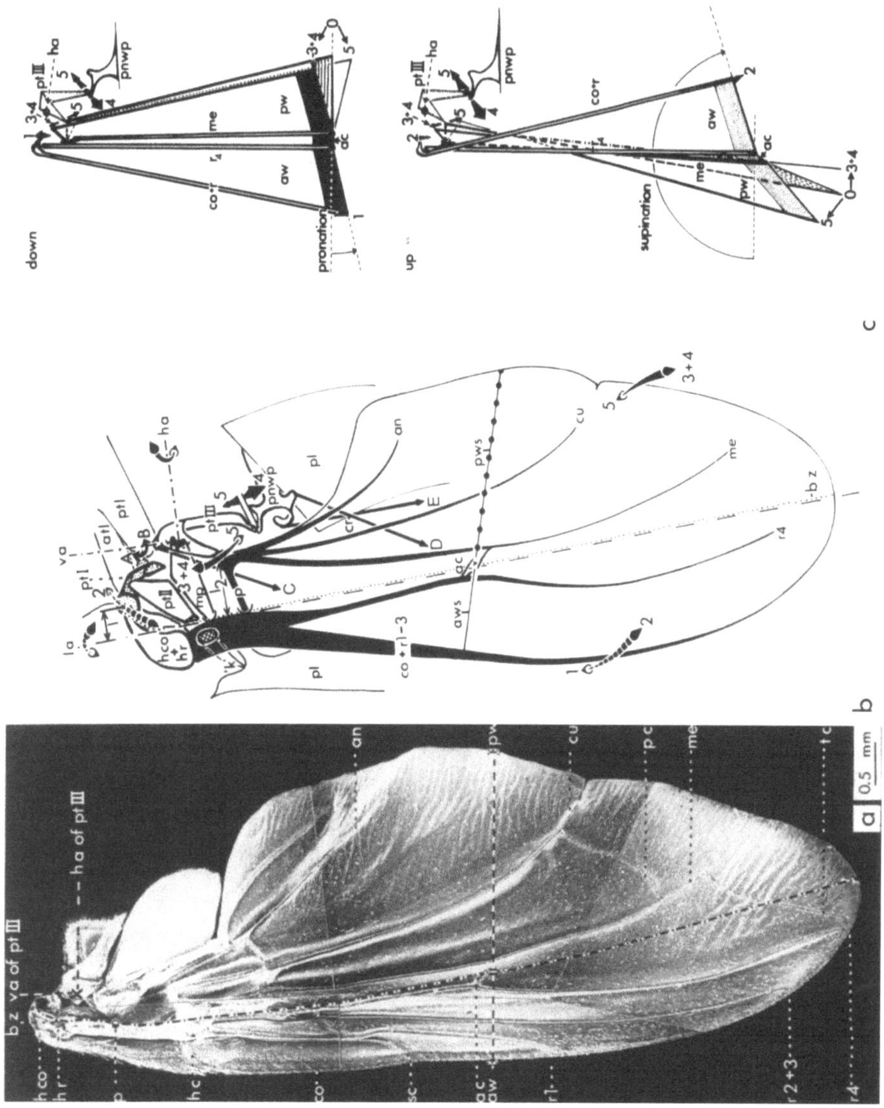

a
0.5 mm
b
c
down
up
pronation
supination

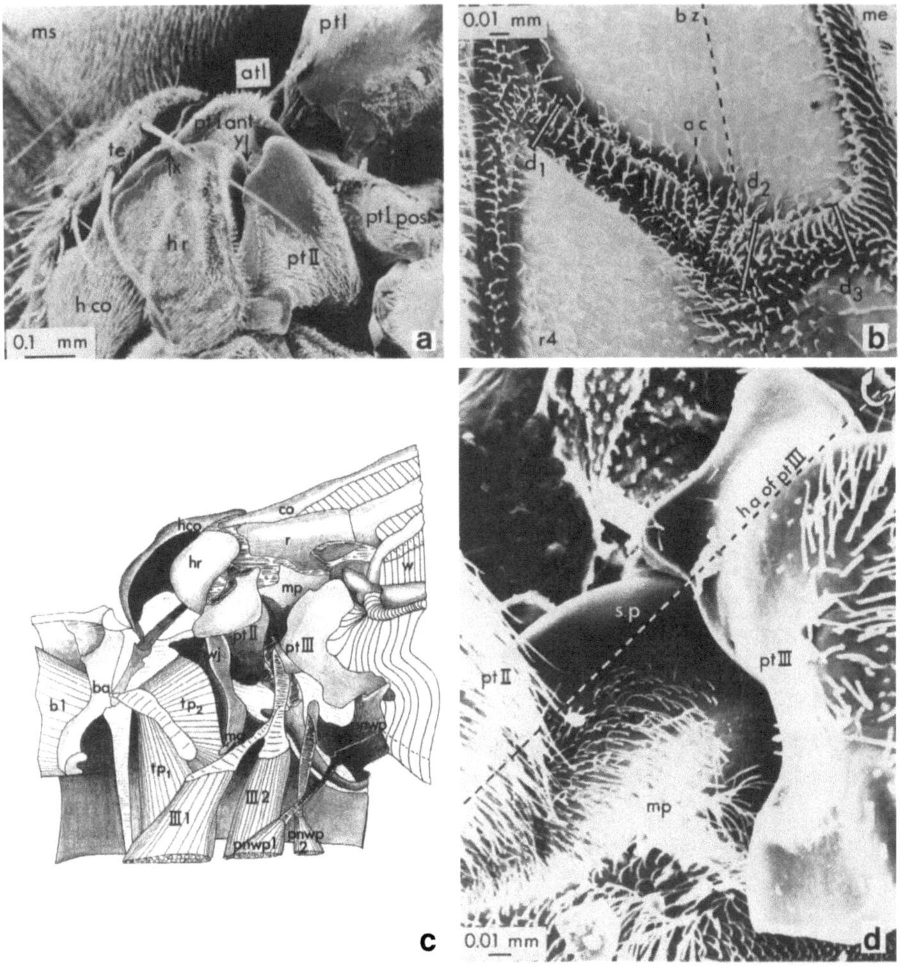

Fig. 4a–d. Morphological details for using the two types of regulating the geometrical angle. **a, b, d** SEM micrographs, *c* 3-D drawing. **a** Dorsal left wing joint, contact regions ($x \rightarrow / y \leftarrow$). **b** Underside of the anterior cross vein (*ac*), flattening of the middle part ($d_1 < d_2 > d_3$). **c** The pterales (scutum and pt I removed), the muscles 1 and 2 of pterale III (*III 1–2*), the muscles 1 and 2 of the posterior notal wing process (*pnwp 1–2*). **d** Semi-circular part (*sp*) and its contact region on the triangular plate of pterale III (*pt III*)

During flight a "*wing-drive-independent*" additional regulation of the geometrical angle of attack exists for the posterior part of the wing (Fig. 3b, c, pw). On the rear edge of the wing joint we find two axillares, the pterale III (pt III) and the posterior notal wing process (pnwp), which inserts longitudinally on the posterior pleural wall. The basal tip of the elongated pterale III is connected to the distal bent end of the posterior notal wing process (Fig. 4c). The anterior basis of the distal triangular plate of pterale III contacts the semi-circular part (sp) of the middle plate (mp, Fig. 4d), thus forming a horizontal and a vertical axis (ha, va) for pterale III movements. On the pterale III four direct muscles end, III 1 together with a common tendon of the muscles III 2–4. On the posterior notal wing process also four direct muscles (pnwp 1–4) insert. The tendon of its first muscle crosses the pathway of the common tendon of its other muscles (Fig. 4c, cr).

The pull of the muscles III 2–4 move their insertion point around the horizontal axis (Fig. 3b, C, ha) inwards. At the same time the first muscle of the posterior notal wing process (pnwp 1) moves the distal bent end of this process around its pleural hinge joint (Fig. 3b, D) outwards. Now the distal tip of the triangular plate of pterale III and the region between the median (me) and anal veins (an), the posterior part of the wing (pw), move upwards (Fig. 3b, *3+4*). During the main phase of the downstroke this part changes to a pronated inclination (Fig. 3c, *3+4*) and the camber of the wing spread decreases. During the main phase of the upstroke the camber of the wing spread increases by decreasing the supinated inclination of the posterior part of the wing (Fig. 3c, *3+4*), if *0* (Fig. 3c) marks the starting point.

The muscles pnwp 2–4 (Fig. 3b, E) move the distal bent end of the posterior notal wing process (pnwp) around its hinge joint and the basis of pterale III inwards. Now the tip of the triangular plate of the pterale III and the posterior part of the wing (pw) move over the horizontal axis (ha) of pterale III downwards (Fig. 3b, *5*). During the main phase of the downstroke the camber of the wing spread increases when the posterior part changes to a supinated inclination (Fig. 3c, *5*). During the main phase of the upstroke the posterior part of the wing (pw) changes to an increased supinated inclination (Fig. 3c, *5*) and the camber of the wing spread decreases.

But here we must indicate the attacking aerodynamic forces which delay and increase or decrease during each stroke phase (Nachtigall 1966) the effects of the "*basic*" and "*wing-drive-independent*" mechanism at the wing joint.

The Pleural Gearbox. The complicated structure of the pleural wing joint (pwj) can only be seen when the insertion region (ir) of the first subunit of the tergo-pleural muscle (tp_1) and the whole ventro-lateral membrane of the joint are removed (Fig. 5a). The two insertion regions of the pterale II (pt II) – on the lateral wing joint its anterior peak articulates in the semi-circular groove of the pleural wing process (pwp) and on the dorsal joint the tip of the dorsal pterale II plate lies in an inlet between the anterior and posterior upper branch of the pterale I (pt I ant+post) – form a hinge-like vertical axis (Fig. 5a) over which the whole wing is turned forwards or backwards. The first subunit of the tergo-pleural muscle inserts on a kidney-shaped vault just in front of the pleural wing process (pwp) and its second unit runs as a small muscle band along the boundary of the pleural apophysis (pa) on the pleural wall behind the insertion of the first subunit. Their resilin-containing tendon (Weis-Fogh 1960) ends on a small sclerotized ledge on the meeting point of the notopleural and transversal ridge (Fig. 5b, asterisk). The shape of the pleural wing joint (pwj) is bent backward and outward, its process consists of a double-walled plate with three upper peaks (p_{1-3}), which are situated morphologically in different positions (Fig. 5c). A cross-section (Fig. 5c, A) through the stem of the pleural wing joint showes its V-configuration (Fig. 5d).

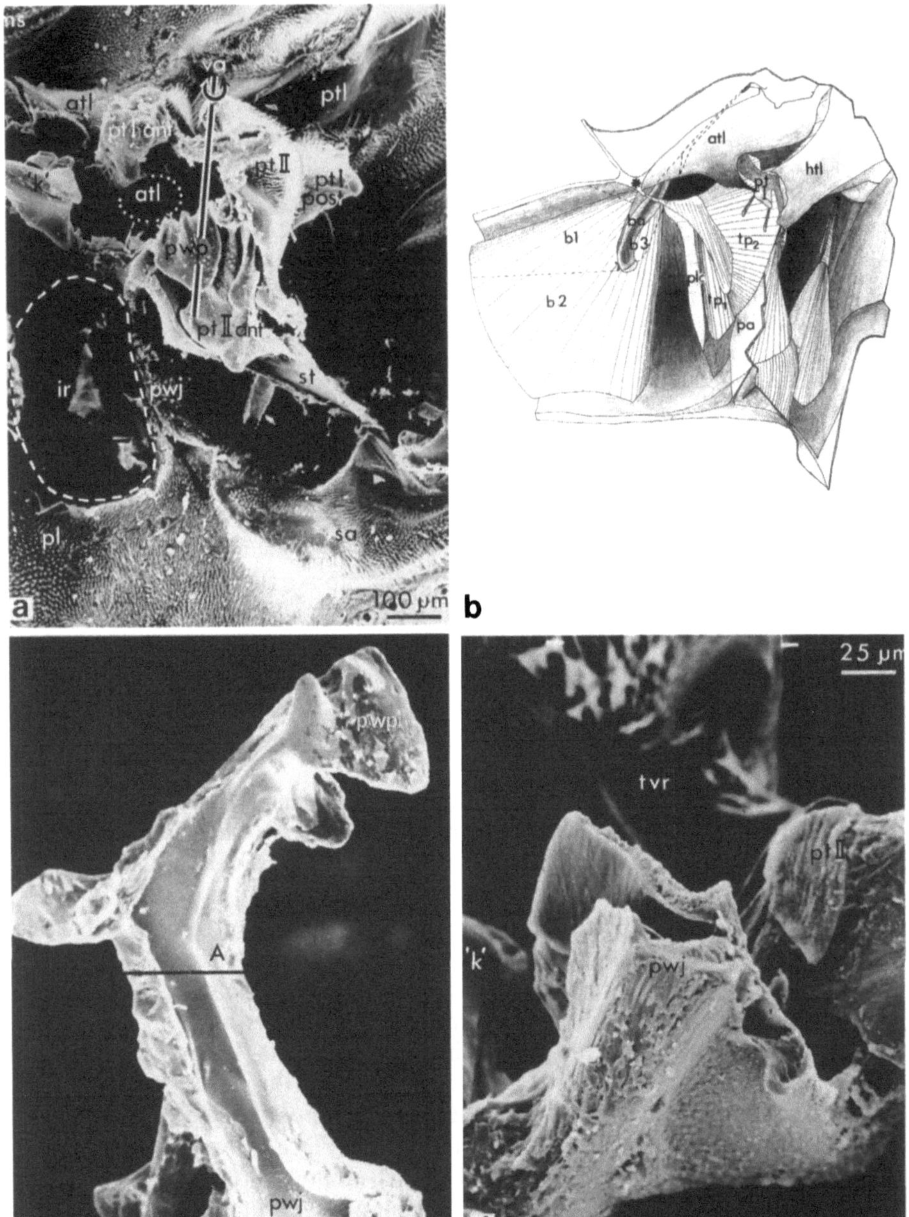

Fig. 5a–d. Structures for using the gearbox. **a, c, d** SEM micrographs, **b** 3-D drawing. **a** Pleural wing joint (*pwj*), pterale II (*pt II*), its vertical axis (*va*), insertion region (*ir*) of the first subunit of the tergo-pleural muscle (tp_1), the whole membrane of the ventral wing base and the distal parts of the wing joint are removed (lateral view). **b** Second basalar muscle (*b2, b1* and *3* are removed), two functional subunits of the tergo-pleural muscle (tp_{1-2}). **c** Pleural wing joint (*pwj*, anterior view). **d** Cross-section through *A*

The tooth of the ventral radial vein (Fig. 6, tvr) inserts distally on the vein's enlarged base. A small groove is to be found on the ventro-lateral side of its tip. When flies were killed with carbon tetrachloride (CCL_4) fumes and the wing was brought into flight position, the groove fitted the anterior (Fig. 6a), middle (Fig. 6b) or posterior peak (Fig. 6c, p_{1-3}) of the pleural wing process (pwp).

The movements of the gearbox elements of the pleural wing joint (pwj) and the wing base are shown in Fig. 7a. The pull of the first subunit of the tergo-pleural muscle (tp_1) pulls the V-shaped stem of the pleural wing joint like a torsion rod outwards (x). Owing to its morphological configuration – the stem is bent backward and outward – the peaks of the pleural wing process (pwp) move into the positions 1′, 2′, 3′. The outward torsion is counteracted by the elasticity of this torsion rod (y) and the pull (z) of the subalar (sa) tendon (st). The tooth of the ventral radial vein (tvr) on the front edge of the wing is turned forwards by the contractions of the first and second basalar muscle (b 1+2) and backwards by the antagonistic contraction of the first pterale III muscle (III 1), supported by the contraction of the third basalar muscle (b 3). The positions *A*, *B*, or *C* are adjusted by the relative strength of the contractions of these muscles. So the tooth of the ventral radial vein (tvr) contacts with its groove peak 1, 2, or 3.

In flight before a wind tunnel the flies normally do not use the gearbox (Fig. 7b). The wing base moves down (1–7) and up (7–1) over its turning axis – in the diagram only the posterior turning point is shown. If the dorso-longitudinal muscles are active the posterior tergal lever (ptl), the pterale I (pt I), and the anterior lever (atl) are moved upwards by the power of these muscles (dotted area), the elastic pleural wall is pushed against the inward pull of the pleuro-sternal muscles (ps) and the tergo-pleural muscle (tp, not shown) outwards (white area) storing potential energy in this spring system until the joint clicks through its bistable position (0 = 4) upwards. After the click, on releasing this energy, the pleural wall turns along the pull of these tension muscles inwards to its starting position. During upstroke, when the dorso-ventral muscles are active, the movements are reversed and the joint clicks downwards.

Sometimes the gearbox is used for several wing beat cycles. There exist two possibilities how to use it:

1. Pfau (1973) discussed a gear-change mechanism. The morphological elements of the wing joint and interruptions in the records during the downstroke phase point to such a mechanism, but this must be modified. In Fig. 7c only one of three possible meetings of the tooth of the ventral radial vein (tvr) with one of the three peaks (p_{1-3}) of the pleural wing process (pwp) is shown during downstroke. The pull of the first subunit of the tergo-pleural muscle (tp_1) turns the pleural wing joint (pwj) outwards like a torsion rod (t). By the combined actions of the basalar muscles 1 and 2 (b 1,2) and the pterale III muscle 1 (III 1), supported by the basalar muscle 3 (b 3), the front edge of the wing with this tooth turns into a meeting position. Now the power input of the downstroke muscles (dotted + broad-striped areas) increases, the pleural wall is pushed further outwards (white + heavily striped areas) against the contracted tension muscles storing more potential energy in this spring system. When the contact between the tooth and a peak is lost in the second downstroke phase, the click occurs. Releasing this energy, the pleural wall turns further inwards (finely lined area) and the

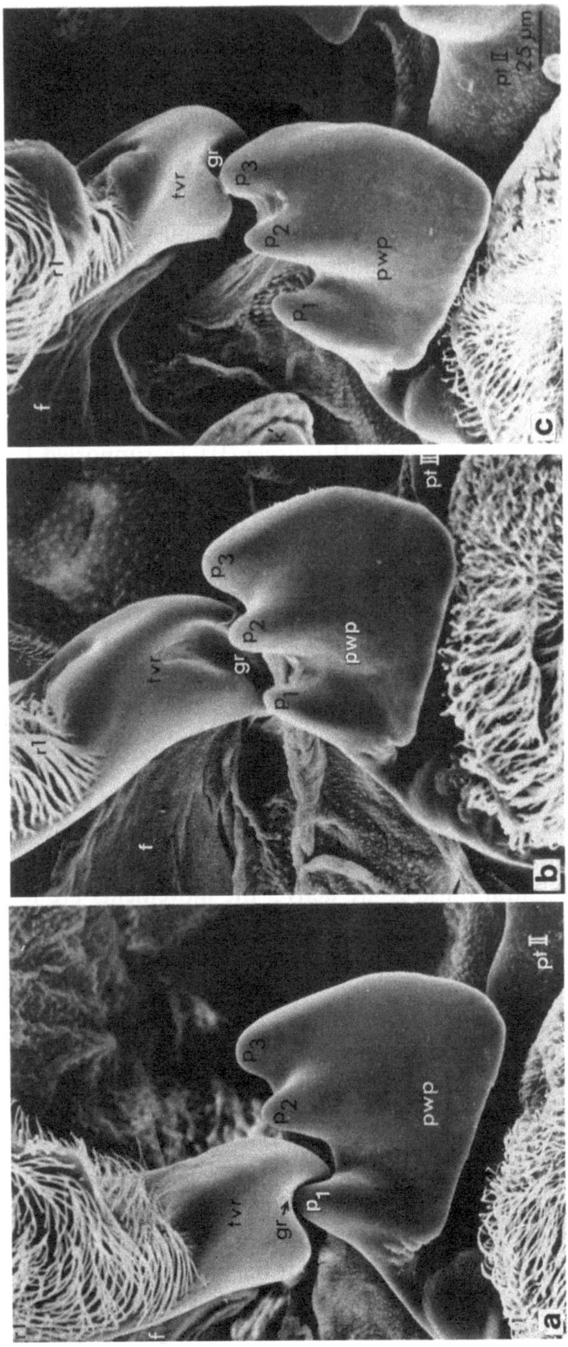

Fig. 6a–c. SEM micrographs of the first (**a**), second (**b**), and third (**c**) gear position (p_{1-3})

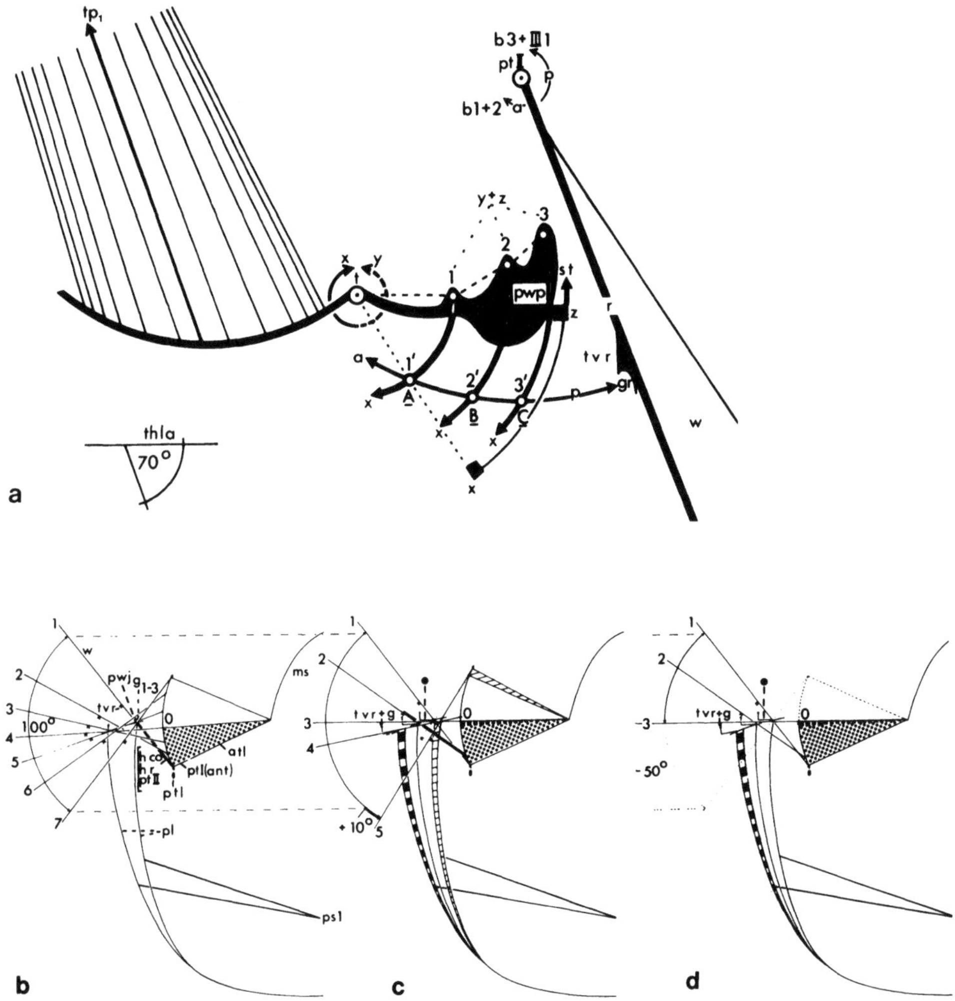

Fig. 7a–d. Functional diagrams of the gearbox. **a** Steering of the pleural gearbox. **b** In normal flight the tooth of the ventral radial vein (*tvr*) does not contact one of the peaks, clicks occur up- or downwards. **c** During downstroke the tooth contacts one of the peaks; after loss of contact the beat amplitude increases; during upstroke (**b, c**) there is never a contact, clicks occur. **d** During downstroke the tooth stops on a peak, the beat amplitude decreases, no click occurs

wing beats down to a greater amplitude. The increasing of the amplitude depends on the gear position, because the peaks stand in different morphological positions on the pleural wing process. The upstroke occurs as in normal flight.

2. Prevailing records show wing beat stops (Miyan and Ewing 1985a, b; Wisser 1988; Nalbach 1989). For several wing-beat cycles the tooth remains during each downstroke on one of the peaks of the pleural wing process (Fig. 7d). Now no click mechanism arises, the stored potential energy may be used for accelerating the upstroke movement of the front wing edge via the tooth of the ventral radial

vein. The downstroke amplitude is decreased, depending on the stop position of the tooth.

These two possibilities can be used synchronously on both sides or unilaterally but independent of the wing drive. Besides an increase (1st possibility) or a decrease (2nd possibility) in the downstroke amplitude, the pronation or supination may also have been altered (Pfau 1973, 1985) because the posterior turning point of the wing beat axis has changed (Wisser 1988). Consequently the fly will turn contralaterally if the first possibility or it will turn ipsilaterally if the second possibility is used unilaterally.

3 Discussion

Boettiger and Furshpan (1951) described a click mechanism in dipteran flight. The wing of flies treated with CCl_4 remain either in an up or down position. Electrical stimulation near the thoracic ganglion of *Sarcophaga* (Boettiger 1951) leads to normal flight movements at stimuli of $15-20\,s^{-1}$. When the frequency of stimulation is increased suddenly, the wing stops either in the down or up position similar to the CCl_4 effects. The model fitted well with the physiological properties of the indirect fibrillar muscles (Pringle 1949). Recently, Miyan and Ewing (1985a, b, 1988) and Ennos (1987) have challenged the model of Boettiger and Furshpan. Miyan and Ewing discussed (1) a wing-beat stop during the whole flight, storing elastic energy in the pleural wing joint and (2) a locking mechanism between the parascutum (= anterior tergal lever) and the pterale I. (1) During flight the wing stroke occurs mainly in a pathway anterior of the pleural wing process (position 0, Wisser 1988 or mode 3, Nalbach 1989). Here potential energy may be stored in the elastic pleural wall and the contracted tension muscles and released during each halfstroke as in a click mechanism (Fig. 2d). (2) There is no necessity for a locking mechanism because the scutum and synchronously the anterior tergal lever (atl) move up over its hinge joint during downstroke. The lateral tip of the anterior tergal lever, connected to the anterior upper branch of the pterale I, moves faster upwards than the lateral margin of the meso-scutum (ms) because the thoracic movements during downstroke are magnified by the posterior tergal lever system, driving the pterale I and, at the same time, over a chitinous bridge the posterior part of the anterior tergal lever upwards (Wisser and Nachtigall 1983). Ennos, like Miyan and Ewing, described a wing-beat stop in his novel model, but he negated their locking mechanism. He disregarded the hinge joint of the anterior tergal lever and his "diagrams also omit the scutellar lever, which surely has a structural role", but this scutellar lever (= posterior scutellar lever) is the main driving element between the thorax vibrating and the wing beating.

In a first approach, Nachtigall (1966) discussed the wing spread as a flat plate during flight. In tests, Vogel (1967) found that cambered wings can produce higher lift than flat ones and often have a higher maximum lift to drag ratio. In locusts the forewings are also able to change their camber during flight. During downstroke a cambered wing produces higher lift at small and medium angles of attack. At high angles it produces less lift than a flat wing. During upstroke lift production of a cambered wing is higher at angles greater than 15° (Nachtigall 1981). Ennos (1988a, b, 1989) gave a good model for cambered wings, where corrugated spars diverge at angles from the front wing spar. He showed that the attacking aerodynamic forces may camber the wing automatically during the main phase of the down- and upstroke. He also said that "there are no muscles in the main wing area of insects to bring about these shape changes, and also in many insects direct muscles could not operate at the high frequencies required". In *Calliphora* there is a modification of his model because we find an anterior and posterior part of the wing (aw, pw) connected by the bending zone (Figs. 3, 4, bz). The muscles of the pterale III 2–4 and of the posterior notal wing process 1–4 change the position of the pterale III influencing over some strokes "*wing-drive independently*" the inclination of the posterior part of the wing (Wisser 1987). This changes the camber of the whole wing during flight. Afferent feedback of the change of camber may be given by campaniform sensillae on the anterior cross vein (Dickinson and Palka 1987). In hover flies a reinforcement, the so-called vena spuria, tends to reinforce the anterior part of the wing and provides linear reinforcement parallel to which the wing must change its inclination (Weis-Fogh 1973). So the anterior and posterior parts are, just as in *Calliphora*, separated functionally.

Pfau (1973, 1977, 1985) described the interactions of an element A (= radial stop = tooth of the ventral radial vein) with an element B (= pleural wing process) in the sense of a gear-change mechanism including two possible gears and used for some wing-beat cycles. "A and B are extremely densely sclerotized, but not padded and can hardly be exposed to constant use" (Pfau 1973). Nachtigall and Roth (1983) found two-peak frequency maxima of the frontal wing-beat angles during steering movements which may correspond to the two-gear positions described by Pfau. The morphological construction of the gearbox and records of fixed flying flies (Wisser 1988) show three possible positions for interaction (Fig. 6). Nalbach (1989) correlated the frontal wing-beat angles with the three positions for interaction. She found a modification of mode 3 which she called mode 3h – an interaction of the radial stop with the "*klöppel*" (k) only. The "*klöppel*" is covered with a strong coating of resilin that has been stained with methylene blue (Weis-Fogh 1960) and is used during the greatest part of the normal flight as anterior turning point of the wing-beat. The later findings prefer an interpretation of this system as a wing-beat stop (Fig. 7d) which can be used only for some wing cycles. But there may be also a gear-change mechanism similar to Pfau's findings which perhaps could not be recorded under the given experimental conditions. The wing-beat stop prevents the joint from clicking, but the movements in normal flight do not render the click mechanism impossible.

Finally, we want to indicate the steady-state or stationary and nonstationary aerodynamic effects onto the wing spread which delay and/or increase/decrease the movements of the steering systems. Weis-Fogh (1973) has found a clap and fling mechanism in the flight of *Encarsia formosa*, a diminutive parasitic wasp, and a flip mechanism in the wing-beat of the Syrphidae, which induce nonstationary aerodynamic effects during the upper or lower turning point. Ellington (1980) presented a vortex theory of hovering animal flight. Nachtigall (1979, 1980) discussed some nonstationary effects during the flight of *Calliphora* like the "*flash-supinating mechanism*" and described the mechanics and aerodynamics, the stationary and nonstationary effects, during insect flight (Nachtigall 1989).

References

Boettiger EG (1951) Stimulation of the flight muscles of the fly. Anat Rec 111:443

Boettiger EG, Furshpan E (1951) Observations on the flight motor of Diptera. Fed Proc 10:17

Boettiger EG, Furshpan E (1952) The mechanics of flight movements in Diptera. Biol Bull Mar Biol Lab Woods Hole 102:200–211

Dickinson MH, Palka J (1987) Physiological properties, time of development and central projection are correlated in the wing mechanoreceptors of *Drosophila*. J Neurosci 7(12):4201–4208

Ellington CP (1980) Vortices and hovering flight. In: Nachtigall W (ed) Instationäre Effekte an schwingenden Tierflügeln. Akad Wiss Lit, Mainz, pp 64–101

Ennos AR (1987) A comparative study of the flight mechanism of Diptera. J Exp Biol 121:355–372

Ennos AR (1988a) The importance of torsion in the design of insect wings. J Exp Biol 140:137–160

Ennos AR (1988b) The inertial cause of wing rotation in Diptera. J Exp Biol 140:161–169

Ennos AR (1989) Inertial and aerodynamic torques on the wings of Diptera in flight. J Exp Biol 142:87–95

Miyan JA, Ewing AW (1984) A wing synchronous receptor for the dipteran flight motor. J Insect Physiol 307:567–574

Miyan JA, Ewing AW (1985a) Is the "*click*" mechanism of dipteran flight an artefact of CCl_4 anaesthesia? J Exp Biol 116:313–322

Miyan JA, Ewing AW (1985b) How Diptera move their wings: a reexamination of the wing base articulation and muscle systems concerned with flight. Philos Trans R Soc London Ser B 311:271–302

Miyan JA, Ewing AW (1988) Further observations on the dipteran flight: details of the mechanism. J Exp Biol 136:229–241

Nachtigall W (1966) Die Kinematik der Schlagflügelbewegungen von Dipteren. Methodische und analytische Grundlagen zur Biophysik des Insektenflügels. Z Vergl Physiol 52:155–211

Nachtigall W (1978) Der Startsprung der Stubenfliege, *Musca domestica* (Diptera: Muscidae). Ent Germ 4:368–373

Nachtigall W (1979) Rasche Richtungsänderungen und Torsionen schwingender Tierflügel und Hypothesen über zugeordnete instationäre Strömungseffekte. J Comp Physiol 133:351–355

Nachtigall W (1980) Rasche Bewegungsänderungen bei der Flügelschwingung von Fliegen und ihre mögliche Bedeutung für instationäre Luftkrafterzeugung. In: Nachtigall W (ed) Instationäre Effekte an schwingenden Tierflügeln. Akad Wiss Lit, Mainz, pp 115–129

Nachtigall W (1981) Der Vorderflügel großer Heuschrecken als Luftkrafterzeuger. I. Modellmessung zur aerodynamischen Wirkung unterschiedlicher Flügelprofile. J Comp Physiol 142:127–134

Nachtigall W (1989) Mechanics and aerodynamics of flight. In: Goldsworthy GJ, Wheeler CH (eds) Insect flight. CRC, Boca Raton, pp 1–29

Nachtigall W, Roth W (1983) Correlations between stationary measurable parameters of wing movement and aerodynamical force production in the blowfly (*Calliphora vicina* R.-D.). J Comp Physiol 150:251–260

Nachtigall W, Wilson DM (1967) Neuro-muscular control of dipteran flight. J Exp Biol 47:77–97

Nalbach G (1989) The gear change mechanism of the blowfly (*Calliphora erythrocephala*) in tethered flight. J Comp Physiol A 165:321–331

Pfau HK (1973) Fliegt unsere Schmeißfliege mit Gangschaltung? Naturwissenschaften 60:160

Pfau HK (1977) Funktionen einiger direkter tonischer Flügelmuskeln von *Calliphora erythrocephala* Meigen. Verh Dt Zool Ges 70:275

Pfau HK (1985) Zur funktionellen und phylogenetischen Bedeutung der "Gangschaltung" der Fliegen. Verh Dt Zool Ges 78:168

Pringle JWS (1949) The excitation and contraction of the flight muscles of insects. J Physiol (London) 108:226–232

Vogel S (1967) Flight in *Drosophila*. III. Aerodynamic characteristics of fly wings and wing models. J Exp Biol 46:431–443

Weis-Fogh T (1960) A rubber-like protein in insect cuticle. J Exp Biol 37:889–907

Weis-Fogh T (1973) Quick estimates of flight fitness in hovering animals, including novel mechanisms for lift production. J Exp Biol 59:169–230
Wisser A (1987) Mechanism of wing-rotating regulation in *Calliphora erythrocephala* (Insecta, Diptera). Zoomorphology 106:261–268
Wisser A (1988) Wing beat of *Calliphora erythrocephala*: turning axis and gearbox of the wing base (Insecta, Diptera). Zoomorphology 107:359–369
Wisser A, Nachtigall W (1983) Funktionelle Gelenkmorphologie und Flügelantrieb bei der Schmeißfliege. In: Nachtigall W (ed) Biona-Report 1. Fischer, Stuttgart New York, pp 29–34
Wisser A, Nachtigall W (1984) Functional-morphological investigations on the flight muscles and their insertion points in the blowfly *Calliphora erythrocephala* (Insecta, Diptera). Zoomorphology 104:188–195

Part III Bioconstruction: Morphogenesis

Ontogenetic Changes Reflected in the Morphology of the Molluscan Shell

KLAUS BANDEL [1]

Summary

The course of ontogeny finds its reflection in the shape and structure of the molluscan shell. The shell becomes mineralized after it has first been secreted. The type of embryonic development is imprinted on the shape of the first mineralized shell and its course can be reconstructed from this shape in modern and fossil cephalopods. The place in the environment causes a reaction in morphology. Gastropod limpets clinging to hard substrates thus have developed in many convergent lines starting out from ancestors that withdrew into their shell. When the later portion of life deviates from that of the ancestors, this may be reflected onto earlier portions of the ontogeny or not. The uncoiled Caecidae may have helically coiled, planspiral, or decoiled larval shells and secondarily coiled caecids may or may not revers this tendency. Larval life in the plankton follows very different pathways of adaptation than adult life in the benthos, as is documented with some mesogastropods. Both adaptations have to be fitted into the same life cycle; but when the life cycle takes a short cut, a major change may be the result, as in the pteropods. It is not easy to differentiate between changes that can be reversed and others that are final. Sometimes pathological cases illustrate a possible way in which phylogeny proceeded, as is demonstrated in an atavistic bivalve embryo.

1 Introduction

The shell of molluscs reflects the course of ontogeny from the first secretion of the shell onward. Shell composition and shape influence the function of the organism during most of its life time, or, with very few exceptions, during parts of their individual ontogeny. Careful analysis of shell production of modern species of cephalopods, bivalves, and gastropods enables us to reconstruct the life history of fossil species.

It is discussed what kind of constraints and benefits shell construction offers to molluscs. The amount of yolk present during the course of early ontogeny is reflected in shell shape, as is exemplified with cephalopods. The mode of life has a strong influence on the shape of the shell and the organization of the body, as is demonstrated with gastropod limpets. During a life cycle, quite different environments may act upon an individual and be reflected in the shape and composi-

[1] Geologisch-Paläontologisches Institut und Museum, Universität Hamburg, Bundesstraße 55, D-2000 Hamburg 13, FRG

N. Schmidt-Kittler and K. Vogel (Eds):
Constructional Morphology and Evolution

tion of its shell, as is the case with higher Mesogastropoda. Here separate evolutionary trends have to be fitted into one and the same ontogeny. When one of these is changed, it may or may not be reflected on the other, as is shown with caecid gastropods. A serious reduction of one or several steps of organ differentiation during an ontogeny have strong impact on the evolution as is shown in the case of the pteropods. In last section, the possibility of reversals of strong morphological impacts on the shell due to the amount of available yolk, as well as the potentials of interpretation of pathological cases, are discussed.

2 Only a Restricted Range and Quality of Materials Are Available for Construction of the Molluscan Shell (Fig. 1)

The shell (skeleton) of molluscs consists of minute crystallites of calcium carbonate intimately associated with organic material. These mineral particles are arranged in structural patterns that are indirectly rather than directly influenced by the cells of the mantle secretory epithelium.

Aragonitic crystallites occur in molluscs in two fundamentally different forms. One form consists of 0.2-μm-wide basic structural units comprising acicular crystallites that in turn form crossed lamellar and a variety of related crossed, helical, and prismatic structures. The other form consists of basic structural units united to form tablets and laminae in nacreous and similar structures. With few exceptions, crossed lamellar and nacreous structures do not grade into one another and they are rarely found together within a single shell. Regarding precursors and derivates of the crossed lamellar structure and those of the nacreous structures, a number of features are observed common to both.

When gastropods, bivalves, and cephalopods first evolved from their common limpet-like ancestors during the Cambrian period, the crossed lamellar and related structures may have evolved as an early alternative to nacreous structure. There is presently no reason to believe that nacreous structure in the Gastropoda and Bivalvia is more primitive than crossed lamellar structure and formed prior to it, as is commonly stated in text books dealing with molluscs.

In terms of their mutual intergradation, aragonitic crossed lamellar, helical, and the scaly structures, are closely related. The mediating structure from crossed lamellar to spherulitic structure is composite prismatic; to helical structure it is crossed acicular; and to scaly structure it is dissected crossed prismatic structure. Presently unknown are direct transitions from crossed lamellar and helical structures to nacreous structure. However, spherulitic prismatic, scaly structure, and dissected crossed prismatic structures may grade directly into nacreous structure. Aragonitic structures deposited without direct mantle influence (opercula, egg shells, darts as copulation aids, pedal deposits) are more simply constructed and lack the basic structural units characteristic of true biominerals (Bandel 1979a, 1990a). It appears that only mantle tissue is capable of forming the crossed lamellar, nacreous, and related structures of the Mollusca. In shell repair and the initial layers of a septum aragonitic crystallites of "inorganic" habit grade into "true" biocrystallites with increasing mantle control on shell secretion (Bandel and Hemleben 1975; Bandel 1977b).

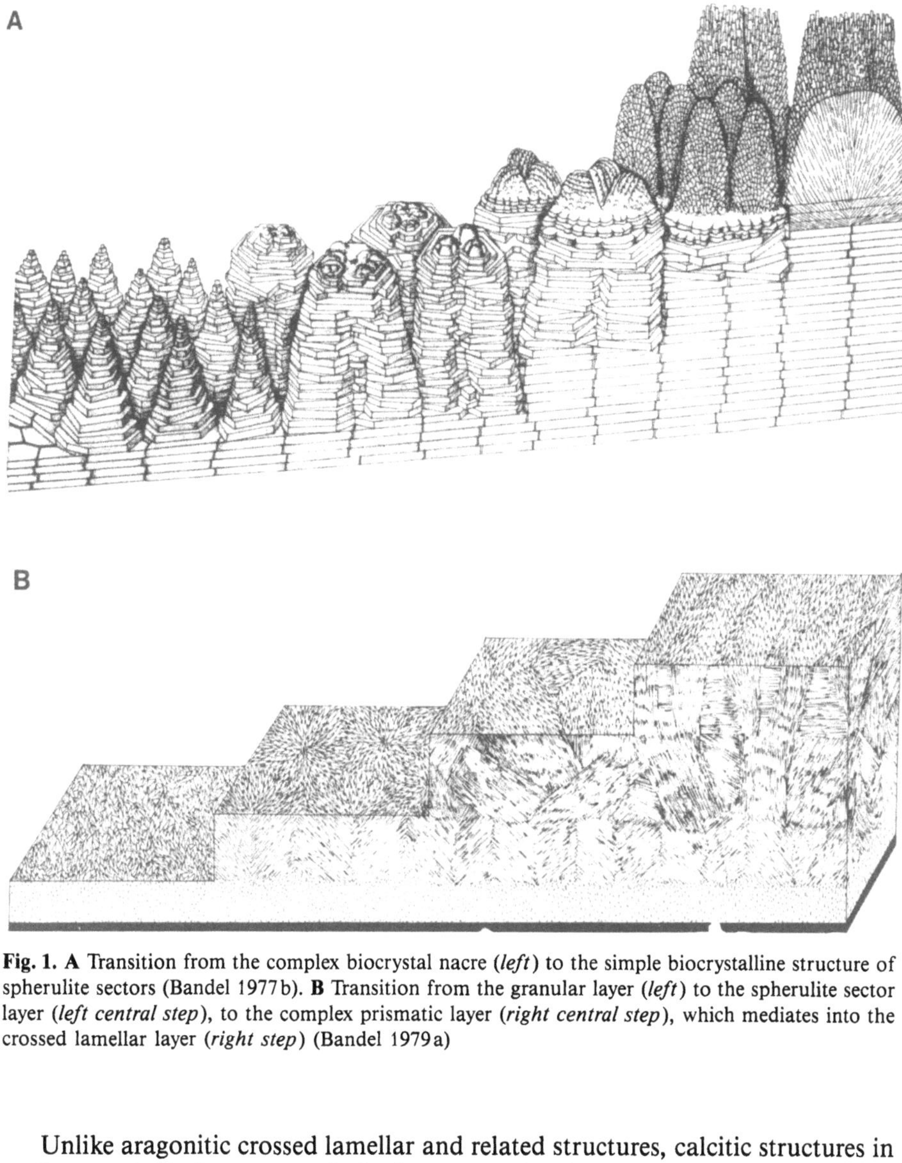

Fig. 1. A Transition from the complex biocrystal nacre (*left*) to the simple biocrystalline structure of spherulite sectors (Bandel 1977b). **B** Transition from the granular layer (*left*) to the spherulite sector layer (*left central step*), to the complex prismatic layer (*right central step*), which mediates into the crossed lamellar layer (*right step*) (Bandel 1979a)

Unlike aragonitic crossed lamellar and related structures, calcitic structures in molluscan shells do not contain basic structural units. Instead, calcitic structures consist of needles, rods, and plates that represent the growth fronts of larger structural units with uniform optical properties. The smallest structural elements visible on growth surfaces show a variety of sizes unlike the uniform size of basic structural units in aragonitic structures. The morphology of these structural units is commonly influenced by associated organic shell deposits. With the exception of some patellogastropods and oysters, calcitic shell layers are generally limited to outer shell walls. Calcite occurs among gastropods in the opercula of many unrelated species, in the egg capsule of some Neritidae (others have aragonite here) and the mineralized eggs, and in the diaphragmata of pulmonates. In cephalopods the phragmocone is secreted by the shell mantle behind the periostracum gland zone and consists of true biomineralizates. External rostra of most belemnites excreted by the muscular mantle, reinforcements of the nautilid beak produced by the lips, brood chambers of *Argonauta* secreted by the arms, and chamber fills (lamellorthoceratids, Bandel and Stanley 1989), as well as fill structures of the siphuncle of endoceratids, are calcitic. These shell deposits have a rather simple internal structure (Bandel 1990b).

While crossed lamellar structure and nacre have been present since early Cambrian and used in practically all mineralized molluscan shells, they are absent in nonmolluscan shells and in molluscs that have been derived from precursors with nonmineralized shells. The presence of crossed lamellar structure in Hyolithida is either doubtful or due to similar calcitic structural types as found in the tube of worms like *Ditrupa* (Bandel 1986b). Nacre and crossed lamellar structure usually exclude each other and form a constraint to those groups that have them. A switch over from one to the other is usually not possible, and each group of molluscs is constrained to use only the one or the other or its derivates, with possible addition of new and more simple structure.

3 Biochemical and Biomechanical Prerequisites Influencing Embryonic Construction of the Shell (Fig. 2)

As in many other molluscs, the structure and morphology of the embryonic cephalopod shell are strongly correlated with the amount of yolk involved in embryogenesis. Von Boletzky (1974) noted that in cephalopods with eggs smaller than 2 mm the longitudinal axis of the animal is predetermined by the longitudinal axis of the egg. Here the embryo covers about half of the yolk length when an outer and an inner yolk mass are differentiated. The mantle secretes an organic shell whose shape reflects the round inner yolk mass. It becomes functional and detached from its secretory epithelium (the periostracum-secreting cells) before it is mineralized. Among modern cephalopods with a calcified shell, only *Spirula* has eggs so small (1.7 mm) that shell gland differentiation and formation of the first shell coincide with the time of yolk rearrangement and morphological lifting of the embryo. Mineralization of the organic primary shell occurs in two phases (Bandel and von Boletzky 1979). During the first phase aragonitic crystallites penetrate and partly replace the organic primary shell. During the second phase, the primary shell is covered by inner-prismatic deposits. Mineralization of the embryonic shell in *Spirula* resembles the mode of mineralization of the ammonitella (the ammonite embryonic shell) in all well-preserved Triassic, Jurassic, and Cretaceous ammonites so far studied (lit. see Bandel 1982, 1986a). As in *Spirula*, the first chamber within the ammonite ammonitella is spherical, ovoid, or lenticular (spindle-shaped) and it probably surrounded the embryonic inner yolk sac. Secretion of the entire ammonitella was evidently completed in contact with the mantle epithelium prior to mineralization, judging from the fact that the ammonitellae are devoid of growth lines, that the shell was flattened and deformed where the head and yolk sac of the developing embryo lay (Bandel 1986a), and the first mineralization affected only the outer walls. A similar pattern is found in the embry-

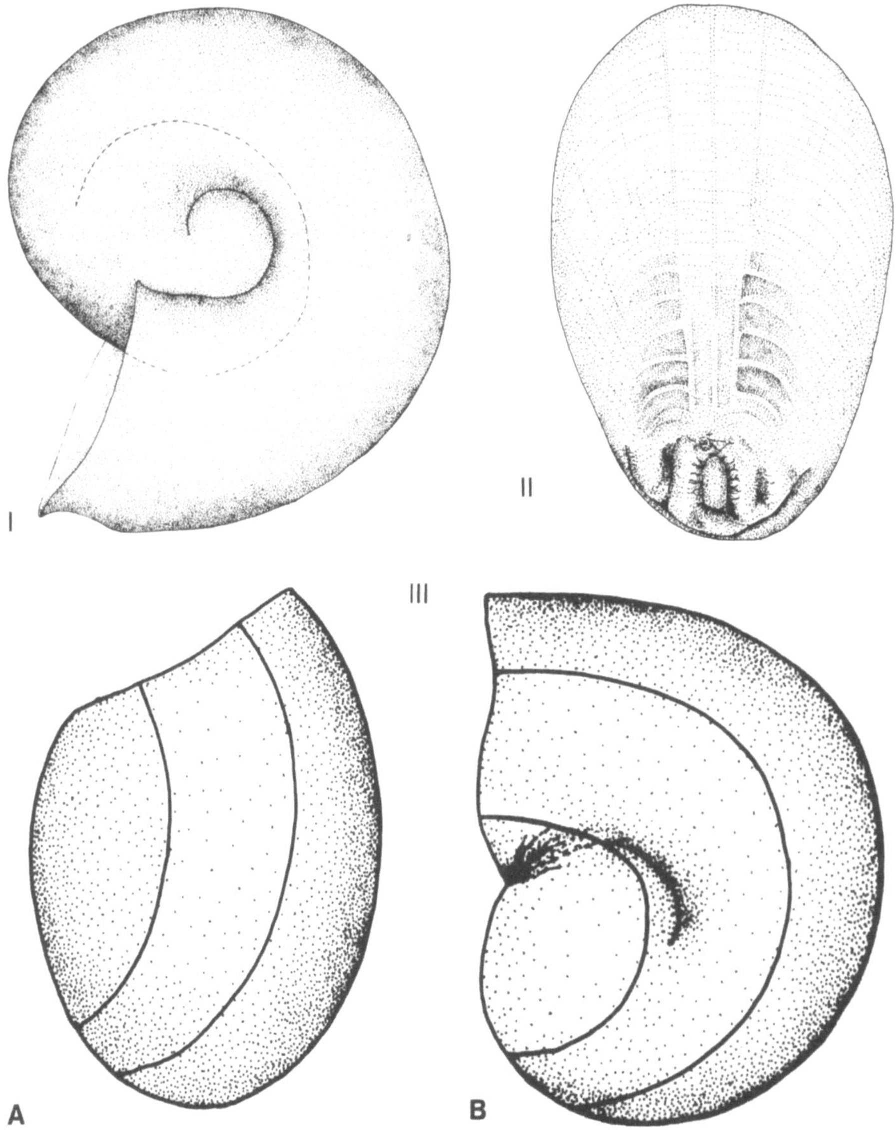

Fig. 2. I The ammonitella as the embryonic shell of ammonites developed from an egg of less than 2 mm in diameter. Here the embryonic shell was mineralized as soon as it had been completely formed (Bandel 1982). **II** The embryonic shell of *Sepia* develops in an egg larger than 2 mm. Here the embryonic shell is deformed before it becomes mineralized much later during embryogenesis. Thus wrinkles form and the hatching young is adult-like. Much yolk retards hatching and the yolk sac is present for a long time (Bandel and von Boletzky 1979, 1988). **III** The embryonic shell of Archaeogastropoda generally is of bilaterally symmetrical cup-like shape at first (**A**) and later becomes deformed (**B**) prior to the beginning of benthic life (Bandel 1982)

onic shell of belemnites and aulacoceratids (Bandel 1985; Bandel et al. 1984b; Bandel and Kulicki 1988), and this probably also occurs in bactritids and orthoceratids, which have embryonic shells like belemnitids in general. In ammonites and belemnites with a spherical first chamber, the first septum differs considerably in structure and morphology from later septa (Bandel 1982; Landman and Bandel 1985). This type of prismatic proseptum, which closes the first chamber to form the siphuncular system, was apparently inherited by the ammonites from Upper Silurian orthoceratid cephalopods. The latter are also believed to have evolved into the first coleoids (Bandel et al. 1983; Bandel and von Boletzky 1988). Belemnoids in the broad sense (including aulacoceratids and belemnitids) have a similar prismatic septum that differs in shape from the later nacreous septa; but belemnites also have an organic septum that is not penetrated by a siphuncle ("closing membrane" of Jeletzky 1966; Bandel et al. 1984b; Bandel and Kulicki 1988). The organic nature of the first septum in belemnites suggests that they were able to hatch and utilize their first chamber as a flotation device before their visceral mass had differentiated a siphuncle.

In *Nautilus* and *Sepia* the eggs are yolk-rich and much larger (5 to 20 mm) than 2 mm in maximum diameter. Here the shell gland develops on a disc-like cell plate resting on the yolk mass. In *Sepia officinalis* the primary shell detaches from the mantle at a length of about 0.9 mm. During reorganization of the yolk, the purely organic conch is deformed and pulled inward along a central line of shell-tissue attachment. Further shell growth occurs with the formation of clear growth increments. At about 2 mm shell length, the first mineralization occurs both outside and inside the shell. Both exterior and interior deposits are organized in the same manner as normal adult shell, being accreted onto an organic periostracal layer. There are no early shell structures comparable to those seen in shell-bearing cephalopods with eggs smaller than 2 mm.

Nautilus develops from a very yolk-rich egg within a large egg capsule measuring up to 20×14 mm (Arnold et al. 1987). Here, as in *Sepia*, the primary organic shell becomes strongly deformed before initial mineralization fixes its shape. The primary organic shell is larger than 2 mm and it sits cap-like on the top of the embryo (Arnold 1987; Arnold et al. 1987). The initiation of mineralization and continued growth of the large embryonic conch are clearly visible in its exterior morphology. The wrinkled, unmineralized primary shell formed by the still connected mantle and shell changes abruptly and quite visibly to the regularly sculptured, evenly formed secondary embryonic shell. The primary shell is stretched and becomes wrinkled around a central fold before it becomes mineralized and fixed in shape (Erben and Flajs 1975). The irregular initial mineral deposition is underlain by nacre not unlike that formed by adults. As in *Sepia*, the embryonic shell does not differ in style of mineralization from the adult shell with the exception of the small portion of primary shell. This primary shell becomes deformed before it was mineralized, in the case of *Sepia* initially from the outside, in the case of *Nautilus* initially from the inside. The first septa of *Nautilus* and *Sepia* are constructed in the same way as their later septa. The embryonic shell of cephalopods is thus a faithful reflection of the mode of development within the egg and of the first mode of free life within the sea after hatching. Ammonites and belemnites on one hand and *Nautilus* and relatives from the Jurassic period onward on the other hand have not varied their reproductive strategies, while Paleozoic nautiloids used, and modern coleoids still use both strategies.

Interestingly, archeogastropods initially mineralize their organic embryonic shell in a manner similar to ammonites and such cephalopods that have an embryonic shell of the dimension of the ammonitella (Bandel 1982, 1986a). Cambrian ancestors of cephalopods and gastropods alike may have developed this mode of shell mineralization in response to a change from a free swimming nonfeeding larva requiring a light shell, to a benthic juvenile in need of protection through mineralization. The very rapid growth of needles 0.2 μm in diameter into the organic shell may represent a primordial mode of initial mineralization in conchiferan molluscs (Bandel 1982, 1985, 1986a). The mineralization of the embryonic shell of the Monoplacophora is not known to date, but it should fit into the presented data. Evolution of similar reproductive strategies in different groups of molluscs could also have resulted in a convergent mode of mineralization of the first shell.

4 Shell Shape as Reflection of the Mode of Life: the Limpet Case (Fig. 3)

Many different and unrelated groups of gastropods modified their shell toward a more or less perfect bilateral symmetry when they acquired limpet shape. The animal inside the conical shell is torted and thus easily differentiated from such nongastropod limpets such as the monoplacophorans. These limpet shells are convergent. It reflects the mode of life clinging to a hard substrate by means of the foot. The shell is pulled down onto the foot when the animal is disturbed and retracted into the shelter of its shell. In contrast, gastropods with coiled shells pull the foot into the shell. The pedal musculature in limpets is attached to the shell above the foot at several places in a more or less symmetrical pattern, while in coiled gastropods there is usually only one retractor muscle asymmetrically attached to the shell columella.

In higher gastropods early ontogeny includes a larval phase and the shell of the prelimpet stage of ontogeny in all marine forms reflects planktotrophic life or its short cut in the egg capsule. Both forms are clearly coiled, the first with an embryonic and larval shell, the second only with an embryonic shell (Bandel 1975b). The larval shells are distinct, so that neritomorph limpets like Recent *Phenacolepas* can clearly be distinguished from other limpets (Bandel 1982) and problematic fossil limpets like the crinoid parasite *Platyceras* can be placed in this group. Very similar mesogastropod limpets of the Calyptraeoidea and Capulidae can be separated from each other by the larval shell, which only in the second case is an Echinospira. Eulimid limpets like *Thyca*, which is parasitic on starfish, and calyptraeoid limpets attached to the spines of sea urchins catch their own food, are easily differentiated by the smooth slender larval shell of the eulimid type (Warén 1984). In the Heterostropha the secondary limpet form leads to greater reduction of the larval shell morphology. In the pyramidellid limpet *Amathina*, the larval shell twists from sinistral to dextral coiling before the limpet shape is acquired. In the archeopulmonate *Williamia*, as well as the notaspidean opisthobranch *Umbraculum*, the sinistral embryonic and larval shell do not twist into the right before the limpet of the teleoconch begins (Bandel 1982). In the freshwater

Fig. 3. Limpets have convergent shell shape due to the same function of the shell as in an eulimid (**a** from Warén 1984) and a calyptraeid (**b** from Bandel 1982). Their position in the gastropod system is revealed by the shape of the larval shell. Thus calyptraeoid larvae from the modern Red Sea (**c**) are quite different from neriomorph larvae as present in the Triassic Platyceratidae from the Tethys Ocean (**d**). When no planktotrophic larva is developed, the early ontogenetic shell can be simplified as in the Caribbean member of the Calyptraeoidea (**e** from Bandel 1982) or strongly reduced as in the freshwater limpet *Ancylus* (**f**)

basommatophoran *Ancylus* and *Acroloxus*, coiling of the embryonic shell is lost and the larval stage is extremely modified into a stage of yolk uptake and no longer reflected in the shape of the embryonic shell.

In Archeogastropoda, the ontogeny is shorter than in higher Gastropoda. A reduction of the characteristic pattern of embryonic shell deformation may easily occur and thus hide any original dextral or sinistral twist of the shell. Both left- and right-handed coiling may be present, but the groups are now fixed to one of these. Within many Patellogastropoda the shell is forced into a sinistral whereas in all other archeogastropods it is pulled into the dextral coil. This fixation seems to have originated early in their evolutionary histories and is connected to morphological changes in the soft body, especially associated with muscle reconstruction. No case is known where a reversion has happed, but lecithotrophic development may simplify muscle reconstruction, which causes problems in regard to the systematic placement of many recently discovered, deep sea, archeogastropod limpets mainly from deep sea hydrothermal vents. Evolutionarily "new" limpets may still retain the operculum during adult life, as is the case in the scissurellid limpet *Incisura lytteltonensis*, that clings to algae in the intertidal rocky coast of New Zealand. The operculum may be formed and later be lost during the intracapsular development, as is the case in the archeopulmonate limpet *Latia neritoides*, that lives on rocks in rapids in freshwater streams in New Zealand. In both cases the operculum represents a relict not needed in the function of the limpet.

5 Combination of Two Separate Evolutions, the Case of the Larval Shell of the Cenogastropods. Constructional Morphology and Systematics (Fig. 4)

Strathmann (1978) postulated that most types of planktotrophic larvae are of extreme antiquity. This author assumed that mechanisms of larval feeding are more readily lost than gained in existing species of many phyla. He thought that present barriers to the acquisition of planktotrophic larval stages were probably absent in ancestral forms with ciliary feeding and more direct development from larva to juvenile.

Only higher bivalves (nonprotobranchs) and higher gastropods have planktotrophic larvae, while all other mollusc larvae rely on the yolk reserves of the egg for swimming in the sea. If one does not accept the unlikely possibility that the simplest living molluscs, Aculifera, Protobranchia, Monoplacophora, Scaphopoda, and Archeogastropoda have all secondarily lost a plankton-feeding larva, we have to reconstruct the early ontogenesis of molluscs without a planktotrophic larval phase. Based on this thesis, it has to be taken into account that bivalves and gastropods have independently acquired planktotrophic larvae, a view contrary to Strathmann's. In contrast to machines, every organism is built up by an ontogenetic development, during which every ontogenetic stage must function to survive. Biomechanical forces are quite different at the embryonic phase, the stage of planktotrophic swimming and feeding, and final benthic crawling and food-searching phase. Many different groups can be differentiated by the morphology and the sculpture of their larval shells. Each level of ontogeny

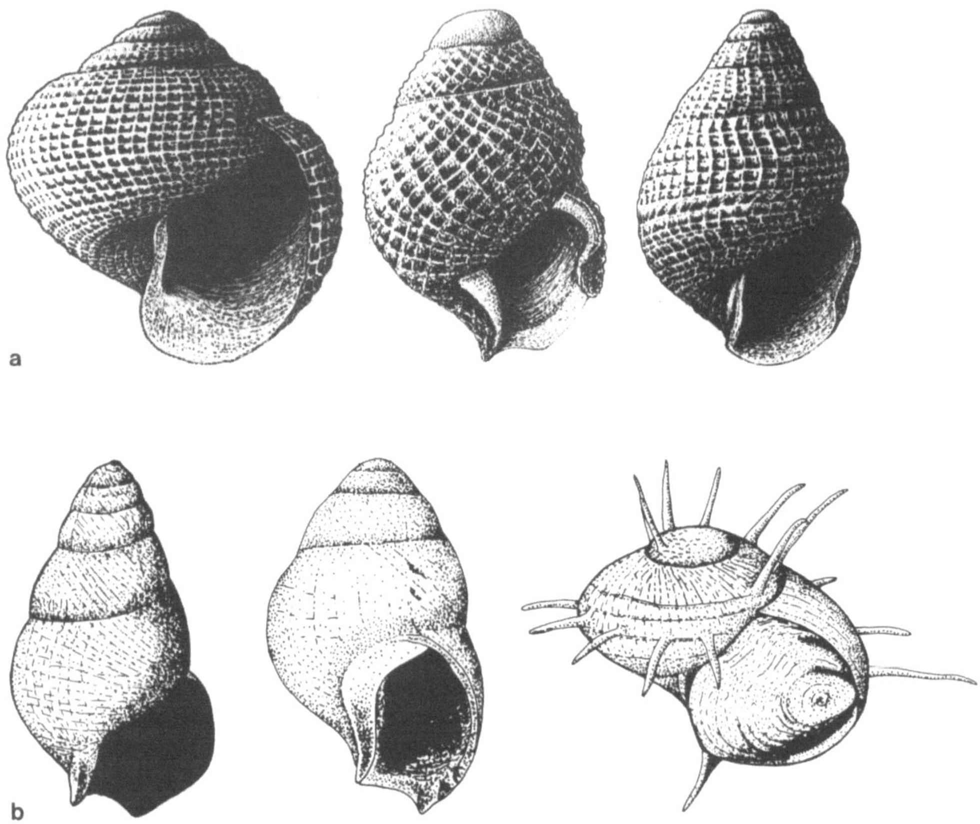

Fig. 4. The larval shells of modern mesogastropoda like the Cypraeoidea look very different from the later Cowrie-shell (**a**). Many tonnoidean larvae cover the apical portion of their shell with an additional organic layer (**b**)

follows its own evolution and has its own phylogeny. Even embryonic shells show group-specific features (Bandel 1975b). It is possible to discriminate different taxa on the basis of their ontogeny. This is true even though the selective forces are practically identical for most developments within the shelter of the egg capsule (Bandel 1982). Comparative ontogenetical studies can thus provide general rules in the interpretation of evolutionary changes.

The larval shell of the Caenogastropoda shows the same direction of coiling as the adult shell, but quite different morphology. Shape, composition, and sculpture of the larval shell are characteristic for certain systematic units, and reflect their own history. As an example, the adult shells of fossil mesogastropod cerithioids and the ctenoglossan epitonioids show numerous cases of convergencies, while the larval shell separates both groups clearly in beds as ancient as the Triassic. As one further example, Caenogastropoda belonging to the groups of the Naticoidea, Tonnoidea, Muricoidea, Cypraeoidea, Lamellaroidea, and Calyptraeoidea each have characteristic planktotrophic larvae showing features that are quite independent of those of the benthic stages.

Cypraea and relatives, the so-called cowries, have a glossy, colorful, ovate shell with a long narrow aperture bounded on both sides by inrolled lips. The last whorl covers all former whorls and is glazed over by additional shell that is produced by the mantle that covers the shell from the outside. When disturbed, the mantle can be retracted into the shell.

Cowries generally live below stones and among coral debris and feed on crust-forming colonial animals that they graze or bite off from hard substrates. Ovulids with similar shell and mantle feed on polyps of octocorals. The planktotrophic larvae, in contrast, feed on unicellular algae that they collect with their ciliated velum while swimming in the open sea. The larval shell is trochospirally coiled and its sculpture consists of a strong pattern of ribs. In the case of the cypraeids, this pattern is rectangular; in the case of the ovulids, it consists of rhomboids. They have been distinct since the Upper Cretaceous. The shell in both larvae is partly covered by mantle when the animal is active and it is light, with thin walls composed of little mineral and much organic material.

The Tonnoidea are even more remarkable. Most species are rather large and hunt different animals, in many cases echinoderms, or, in other examples, molluscs and worms. Adult shells show a variety of shapes; some closely resemble those of some Neogastropoda. Many of these species that live in near-shore tropical environments brood their spawn until the young hatch and swim off into the open sea (Bandel 1976b). The larvae also feed on unicellular algae and show a very different shell and developmental pattern from their adults. After the larvae have left their egg capsules, they usually construct a shell of a pattern similar to that found in the cypraeoids. Later, the shell produced by the mantle edge is purely organic and does not mineralize from below (Bandel 1981a). The mantle margin, and with it the periostracal gland zone, grows into a trunk that is large enough to cover the shell from the apertural margin to the apex of the shell. New shell is not only added to the apertural margin but also, in continuation of it, onto former shell up to the apex. In this way, the veliger can cover former shell with new shell and thus cover organisms that have attached to the shell. The larva may also change style of the sculptural pattern of its shell during growth. Some species add spines to the formerly smooth whorls, others cover an originally spiny shell with a smooth layer. When larvae are ready to change to benthic life after having stayed in the plankton for as much as a year (Simroth 1895; Scheltema 1971), the metamorphosis is quite drastic.

The growth of epiphytes and animal crusts on the shell of the planktotrophic larva is a commonly seen feature, but not in lamellaroids. These have a gelatinous cover, the echinospira, which surrounds the actual shell. It is secreted by the epithelium of the muscular mantle in front of the periostracal gland zone and is usually lost when the pediveliger changes into the benthic animal. A special form has developed in the Antarctic genus *Marseniopsis*, where the muscular mantle surrounds the larval shell. This is the only veliger that swims for long periods in the cold Antarctic waters and uses its living echinospira shell as a buoyancy control and defence system (Bandel et al. 1990). The lamellaroideans demonstrate the difference in evolution of the larval life in comparison with the benthic life particularly well. From a common early Cretaceous ancestor that could also have been morphologically close to the ancestor of the cypraeoids, naticoids and tonnoids a first echinospira larva developed. In forms with normal coiling it retained its normal trochospiral shape, as still exhibited in some trichotropids (Pilkington 1974). But the mode of life of the benthic phase evolved forms like *Trivia* and *Erato* in which the mantle covers the shell when active, and is retracted when disturbed, much as in the cowries. These two groups have commonly been mixed and many fossil species are still not clearly referable to either group. *Lamellaria* and related forms cover their shell partly or totally by mantle and can no longer retract into their shell. Most species feed on tunicates like *Lamellaria, Trivia*, etc. In contrast, species of *Trichotropis, Capulus*, and related taxa filter food from the sea water. Some are mobile snails with an operculum (trichotropids), while others are sessile limpets (capulids). *Capulus* species are often mistaken for calyptraeids, which have a very similar mode of life but totally different larvae. The three related groups Cypraeoidea, Tonnoidea, and Lamellaroidea can be compared with the naticoids, which have a large and almost spherical larval shell, and the calyptraeoids, with highly sculptured and spherical larval shell. They clearly demonstrate that evolution in the early part of the ontogeny differs from that of later parts of the ontogeny. Both portions of ontogeny are connected to each other in multiple ways, but selectional forces acting on the early part of the ontogeny happening in the open sea are quite different from those acting on the benthic juveniles and adults. To understand the evolution of the members of these groups of gastropods, ecological processes of the open sea affecting the early portion of individual life, as well as factors of benthic ecology acting on the later portion of life, have to be taken into consideration.

6 Changes in Life Style in One Part of the Ontogeny Can Affect Other Ontogenetic Stages (Fig. 5)

Variations without impact on selection are demonstrated in the case of the Caecidae. The caecids appear first in the Eocene and have probably evolved from normally coiled truncatelloids with a low spire similar to Recent vitrinellids. The earliest-known caecids had larval shells of low trochospiral shape, as are still found in some living caecid species. They did not cast off the early part of their shell during later life (Beu and Maxwell 1990). After larval life, the benthic *Caecum* secretes an uncoiled shell. It then closes off the juvenile shell portion with a septum and usually discards these parts of the shell. The tiny (1 to 3 mm long) animal with tusk-shaped shell plows its way through soft sediment, algal thickets, and across pebble surfaces. They feed on unicellular algae. The shape of the embryonic and larval shell is of little significance to the benthic animal, since it is discarded soon after metamorphosis. To a veliger swimming with a ciliary apparatus, the morphology of the shell is also of little importance, as soon as evolution had produced an uncoiled adult shell. The shell is here utilized as protective exoskeleton and its shape is not very significant in regard to the fitness of the larva. New free paths have thus been opened to the morphology of the embryonic and larval shell of caecids. Their presence is documented by (1) normally trochospirally coiled larval shells in some species; (2) planspirally coiled larval shells in other species; (3) uncoiled larval shells in members of the genus

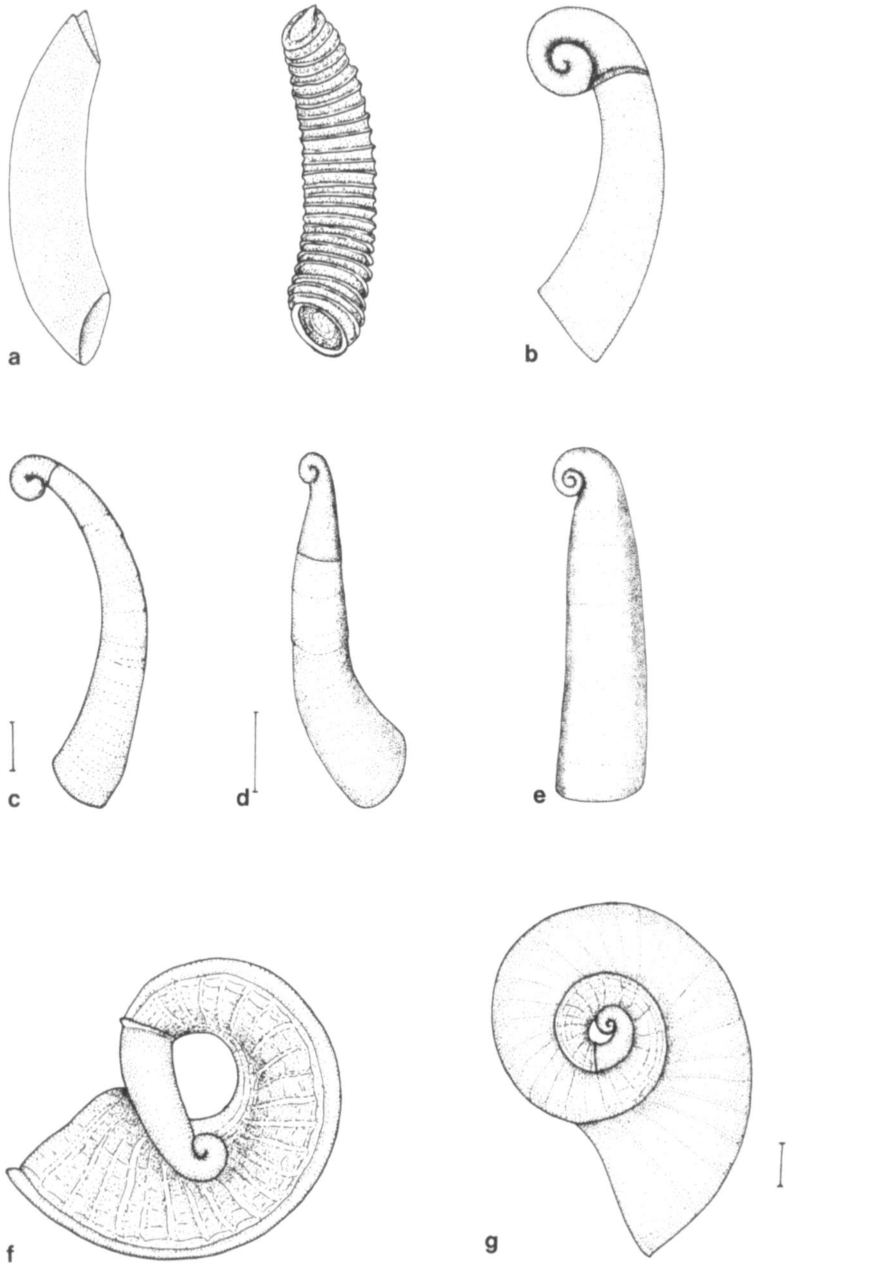

Fig. 5. The Caecidae have developed an uncoiled shell and often cast off the early shell portion (**a**). The larval shell of *Caecum* may still form a normal coil (**b**), that can be reduced by one whorl when the young hatches crawling from the egg (**c**). In the genus *Parastrophia*, the larval shell may be uncoiled (**d**). These different forms can be found from first records of caecids onward in the Eocene (**e** redrawn from fossil form of New Zealand figured by Beu and Maxwell 1990). Very strange is the reversal of this tendency in the Ctiloceridae of the north Australian shallow sea with *Parastrophia* like larval shell (**f**) that shows the begin of coiling (**g**) in another species

Parastrophia; and (4) uncoiled larval and embryonic shell in the genus *Thecopsella*. In the peculiar side branch of northern Australian caecids surrounding the genus *Ctiloceras*, the adult shell is coiled again while the larval shell is uncoiled the *Parastrophia* way. In one species the trend regarding the larval shell seems to have been reverted again and it is converted into a still open spiral, but evenly curved form.

Similar changes in morphology of the larval shell showing all kinds of morphological differences along with planspiral coiling and uncoiling are exhibited by pelagic Heteropoda, where the adult shell is either planispirally coiled, reduced in size, or lost (Bandel and Hemleben 1987). These examples demonstrate that constraints act on the morphology of the larval shell that are relieved when changes occur in the life style of the adult. Constraints force the preadult shell to be spirally coiled. These no longer prevail when the larval concessions of shell function are solely active, as is also evident in the following case of the pteropods.

7 Ontogenetic Abbreviation Resulting in Great Changes of Large Systematic and Phylogenetic Importance and Connected with Irreversibility (Macro- and Microevolution)

In order to understand the origin of the pteropods we have to consider their probable closest relatives among benthic gastropods, the cephalaspidean opisthobranchs. In the course of their ontogeny, these, as well as the Allogastropoda (in the sense of Haszprunar 1985) and the Archaeopulmonata, may produce an embryonic shell (shell secreted by the embryo while still within the egg capsule) and a larval shell (shell secreted by the veliger swimming in the sea) with a direction of coiling different to that of the adult shell (shell produced by the gastropod after its metamorphosis from the veliger larva). Allogastropoda, Opisthobranchia, and Pulmonata united in the systematic unit Heterostropha Fischer, 1885, were originally separated from all other gastropods by this change in the direction of coiling during ontogeny of the shell, usually from early left-hand to later right-hand coiling. The switch from sinistral to dextral coiling in the larval shell secreted by the pediveliger (veliger larva ready to metamorphose) can be traced back in the fossil record to the Lower Carboniferous (Kollmann and Yochelson 1976) and is well established in Allogastropoda and Opisthobranchia in the Triassic (Bandel 1988). Even though the necessity for heterostrophic shell twisting in the Heterostropha (Allogastropoda, Opisthobranchia, and Pulmonata) is not understood; its sustained existence from at least the early Carboniferous onward in three quite different major groups is strong evidence for its selective value in all such cases where a free planktotrophic veliger is present.

Pteropods differ from the Heterostropha by retaining the larval, originally left-hand coiled shell throughout adult life (Bandel et al. 1984a). The total disappearance of the adult shell with its dextral coil is probably connected with the origin of the pteropods in general. It represents a major change in the course of ontogeny during which the program of construction of the adult shell was eradicated. Constraints connected to construction of the adult shell were also eliminated, and the program of shell production to continue after metamorphosis. Early

pteropods may have become pelagic by production of a mucus raft (Gilmer and Harbison 1986) and had thus left behind the need to establish themselves as epi- or endobenthic species settling on or in the bottom sediment. This, together with the new freedom of shell production of the growing young, may have resulted in the unusual shell shapes found among pteropods more of less from the time of their origin onwards.

The only apparent difference from other gastropods is seen in the shape of the embryonic and larval shell, which in many pteropods is not coiled; but this is not a feature that is confined to the pteropods. Some benthic opisthobranchs produce a veliger shell very much like that of a pteropod (personal observation 1987) and the embryonic and larval shell of the gymnosomate pteropods is uncoiled. Uncoiled shells occur as pathologic cases in many marine gastropods with a coiled shell (Bandel 1975b, 1982), but here lead to a nonfunctioning teleoconch and are thus exterminated during ontogeny. This restriction does not apply to the pteropods, where an uncoiled shell is no disadvantage. This has been selected out at the very beginning of recognizable pteropod evolution in the Eocene, where straight early ontogenetic shells are present just as in modern times. Thus, we have to take into account a mode of development that favored the pathological case of benthic shelled gastropods and selected out uncoiled forms rapidly for the cavolinid line of pteropod evolution. The case demonstrates that pathology sometimes is only a relative term, analogous to adaptation.

Loss of the teleoconch in pteropods can be seen as a model for the loss of the larval shell in some nudibranchs, the loss of a functional larval shell in some dorids, and the disappearance of the shell as a whole in only a very small group of opisthobranchs within the genus *Rhodope*. The use of the larval shell in adult life may have resulted in the appearance of a type of shell-building program that is not tightly integrated with other systems of the adult organization and thus decoupled and free to change independently. The new structure of the shell (helical structure), its rebuilding during ontogeny, and unconventional shape may be due to this new freedom acquired by the use of a new program to the construction of the adult shell.

8 Open Question of Reversibility of Loss of Larval Shell and its Implication on Interpretation of the Shell (Fig. 6)

Is the duration of a species in time correlated with the mode of larval development? This has been postulated by Strathmann (1978, 1985) and been picked up by Jablonsky and Lutz (1983) and taken to be the case in gastropods. They suggest that a prolonged stay in the plankton would bring about a ready flow of genes and a wide geographic distribution. This is supposed to be of positive influence to the time of existence of a species (Jackson 1974). No convincing proof can be presented that mollusc species with a planktotrophic larva have been in existence longer than those without such a larva (contrary opinion see Hansen 1980). Examples can be presented for both cases. Small adult size was supposed to result in a nonplanktotrophic larva by Hoagland (1984) in the case of the genus *Crepidula*. A group of very small allogastropod species, including the omalogyrids and

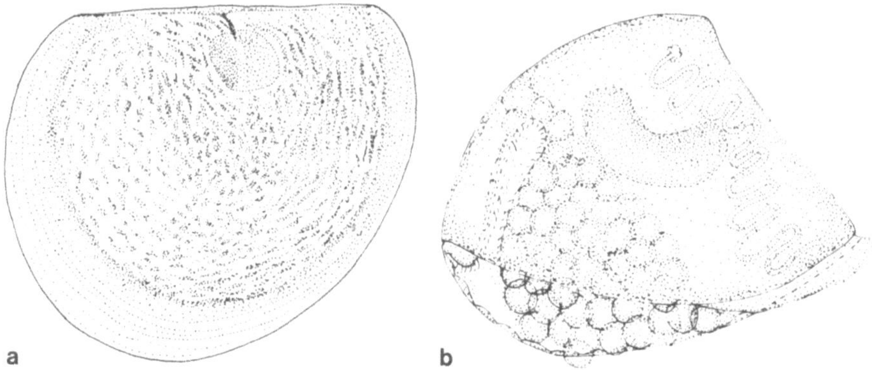

Fig. 6. When the normal bivalve embryonic shell does not mineralize separately in the two valves (**a**) but everywhere the outcome is a pathologic shell which reflects a shape of the shell of the possible ancestor to bivalves among the monoplacophoran limpets that may have lived in the Cambrian (**b** Bandel 1988)

the orbitestellids, can be traced into the Triassic time, and throughout their histories they show some species with both planktotrophic and others with lecithotrophic development (Bandel 1988).

Strathmann (1985) expressed the opinion that once a species has evolved a nonfeeding larval development, planktotrophy is no longer an option. Observations in molluscs cannot support this view. The example of the wood-boring bivalve *Teredo* suggests a different conclusion. Here the offspring may be planktotrophic or not, according to the mode of parental care of the eggs (Bandel 1988). According to Strathmann (1978, 1985), Jägersten (1972) and Thorson (1946) have presented morphological evidence that species with nonfeeding larvae have developed from species with feeding larvae much more frequently than the reverse, and that this has occurred many times in many of the classes of marine invertebrates. The second part of this statement can be shown to be true in all marine gastropods. But the first part is an assumption that needs testing and much more research.

Observations on the spawn of *Tonna* from the Caribbean Sea indicated the possibility that within the egg mass produced by one and the same female, long-term veligers may hatch from one capsule, while in a neighboring capsule one individual may utilize all other embryos as nurse eggs and hatch as crawling young (Penchaszadeh 1981). Thus genotype expression into a phenotype may here be related to an influence coming from the external environment. In this case the availability of larval foods in the form of nurse eggs triggers a shell program totally different from the usual.

9 "General Rules" Extracted from Shell Morphology and Shell Construction, Constraints, and Free Ways and Possible Experiments to Test Bioconstruction

All gastropods, with the possible exception of *Rhodope*, produce an embryonic shell, even those that hatch from their egg masses naked, like some species of

nudibranchs. It is not known why this is so, since the mantle connected to the shell is there before the shell and secretes it. It can be interpreted as "vestigial" character, indicating that they evolved from shelled ancestors and there is no selection against its presence in the capsule. But it could be seen to provide some kind of structural support or template for the later development of the embryo.

Stages in the ontogenetic development are a sequence of musts; one is not possible unless the one before it had occurred. Within each stage of ontogeny organ rudiments may either develop rapidly into functional organs, or may be retarded to varying degrees in development. Early portions of the ontogeny of several molluscs have been studied well with experimental approach by the group of Biggelaar from Utrecht (van den Biggelaar et al. 1981; van den Biggelaar 1977; van den Biggelaar and Serras 1988), but most of the ontogeny is quite unknown in this respect. Strathmann (1978) expressed the hypothesis that major body plans of marine invertebrates decreased in number with time. In my view this statement cannot be supported on the evidence of gastropods or bivalves, where we can observe a steady increase in complexity from the end of the Cambrian onward. The complexity may be coupled with a decrease of the potential to evolve in many directions, but overall complexity has certainly become much greater, even including the setbacks of major extinction events.

Pathological cases commonly found in the egg masses brooded by *Littorina saxatilis* have been studied by Bandel (1982), and are currently being studied by us. Here a close interaction between uncoiling of the shell and uncoiling of the operculum was noted that could be related to Recent forms such as *Pseudomalaxis* species with disjunct whorls and trochospiral operculum or fossil relation of *Neritopsis* from the Triassic and Paleozoic times with whorls in disjunct spirals and cow-horn-shaped opercula. *Littorina* embryos with defects such as sinistral coiling of the shell can be cultured and in this way models of "hopeful monsters" can be studied in detail. Where is the end of pathology and the begin of evolution to new forms? Also, experiments with decoiling stylommatophoran gastropods, such as that carried out by Moor (1978) can be reproduced and extended to other species to explain the story of the evolution of *Gyraulus* in the lake of the Steinheim impact crater or the production of numerous "species" of fossil and modern *Melanopsis* connected to the Mediterranean area, or *Paramelania* of Lake Tanganyika and *Pyrgulifera* of the estuaries of the Gosau (Riedel 1990) (cf. also the effects of uncoiling in rudists as described by Skelton 1985).

Pathological cases as in the larval shell of *Anodonta* reflect changes in the ontogeny of the bivalves that must have occurred as far back in time as the Cambrian (Bandel 1988). Perhaps studies on such atavisms can help to unravel the evolution of the major features of mineralization and shell shape within different monophyletic lines of the conchiferan molluscs.

References

Arnold JM (1987) Reproduction and embryology of *Nautilus*. In: Saunders WB, Landman NH (eds) *Nautilus* – biology and paleobiology of a living fossil. Plenum, New York London, pp 353–372

Arnold JM, Landman NH, Mutvei H (1987) Development of the embryonic shell of *Nautilus*. In: Saunders WB, Landman NH (eds) *Nautilus* the biology and paleobiology of a living fossil. Plenum, New York London, pp 373–400

Bandel K (1975a) Entwicklung der Schale im Lebenslauf zweier Gastropodenarten; *Buccinum undatum* and *Xancus angulatus* (Prosobranchier, Neogastropoda). Biomineralisation 8:67–91

Bandel K (1975b) Embryonalgehäuse karibischer Meso- und Neogastropoden (Mollusca). Akad Wiss Lit Abh Math Naturwiss Kl 1975 (1):1–133

Bandel K (1976a) Observations on spawn, embryonic development, and ecology of some Caribbean lower Mesogastropoda (Mollusca). Veliger 18:249–271

Bandel K (1976b) Die Gelege karibischer Vertreter aus den Überfamilien Strombacea, Naticacea und Tonnacea (Mesogastropoden, Mollusca). Mitt Inst Colombo Aleman Invest Cient Santa Marta (Columbia), Gießen 8:105–139

Bandel K (1977a) Die Herausbildung der Schraubenschicht bei Pteropoden. Biomineralisation 9:28–47

Bandel K (1977b) Übergänge von der Perlmutter-Schicht zu prismatischen Schichttypen bei Mollusken (Transitions from the nacreous layer to prismatic type layers in molluscs). Biomineralisation 9:28–47

Bandel K (1979a) Übergänge von einfacheren Strukturtypen zur Kreuzlamellenstruktur bei Gastropodenschalen. Biomineralisation 10:9–37

Bandel K (1979b) The nacreous layer in the shell of the gastropod family Seguenziidae and its taxonomic significance. Biomineralisation 10:49–61

Bandel K (1981a) Struktur der Molluskenschale im Hinblick auf ihre Funktion. Paläontol Kursbücher 1:25–48

Bandel K (1981b) The structure and formation of the siphuncular tube of *Quenstedtoceras* compared with that of *Nautilus* (Cephalopoda). N Jahrb Geol Paläontol Abh 161:153–171

Bandel K (1982) Morphologie und Bildung der frühontogenetischen Gehäuse bei conchiferen Mollusken. Facies 7:1–198

Bandel K (1985) Composition and ontogeny of *Dictyoconites* (Aulacocerida, Cephalopoda). Paläontol Z 59:223–244

Bandel K (1986a) The ammonitella: a model of formation with the aid of the embryonic shell of archaeogastropods. Lethaia 19:171–180

Bandel K (1986b) The reconstruction of *"Hyolithes kingi"* as annelid worm from the Cambrian of Jordan. Mitt Geol Paläontol Inst Univ Hamburg 61:35–101

Bandel K (1988) Stages in the ontogeny and a model of the evolution of bivalves (Mollusca). Paläontol Z 62:217–254

Bandel K (1990a) Shell structure of the Gastropoda excluding Archaeogastropoda. In: Carter JG (ed) Skeletal biomineralization: patterns, processes, and evolutionary trends. Van Nostrand Reinhold, New York, pp 117–134

Bandel K (1990b) Cephalopod shell structure and general mechanisms of shell formation. In: Carter JG (ed) Skeletal biomineralization: patterns, processes, and evolutionary trends. Van Nostrand Reinhold, New York, pp 97–115

Bandel K (in preparation a) Allogastropoden aus der triassischen St. Cassian Formation (Gastropoda, Dolomiten)

Bandel K (in preparation b) Trochoidea aus der triassischen St. Cassian Formation (Gastropoda, Dolomiten)

Bandel K, Boletzky S von (1979) A comparative study of structure, development and morphological relationships of chambered cephalopod shells. Veliger 21:313–354

Bandel K, Boletzky S von (1988) Features of development and functional morphology required in the reconstruction of early coleoid cephalopods. In: Wiedmann J, Kullmann J (eds) Cephalopods – present and past. Schweizerbart, Stuttgart, pp 209–246

Bandel K, Almogi-Labin A, Hemleben C, Deuser WG (1984a) The conch of *Limacina* and *Peraclis* (Pteropoda) as a model for the evolution of planktonic gastropods. N Jahrb Geol Paläontol Abh 168:87–107

Bandel K, Engeser T, Reitner J (1984b) Die Embryonalentwicklung von *Hibolithes* (Belemnitida, Cephalopoda). N Jahrb Geol Paläontol Abh 167:275–303

Bandel K, Hemleben C (1975) Anorganisches Kristallwachstum bei lebenden Mollusken. Paläontol Z 49:298–320

Bandel K, Hemleben C (1987) Jurassic heteropods and their modern counterparts (planktonic Gastropoda, Mollusca). N Jahrb Geol Paläontol Abh 174:1–22
Bandel K, Hemleben C (1990) Observation on the ontogeny of the thecosomate pteropods during cruise 5/5 of the Meteor in the southern Red Sea substituted by observations from Bermuda (in preparation)
Bandel K, Kulicki C (1988) *Belemnoteuthis polonica*: a belemnite with an aragonitic rostrum. In: Wiedmann J, Kullmann J (eds) Cephalopods – present and past. Schweizerbart, Stuttgart, pp 303–316
Bandel K, Reitner J, Stürmer W (1983) Coleoids from the Lower Devonian Black Slate ("Hunsrück-Schiefer") of the Hunsrück (West Germany). Neues Jahrb Geol Paläont Abh
Bandel K, Stanley GD (1989) Reconstruction and biostratinomy of Devonian Cephalopods (Lamellorthoceratidae) with unique cameral deposits. Senckenb Lethaea 69:391–437
Bandel K, Wedler E (1987) Hydroid, amphineuran and gastropod zonation in the littoral of the Caribbean Sea, Colombia. Senckenb Mar 19:1–129
Bandel K, Hain S, Riedel F, Tiemann H (1990) Limacosphaera, an unusual mesogastropod larva of the Weddell Sea (Antarctica). Nautilus (in press)
Beu AG, Maxwell PA (1990) Cenozoic Mollusca of New Zealand. Drawings by RC Brazier. NZ Geol Surv Paleontol Bull 58:1–518
Biggelaar JAM van den (1977) Development of dorsoventral polarity and mesentoblast determination in *Patella vulgata*. J Morphol 154:157–186
Biggelaar JAM van den, Serras F (1988) Determinative decisions and dye-coupling changes in the molluscan embryo. Gap junctions. Liss, New York, pp 483–493
Biggelaar JAM van, Dorresteijn AWC, de Laat SW, Bluemink JG (1981) The role of topographical factors in cell interaction and determination of cell lines in Molluscan development. Int Cell Biol 1980–1981:526–538
Boletzky S von (1974) The "larvae" of Cephalopoda: a review. Thalassia Jug 10:45–76
Erben HK, Flajs G (1975) Über die Cicatix der Nautiloideen. Mitt Geol Paläontol Inst Univ Hamburg 44:59–68
Gilmer RW, Harbison GR (1986) Morphology and field behavior of pteropod molluscs: feeding methods in the families Cavoliniidae, Limacinidae and Peraclididae (Gastropoda: Thecosomata). Mar Biol 91:47–57
Hansen TA (1980) Influence of larval dispersal and geographic distribution on species longevity in neogastropods. Paleobiology 6:193–207
Haszprunar G (1985) Zur Anatomie und systematischen Stellung der Architectonicidae (Mollusca, Allogastropoda). Zool Scr 14:25–43
Haszprunar G (1988) On the origin and evolution of major gastropod groups, with special reference to the Streptoneura. J Mollusc Stud 54:367–441
Hoagland KE (1984) Use of molecular genetics to distinguish species of the gastropod genus *Crepidula*. Malacologia 25:607–628
Jablonski D, Lutz R (1983) Larval ecology of marine benthic invertebrates: paleobiological implications. Biol Rev 58:21–89
Jackson JBC (1974) Biogeographic consequences of eurytopy and stenotypy among marine bivalves and their evolutionary significance. Am Nat 108:541–560
Jägersten G (1972) Evolution of the Metazoan life cycle. Academic Press, New York London, 282 pp
Jeletzky JA (1966) Comparative morphology, phylogeny, and classification of fossil coleoidea. Univ Kansas Paleontol Contrib Mollusca, Art 7:1–162
Kollmann HA, Yochelson EL (1976) Survey of Paleozoic gastropods possibly belonging to the subclass Opisthobranchia. Naturhist Mus Wien Ann 80:207–220
Landman NH, Bandel K (1985) Internal structures in the early whorls of Mesozoic ammonites. Am Mus Nov 2823:1–21
Moor B (1978) Die Embryonalschale von *Theba carthusiana* Müller (Gastropoda, Pulmonata, Stylommatophora). Ein Beispiel milieuabhängiger Variation der Schalenform. Zool Anz Jena 201:353–363
Penchaszadeh P (1981) A peculiar development pattern in *Tonna galea* from Venezuela. Int J Invertebrate Reprod Dev 4:209–212
Pilkington MC (1974) The eggs and hatching stages of some New Zealand prosobranch Molluscs. J R Soc N Z 1974 (4):411–431

Riedel F (1990) Die Bedeutung der Embryonalschale für die systematische Einstufung rezenter und fossiler Süßwasserschnecken. Masters Thesis, Univ Hamburg
Scheltema RS (1971) Larval dispersal as means of genetic exchange between geographically separated populations of shallow-water benthic marine gastropods. Biol Bull 140:284–322
Simroth H (1895) Die Gastropoden der Plankton-Expedition. Ergeb Planktonexped Humboldt Stift 2:1–206
Skelton PW (1985) Preadaptation and evolutionary innovation in rudist bivalves. In: Cope JCW, Skelton PW (eds) Evolutionary case histories from the fossil record. Spec Pap Palaeontology 33
Strathmann RR (1978) Progressive vacating of adaptive types during the Phanerozoic. Evolution 32:907–914
Strathmann RR (1985) Feeding and nonfeeding larval development and lifehistory evolution in marine invertebrates. Annu Rev Ecol Syst 16:339–361
Thorson G (1946) Reproduction and larval development of Danish marine bottom invertebrates, with special reference to the planktonic larvae in the Sound. Meddr Komm Dan Fisk Og Havunders Ser Plankton 4:1–523
Warén A (1984) A generic revision of the family Eulimidae (Gastropoda, Prosobranchia). J Mollusc Stud Suppl 13:96 pp

A Functional Approach to the Development of the Head of Teleosts: Implications on Constructional Morphology and Constraints

KAREL F. LIEM[1]

Summary

The relative roles of intrinsic and extrinsic determinants of form in teleost skulls is assessed on the basis of a functional approach to development. The patterns of emerging kinematic connections, changing proportions of structural elements, and their probable functions have been analyzed in phylogenetically closely related lineages with strikingly different modes of development: (1) the oviparous Pomacentridae with free-living larvae subjected to intense environmental selection pressures; (2) the viviparous Embiotocidae which give birth to highly developed juveniles that have been shielded from environmental influences. The developmental patterns in the two groups are very similar. Thus it is suggested that transformations in functional components, changes accompanying the proportional reduction in eye size, and the changes in the functional "hot spots" are governed by internal rules inherited from a common ancestor. It is concluded that internal programmatic rules governing the emergence of morphological variations and transformations and their vitally adaptive functions are inseparably interconnected at the level of the generative process.

1 Introduction

One of the most fundamental questions in evolutionary biology is: why is the available morphospace so discontinuously occupied by organisms? Proponents of the supremacy of intrinsic determinants of form have argued that discontinuity in morphological variation proves that the realm of possible forms is governed by internal developmental canalizations and not by extrinsic agents of natural selection. In sharp contrast, advocates of the dominance of extrinsic determinants of form embrace the concept of the adaptive landscape, which may be considered a phenotypic space with each occupying phenotype having a fitness value. Organisms evolve by "selective entrainment", occupying the tip of the nearest adaptive peak. In this landscape the adaptive peaks and valleys are irregularly interspersed. Because the peaks are discontinuous, it leads to the discontinuity of the ways and means by which organisms derive their energy from the environment. In this view, natural selection reigns in creating discontinuity in the variation of designs of organisms.

[1] Museum of Comparative Zoology, Harvard University, 26 Oxford Street, Cambridge, Massachusetts 02138, USA

N. Schmidt-Kittler and K. Vogel (Eds):
Constructional Morphology and Evolution

Thus proponents of the supremacy of intrinsic determinants argue that in the absence of natural selection certain designs of organisms would be clustered and gaps would be prominent in the distribution of designs. They present evidence that the evolutionary reduction of digits has occurred independently in several lineages of amphibians. It involves the loss of the innermost digit in frogs, but the outermost one in salamanders. When a generalized salamander limb bud was treated with a mitotic inhibitor, the outermost digit was lost. The same experiment on the frog would yield the loss of the innermost digit. Thus developmental differences have evolved in the two groups that determine which variations can evolve (Alberch and Gale 1985).

Proponents of the supremacy of extrinsic determinants of form use the adaptive landscape as the central metaphor correlating the diversity of organisms with the great variety of environments and niches. They argue that the preponderance of convergences and parallelisms in nature reflect the molding forces of natural selection.

In this chapter I will attempt to examine the relative roles of intrinsic and extrinsic determinants of organismic design by combining functional and comparative studies of adults with functional morphological analyses of ontogenetic stages. I propose that the focus on intrinsic determinants in early life history stages is not mutually exclusive with the functional morphological approach. A simultaneous analysis of both the intrinsic and extrinsic determinants seems to be the best approach to explain the emergence of variation in design, to elucidate the possible causal factors underlying the variation, and to demonstrate the evolutionary consequences of the variations in design (Lauder et al. 1989).

2 Networks of Interacting Constraints

Internal constraints in the skulls of vertebrates are the result of various interactions between structural components (e.g., Dullemeijer 1974; Liem 1980; Lauder 1981). When components interact mechanically, developmentally, or genetically, it will be difficult to change one component independently of the others. Such a linked system is thought to be coupled (Liem 1980; Lauder et al. 1989). For example, in the adult teleost fish, the various structural elements interact upon each other by their mechanical, structural, functional, and topographical properties, the nature of which can be determined by the methods of functional morphology (Fig. 1). In a coupled network, as found in the head of a primitive percoid fish, jaw protrusion and opening are closely linked. Thus the maxilla is used to perform protrusion of the premaxilla, to open the mouth, and also to close the jaws! Such a tight coupling is the result of a mechanical linkage between the premaxilla and maxilla (Fig. 2). This coupling dictates the very limited range of independently rotating and hinge movements that the maxilla and premaxilla can make. If a pull is exerted part way along the maxilla, the maxilla rotates along its long axis. This rotation of the maxilla has one predictable effect on the premaxilla: it is pushed forward by the mechanical linkage between the two bones.

In the cichlid fish, the maxilla and premaxilla mechanically are decoupled (Fig. 3). Each of the two elements is now available for independent specialization.

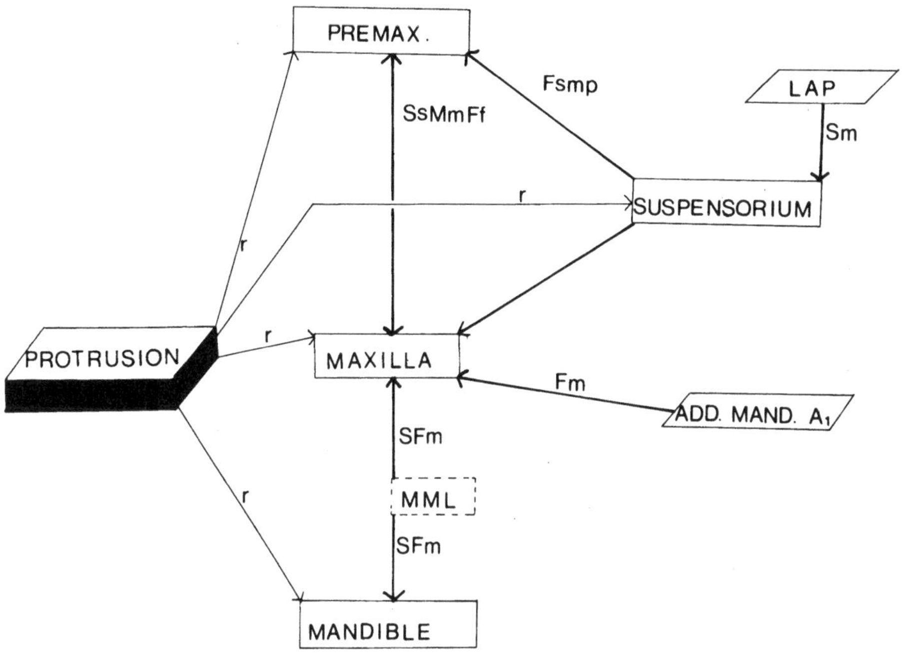

Fig. 1. Pattern of mutual influences between structural elements in the head of a generalized percoid fish, involved in upper jaw protrusion. *Rectangles drawn with solid and broken lines* stand for, respectively, bony and ligamentous morphological elements, and *parallelograms* for muscular morphological elements. The *three-dimensional rectangle* represents an activity having a major significance for the form of the head. The *heavy arrows* run in the direction of the influenced element. The *light arrows* indicate realization. The *capital letters* indicate the property that is influenced, the small letters the influencing property. *f* function; *m* mechanical influence; *p* position; *s* shape and structure; *r* relation between form and function. *ADD. MAND. A_1* adductor mandibulae A_1, muscle; *LAP* levator arcus palatini; *PREMAX* premaxilla. (Liem 1980)

It can be considered as a release of a constraint. As a result, the network of interacting constraints underlying jaw protrusion has changed drastically (Fig. 4). The release of the constraint has opened four pathways by which jaw protrusion can be realized: (1) lifting of the neurocranium by the epaxial muscle (EM); (2) abduction of the suspensorium by the levator arcus palatini (LAP); (3) by depressing the mandible as a result of action by the levator operculi (LO); (4) depression of the mandible by action of the sternohyoideus (SH). Lauder et al. (1989) have emphasized the importance of multiple mechanical pathways performing the same or similar functions as setting the stage for one of the pathways to be modified to acquire an improved or new function. Thus in functional and constructional morphology, adaptation should not be considered exclusively as the response of a trait to external environment. The adaptive potential and sensitivity to changes in the external environment depends on the nature of the network of interacting constraints.

An increase in the number of independent mechanical pathways underlying a biological role can be hypothesized to favor the evolution of morphological and

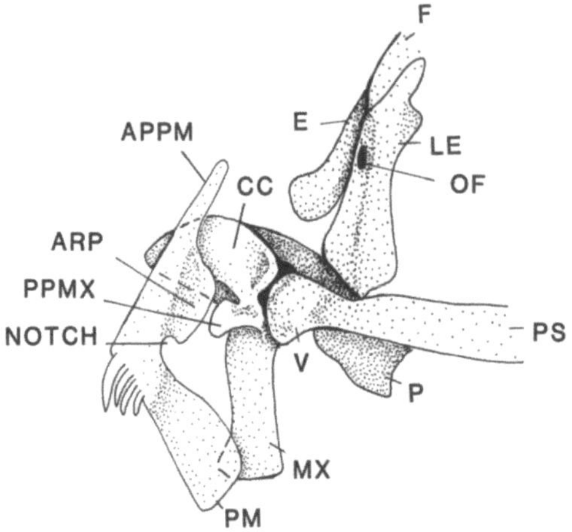

Fig. 2. Coupled premaxilla and maxilla in *Embiotoca lateralis.* Medial view of the right premaxilla, maxilla, and palatine, to show abutting articular process of the premaxilla (*ARP*) and the premaxillary process of the maxilla (*PPMX*). *APPM* ascending process of the premaxilla; *ARP* articular process of the premaxilla; *CC* cranial condyle of the maxilla; *E* ethmoid; *F* frontal; *LE* lateral ethmoid; *OF* olfactory foramen; *P* palatine; *PM* premaxilla; *PPMX* premaxillary process of the maxilla; *PS* parasphenoid; *V* vomer

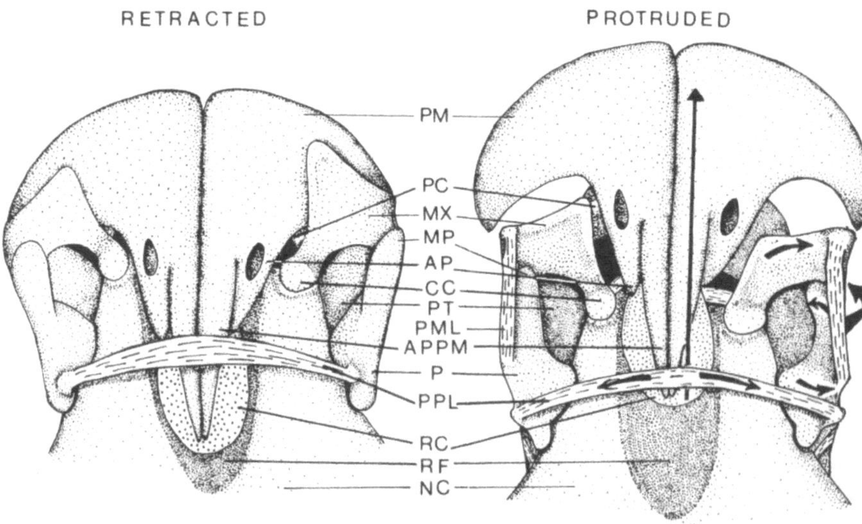

Fig. 3. Dorsal view of principal components of the apparatus underlying upper jaw protrusion in *Cyathopharynx furcifer.* Position of the elements determined from carefully fixed specimens with retracted and closed jaws, and protracted open jaws. Principal movements depicted by ***heavy arrows.*** Tension in palatopalatine depicted by ***arrows in opposite directions.*** *AP* articular process of the premaxilla; *APPM,* ascending process of the premaxilla; *CC* cranial condyle; *MP* maxillary process; *MX* maxilla; *NC* neurocranium; *P* palatine; *PC* premaxillary condyle or process of the maxilla; *PM* premaxilla; *PML* palatomaxillar ligament; *PPL* palatopalatine ligament; *PT* pterygoid portion of the suspensorium; *RC* rostral cartilage; *RF* rostral fossa (Liem 1980)

functional diversity in several systems when exposed to the appropriate environmental conditions (Lauder 1981; Lauder et al. 1989; Liem 1987, 1989). Thus, a clade possessing a decoupled network of interacting constraints will exhibit an increase in morphological and functional diversity when compared to sister clades with a coupled network of interacting constraints which serve as control groups

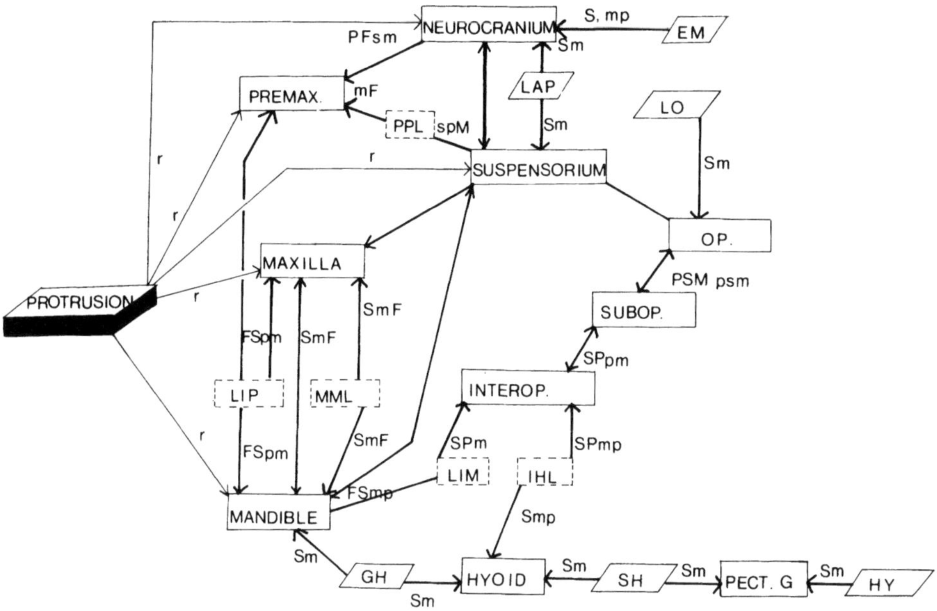

Fig. 4. Pattern of influences between structural elements in the head of *Petrotilapia tridentiger.* For further explanation see caption of Fig. 1. *ADD. MAND. A_1* adductor mandibulae part A_1; *EM* epaxial muscles; *GH* geniohyoideus; *HY* hypaxial muscles; *IHL* interoperculohyoid ligament; *INTEROP.* interoperculum; *LAP* levator arcus palatini; *LIM* interoperculomandibular ligament; *LO* levator operculi; *MML* maxillomandibular ligament; *OP* operculum; *PECT. G.* pectoral girdle; *PPL* palatopalatine ligament; *PREMAX.* premaxilla; *SH* sternohyoideus; *SUBOP.* suboperculum (Liem 1980)

for a natural experiment (Lauder 1981; Skelton 1985; Stiassny and Jensen 1987). Such a hypothesis has been tested recently by Lauder et al. (1989) as illustrated in Fig. 5. Actinopterygian fishes primitively have only one biomechanical pathway to lower the mandible. Among this radiation, a second biomechanical pathway for lowering the mandible evolved at the halecostome level. Thus two independent pathways underlying jaw opening evolved, decoupling the second system from the mechanical constraints imposed on the coupled primitive system. The effect of this decoupling is significant as a primitive clade of teleost fishes with the two independent pathways, the Osteoglossomorpha, has greatly modified the primitive hyoid pathways for mastication (Fig. 5). The decoupling of the hyoid pathway from its primitive mouth-opening function is predicted to have permitted the evolution of a morphologically and functionally specialized masticatory system, provided the appropriate environmental conditions are present (Lauder et al. 1989).

3 Ontogeny and Constructional Morphology

Functional analyses on the feeding apparatus of teleost fishes have yielded a thorough understanding of the kinematics, muscle functions, and the output performance of the apparatus (e.g., Alexander 1967; Osse 1969; Liem 1978; Lauder

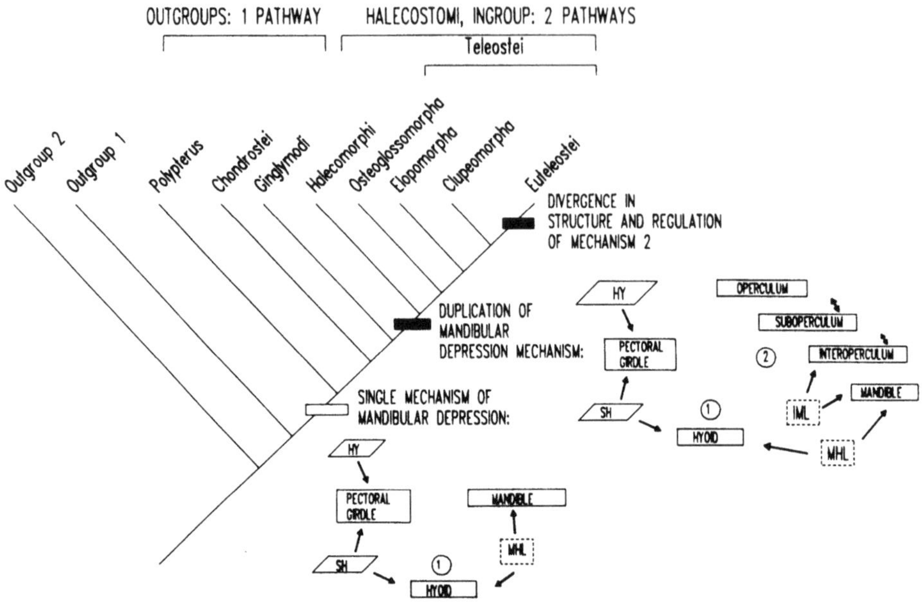

Fig. 5. Diagram of the duplication in biomechanical pathways in the feeding mechanism of ray-finned fishes. The diversity of jaw structure and function in the skull of lower vertebrates is causally related to the primitive duplication of biomechanical pathways. Primitively in ray-finned fishes, there is one biomechanical pathway for depressing the mandible when the mouth is opened. This mechanism involves co-contraction of the hypaxial muscles (*HY*), the sternohyoideus muscle (*SH*), and the application of a force to the mandibulohyoid ligament (*MHL*). The halecostome clade (the ingroup) is characterized primitively by two biomechanical pathways, both of which act to open the mouth. One of these (*1*) is identical to the primitive system, while the second (*2*) is a novel functional linkage between the interoperculum and the mandible via the interoperculomandibular ligament (*IML*). In the diagrams of the biomechanical pathways, muscles are shown in *parallelograms*, bones in *solid rectangles*, and ligaments in *dashed rectangles*. Note that much subsequent divergence of function and structure in the hyoid linkage system occurs in teleost fishes. This diversification is causally related to the duplication of biomechanical pathways that allowed specialization to occur (Lauder et al. 1989)

1985; Sibbing 1986; Van Leeuwen and Muller 1984). These findings have enabled Otten (1983) to formulate a mathematical computer model to deal with the kinematics and the statics of the jaw apparatus of teleosts. Otten devised a calculation procedure to determine the importance of strategic anatomical points ("hot spots") in the mechanism for force transmission during biting and suction (Fig. 6). In the model it is possible to change the "hot spots" so that the efficiency (or performance) of the system for the functions of biting or suction is altered. The relative positions of the "hot spots" and proportions of musculoskeletal elements change during ontogeny. Thus the functional design of the teleost skull undergoes important transformation during the early life history of fishes. Empirical data on the changes of the "hot spots" are based on studies for the Cichlidae, Pomacentridae and Embiotocidae.

In the cichlid *Haplochromis elegans*, Otten (1983) has reported the following changes in the "hot spots" during ontogeny from the free-living larva to the adult:

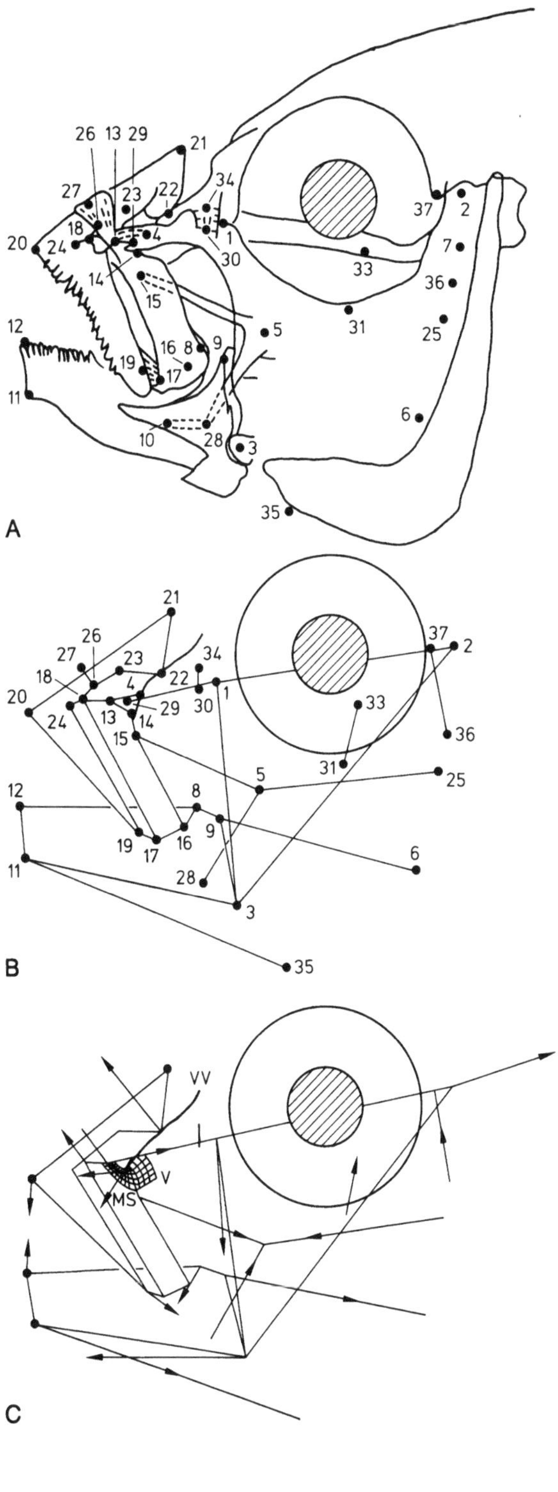

Fig. 6. "Hot spots" in the cichlid *Astatotilapia elegans* (Otten 1983). Anatomical positions and proportions are indicated on the skull components. They are selected on the basis of functional and anatomical data of the adult fish. Forces act on each of these spots and movements occur on several of the positions, which have been measured. Comparative studies on forces, velocities and prey capture patterns have revealed that shifts of proportions and positions of these spots are correlated with changes in function. These changes can be derived from the forces necessary to balance the bony elements during a bite. Biters would have differently positioned hot spots than suction feeders. **A** The head with the selected anatomical points. **B** Line diagram of the jaw mechanism. **C** Line diagram of the jaw mechanism with *arrows* indicating the forces necessary to balance bony elements during biting

1. Lengthening of the premaxillary ascending process.
2. Position of the premaxillary ascending process becomes less steep.
3. The palatopalatine ligament moves caudad relative to the maxilla.
4. Lengthening of the coronoid process of the dentary.
5. Shortening of the maxillary process of the palatine.

Otten (1983) concludes that because of the more favorably positioned "hot spots" in the free-swimming larva, the apparatus is better adapted for snapping the jaws ("biting" or suction) than the adult. These mechanical advantages in the larva over the adult are needed to counter the disadvantages resulting from the proportionally much larger eyes. The proportionally enormous eyes displace the tendons of the adductor mandibulae A_1 ventrally and the insertion sites on the maxilla anterodorsally. During ontogeny, the jaw-closing mechanism is optimized by having five of the seven "hot spots" in favorable positions to compensate for the spatial demands imposed by the proportionally enormous eyes, which prevent the optimizations of the adductor mandibulae complex. During the early life history of the fishes, absolute eye size is highly critical for visual acuity (Otten 1983). A free-swimming larva reserves virtually all the retinal space for visual acuity. It can be hypothesized that the eye is a dominant organ for survival of the larva. However, a high selective premium also applies to the feeding apparatus. The two competing demands, i.e., visual acuity and high feeding performance, are accommodated by the constructional morphology of the larval design in cichlids.

To test the constructional morphological notion that compensatory design in the feeding apparatus will neutralize the adverse effects of a proportionally large eye, analyses have been performed on free-swimming larvae of members of the pomacentrids, which are related to cichlids, *Amphiprion frenatus* and *A. polymnus.* The data obtained by Lisa Dilling in my laboratory do not always agree with the findings in the cichlid *H. elegans.* "Hot spot" # 1 predicts a lengthening of the premaxillary ascending processes. Instead the premaxillary process lengthens only during the first 8 days after hatching and then remains constant in proportion to overall head length (Fig. 7). "Hot spot" # 2 predicts that the angle of the ascending process will become "less steep" as ontogeny progresses in *H. elegans.*

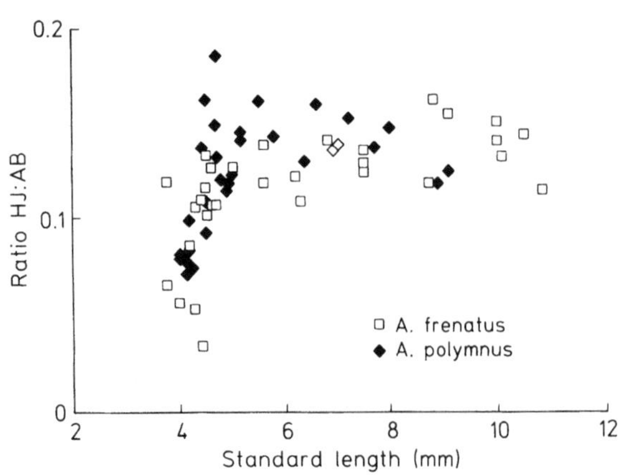

Fig. 7. Ratio ascending process of premaxilla (*HJ*): head length (*AB*) versus standard length (Dilling 1989). Note decreasing variance of ordinate values as ontogeny progresses, indicating possible stabilizing selection

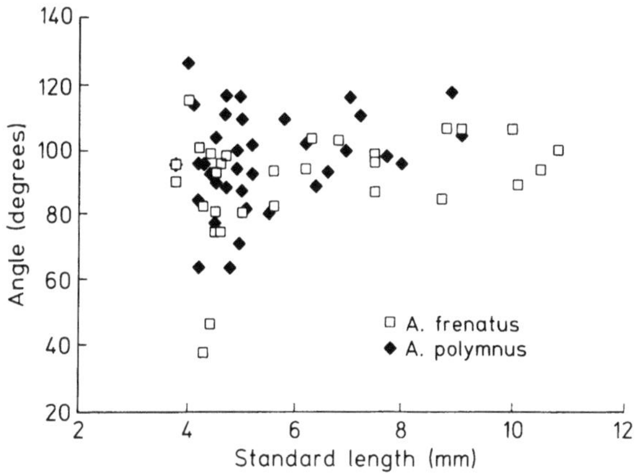

Fig. 8. Angle of premaxilla ascending process versus standard length (Dilling 1989). See note for Fig. 7

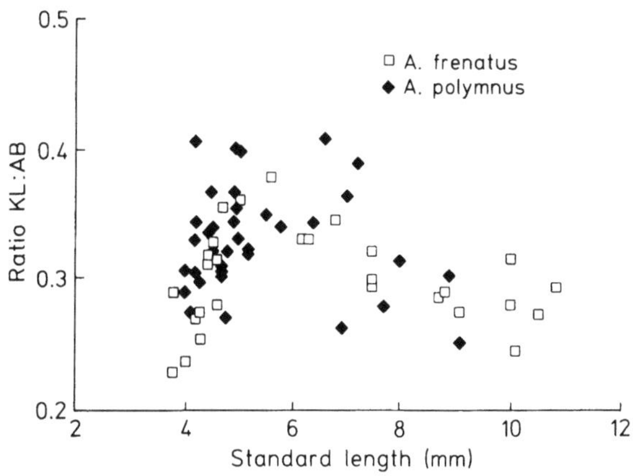

Fig. 9. Ratio coronoid process of the dentary (*KL*): head length (*AB*) versus standard length (Dilling 1989)

However, in *Amphiprion* the angle remains constant during development even if we break down the curve into early and late development (Fig. 8). The data in Fig. 7 and 8 seem to show higher variances of ordinate values at low values of standard length. This could reflect stabilizing selection in free-living larvae, with decreasing variance around mean values as the larvae grow. In regard to "hot spot" # 4, the coronoid process lengthens in the first 8 days in relation to head length, but begins shortening in relation to head length as the rest of the head catches up development (Fig. 9). The palatopalatine ligament ("hot spot" # 3) moves caudad relative to the maxilla in the first 8 days, but remains relatively fixed in this position for the rest of the ontogeny. A shortening of the maxillary process of the palatine ("hot spot" # 5) is measurable in the first 8 days of development. It seems that the design changes during approximately the first 8 days after hatching, accompanied by a selective "narrowing" of variance.

Thus, it is in the first 8 days that we can observe a drastic change in skull design in the developing larva in *Amphiprion.* These changes coincide with functional changes described by Otten (1983). The pattern of changes can indeed be interpreted as compensatory adjustments to minimize the disadvantages imposed by the relatively large eyes in earlier ontogenetic stages. The necessity of relatively large-sized eyes can be considered a constraint on the design of the feeding apparatus during early ontogeny and in adults (Barel 1983; Otten 1983; Dullemeijer 1989).

Throughout the early ontogeny of both cichlid and pomacentrid larvae, the ratio of the depth of the head and the head length increases (Fig. 10). An increase in head depth is accompanied by a switch from ram feeding to suction feeding. After hatching the larva has a shallow suspensory apparatus, a large gape and a cylindrical buccopharyngeal cavity (Fig. 11). Such a cylindrical design is not con-

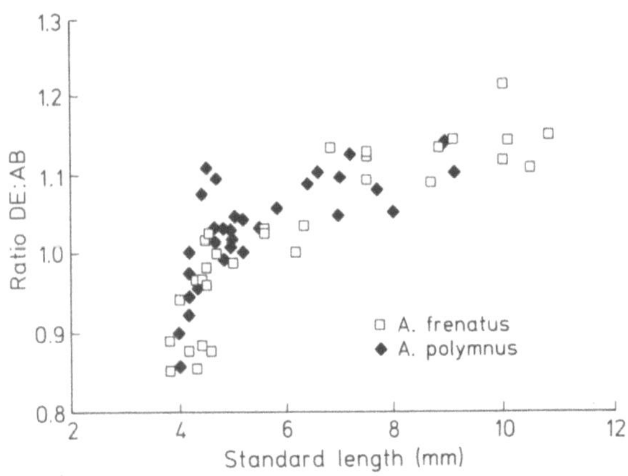

Fig. 10. Ratio depth of head (*DE*): head length (*AB*) versus standard length (Dilling 1989)

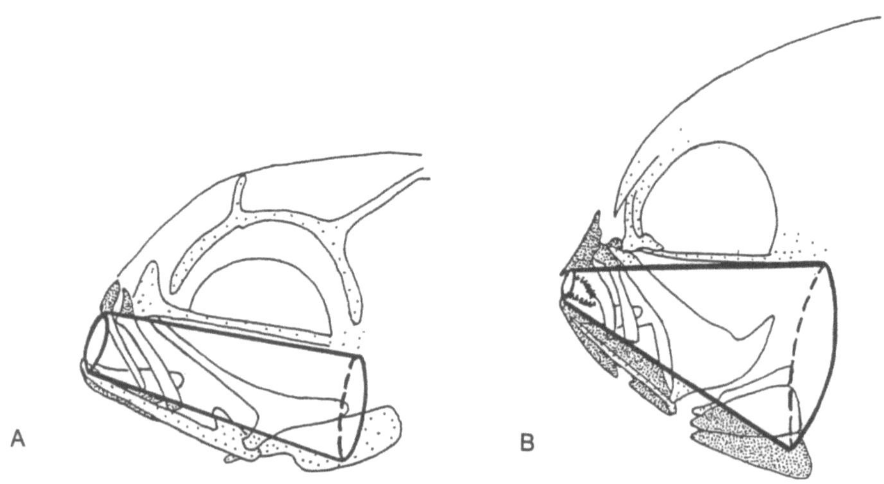

Fig. 11. A Early larva of *Amphiprion* with cylindrical buccopharyngeal cavity and large gape. **B** Late larva with conical buccopharyngeal cavity and gape (Dilling 1989)

ducive for suction feeding (Brainerd 1985). High speed video analysis shows that early larvae with cylindrical buccal cavities invariably feed by ram feeding, i.e., they feed by approaching the prey, protruding and opening their jaws over the prey while moving forward and engulfing the prey, followed by adduction and retraction of the jaws. When the skull changes from a cylindrical to a truncated cone design (Fig. 11) the larva switches from ram feeding to a predominantly suction feeding strategy. The skull is characterized by a greater depth, a deep suspensory apparatus, relatively small gape, a mandible with a longer ascending process of the dentary, and relatively shorter jaw bones. At this stage, the larva becomes a very efficient suction feeder, as extrapolated from its feeding mode and prey capture success. Suction is hypothesized to be generated by a very rapid expansion of the cone-shaped buccopharyngeal cavity (Osse and Muller 1980; Lauder 1985).

Otten (1982) has demonstrated a threshold ontogenetic stage of cichlids during which there is a very rapid change in the mouth-opening mechanism. It involves: (1) the establishment of an articulation of the maxilla with the palatine; (2) the differentiation of a coronoid process (ascending process of the dentary) to form a new corner of the gape of the mouth; (3) the differentiation of the levator operculi muscle, and ligament between the interoperculum and the lower jaw thereby establishing the levator operculi coupling to open the mouth (Liem 1970). This coupling to open the mouth is activated as soon as the working line of the geniohyoideus muscle has moved dorsal to the jaw joint (between quadrate and mandible). Just prior to this stage, the working line of the geniohyoideus muscle is ventral to the jaw joint and represents the primary jaw-opening mechanism so characteristic of the initial ontogenetic stages. Contraction of the geniohyoideus in the position ventral to the jaw joint would pull the mandible down, especially since the very large hyoid is stabilized by the sternohyoideus muscle (Fig. 12). However, during growth, the working line of the geniohyoideus muscle moves dorsal to the jaw joint (Fig. 12) and consequently can no longer operate as a jaw opener. It is at this threshold stage that the levator operculi coupling becomes operational to open the mouth. Without an operational levator operculi coupling the larva would not be able to open its mouth to breathe. Thus the relative position of the jaw joint is very important and is a highly programmed feature of the emergence of kinematic mechanisms constrained by changes in the proportions of the head. Similarly exact synchronizations of developmental patterns of the changes in the position of the jaw joints and emergences of the levator operculi couplings are found in pomacentrids and other cichlids. Thus ontogenetic change is finely tuned to the execution of vital functions.

As ontogeny progresses, new couplings evolve through the differentiation of ligaments, i.e., the ligament between the hyoid and interoperculum, the palatopalatine ligament, and other linkages between bony elements. The result is a network of interacting constraints with built-in redundancies in mechanical pathways (Fig. 4). These couplings may become weaker or stronger depending on the changes of the proportions of the structural elements, which in turn depend on changes in the functions of the components. These data support the notion that the sequence of ossifications, appearance of ligaments, and establishment of articulations reflect adaptations to specialized and vital larval requirements. For example, a delayed onset of the opercular coupling when the working line of the

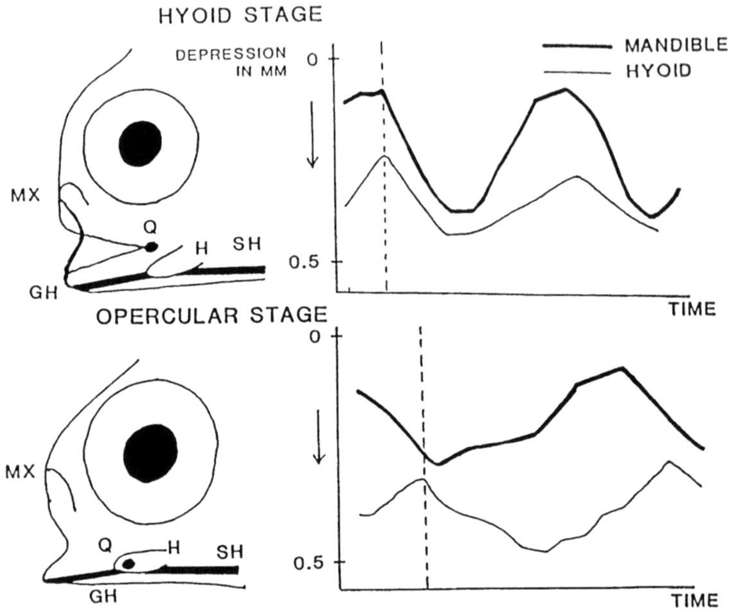

Fig. 12. The hyoid and opercular stages in *Cichlasoma*. Larva begins with the hyoid stage with the hyoid (*H*) positioned in such a way that the working line of the sternohyoideus (*SH*) and geniohyoideus (*GH*) muscles remain below the jaw joint *Q*. *MX* head of maxilla. The graph shows that mandibular depression is synchronous with that of the hyoid, indicating that mouth opening is accomplished by action of the sternohyoideus and levator operculi muscles. In the subsequent opercular stage, the hyoid has shifted position in such a way that the working line of the sternohyoideus and geniohyoideus muscles moves above the jaw joint *Q* rendering it ineffective in opening the mouth. At this stage another mouth-opening mechanism emerges (the levator operculi coupling discussed in the text) and one can see from the kinematic analysis that mandibular depression occurs when the hyoid moves upward

geniohyoideus muscle has moved above the jaw joint would result in a very high risk of asphyxiation of the larva.

If developmental constraints are defined as biases on the production of variant phenotypes or limitations on phenotypic variability caused by the structure, character, composition, or dynamics of the developmental system (Smith et al. 1985), the synchronization of the proportional changes in the developing fish skull (including the shift of the jaw joint and the differentiation of the levator operculi coupling) constitutes a constraint, since it is quite binding for the survival of the larva. Constructional morphology has also revealed the necessary compensatory features which invariably accompany the large eyes of fish larvae. The co-occurrence of large eyes and the positions of kinematic "hot spots" can also be interpreted to constitute developmental constraints. Such developmental constraints may well canalize the subsequent developmental trajectories of the components.

4 Intrinsic Versus Extrinsic Determinants of Form

Emerging structures often appear in a finely timed sequence thereby establishing functions without which the evolving larva is not viable. Constructional morphological principles underlie further differentiation of the teleost skull, resulting in a design that compensates for the adverse effects of the proportionally large eyes so characteristic of early ontogenetic stages. Thus ontogenetic features are structurally and temporally integrated and perform essential functions for the survival of the larva. Such structural, functional, and temporal integrations constitute constraints, giving the emerging systems directionality. Functional demands may well dictate the precise timing, mechanically adaptive shapes and strategic kinematic connections of emerging structures during ontogeny. How can these functional findings be reconciled with the structuralist (developmentalist) perspective emphasizing the internal rules of assembly (intrinsic determinants) rather than functional demands (extrinsic determinants)?

In order to assess the relative roles of intrinsic and extrinsic determinants of form, I have compared the patterns of development of emerging kinematic connections, proportions and their probable functions in phylogenetically closely related lineages (Stiassny and Jensen 1987) with strikingly different modes of development. In the Pomacentridae, development proceeds outside the mother, with the various stages of the free-living larvae occupying radically different niches. Selection pressures (see Figs. 7, 8) seem to be particularly intense (Balon 1985; Kendall et al. 1984; Lasker 1981) as the developing fish passes through a sequence of steady states or thresholds, interrupted by short unsteady phases when several organ systems pass at accelerated rates to become functional, thereby enabling the fish to make internal adjustments to changing life strategies. For example, the switch from the hyoid to the opercular stages of mouth opening is critical and even the slightest error in timing results in considerable mortality due to asphyxiation. In many teleosts, larvae suffer the greatest mortality during the switch from hyoid to opercular stage. In sharp contrast, the Embiotocidae exhibits not only highly specialized internal fertilization and viviparity, but also an extremely prolonged period of development within the ovary, and birth does not occur until the juveniles are in a highly advanced stage of development, exiting from the mother as small replicas of the adults in all major features. Within one hour of birth, the young begin to feed in the same way on the same prey as the adults! (Schmitt and Holbrook 1984; Wiebe 1968).

In embiotocids, head structures differentiate without the functional demands for aquatic respiration and feeding. Respiration by means of gas exchange through the surface of characteristically enlarged and vascularized median fins (Webb and Brett 1972) and intraovarian feeding does not require the demands of speed, suction, and biting seen in free-swimming fish larvae. A comparative functional analysis of ontogenetic trajectories of the major skull components in the two groups may reveal the relative roles of programmatic intrinsic factors, as exhibited in the viviparous embiotocids, and the extrinsic or functional molding forces, as exhibited in pomacentrids, in which all development must be flawlessly designed and scheduled for entry into a competitive world.

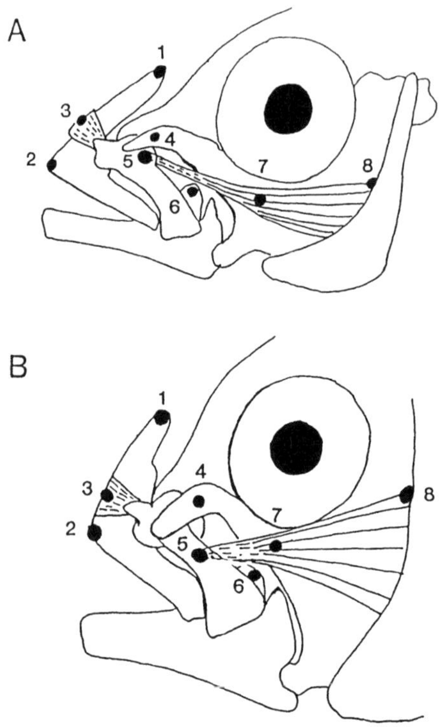

Fig. 13. "Hot spots" in two developmental stages in *Amphiprion*. **A** "Hot spots" in the younger ram feeding larva. Proportions and positions are conducive to rapid expansion and protrusion of the mouth as determined by high speed motion analysis. These proportions would be ineffective in "biting". **B** "Hot spots" in the subsequent suction feeding larva. Proportions and positions have changed and are conducive to the functional shift to suction feeding and "biting" as revealed by high speed motion analysis

The switch from the hyoid stage to the opercular stage of jaw opening accompanying the proportional changes during the growth of the skull do occur quickly and synchronously with the movement of the jaw joint below the working line of the geniohyoideus muscle in both the cichlids (Otton 1982; pers. observ. of *Pterophyllum, Cichlasoma*) and pomacentrids (Fig. 12). This synchronously co-occurring pattern is functionally adaptive since it enables the larva to breathe and therefore to survive. The simultaneous shift in jaw joint position and switch from the hyoid to opercular mode of jaw opening are rigidly coupled and constitute a constraint. In the viviparous embiotocids there is no need for the synchronously co-occurring shift to emerge since the developing larva does not use the mouth for either breathing or feeding. Yet the shift occurs synchronously albeit at a greatly reduced speed (Figs. 14, 15). The jaw joint moves anteriorly and so does the relative position of the hyoid. The working line of the geniohyoideus moves dorsal to the jaw joint. At this stage the levator operculi muscle becomes functional, acting on gradually ossifying opercular bones and a prominent interoperculomandibular ligament. Thus the very slowly emerging synchronously co-occurring shift seems to be internally generated on the basis of programmatic properties. Even in the absence of functional demands, the synchronously co-occurring shift takes place programmatically. It proceeds according to an internally determined order. This internally generated pattern in the embiotocids may have been inherited from its plesiomorphous nonviviparous ancestors. Phylogenetical-

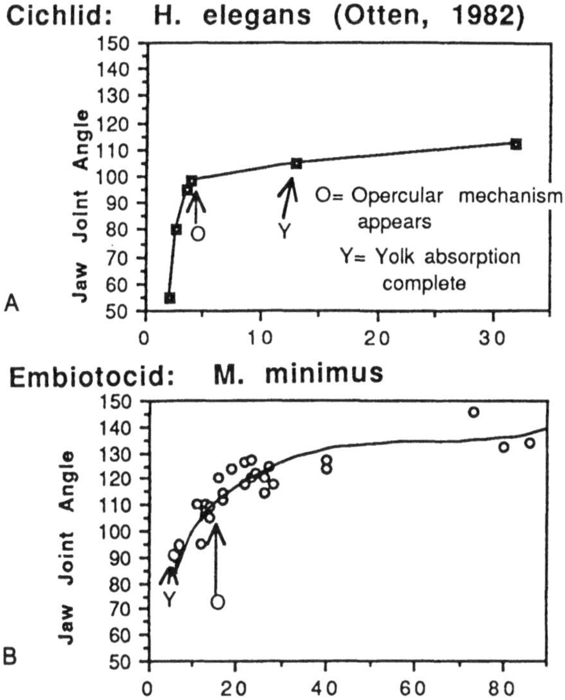

Fig. 14. Jaw joint position. The position of the joint of the mandible was determined by measuring the angle between lines from the center of the eye to the jaw joint and to the middle of the portion of the chorda dorsalis under the cleithrum. The angle is plotted on the *vertical axis* and age in days on the *horizontal axis*. **A** The cichlid *Haplochromis elegans* (After Otten 1982). **B** The embiotocid *Micrometrus minimus* (Golomb 1990). *O* opercular mechanism appears; *Y* yolk absorption completed. Note change in angle on day 6 in *H. elegans* and on day 12 in *M. minimus*

ly, the evidence (Stiassny and Jensen 1987) favors the hypothesis that the pattern in embiotocids is indeed inherited from nonviviparous ancestors. Hence the implication is that a residual canalizing gene complex has not yet decayed through mutation. Since a change in the pattern would have been so crucial in the ancestors, it is very likely that a fairly large gene complex (with much homeostatic "backup") might have evolved, and so would decay only very slowly under normal point mutation or drift.

Likewise the transformation in designs from ram feeding to suction feeding and/or biting are accompanied by a programmatic sequence of events in both the oviparous cichlids and pomacentrids (Figs. 6, 11, 13) as well as the viviparous embiotocids (Fig. 15). Even in the embiotocid larvae that neither ram feed nor suction feed, the buccopharyngeal cavity changes from a cylindrical to a truncated cone design. Early in ontogeny the larva has a shallow suspensory apparatus and a large gape. As development proceeds the suspensory apparatus becomes deeper and the gape becomes relatively smaller. Intraovarian embiotocids do move their jaws, but they do not bite or suction feed. Yet, earlier larvae with proportionally large eyes possess shorter ascending processes of the premaxillae, steeply positioned premaxillary ascending processes, forward positioned "palatopalatine" ligaments, shorter coronoid processes of the dentary, and longer maxillary processes of the palatine. Later in ontogeny, when the eyes become relatively smaller, all these features change so that the ascending premaxillary processes become longer and less steep, the "palatopalatine" ligaments move backward, the coronoid pro-

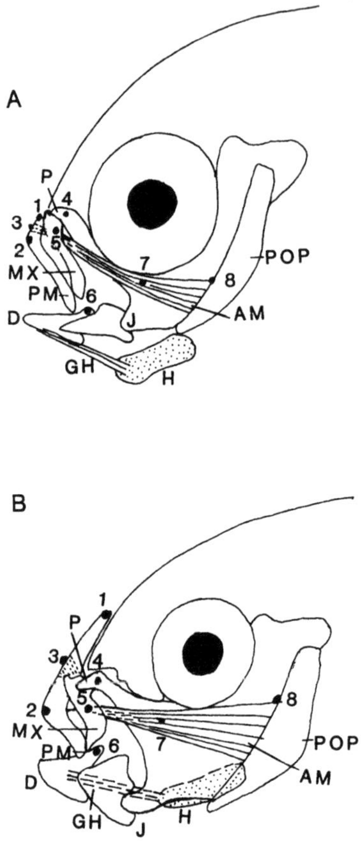

Fig. 15. The hyoid (**A**) and opercular (**B**) stages and "hot spots" in *Micrometrus minimus*. Larva begins with the hyoid (*stippled, H*) positioned in such a way that the working line of the geniohyoideus (*GH*) is positioned below the jaw joint. In later stages the jaw joint is positioned ventrally to the working line of the geniohyoideus. In the hyoid stage (**A**), the design is comparable to that of the ram feeding larva of cichlids and pomacentrids (see Fig. 13). In the opercular stage (**B**) the proportions and positions of the "hot spots" (*1–8*) have shifted to the pattern found in typical suction feeding cichlids and pomacentrids. AM_1 part A_1 of the adductor mandibulae complex; *D* dentary; *GH* geniohyoideus; *H* hyoid; *J* jaw joint; *MX* maxilla; *P* palatine; *PM* premaxilla; *POP* preoperculum

cess of the dentary becomes higher and the maxillary process of the palatine becomes proportionally shorter (Fig. 15). Thus the adductor mandibulae muscle complex is optimized in the early larval stages when the larva does not even bite. We may conclude that the changes accompanying the proportional reduction in eye size proceed programmatically. The changes in "hot spots" (Figs. 13, 15) seem also to be governed by internal rules that order the appearance of morphological transformations even in the absence of functional demands. These internal rules are most likely inherited from the ancestors in which the developmental pattern was of great functional significance.

The comparative functional analysis of the developmental trajectories of skull features in oviparous and viviparous fishes has revealed that intrinsic and extrinsic determinants of form are not mutually exclusive. Internal rules do govern the appearance of morphological transformations that are of great functional significance and of great survival value. Actually, it is impossible to assign a dominance hierarchy of form over function or vice versa. Internal rules of assembly of form and vitally adaptive functions are interconnected at the level of the generative process.

5 Conclusions

As eloquently argued by Alberch (1989) in his studies of monsters and teratologies, the theory of form has to be based on the properties of the complete network of interacting constraints that characterize development. In this chapter I have illustrated the nature of interactions between emerging elements resulting in integrated components that underlie vitally important functional adaptations. Integrations result in various kinds of contraints. Both intrinsic and extrinsic determinants of form are tightly linked. Internal programmatic rules governing the emergence of morphological variations and transformations and their vitally adaptive functions are inseparably interconnected at the level of the generative process. As a result, each integrated system has an intrinsically defined potentiality allowing only a very limited set of possible transformations. Selection determines the probability of survival of each of the possible transformations and indirectly affects the governing internal morphogenetic programs (Alberch 1989). Thus the realm of possible phenotypes is internally canalized. Internal canalization combined with selection would produce the discontinuous variation in morphology and the empty spaces in the phenotypic space.

Acknowledgments. I am greatly indebted to Dr. Eric Schultz for collecting the splendid series of embiotocid larvae, to Dr. Les Kaufman for the pomacentrid larvae, and to Lisa Dilling and Meredith Golomb, who have provided most of the research data on respectively the pomacentrids and embiotocids. I thank Dr. Peter Skelton for greatly improving the manuscript and providing me with conceptual ideas. The following individuals have contributed significantly: Chris Rose, Karsten Hartel, Jeff Jensen, Ernest Wu, Bruce Miller, John Wourms, Lisa Dilling, Meredith Golomb, Chien Chen, Colleen Boggs and Les Kaufman. To them I extend my gratitude. This research is supported by a William Milton Fund Grant and the National Science Foundation Grant # BSR 88-18014.

References

Alberch P (1989) The logic of monsters: evidence for internal constraint in development and evolution. Geobios Mém Spéc 12:21–57

Alberch P, Gale E (1985) A developmental analysis of an evolutionary trend: digital reduction in amphibians. Evolution 39:8–23

Alexander RM (1967) The functions and mechanisms of the protrusible upper jaws of some acanthopterygian fish. J Zool (London) 151:43–64

Balon EK (1985) Early life histories: new developmental, ecological and evolutionary perspectives. Junk, The Hague, 280 pp

Barel CDN (1983) Towards a constructional morphology of cichlid fishes (Teleostei, Perciformes). Neth J Zool 33:357–424

Brainerd EL (1985) An experimental analysis of prey capture mechanisms in teleost fishes with implications for present hydrodynamic models of suction feeding. Undergraduate Honors Thesis, Department of Organismic and Evolutionary Biology, Harvard University

Dilling L (1989) An ontogenetic study of the jaw mechanism and feeding modes in *Amphiprion frenatus* and *A. polymnus*. BSc Honors Thesis Biol, Harvard Univ, 49 pp

Dullemeijer P (1974) Concepts and approaches in animal morphology. Van Gorcum, Assen, Neth

Dullemeijer P (1989) On the concept of integration in animal morphology. In: Splechtna H, Hilgers H (eds) Trends in vertebrate morphology. Fischer, Stuttgart, pp 18–33

Golomb MR (1990) An analysis of the relative roles of internal and external factors underlying morphogenesis: Patterns of development in labroid fishes. BSc Honors Thesis Biol, Harvard Univ, 70 pp

Gould SJ (1982) Change in developmental timing as a mechanism of macroevolution. In: Bonner JT (ed) Evolution and development. Springer, Berlin Heidelberg New York, pp 33–346

Ismail HM, Verraes W, Huysseune A (1982) Developmental aspects of the pharyngeal jaws in *Astatotilapia elegans* (Trewavas, 1933) (Teleostei: Cichlidae). Neth J Zool 32:513–543

Kendall AW, Ahlstrom EH, Moser HG (1984) Early life history stages of fishes and their characters. In: Moser HG et al. (eds) Ontogeny and systematics of fishes. pp 11–22

Lasker R (1981) Marine fish larval. Morphology, ecology and relation to fisheries. Univ Press, Washington

Lauder GV (1981) Form and function: structural analysis in evolutionary morphology. Paleobiology 7:430–442

Lauder GV (1985) Aquatic feeding in lower vertebrates. In: Hildebrand M, Bramble DM, Liem KF, Wake DB (eds) Functional vertebrate morphology. Harvard Univ Press, Cambridge, MA, pp 210–229

Lauder GV, Crompton AW, Gans C, Hanken J, Liem KF, Maier WO, Meyer A, Presley R, Rieppel OC, Roth G, Schluter D, Zweers GA (1989) Group report: how are feeding systems integrated and how have evolutionary innovations been introduced? In: Wake DB, Roth G (eds) Complex organismal functions: integration and evolution in vertebrates. John Wiley & Sons, New York, pp 97–115

Leeuwen JL van, Muller M (1984) Optimum sucking techniques for predatory fish. Trans Zool Soc London 37:137–169

Liem KF (1970) Functional anatomy of the Nandidae (Pisces, Teleostei). Fieldiana Zool 56:1–66

Liem KF (1978) Modulatory multiplicity in the functional repertoire of the feeding mechanism in cichlid fishes. Part I: Piscivores. J Morphol 158:323–360

Liem KF (1980) Adaptive significance of intra- and interspecific differences in the feeding repertoires of cichlid fishes. Am Zool 20:295–314

Liem KF (1987) Functional design of the air ventilation apparatus and overland excursions by teleosts. Fieldiana NS 37:1–29

Liem KF (1989) Functional morphology and phylogenetic testing within the framework of symecomorphosis. Acta Morphol Neerl Scand 27:119–131

Meyer A (1987) Phenotypic plasticity and heterochrony in *Cichlasoma managuense* (Pisces, Cichlidae) and their implications for speciation in cichlid fishes. Evolution 4:1357–1369

Morris SL, Gaudin AJ (1982) Osteocranial development in the viviparous surfperch *Amphistichus argenteus* (Pisces: Embiotocidae). J Morphol 174:95–120

Osse JWM (1969) Functional anatomy of the head of the perch (*Perca fluviatilis* L.): an electromyographic study. Neth J Zool, 19:389–392

Osse JWM, Muller M (1980) A model of suction feeding in teleostean fishes with some implications for ventilation. In: Ali MA (ed) Environmental physiology of fishes. Plenum, New York, pp 335–352

Otten E (1982) The development of a mouth-opening mechanism in a generalized *Haplochromis* species: *H. elegans* Trewavas, 1933 (Pisces, Cichlidae). Neth J Zool 32:31–48

Otten E (1983) Vision and jaw mechanism during growth of the cichlid fish *Haplochromis elegans*. Doct Diss, Univ Leiden, Neth

Schmitt RJ, Holbrook SA (1984) Ontogeny of prey selection by black surfperch *Embiotoca jacksoni* (Pisces: Embiotocidae): The role of fish morphology, foraging behavior, and patch selection. Mar Ecol Prog Ser 18:225–239

Sibbing FA (1986) Food handling in the carp (*Cyprinus carpio*): its movements patterns, mechanisms and limitations. J Zool London A 210:161–203

Skelton PW (1985) Preadaptation and evolutionary innovation in rudist bivalves. In: Cope JCW, Skelton PW (eds) Evolutionary case histories from the fossil record. Spec Pap Paleontology 33:159–173

Smith JM, Burian R, Kauffman S, Alberch P, Campbell J, Goodwin B, Lande R, Raup D, Wolpert L (1985) Developmental constraints and evolution. Q Rev Biol 60:265–287

Stiassny MLJ, Jensen JS (1987) Labroid intrarelationships revisited: morphological complexity, key innovations, and the study of comparative diversity. Bull Mus Comp Zool 151:269–319

Strauss R (1984) Allometry and functional feeding morphology in haplochromine cichlids. In: Echelle AA, Kornfield I (eds) Evolution of species flocks. Univ Press, Maine, pp 217–229

Triplett EL (1959) Note on the life history of the barred surfperch *Amphistichus argenteus* Agassiz, and a technique for culturing embiotocid embryos. Cal Fish Game 90:433–439

Webb PW, Brett JR (1972) Oxygen consumption of embryos and parents, and oxygen transfer characteristics within the ovary of two species of viviparous seaperch, *Rhacochilus vacca* and *Embiotoca lateralis*. J Fish Res Bd Can 29:1543–1553

Wiebe J (1968) The reproductive cycle of the viviparous seaperch *Cymatogaster aggregata* Gibbons. Can J Zool 46:1221–1234

Wourms JP (1981) Viviparity: The maternal-fetal relationship in fishes. Am Zool 21:473–515

Self-Organizing Mechanisms in Morphogenesis and Evolution

ADOLF SEILACHER[1]

Summary

In the framework of morphodynamics (new term) self-organizing processes in the form of mechanical regulation and synergetic pattern formation are important elements of developmental and evolutionary transformation. Generally available mechanisms may first be randomly adopted and then become tamed by selection towards specific internal and external functions. Even in the well-adapted state, however, the autonomy of such processes is expressed by teratological cases and the unpredictability in details that do not infringe on fitness. The synergetic model also applies to systems in which a symbiotic partner, rather than a physical mechanism, has become adopted.

1 Constructional Morphology, Morphodynamics, and Evolution

The term "constructional morphology" was coined by the zoologist Hermann Weber (1955). It became later adopted by paleontologists (Seilacher 1970; Reif et al. 1985) for a morphological research program in which emphasis is placed on *morphogenetic* (or fabricational) constraints, in addition to the *phylogenetic* and *functional* factors, as determinants of biologic form. This view was graphically expressed by the constructional morphology triangle.

From the literature (e.g., Barel et al. 1989), from discussions at the Bad Homburg symposium, and from the very title of this book it became clear, however, that some zoologists use the term in a different and more restricted sense. For them, constructional morphology is the correlate to functional morphology – with the difference that it refers to internal and architectural, rather than external, functions (Nachtigall this Vol.). In order to avoid further confusion, a new term, *morphodynamics* (or biomorphodynamics, to avoid homonymy with a term in physical geography), is used here to express the broader relationships (Fig. 1).

At this point it is first necessary to emphasize that the scope of the field has amplified during the years. The original triangle (Seilacher 1970), still expressed by the front face in the pyramid of Fig. 1, was meant as a simple working procedure for the analysis of individual organisms. In the meantime, the approach has gradually gained a conceptual quality and expanded into the evolutionary dimen-

[1] Geol.-Paläontologisches Institut, Universität Tübingen, Sigwartstr. 10, D-7400 Tübingen, FRG and Dept. Geology and Geophysics, Yale University, Kline Geology Laboratory, P.O. Box 6666, New Haven, Connecticut 06511/8130, USA

N. Schmidt-Kittler and K. Vogel (Eds):
Constructional Morphology and Evolution

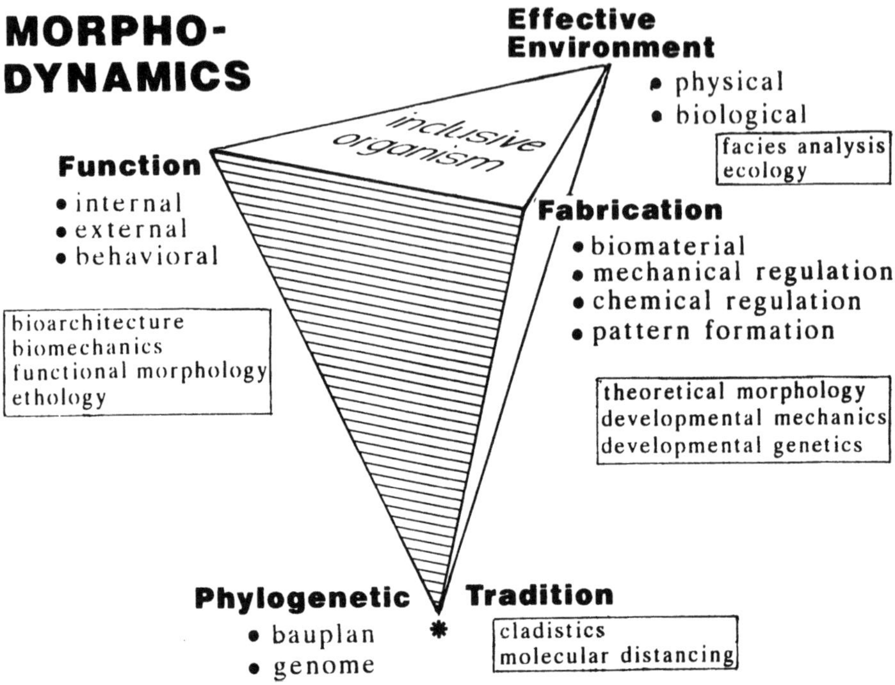

Fig. 1. The conceptual framework of (bio-)morphodynamics and some relevant approaches (*in boxes*). Note that the vertical time axis can be viewed at the ontogenetic as well as the phylogenetic scale

sion (Reif et al. 1985). In the perspective of *evolution*, however, the original three corners lose their equivalence. Darwinian selection, and thereby the basic driving force of evolutionary change, can be understood primarily at the level of fabrication and function and the resulting fitness of the phenotype. Phylogenetic constraints, in contrast, have a historical quality and tend to be more stringent and inclusive (for this reason, mass extinctions have their most effective lever at this level).

In a more encompassing concept, we must also reconsider the role of the *environment*. At first glance, it appears to be an extrinsic factor clearly separated from the organism. In biological reality, however, it is not an independent entity but consists of a specific set of parameters that are relevant to the life of the organism, including other organisms (von Üxküll 1928). This "effective environment" differs quantitatively and qualitatively from one species to another. It also changes through time – not only by changing physical conditions and evolution in the vicinity of an organism, but also by its own adoption of new environmental parameters through the establishment of new sensory and behavioral linkages (Peters this Vol.; hermit crab example in present chapter). In this perspective, the effective environment becomes an intrinsic part of the *inclusive organism*, which is the true unit of evolution. Only because naturalists are accustomed to describe and name only the collectible part of it, the term "organism" is colloquially understood in the restricted sense. In a truly holistic view, the effective environment must thus be included in the diagram and at the same level with function and fabrication (A. Gorthner 1990). As a result, we have a tetrahedral pyramid with a vertical time axis and the final inclusive organism at the top of the historical pile of earlier stages.

Since every feature of a Bauplan has once originated at the fabrication/function/environment level of selectional testing, evolutionary pathways can thus be reconstructed in a step-by-step analysis that leads back to the origination of the most basic designs (such as four-leggedness). In principle, morphodynamic backtracing resembles the procedure of cladistics, but with the important difference that fabricational, functional, and environmental explanations are required at every step. The pyramid of Fig. 1 may be adapted to the ontogenetic as well as the evolutionary time scales, because all transformations are controlled by a similar set of constraints. Thus it should be made clear in every case which scale is being considered.

A variety of *approaches* can be used, inductive as well as deductive ones, to reveal morphodynamic relationships. Genetic distance analysis, developmental genetics, biomechanics, theoretical morphology, and comparative functional morphology with regard to internal or external functions are represented in this volume, but others may emerge with the progress of science. In emphasizing self-organization, the present chapter will focus on processes of mechanical regulation (pneu structures), pattern formation (zebra designs and hierarchic fractals), and the adoption of symbionts, in relation to architectural and external functions. Methodologically, we will follow the good old tradition of comparative morphology. This approach, unduly discredited in modern biology, has the advantage that modifications within a population, or transformation within a clade and between convergent taxa, serve as natural experiments, against which conceivable models can be tested on the spot. Additional, but more expensive tests in terms of biomechanics (Wisser and Nachtigall this Vol.) or theoretical morphology (Weishampel this Vol.) have not been performed, but could probably be designed for further corroboration in each case.

2 Pneu Morphologies in Rigid Skeletons

2.1 Echinoid Tests as Rigified Balloons

Balloons and pressure capsules belong to the same family of structures, both in a morphological and a static sense. Ideally, they are spherical, but under load they transform into rotational ellipsoids and domes. The striking similarity in shape between a fluid-filled rubber balloon resting on a table and a regular echinoid test (Fig. 2) can also be considered as a kind of functional symmetry, because in the one case tensional forces are transmitted through an elastic skin, and in the other compressive forces through a rigid shell, with a minimum of buckling. In a fabricational sense, however, there is an important difference: balloon shapes do form automatically whenever a tensional membrane encloses a denser or pressurized space (water drops, bread loaves, melons), while rigid domes must be constructed. Therefore origin in a tensional state is likely whenever pneu morphologies occur in the world of hard skeletons.

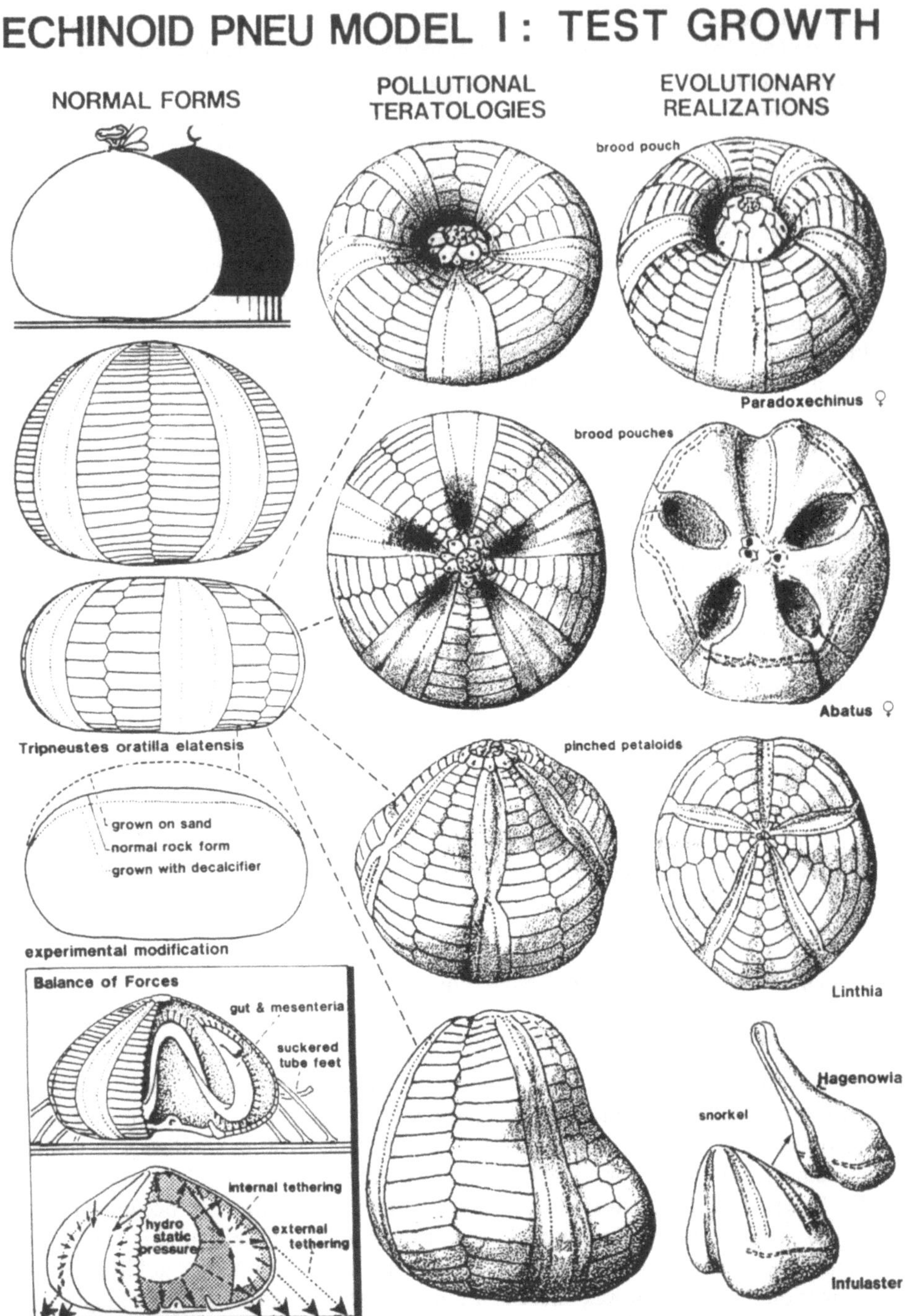

Fig. 2. Pollutional modification reveals that the normal growth of a regular echinoid test depends on the balanced interaction of external and internal mechanical forces. In other species (particularly irregular echinoids) similar deformations of the simple balloon shape have become a constant and functional feature (After Dafni 1986)

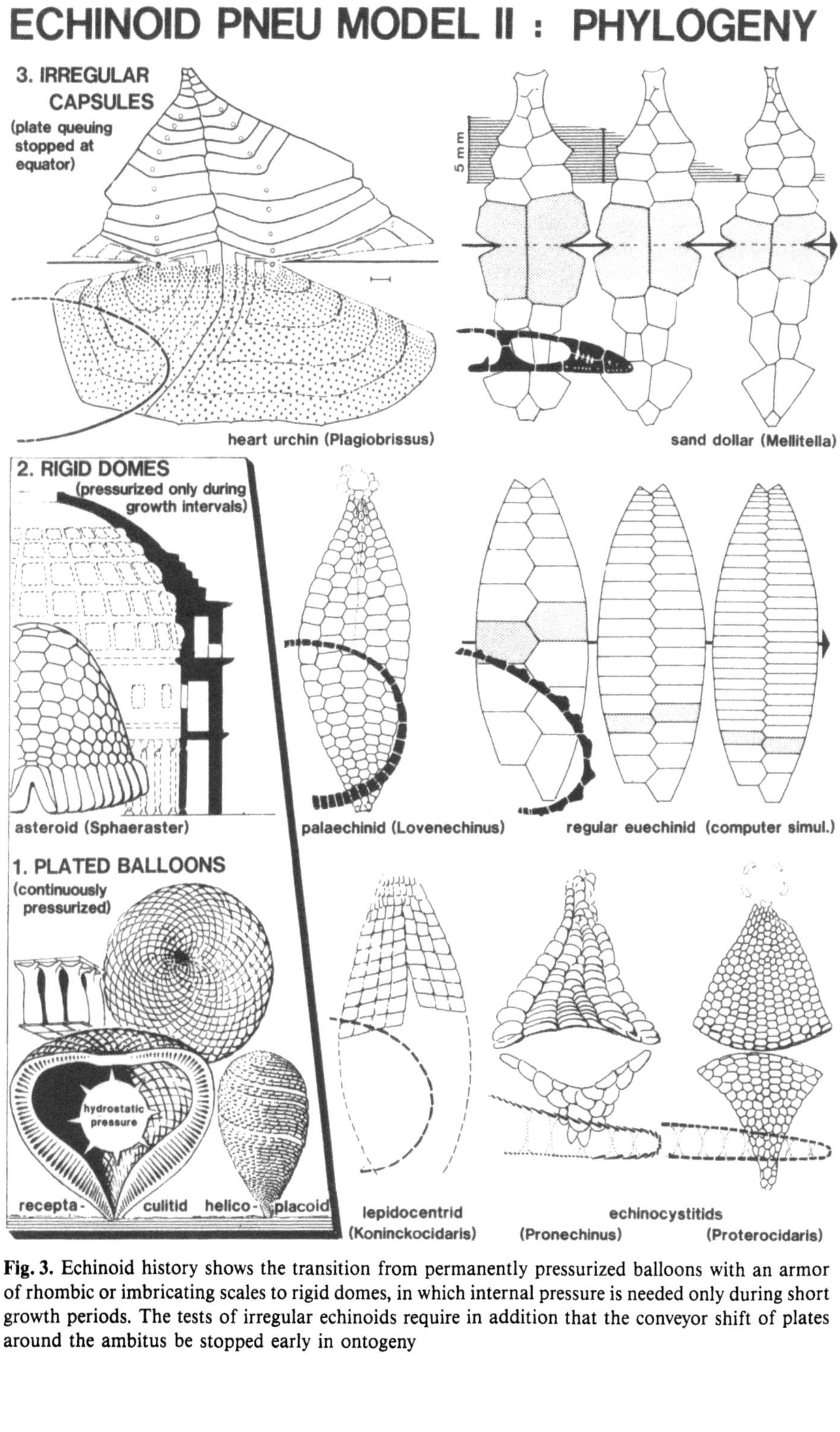

Fig. 3. Echinoid history shows the transition from permanently pressurized balloons with an armor of rhombic or imbricating scales to rigid domes, in which internal pressure is needed only during short growth periods. The tests of irregular echinoids require in addition that the conveyor shift of plates around the ambitus be stopped early in ontogeny

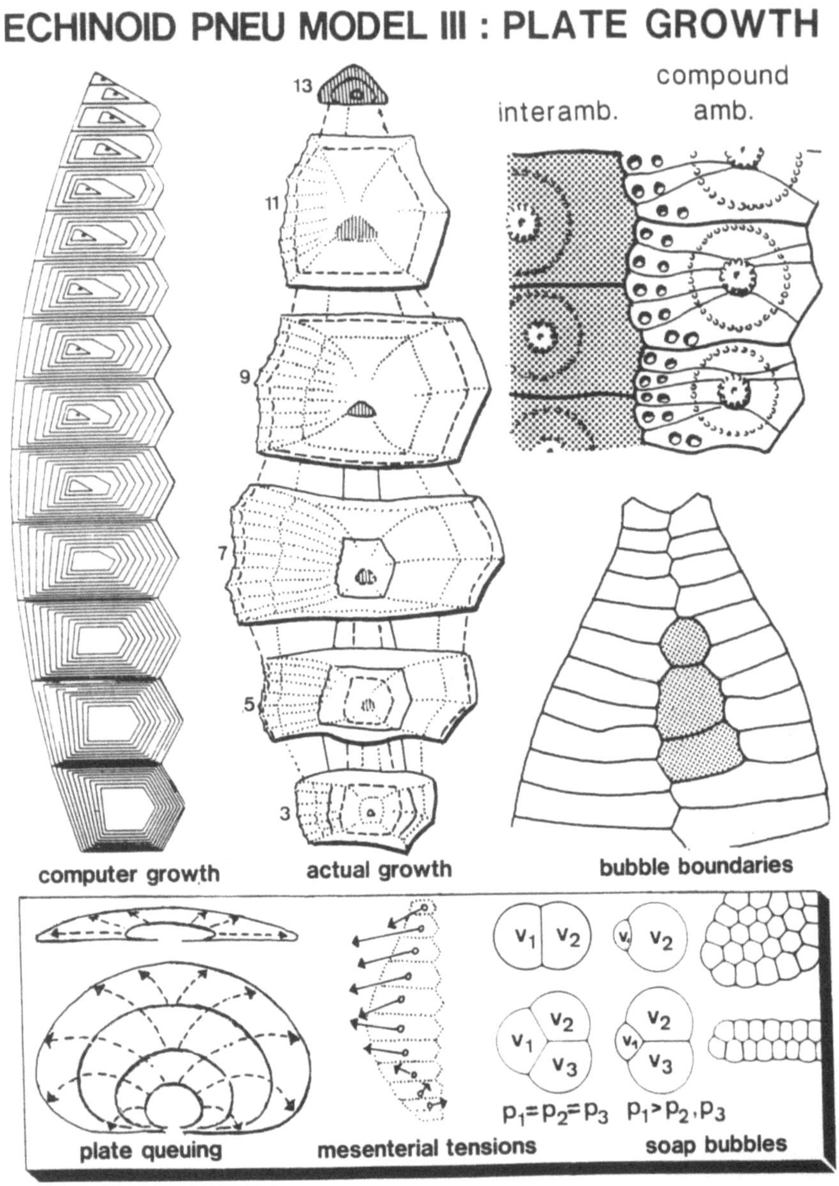

Fig. 4. Pneu morphology is also expressed in the outlines of individual interambulacral plates. In contrast to the computer model, their borders are indented according to the soap bubble rule. The serrated ambulacral border shows that in the compound ambulacral plates only stereoms have merged, while the constituent syncytial stroma cells remain separated

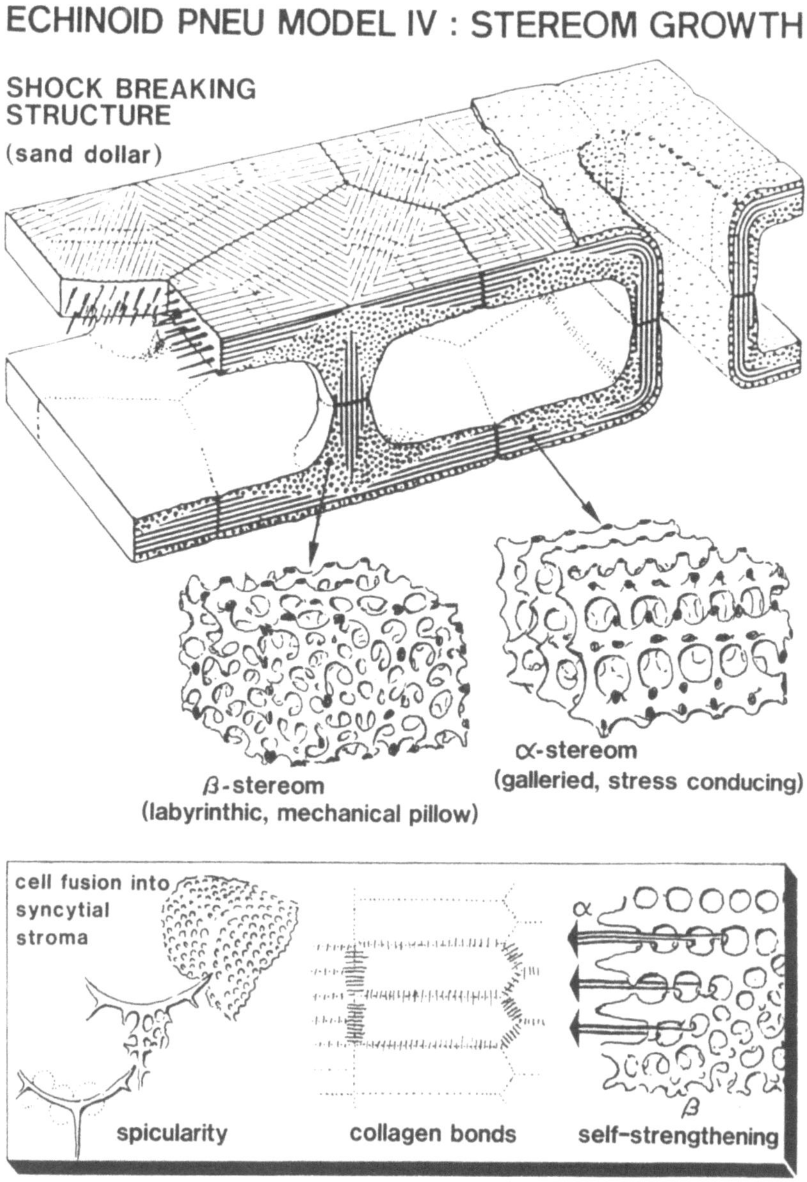

Fig. 5. Tensional forces of the whole test pneu translate to the collagen fibers that connect the test plates. Since these fibers also control sutural plate growth, the resulting distribution of galleried α-stereom and of labyrinthic β-stereom is automatically optimized to distribute load forces within the rigid test wall

The fabricational conversion of a tensional pneu (Otto 1977) into a rigid structure is easy in an egg shell, but its application to a plated and growing echinoid capsule incurs difficulties. Since the test must grow, can it simultaneously act as a dome *and* a balloon?

The bizarre morphologies that Dafni (1986) observed in populations of a regular echinoid surviving in polluted waters near Eilat (Israel) serve as a natural test for the pneu model (Fig. 2): though dramatically deviating from evenly vaulted domes, they still correspond to the internally tethered pneu shapes of gourds and irregular echinoids. The phylogenetic history of echinoids also confirms the pneu model (Fig. 3): Paleozoic forms commonly have rhombic or scale-like test plates – in contrast to the hexagonal plates in modern representatives. In analogy with scale patterns in fishes (where hexagonal scales occur in box fish), rhombic plate configurations suggest that the tests of primitive echinoids were flexible and needed permanent support by a slightly pressurized coelomic fluid. Transition from flexibly plated balloons to rigid mosaic tests could simply mean that the expensive pressurized, or denser, state (possibly by osmosis) was restricted to short episodes of growth – in a way like in arthropods, but without molting.

Interestingly, echinoid plate configurations themselves are also controlled by the pneu principle, because in echinoderms each ossicle is contained in a giant, syncytial stroma cell. This is why the close-packed plates do grow hexagonally and why smaller plates indent larger neighbors along the sutures (Fig. 4) – just as smaller bubbles do in a raft of soap bubbles. This rule is particularly well expressed in compound ambulacral plates (Fig. 4), where identical crystallographic orientation has allowed the stereom of adjacent plates to fuse, while the suture with a larger interambulacral plate still shows separate indentations of the constituent stroma cells.

At the still smaller, intracellular scale of stereom structure (Fig. 5), the simple pneu model fails. Still tensional forces do play a role in the scaffolding of the rigid calcite meshwork in its α and β versions and the self-organized establishment of a highly optimized architecture.

In summary we can say that vaulting, as well as plate patterns of echinoid tests, are basically controlled by physical laws, with differential growth rates as the major biological determinants.

2.2 Coral Skeletons and Mollusc Shells: Marginal Pneu Accretion

Coral skeletons are made by polyps, whose general pneu character is expressed in the circular outlines of solitary corallite walls and their hexagonal configurations in closely packed colonies. Another pneu-related feature is the straightness of the radial septa, in sharp contrast to equivalent structures in the stress-free columellar area. This is because the folds of the body wall, in which sclerosepta form by accretion, are tied to the thecal wall on one side and to the columellar margin on the other. In side view, septal edges curve smoothly – as bivalve and brachiopod shells do between their tie points at the hinge edges. The septal edge, however, bulges towards the center of the body pneu, indicating a counter-pneu along the growing septal margin! In reality, the septal pneu consists of a chain of minute blisters, in which sclerodermites are crystallizing. The edge of the calcified sep-

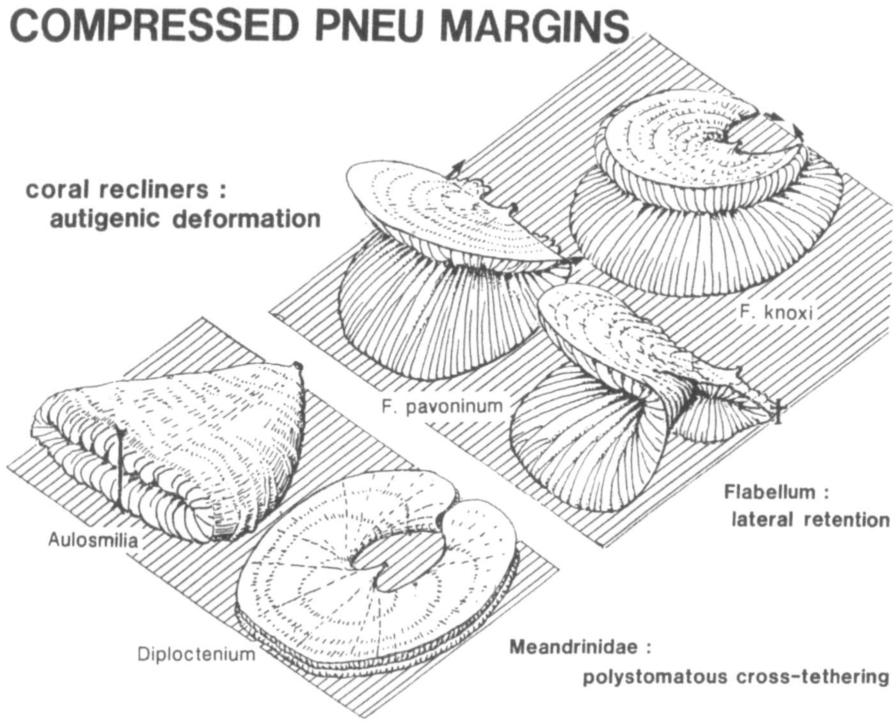

Fig. 6. The flattening of corallites required by a recumbent mode of life may be induced either by tie points along the originally circular growth margins or by soft cross-tetherings in polystomatous species. Note that the edges of the septa also reflect control by a tied pneu line but with curvatures directed against the pressure of the body pneu

tum, however, is not only the product of a chain of pneu blisters; it also forms the base for the next increment. Therefore, in order to minimize the size of the new blisters (and thereby increase their pressure relative to that of the body pneu), the septal edge must remain sharp as a knife. This is in marked contrast to the septa of Early Paleozoic "corals" such as *Spatangopsis*, whose septa were formed not by ectodermal carbonate, but by sand packed into the mesogloea and therefore could afford blunt edges (Seilacher 1990).

The bivalve analogy also helps understanding deformation of the thecal wall in some solitary scleractinian genera. In the pleurothetic recliner *Flabellum pavoninum* (Fig. 6) the conical theca has, by introduction of a pair of protoseptal tie points, become transformed into two flat semi-circular flaps that meet along a straight edge. Teratological loss of such a tie point leads to a conical thecal wall. Another way to flatten the conical corallite for pleurothetic reclining is found in Cretaceous meandrinids (Fig. 6). Here incomplete budding produces a series of mouths, whose boundary tissues hold the corallite edges together like stitches across a wound. This is also the principle that has allowed the development of meandroid patterns in modern colonial relatives.

In a similar way, tensional forces control the curvature of *molluscan* mantle edges that act as generating curves for conical and spiral shell growth. Since the

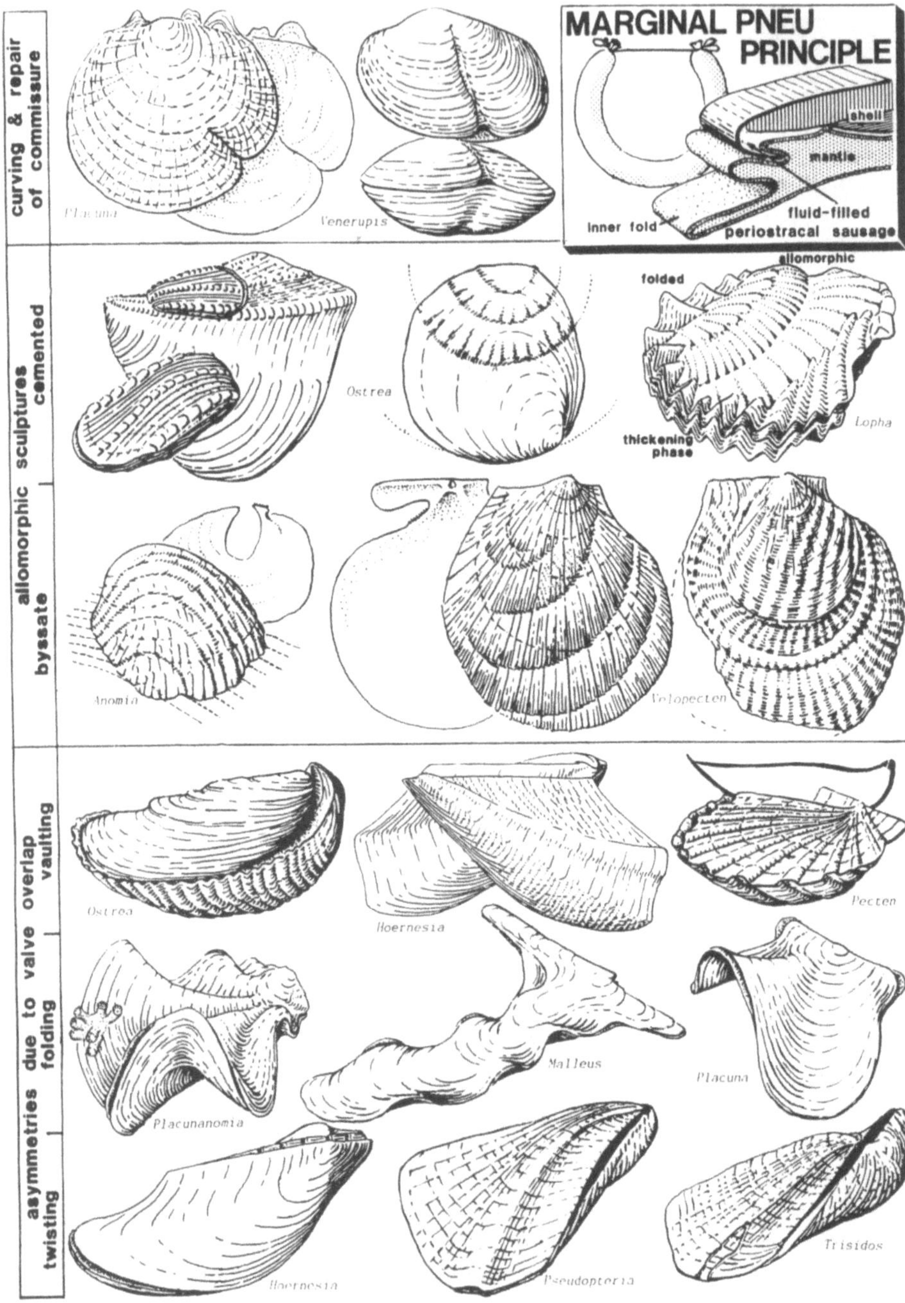

Fig. 7. The geometry of bivalve shells is defined by the rules of logarithmic spiral growth and the shape of the mantle margin (generating curve). The pneu character of this margin is expressed in teratological specimens, in upper-valve xenomorphic copies of the substrate and in folded and twisted morphologies

mantle itself is a fluid-filled duplicature and since it produces at its edge the flexible periostracum, we again deal with a linear, but in this case continuous, marginal pneu (Fig. 7). Pressed against either the opposite valve or a hard substrate, this sausage not only causes the smoothly recessed festooning of injured shell margins, but also allomorphic sculptures and inequivalve shell curvatures, folds, and sculptures (Fig. 7).

It is due to the ubiquity of pneus and accretional growth that the theoretical morphospace of spiral cones (Raup 1966) applies to so many kinds of shells, beaks (see Peters this Vol.), elephant tusks, horns, and other skeletal structures – in spite of the fact that a simple circle has originally been used as a generating curve in the computer simulation.

3 Zebra Patterns

3.1 Zebra Patterns in Skins

To study zebra patterns, we do not have to go to the zoo, because our own fingerprints provide a handy example. Formally, such patterns correspond to the "Bénard Figures" (Fig. 8) in fluid films, as well as to other dissipative structures in which equidistant instabilities form spontaneously under certain boundary conditions. In such chaotic systems (Murray 1981) regularity is inherently combined with uniqueness: in repeated experiments the pattern will never turn out to be exactly the same. Likewise, finger prints are distinctive not only between persons, but also between every one of our fingers! Such asymmetry is found in all biological patterns of this category, whether expressed as linear or spotty designs and whether in color (salamanders, fish, mammals, arthropods) or sculpture (ostracods, fish scales, stegocephalian skulls). Certainly, zebra patterns are distinctive enough to be taxonomically useful; but their elements can be homologized only to a certain point, beyond which the principle is free to go its own ways (Fig. 9).

3.2 Zebra Patterns in Accretionary Shells: the Divaricate Principle

Color and sculpture patterns of mollusc shells correspond to the true zebra principle only where extended tissue folds lay down thin sheets of carbonate on the *outside* of a shell (cowries, opercula of *Turbo*).

If combined with marginal accretion, zebra patterns acquire the shapes of "*divaricate*" triangles, chevrons, or oblique lineations. This is so because the actual dissipative rhythm is at any moment restricted to the mantle margin (Seilacher 1973). What we see as divaricate ornaments is thus the record of a repetitive marginal zebra pattern that shifts through growth time and is presumably induced by the interaction of morphogenetic substances (activator and inhibitor, Meinhardt and Klingler 1987) within cell families of the mantle edge.

Divaricate patterns in bivalve shells show contrasting levels of variability (Figs. 9, 10). Some species have them only in rare cases. In *Tapes philippinarum*, divaricate coloration has become a common feature, but occurs in diverse modes

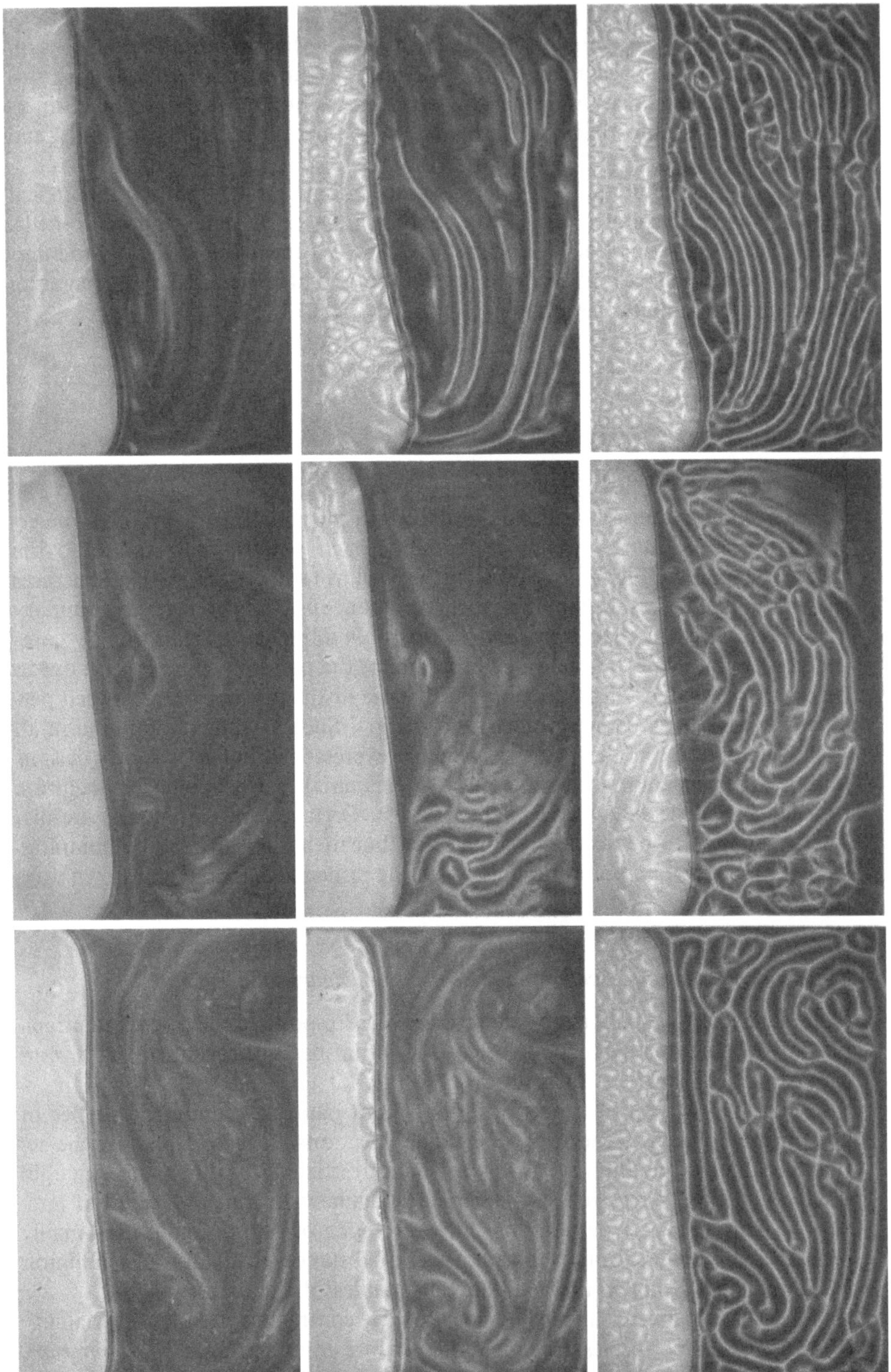

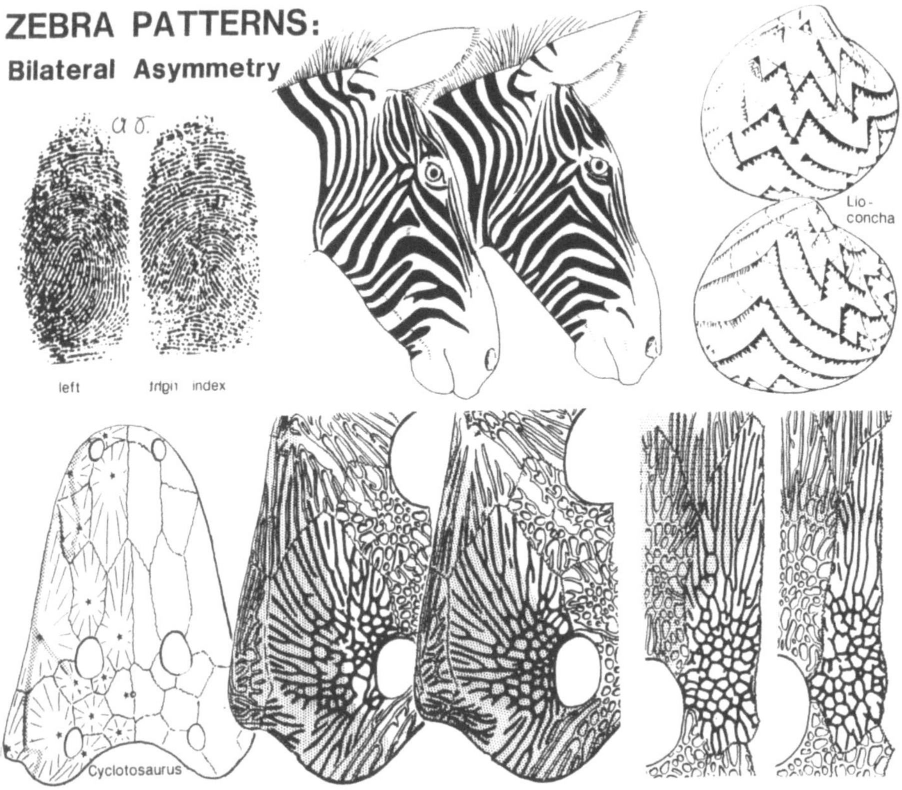

Fig. 9. The autonomy of zebra pattern formation is expressed by right/left asymmetries (right parts optically reversed). The near-identity in parts of the stegocephalian skull is not the work of nature, but of a preparator that modeled missing parts (*shaded*) as a mirror image of the other side

within the same population (Seilacher 1972, Fig. 9). Since *T. philippinarum* is an infaunal species, its divaricate color pattern can be considered as "fabricational noise" without any functional significance other than getting rid of metabolic waste products. In *Pteria zebra*, however, which lives attached to hydrozoan colonies (Seilacher 1972, Fig. 10), the divaricate color pattern has acquired a mimetic function. Accordingly, its variability is very much reduced, although noncolored morphs still do occur.

As another functional example, the cockle *Corculum* uses divaricate "windows" to supply its endosymbiotic algae with sunlight (Fig. 10 and Seilacher 1973, Fig. 6). All individuals do have "*windows*", but these are made also on the lower (anterior) side of the shell that is never exposed to sunlight. Presumably,

Fig. 8. The spontaneously formed pattern of convection cells in a flat fluid chamber heated from below (Bénard figures) is influenced by equally random flow vortices present at the onset. Note also that in all three experiments more reticular patterns did form in the upper part of the picture, where the fluid layer happened to be thinner because of an enclosed air bubble. (Selected time-lapse pictures taken by W. Wetzel, Tübingen; fluid chamber provided by E. W. Bolton, Yale University)

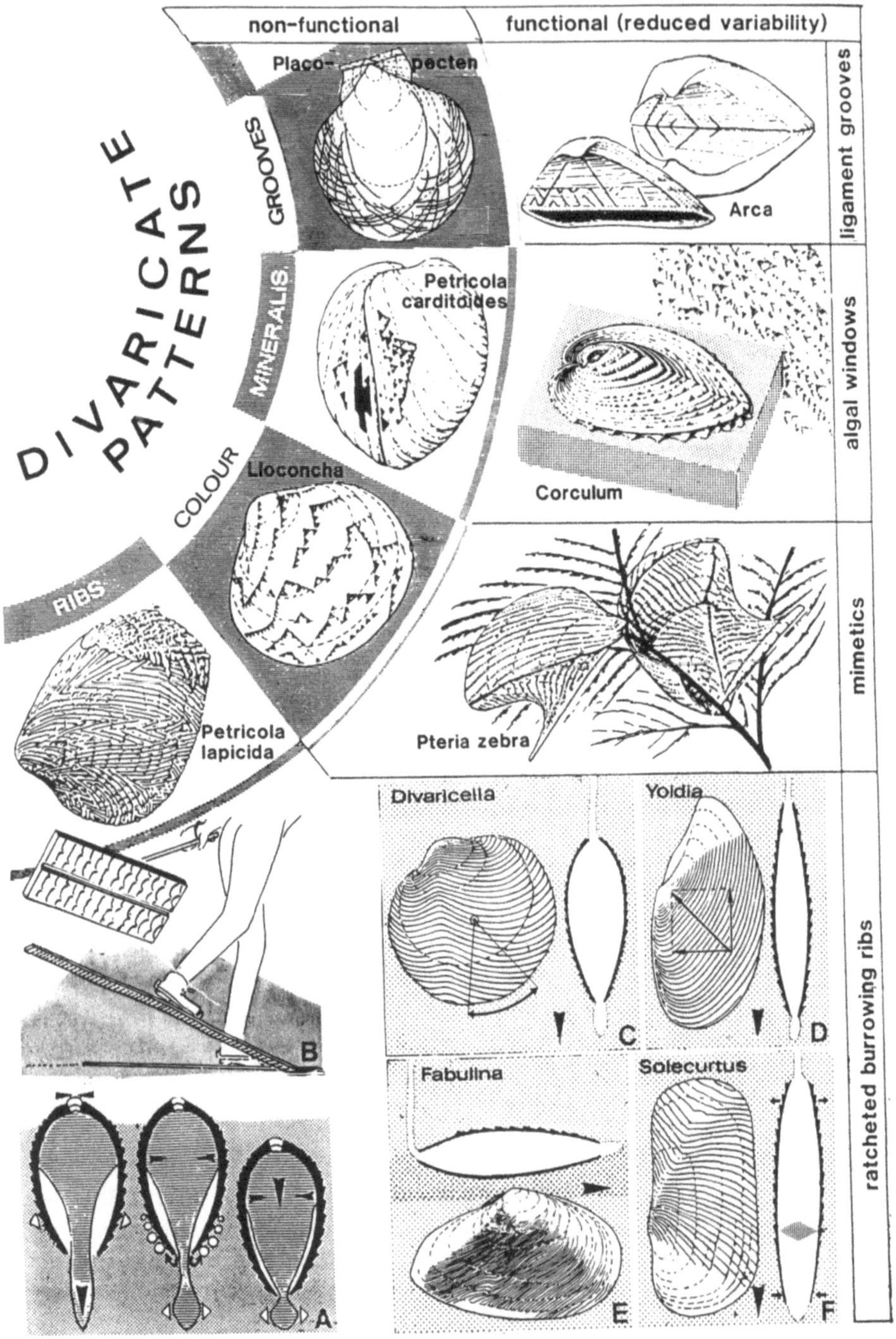

Fig. 10. Divaricate shell patterns record the generation and lateral shift of zebra patches along the soft tissues of the mantle edge (Fig. 7) through growth. Expressed in various modes, such patterns may be an accidental feature that appears only sporadically and with a high degree of variability. In other species they have become tamed for specific functions. Still the original triangular pattern shows up in repair situations (After Seilacher 1972)

turning on window fabrication along the whole mantle edge comes cheaper than restricting it to the relevant sectors. Likewise, the triangularity of the windows is pure fabricational noise without any functional significance.

Finally, bivalves employ the divaricate principle to produce highly functional *burrowing ribs* (Seilacher 1972; Savazzi et al. 1982). Comparable to the ratcheted profile of a crosscountry ski, these sculptures reduce backslippage and are convergently tuned to shell shape, substrate grain size and burrowing position in many unrelated groups. Nevertheless, the original triangular design does surface again where the mantle margin had been wounded and the functional file pattern needed to be restored.

At this point we may draw a first conclusion with regard to the *evolutionary* scenario. A self-organizing physical process of pattern formation is generally available and may at first occur only as a developmental accident under minimal genetic input. The crucial second stage is the genetic assimilation (Waddington 1953) or "*adoption*", of this mechanism. This step guarantees the reproduction of the pattern, but as yet with a highly unpredictable outcome. As selection discovers a specific function for the adopted character (third stage), Darwinian selection will "*tame*" it by reducing the inherent variability towards the required design. Still, even in the most perfectly adapted stage, individual mistakes or right/left asymmetries will occur, in which the autonomy of the mechanism expresses itself beyond the adaptational straight-jacket.

4 Cyclic Intercalation Patterns as Hierarchic Templates

4.1 Nondivaricate Shell Ornaments

The divaricate principle is only one of several alternative modes in which repetitive patterns can be programmed into marginally growing shells. Besides their obliquity to growth direction, divaricate elements tend to maintain uniform distances (negative allometry) and equal ranks, independent of their relative ages. The serially introduced metasepta of rugose corals are another familiar expression of the divaricate principle.

Other fabricational categories of shell ornaments include:

a) *Commarginal ribs* ("concentric" in bivalves, "radial" in ammonites). They result from a cyclic state change of all cells along the mantle margin. If the triggering signal is diffusion-controlled, distancing will be uniform, while in the case of an external (daily, monthly, seasonal) control it should vary with growth rates during life time. Interestingly, commarginal *color* bands are rare in mollusc shells.

b) *Normal ribs* (growing normal to the shell edge) are established early in ontogeny. They fail to secondarily increase in number and grow isometrically with the whole shell. For geometric reasons, however, they show a size gradient towards the two tie points. Otherwise there is no rank difference, but ribs and interrib spaces behave differently with respect to secondary ornamentations (spines, tubercles, divaricate windows, color). Examples are some cockle genera, where rib numbers can be used to distinguish species.

c) *Normal folds* differ from normal ribs by high variability and the equivalence between ridges and valleys (therefore they are never spinose). There is also a tendency of the amplitudes to increase during growth (positive allometry) within the limits set by distances from the hinge tie points. Such folds may originate relatively late in ontogeny (oysters, malleids, anomiids, *Plicatula*) and appear to be induced by the local mechanical interaction of the opposed marginal pneus ("overbiting") rather than by chemical signals (Fig. 7).

d) *Bifurcating ribs* are common in brachiopods, but rare in molluscs (some *Pecten*). This is astonishing in view of the fact that bifurcation is used as a means of fractal differentiation (Mandelbrot 1983) in many other biological structures (blood vessels, fern leaf venation, heterocoral septa) and is a dominating synergetic feature in the inorganic world (lightnings, dendrites, river systems, viscosity figures, etc.).

e) *Intercalation ribs* appear, in a formal sense, to represent only the mirror image of bifurcation: instead of the ribs it is the valleys that bifurcate. Since, however, the ribs are the elements that develop secondary ornamentations, the difference becomes morphogenetically fundamental. While bifurcating elements have equal ranks at any time level, intercalation produces *hierarchy*. The principle is best illustrated by juvenile crinoid stems, whose length increases by cyclic introduction of new internodals. As a result, 1, 3, 7, etc. elements of decreasing size order appear between the original nodals.

With respect to mollusc shells, it is noteworthy that the hierarchic order induced by intercalation may control not only the relative sizes of the elements, but also their response to secondary developmental signals. The spines of *Spondylus* (Fig. 11) and the windows of *Corculum* (Fig. 10), for example, correspond to the hierarchic order of the parent ribs, i.e., they are larger and more widely distanced along higher-order ribs. Similarly, the delicate spines that *Murex pecten* produces at every varix correspond to the rank of their generating spiral rib not only in size, but also in angular position. Probably they respond to the same morphogenetic signal in the sequence of their hierarchic order (Fig. 11; see also Gorthner 1990) and thus form a cage for mobile prey (Paul this Vol.).

Another common feature is that a senior rib may dominate, or "enslave", its junior neighbor by size suppression (*Neithea*), or with respect to spine distancing. This is how *Spondylus mirabilis* has managed to produce trident spines with the risk that one of the slaves may disobey and fire its spines prematurely (Fig. 11).

Among *corals*, the Scleractinia have chosen cyclic intercalation to allometrically increase the number of metasepta. As the divaricate method does in rugose and bifurcation in heterocorals, it serves them to solve the scaling problem and to allometrically increase the polyp's digesting endodermal surface. In most cases, the intercalational hierarchy is eventually lost. In some flabellids, however, it is maintained and serves as a template for the secondary morphologic differentiation of the columellar septal edges.

4.2 Ammonite Suture Lines

In a *formal* sense, the dendroid design of ammonitic sutures resembles the inorganic patterns of dendrites on bedding planes and of viscous fingering in fluids.

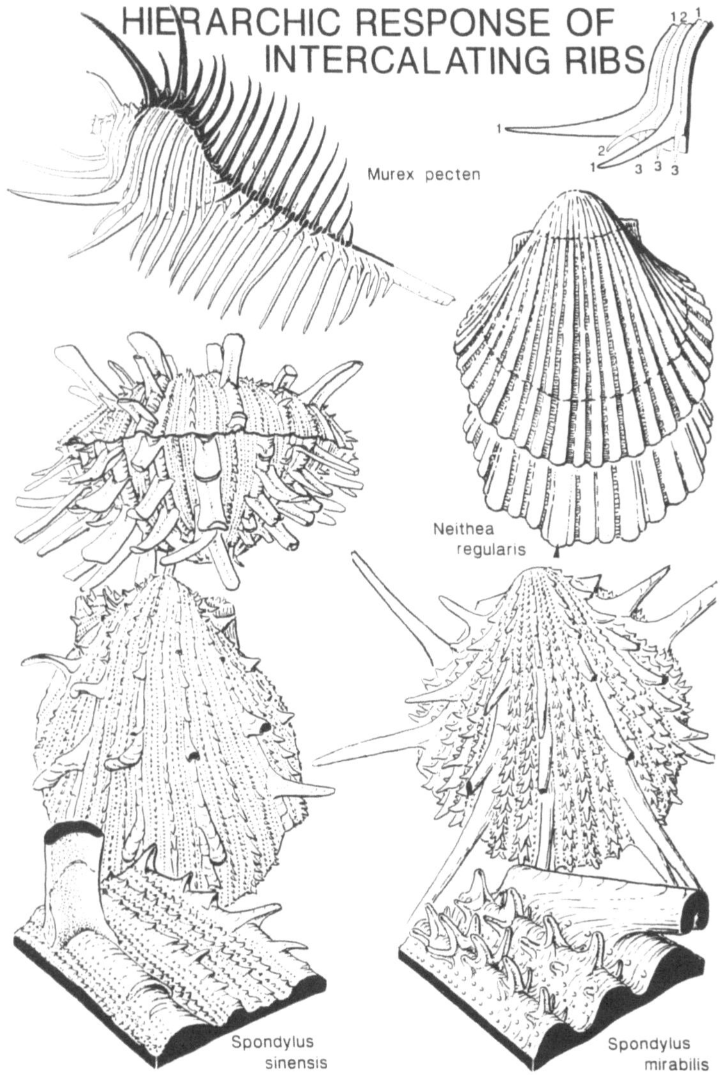

Fig. 11. Different ranks in growth-normal ribs reflect the cyclic intercalation of new elements in early stages. In species of *Neithea*, subordinate ribs may become "enslaved" by first-order neighbors. In the other examples the intercalational hierarchy serves as a template for the differential formation of spines with regard to size, distancing, and angularity

In this sense, ammonite sutures may be called fractals, since they are self-similar and result from the repetition of a simple differentiation mechanism. In a *morphogenetic* sense, however, it is more adequate to compare them to intercalation patterns combined with the differential deformation of a tensional membrane. More precisely, the transition from nautiloid to ammonoid septal configurations must have started with the differentiation of discrete *tie points*, with which the muscular body membrane became initially attached to the shell wall at a new septal position. As lateral distances widened during growth, new tie points became introduced according to the intercalation rule. On the one hand, the fabricational transformation of this linear template into a dendroid suture can be explained by the sequential attachment of the tie points in the order of their intercalation origin. On the other hand, the ultimate configuration of each suture results from bulging of the intervening portions of the body membrane in an apertural direction, i.e., by a pneu-like mechanism.

In a *functional* sense, this transformation has probably allowed the septal membrane of the ammonite body sack to "crawl" to the new septal position without ever giving up its tensional purchase (Seilacher 1988).

In spite of their high specifity and functional importance, ammonitic sutures are not determined to the last details. Like finger prints, equivalent sutural elements on the right and left side of the same shell are never symmetrically isomorphic. Once established, however, the personalized pattern was maintained in subsequent sutures, unless local changes in shell diameter (e.g., strong ribs) modified the stride between tie points of successive ranks.

5 Symbiotic Couples as Cybernetic Systems

Space permitting, the autonomous origin of synergetic, self-organizing pattern formation could be documented with many more examples, such as insect wings, leaf venations, and protist skeletons (Seilacher and Chinzei 1989). Instead, I shall finish this overview with a case, in which not a morphogenetic process, but another organism is being adopted.

Throughout their geologic history, *hermit crabs* have been involved in symbioses with sessile partners. In the case of sponge, colonial coelenterate, or bryozoan encrusters, the commensal relationship may develop into an obligatory mutualism: the host gains a substitute spiral shell that continues to grow on, as soon as its gastropod starter home becomes too small, while the sessile partner takes advantage of increased feeding surface, hitchhiking facilities and some breadcrumbs from the host's meals (Taylor et al. 1989).

In several cases this symbiotic relationship has induced distinctive modifications in both partners (Fig. 12). The crustacean host may change the coiling of its tail from a helicoidal to a planispiral mode after the gastropod shell is outgrown, while the encruster develops, in addition to the substitute shell, winglike lateral extensions and in the mature stage sometimes also a tail. The new structures are well balanced to the point that the two wings are either slender or obese, but always similarly so on either side.

The advantage of this transformation for both partners is obvious. Nevertheless the change can be accomplished almost entirely on the epigenetic level. Once the larva of the encruster has genetically fixed its *behavior* to settle only on pagurized shells, the rest is a consequence of the new cybernetic environment without a need to add further morphogenetic information. Preferred growth of the colony towards where the host supplies food particles produces the shell substitute. On the other hand, preferred growth in places where mud suffocation is minimal results in the formation of wings and eventually the tail. Balance between

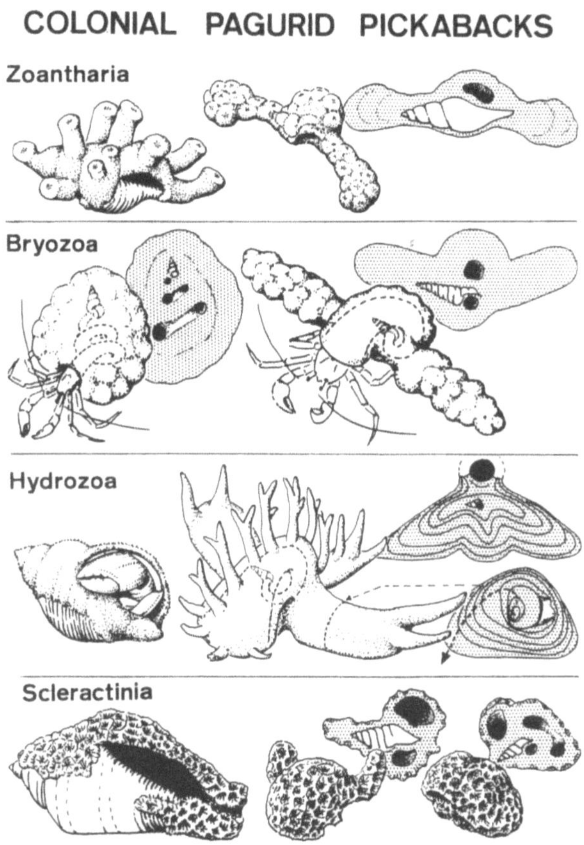

Fig. 12. In mutualistic symbiosis with hermit crabs, various colonial encrusters enlarge the shelter for their host and increase their own feeding surface by the formation of well-balanced horns, wings, and tails. In spite of their highly adaptive appearance, these shapes can be explained by synergetic response to the growth of the adopted partner

the wings is also automatically restored: if one wing becomes lopsided and sinks into the mud, its growth will be reduced in favor of the other. There is no reason to believe that a colony reared without a hermit crab would ever develop such shapes.

Thus the morphological similarity between examples shown in Fig. 12 is not what we usually mean by convergence, but rather the expression of cybernetic feedback between a helicospiral starter shell, its pagurid inhabitant, and a colonial encruster. More truly convergent is the case of the deep sea actinian *Adamsia*. Its basal membrane enlarges the original snail shell so perfectly that the nongastropod part can be recognized only from the darker chitinous shell material. Since the encruster is a solitary organism rather than a colony in which parts can temporarily tolerate suboptimal positions, the cybernetic feedback mechanism must be different. As the growth of the hermit crab rotates the shell, the actinian, which also grows, must readjust its top position and thereby automatically enlarges the shell in the mode of an accretionary spiral cone.

All these cases imply self-organizing mechanisms. In order to turn them on, however, innovation was required at the level of behavior rather than morphogenesis: the preference of larvae to settle on pagurized shells or, vice versa, the preference of hermit crabs to choose shells already encrusted by the suitable symbiont. This supports the view of Peters (this Vol.) that behavior plays a critical role in the origin of morphological adaptations – or (following a formulation by B. Hassenstein) that morphological evolution is commonly dragged along in the wake of physiological innovations.

6 Conclusions

1. The approach of morphodynamics represents an integrated view of biological form by focusing on functional, fabricational, and environmental constraints in the evolutionary transformation of given Bauplans. It amplifies the concept of constructional morphology sensu Seilacher (1970).

2. In this view, morphogenetic mechanisms are not only passive executors of genetic commands, but can also be seen as creative elements and a source of "pro-morphological", but structured, raw material.

3. In the evolutionary process, self-organizing epigenetic mechanisms serve as generally available "processors" to be picked up randomly or under weak selective pressure by genomic adoption.

4. In the further course of evolution, the adopted mechanisms may become tamed towards a specific function by Darwinian selection. Nevertheless, their basic autonomy is still expressed in the outcome of morphogenetic accidents or by random variation in details that do not infringe on functional fitness.

5. A similar model may be applied to the evolution of symbioses from originally commensal or parasitic relationships, only that another organism, rather than a fabricational principle, is being adopted. Obligatory symbiosis, however, may also have a negative side effect: since the new systems depend on two genomes, their probability of becoming extinct during unusual environmental perturbations will be increased.

Acknowledgments. The ideas presented here grew from projects and many discussions in cooperative research programs, SFB 53 (*Palökologie*, Teilprojekt: Konstruktions-Morphologie) and SFB 230 (*Natürliche Konstruktionen*), funded by the Deutsche Forschungsgemeinschaft. W. Maier, W. E. Reif, R. Thomas and D. Weishampel have helped with critical remarks and own ideas. So did participants of an informal workshop in the Heidelberg Academy, summer 1989.

References

Barel CDN, Anker G CH, Witte F, Hougerhoud RJC, Goldschmidt PT (1989) Constructional constraints and its ecomorphological implications. Acta Morphol Neerl Scand 27:1–2, 83–110

Dafni J (1986) A biomechanical model for the morphogenesis of regular echinoid tests. Paleobiology 12 (2):143–160

Gorthner A (1990) Bau, Funktion und Evolution komplexer Gastropodenschalen in Langzeit-Seen. Diss, Univ Tübingen

Gould SJ, Lewontin RC (1979) Spandrels of San Marco and the Panglossian paradigm: a critique of the adaptionist program. Proc R Soc London Ser B 205 (1161):581–598

Mandelbrot BB (1983) The fractal geometry of nature. Freeman, San Francisco

Meinhardt H, Klingler M (1987) A model for pattern formation on the shells of molluscs. J Theor Biol 126:63–89
Murray JD (1981) On pattern formation mechanisms for lepidopteran wing patterns and mammalian coat patterns. Philos Trans R Soc London Ser B 1078:473–496
Otto F (ed) (1977) Pneus in Natur und Technik. IL 9. Mitt Inst Leichte Flächentragwerke, Stuttgart, 334 pp (in German and English)
Raup DM (1966) Geometric analysis of shell coiling: general problems. J Paleontol 40:1178–1190
Reif WE, Thomas RDK, Fischer MS (1985) Constructional morphology: the analysis of constraints in evolution. Acta Biotheor 34:233–248
Savazzi E, Jefferies RPS, Signor PW (1982) Modification of the paradigm for burrowing ribs in various gastropods, crustaceans and calcichordates. N Jahrb Geol Paläontol Abh 164:206–217
Seilacher A (1970) Arbeitskonzept zur Konstruktions-Morphologie. Lethaia 3:393–396
Seilacher A (1972) Divaricate patterns in pelecypod shells. Lethaia 5:325–343
Seilacher A (1973) Fabricational noise in adaptive morphology. Syst Zool 22 (4):451–456
Seilacher A (1988) Why are nautiloid and ammonite sutures so different? N Jahrb Geol Paläontol Abh 177 (1):41–69
Seilacher A (1990) Lost constructions: Vendozoa and Psammocorallia. Abstr GSA Meet, Dallas
Seilacher A, Chinzei K (1989) Selbstbildungs-Mechanismen als Prozessoren organismischer Entwicklung und Evolution. Nat Konstrukt Mitt SFB 230, 3:55–64
Taylor PD, Schembri PJ, Cook PL (1989) Symbiotic associations between hermit crabs and bryozoans from the Otago region. S N Z J Nat Hist 23 (5):1059–1085
Üxküll J von (1928) Theoretical biology. Springer, Berlin
Waddington CH (1953) Genetic assimilation of an acquired character. Evolution 7:118–126
Weber H (1955) Stellung und Aufgaben der Morphologie in der Zoologie der Gegenwart. Zool Anz Suppl 18:137–159

Genetic and Molecular Analysis of Pattern Formation Processes in *Drosophila*

DIETHARD TAUTZ[1]

Summary

A detailed understanding of the processes controlling *Drosophila* development has become possible by a systematic genetic analysis of mutations leading to pattern defects during embryogenesis. It was possible to obtain definitive proof for the involvement of morphogenetic gradients and to define a hierarchy of genetic interactions, which determine the major outlines of the body pattern. Many of the genes involved in this process have been cloned and their molecular functions have been revealed. It could be shown that a variety of different molecular processes can play a role in pattern formation events. It is also becoming increasingly clear that a number of processes are specified in redundant pathways which can partly substitute each other. This has consequences for our understanding of evolution. Redundantly specified pathways can provide a versatile frame for the evolution of developmental traits, since they allow the divergence of one component, while the other remains functional.

1 Introduction

The results from the experimental embryology in insect systems have suggested that the processes of the formation of the polarity and of the initial body pattern are determined at very early stages during embryogenesis. Ligation experiments, experimental manipulation of posterior determinants, and selective destruction of anterior determinants have shown that long-range acting factors, which may reflect morphogenetic gradients, are effective in controlling the basic body plan (reviewed in Sander 1976). However, these experiments did not make it possible to identify the biochemical components of the morphogenetic gradients, nor to predict how the gradients are interpreted by the cells. Such an analysis has now become feasible by the genetic approach to embryology (Nüsslein-Volhard and Wieschaus 1980). The rationale of this approach was the idea that it should be possible to obtain mutants for each of the components which regulate the different developmental processes. Mutations of this kind should lead to specific pattern defects, which can be scored and grouped according to their phenotype. Further analysis of these mutations should then permit learning something about the

[1] Institut für Genetik und Mikrobiologie der Universität München, Maria-Ward Str. 1 a, D-8000 München 19, FRG
New address: Institut für Zoologie der Universität München, Luisenstr. 14, D-8000 München 2

N. Schmidt-Kittler and K. Vogel (Eds):
Constructional Morphology and Evolution

principles of the pattern formation processes. The approach is, however, only sensible if one tries to identify all genes which are required for the process. This can be achieved by performing a systematic saturation mutagenesis screen. Such a screen can only be done in a genetically well-characterized system. This led to the decision to use *Drosophila* as a model system, although *Drosophila* had not been an obvious choice for experimental embryology before. Nonetheless, the approach turned out to be highly successful and it is now possible to describe the processes of pattern formation in *Drosophila* in great detail; in particular, we begin to understand the functional interactions at the molecular level (reviewed in Akam 1987; Nüsslein-Volhard et al. 1987; Ingham 1988).

2 Genes Involved in Pattern Formation

The detailed analysis of the mutants which have been obtained in a number of laborious screens has revealed a genetic hierarchy of interactions which serve to provide the embryo with an anterior-posterior, as well as a dorso-ventral polarity, and which lead to an increasingly refined subdivision of the embryo. The maternal genes are at the top of this hierarchy. They are necessary to organize the oocyte and to provide it with the essential positional cues. These cues are interpreted by the zygotic segmentation genes, which cause a metameric subdivision along the anterior-posterior axis and a regional determination of cell fates along the dorso-ventral axis. A third group of genes, the homoeotic genes, are finally necessary to specify the identity of the metameric units. Each of these groups of genes can be further subdivided into further groups according to their general function (Fig. 1).

2.1 Maternal Segmentation Genes

Three functionally nonoverlapping groups of maternal genes determine the pattern along the anterior-posterior axis. Key genes for each of the groups are *bicoid* for the anterior region, *nanos* for the posterior region, and *torso* for the terminal regions (Nüsslein-Volhard et al. 1987). The molecular functions of each of these three groups are very different, although all three act eventually via localized determinants. However, the involvement of a morphogenetic gradient can so far only be established for one of the three groups, namely for the anterior determinant *bicoid*. Its gene product behaves exactly as one would expect for a classical morphogenetic molecule, which acts via a source-diffusion mechanism. Its RNA is localized at the anterior pole of the embryo, while the translated protein diffuses towards posterior and forms an exponential gradient (Driever and Nüsslein-Volhard 1988). The protein itself is a transcriptional regulator and causes a localized transcription of its target genes in dependence of its relative concentration (Fig. 2; Driever and Nüsslein-Volhard 1989). Thus the distribution of the *bicoid* protein determines directly positional values. If *bicoid* RNA is injected at lateral or posterior positions of an early embryo, it can induce new anterior centers. This proves that the *bicoid* gene product is in a classical sense a morphogenetic molecule.

The determination of the posterior region of the embryo is more complex and its analysis has led to some surprising insights. The key gene for the posterior

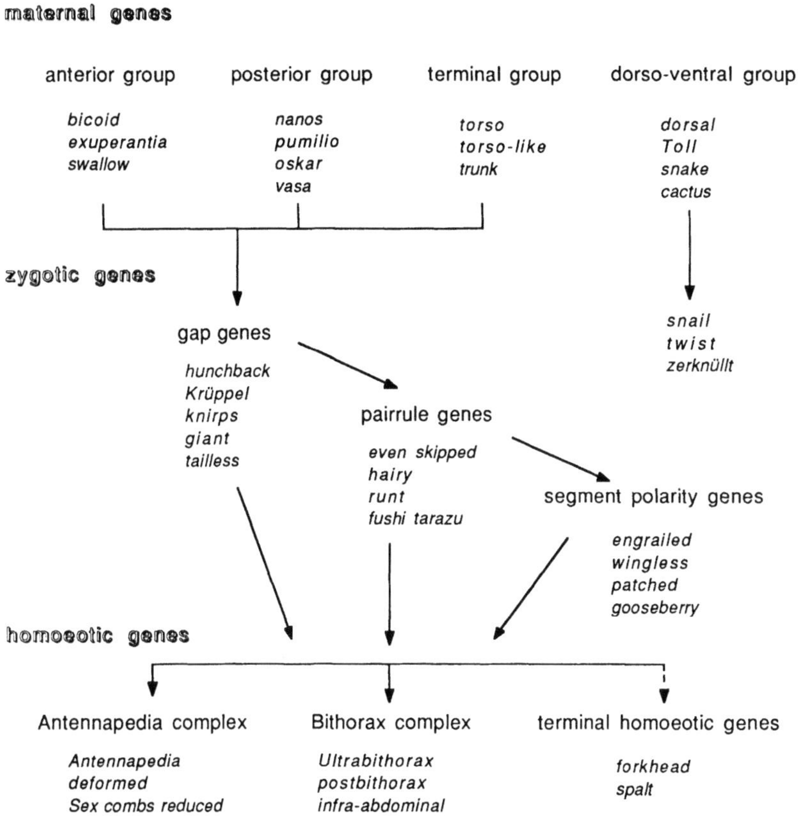

Fig. 1. The genetic hierarchy of genes involved in pattern formation. The major groups of genes are shown as they can be defined by the genetic analysis of the respective mutants. At the top of the hierarchy are the maternal genes, whose products regulate the zygotic segmentation genes, which in turn regulate the homoeotic genes. Each of these major groups of genes can be further subdivided and the regulatory interactions among these groups are symbolized by *arrows*. A number of characteristic loci are given for each of these groups; the list is, however, not complete (see reviews by Akam 1987; Nüsslein-Volhard et al. 1987; Ingham 1988 for more complete lists)

centre is *nanos* and its RNA product is tightly localized at the posterior pole. The protein product is apparently actively transported towards the anterior, but acts neither as a direct transcriptional regulator, nor as a concentration-dependent morphogen. Instead, its sole function appears to be the translational repression of a gene product, which would otherwise inhibit abdominal development. This gene product is the maternally expressed *hunchback* mRNA, which is itself not localized, but homogeneously distributed throughout the embryo. The *hunchback* protein has several functions: it inhibits abdominal development, it triggers the expression of a zygotic target gene, *Krüppel*, in the middle of the embryo and it is involved in setting the frame for the expression of a posterior acting zygotic gene, *knirps* (Hülskamp et al. 1990). Thus, abdominal development need not be organized by a maternal morphogenetic gradient, but is solely determined at the

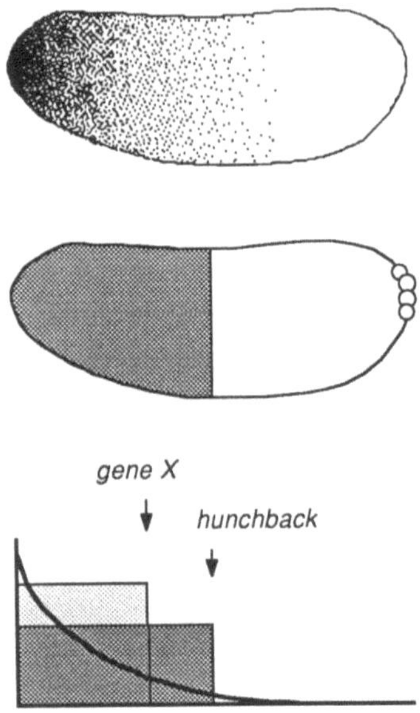

Fig. 2. Effect of the morphogenetic gradient of the *bicoid* gene product. The *bicoid* product forms an anterior-posterior gradient (*upper embryo*), which is directly involved in regulating other genes. This regulation is concentration dependent and the respective target genes are therefore switched on until a certain threshold level of the *bicoid* protein is reached. This is shown for the gap gene *hunchback* (*lower embryo*) and for a hypothetical gene X (Driever et al. 1989), which may be active in the head region

zygotic level. All that is necessary is a boundary and a polarity. Setting up this boundary as well as providing the polarity is exactly the role of the interplay between *nanos* and the maternally expressed *hunchback* mRNA. The embryo is divided into two fields, an anterior one, where the *hunchback* protein is produced and where it serves to inhibit posterior development, and a posterior one, which is free of *hunchback* protein and which allows posterior development (Fig. 3). This subdivision may reflect an archaic process, which may predate the evolutionary "invention" of the anterior organizer *bicoid*. One reason to assume this is the fact that the very first regulatory event which is caused by *bicoid*, is to trigger the zygotic expression of *hunchback* in exactly the same region, which is already occupied by the maternal protein (compare Fig. 2). The zygotic expression of *hunchback* can thus replace the maternal one. This leads to a functional redundancy, which in *Drosophila* is so efficient that one can genetically remove the *nanos*/maternal *hunchback* system and still obtain normally formed embryos and larvae (Hülskamp et al. 1989; Irish et al. 1989; Struhl 1989). This raises the question why a complicated posterior system, which involves the specific interaction of up to ten genes, is present in the *Drosophila* embryo at all, when it can be artificially removed without causing apparent detrimental effects. The answer to this question must simply be that laboratory conditions do not match wild-type conditions and that partly redundant pattern formation systems are apparently necessary to ensure proper development under adverse conditions. Although this answer is simple, it is important to realize that redundant pattern formation systems exist. This has consequences for the understanding of the evolution of developmental traits, which are discussed below.

The terminal system with its key gene *torso* functions in a different way. The *torso* gene product apparently provides a receptor, which senses a signal emanating from specialized follicle cells at both ends of the developing oocyte (Sprenger

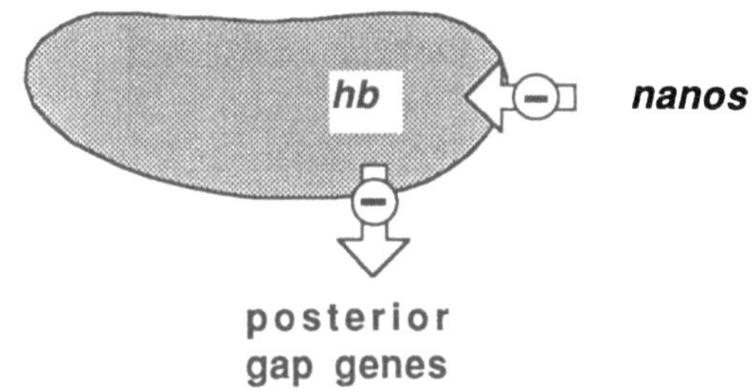

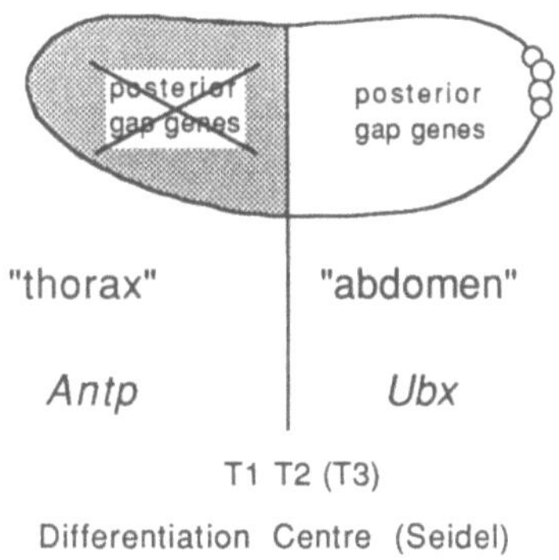

Fig. 3. Function of the posterior morphogenetic centre. The role of the *nanos* gene product is to repress the translation of the maternally provided *hunchback* RNA. It thereby allows a posterior segmentation pathway to proceed in the posterior half of the embryo, which is at the same time repressed by the *hunchback* protein in the anterior half of the embryo. This leads to a very early subdivision of the embryo. The border of this subdivision coincides spatially and functionally with the border between the action of the two major homoeotic gene clusters, the *Antennapedia* and *Bithorax* complexes. The border lies also at the place which was recognized as a "differentiation centre" by Seidel (1934)

et al. 1989). Thus *torso* is itself not a localized molecule, but transmits only a localized signal, which comes from outside the embryo. The cells which are involved in generating this signal are in fact not germline-derived cells, but derive from the somatic cells of the mother.

A similar principle is apparently utilized by the maternal group of genes, which determine the dorso-ventral axis of the embryo. A localized signal emanating from spezialized follicle cells is transmitted via a signaling cascade and is eventually transformed into the asymmetric distribution of a morphogenetic molecule, which is the product of the *dorsal* gene. This gene product then forms a gradient of differential uptake into the nuclei along the dorso-ventral axis and is thereby apparently directly involved in regulating the expression of the zygotic target genes (Rushlow et al. 1989; Steward 1989; Roth et al. 1989).

2.2 Zygotic Segmentation Genes

The primary zygotic genes, both along the dorso-ventral and the anterior-posterior axis, are directly regulated by the differential distribution of regulatory maternal molecules. Best understood is the system along the anterior-posterior axis. Three groups of genes exist, which sequentially divide the embryo into successively smaller units. The first group is formed by the gap genes, which specify large, partly overlapping regions of the embryo; the second group are the pair-rule genes, which specify double segmental units and the third group are the segment-polarity genes, which eventually specify the segmental units (Nüsslein-Volhard

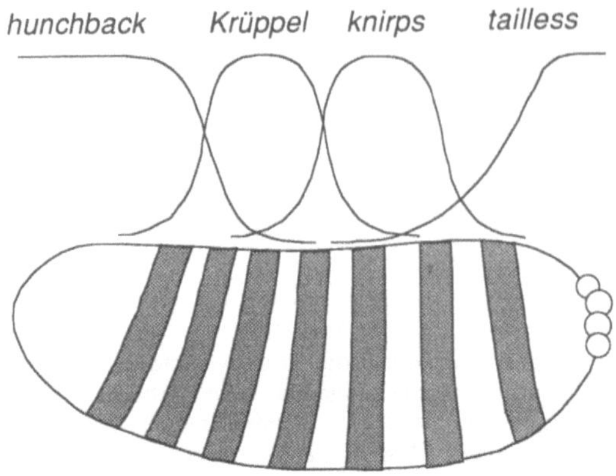

Fig. 4. Gap gene gradients regulate pair-rule gene expression. The gap gene products form short-range gradients, which determine the spatial expression of the pair-rule gene stripes. It is assumed that each of the stripes is generated by a combinatorial regulatory input from at least two gap gene products, whereby again concentration-dependent effects play a crucial role. The details of the process are, however, not yet understood (for review see Carroll 1990). Note that the figure is highly schematical, in particular with respect to the spatial extent of the gap gene expression domains

and Wieschaus 1980). Members of all three groups of genes have been cloned and analyzed at the molecular level. Best understood is the group of gap genes. Thc three classical members of this group of genes, *hunchback*, *Krüppel*, and *knirps* code for transcription factors (Rosenberg et al. 1986; Tautz et al. 1987; Nauber et al. 1988) which are directly involved in the regulation of the genes at the next level of the hierarchy. They are initially expressed in largely nonoverlapping domains, but their translation products diffuse from their primary sites of synthesis and form short-range overlapping gradients in the embryo (Pankratz et al. 1989; Hülskamp et al. 1990). These gradients most likely provide positional signals for the primary pair-rule genes and cause eventually their expression in defined stripes (Fig. 4; Stanojevic et al. 1989; Pankratz et al. 1990; reviewed in Carroll 1990). These stripes are in turn the signals for the secondary pair-rule genes as well as for the segment-polarity genes, which are eventually responsible for the metameric subdivision of the embryo. A number of cross-regulatory interactions exist at the level of the pair-rule and segment polarity genes, which are not yet understood in detail. Both direct diffusion of transcription factors forming microgradients and intercellular signalling (van den Heuvel et al. 1989) may be involved in this process.

2.3 Homoeotic Genes

This class of genes is ultimately responsible for giving the segments their identity. Accordingly, if mutant, they do not produce defects in the segment pattern, as is the case for the previous classes of genes, but lead instead to transformations of segment qualities. This can have drastic consequences, as for example flies with a second pair of wings instead of the halteres, or flies which produce legs at the head instead of the normal antennae (Lewis 1978; Struhl 1981). Two major clusters of homoeotic genes exist on the chromosomes, the *Antennapedia* complex, which specifies the development of anterior segments, and the *Bithorax* complex, which specifies the posterior segments. Most of the homoeotic genes have been cloned and analyzed in molecular detail. They turned out to be unusually complex-regulated genes, which eventually code for specific transcription factors. They have a common motif for DNA binding, which is known as the helix-turn-helix motif from the studies on the lambda repressor. This is encoded within a region of about 60 amino acids, which share considerable homology among the different genes. Since this region was first discovered in homoeotic genes, it was named homoeobox even though it is now clear that it occurs equally frequent in nonhomoeotic genes, as for example in the group of maternal and zygotic genes mentioned above (reviewed in Gehring and Hiromi 1986). The presence of the homoeobox indicates, however, that the homoeotic genes are themselves regulators of other genes, which are then responsible for producing the three-dimensional structures which are specific for each segment. The molecular details of how these last steps are achieved are, however, still completely obscure. All that is known is that the homoeotic genes are responsible for switching between the different programs. This is, for example, clear from studies on the *Antennapedia* gene. A certain mutation at this locus is caused by an inversion within the chromosome, which brings the expression of the normal *Antennapedia* gene under a different regulatory control. It is now expressed in the head instead of the thorax, where it should normally be expressed. This abnormally expressed protein can now reprogram the cells, which would form the antennae, to form leg structures (Frischer et al. 1986). The same effect can be produced artificially, by constructing an *Antennapedia* gene, which responds to a simple temperature shift ("heat-shock control"). If one subjects the developing organisms to such a temperature shift during the right phases of development, the flies show again legs instead of antennae (Schneuwly et al. 1987). These examples demonstrate that the homoeotic genes have to be tightly regulated in time and space during normal development to produce the correct decisions. This regulation is achieved by a concerted action of the previous acting segmentation genes, which provide the body with the correct positional information (see above).

3 Discussion

The genetic and molecular analysis of the developmental processes in *Drosophila* has provided an excitingly detailed insight into the question of how pattern formation works. One old question as to whether morphogenetic gradients play a role could be solved. The *bicoid* gene product undoubtedly provides a morphogenetic molecule which determines positional values in dependence on its relative concentration. However, it does not control the entire anterior-posterior pattern directly, but rather hands over information to secondary systems, the gap genes, which then form short-range morphogenetic gradients. This principle of "subdivided" gradients explains how a large number of different pattern elements can eventually be specified.

A second major outcome of these studies is that metameric subdivision and specification of segment identities are independent processes. The homoeotic genes influence only the latter, not the former. Mutations within a single homoeotic gene can lead to gross alterations of functional parts of the fly, without necessarily changing the viability, at least under laboratory conditions. Such events could therefore be seen as macro-mutational events, which have sometimes been invoked as important for the evolutionary processes. However, macro-mutations like the transformation of antennae into legs obviously disrupt other functions within the fly and therefore it is unlikely that such alleles could be sufficiently stable in a population to allow selection working on them. This may be different for homoeotic mutations with more subtle effects, which could then indeed be the basis for biological novelties.

A completely unexpected outcome of the genetical analysis of pattern formation in *Drosophila* is the insight into the function of the posterior system. It does not operate directly via a morphogenetic gradient, but is characterized by the interplay between two gene products, the maternally derived *hunchback* RNA, which is homogeneously distributed throughout the embryo, and the *nanos* gene product, which is localized at the posterior pole. These two gene products eventually cause a subdivision of the embryo, which separates it into two fields, one which allows abdominal development and one which does not. The spatial organization of the abdomen is then achieved only at the zygotic level. Most intriguingly, however, the whole posterior system can be genetically removed in *Drosophila* without detrimental effects on the embryo. This is only possible since the same molecule (the *hunchback* product), which is the target of the posterior system in the subdivision process, is also utilized by the anterior system (cf. Figs. 2 and 3). It causes an expression domain of *hunchback* which overlaps completely with the one determined by the posterior system and can therefore functionally replace it. This is an example of a truly redundant system in early development. However, although the posterior system can be genetically removed under laboratory conditions, it is apparently not dispensible under wild-type conditions. This can be concluded from the fact that the posterior system and the maternal *hunchback* expression are conserved features, at least among dipteran insects (Sommer and Tautz unpubl.). Thus one can conclude that redundant functions can have a positive selection value. In fact, it seems likely that this example is only the top of the iceberg. The detailed analysis of the pattern formation processes provides an increasing number of indications that many of them are more or less redundant (Elkins et al. 1990; Hülskamp et al. 1990).

It is intuitively clear and also well known from information theory that redundant processes can stabilize the flow of information which is necessary for building a pattern and it should therefore not be surprising to detect such redundancies. Accepting this line of arguing has, however, important evolutionary consequences: *redundant developmental pathways can provide a frame within which morphological evolution can take place.* As long as one of two pathways is basically sufficient to determine a defined process, the "sister" pathway has more freedom to acquire mutations and to diverge. However, this can happen only within certain limits, since the pathways must be kept compatible and since there will still be a defined, though weaker, selection to keep both functioning. Considering further that the genes which are involved in specifying these pathways are usually represented by multiple alleles in a population, it is easy to imagine how a continuum of morphological traits could be generated within certain limits. The variability within this framework could be a basis for selection processes which could thus initiate a gradual transition from one function to another. Furthermore, redundancies could also help to circumvent adaptive peaks which could otherwise be an obstacle for further evolution. The details of such an evolutionary process have as yet to be revealed, but the current data on *Drosophila* development suggest that redundancies may provide an important framework for evolutionary play.

References

Akam M (1987) The molecular basis for metameric pattern in the *Drosophila* embryo. Development 101:1–22

Carroll SB (1990) Zebra patterns in fly embryos: activation of stripes or repression of interstripes? Cell 60:9–16

Driever W, Nüsslein-Volhard C (1988) A gradient of *bicoid* protein in *Drosophila* embryos. Cell 54:83–93

Driever W, Nüsslein-Volhard C (1989) The *bicoid* protein is a positive regulator of *hunchback* transcription in the early *Drosophila* embryo. Nature (London) 337:138–143

Driever W, Thoma G, Nüsslein-Volhard C (1989) Determination of spatial domains of zygotic gene expression in the *Drosophila* embryo by the affinity of binding sites for the *bicoid* morphogen. Nature (London) 340:363–367

Elkins T, Zinn K, McAllister L, Hoffmann FM, Goodman CS (1990) Genetic analysis of a *Drosophila* neural cell adhesion molecule: Interaction of fascilclin I and abelson tyrosine kinase mutations. Cell 60:565–575

Frischer LE, Hagen FS, Garber RL (1986) An inversion that disrupts the *Antennapedia* gene causes abnormal structure and localization of RNAs. Cell 47:1017–1023

Gehring WJ, Hiromi Y (1986) Homoeotic genes and the homoeobox. Annu Rev Genet 20:147–173

Heusel M von den, Nusse R, Johnston P, Lawrence PA (1989) Distribution of the *wingless* gene product in *Drosophila* embryos: a protein involved in cell-cell communication. Cell 59:739–749

Hülskamp M, Schröder C, Pfeifle C, Jäckle H, Tautz D (1989) Posterior segmentation of the *Drosophila* embryo in the absence of a maternal posterior organizer gene. Nature (London) 338:629–632

Hülskamp M, Pfeifle C, Tautz D (1990) A morphogenetic gradient of *hunchback* protein organizes the expression of the gap genes *Krüppel* and *knirps* in the early *Drosophila* embryo. Nature (London) 346:577–580

Ingham P (1988) The molecular genetics of embryonic pattern formation in *Drosophila*. Nature (London) 335:25–34

Irish V, Lehmann R, Akam M (1989) The *Drosophila* posterior-group gene *nanos* functions by repressing *hunchback* activity. Nature (London) 333:646–648

Lewis EB (1978) A gene complex controlling segmentation in *Drosophila*. Nature (London) 276:565–570

Nauber U, Pankratz M, Kienlin A, Seifert E, Klemm U, Jäckle H (1988) Abdominal segmentation of the *Drosophila* embryo requires a hormone receptor-like protein encoded by the gap gene *knirps*. Nature (London) 336:489–492

Nüsslein-Volhard C, Wieschaus E (1980) Mutations affecting segment number and polarity in *Drosophila*. Nature (London) 287:795–801

Nüsslein-Volhard C, Frohnhöfer HG, Lehmann R (1987) Determination of anteroposterior polarity in *Drosophila*. Science 238:1675–1681

Pankratz M, Hoch M, Seifert E, Jäckle H (1989) *Krüppel* requirement for *knirps* enhancement reflects overlapping gap gene activities in the *Drosophila* embryo. Nature (London) 341:337–340

Pankratz M, Seifert E, Gerwin N, Billi B, Nauber U, Jäckle H (1990) Gradients of *Krüppel* and *knirps* gene products direct pair-rule stripe patterning in the posterior region of the *Drosophila* embryo. Cell 61:309–317

Rosenberg UB, Schröder C, Preiss A, Kienlin A, Cote S, Riede I, Jäckle H (1986) Structural homology of the product of the *Drosophila Krüppel* gene with *Xenopus* transcription factor III A. Nature (London) 319:336–339

Roth S, Stein D, Nüsslein-Volhard C (1989) A gradient of nuclear localization of the *dorsal* protein determines dorsoventral pattern in the *Drosophila* embryo. Cell 59:1189–1202

Rushlow CA, Han K, Manley JL, Levine M (1989) The graded distribution of the *dorsal* morphogen is initiated by selective nuclear transport in *Drosophila.* Cell 59:1165–1177

Sander K (1976) Specification of the basic body pattern in insect embryogenesis. Adv Insect Physiol 12:125–238

Schneuwly S, Klemenz R, Gehring WJ (1987) Redesigning the body plan of *Drosophila* by ectopic expression of the homoeotic gene *Antennapedia.* Nature (London) 325:816–818

Seidel F (1934) Das Differenzierungszentrum im Libellenei I. Die dynamischen Voraussetzungen der Determination und Regulation. Wilh Rouxs Arch Dev Biol 131:134–187

Sprenger F, Stevens LM, Nüsslein-Volhard C (1989) The *Drosophila* gene *torso* encodes a putative receptor tyrosine kinase. Nature (London) 338:478–483

Stanojevic D, Hoey T, Levine M (1989) Sequence specific DNA-binding activities of the gap proteins *hunchback* and *Krüppel* in *Drosophila.* Nature (London) 341:331–335

Steward R (1989) Relocalization of the *dorsal* protein from the cytoplasm to the nucleus correlates with its function. Cell 59:1179–1188

Struhl G (1981) A homoeotic mutation transforming leg to antenna in *Drosophila.* Nature (London) 292:635–638

Struhl G (1989) Differing strategies for organizing anterior and posterior body pattern in *Drosophila* embryos. Nature (London) 338:741–744

Tautz D, Lehmann R, Schnürch H, Schuh R, Seifert E, Kienlin A, Jones K, Jäckle H (1987) Finger protein of novel structure encoded by *hunchback*, a second member of the gap class of *Drosophila* segmentation genes. Nature (London) 327:383–389

Design Elements Employed in the Construction of Animal Skeletons

R.D.K. THOMAS[1] and W.-E. REIF[2]

Summary

Patterns recur in nature because the possibilities of structure are not unlimited. Among organisms, like forms may be linked by common ancestry. Alternatively, their similarities may reflect identical architectural constraints on structures that must grow and function if the organism is to be fruitful and multiply. Ubiquitous convergence of design observed in the skeletal structures of unrelated living and extinct organisms is usually explained in terms of adaptation. Organic form is assumed to approach optimal functional paradigms, to the extent permitted by ancestral inheritance and the usual necessity to accommodate more than one function. Such explanations are generally not incorrect, but they are incomplete. The range of possible organic structures is determined by the properties of a limited number of suitable materials, and by the dynamic geometry of growth processes which define the topological relationships of increments added during ontogeny. Simple geometric rules define a multivariate skeleton space, defined here in terms of alternate character states of seven structural variables. Among 186 possible pairs of these variables, 12 are disallowed because their specifications are mutually contradictory or dysfunctional. More than half of the remaining pairs are abundantly represented by skeletal structures of known organisms, while fewer than one third are rare or unknown. The number of viable design paradigms prescribed by the skeleton space is not enormous and this abstract morphospace is rather well filled by structures that have been realized in nature. These observations show that the organizational themes of skeletal form are strongly determined by the possibilities of structure and that such possibilities have been rather fully exploited. Thus, organic design is not merely contingent upon the course of history.

1 How Many Kinds of Organisms?

Is it more remarkable that swordfish and narwhals, rudists and richtofeniids, or pterodactyls and hang-gliders are so different or that they are so much alike? The adaptive opportunism and diversity of living and extinct organisms are impres-

[1] Department of Geology, Franklin and Marshall College, P.O. Box 3003, Lancaster PA 17604-3003, USA

[2] Institut für Geologie und Paläontologie, Universität Tübingen, Sigwartstraße 10, D-7400 Tübingen 1, FRG

N. Schmidt-Kittler and K. Vogel (Eds):
Constructional Morphology and Evolution

sive. So great is this diversity and so clear the argument that the distributions of particular taxa in time and space are accidents of history that most modern biologists, working in the NeoDarwinian tradition, have come to regard chance and historical contingency as the primary determinants of patterns of evolution. This view is asserted most forcefully in the writing of Stephen Gould, notably in his recent book *Wonderful Life* (1989). Certainly, the variety of possible organisms is virtually infinite, given the complexity of even the simplest living things. Yet, similar forms recur, and not infrequently or at random. These common themes reflect a modest number of conceivable geometries and growth patterns, the use of suitable materials, and the approach of adaptation to limits set by physical laws. Where convergence results from adaptation to a common mode of life, it speaks of similar opportunities presented and taken up at different times and places. However, convergence results more fundamentally from the fact that a given function or growth pattern prescribes a limited range of viable designs, as Thomas (1988) has shown for bivalved shells.

Recurrent patterns of form are biological universals, based upon immanent properties of this Universe. D'Arcy Thompson (1917) focused his attention on these universals. A thorough-going mechanist, he saw a common mathematical and physical basis for order and discontinuity in both inorganic and organic structures. Earlier, Galton (1869, 1892) had observed the recurrence of distinctive variations, notably three very different categories of human fingerprints that are distributed across lines of race and social class. This led him to argue that evolutionary change must occur in rapid jumps between discrete states, "definite positions of organic stability" that constrain variation. Similarly, the Russian geneticist, Vavilov (1922) equated paralled sets of variants, that recur in different species, genera, families, and even distantly related taxa of flowering plants, with the ordered classes of chemistry and crystallography. However, no formal or causal basis for the supposed transcendent regularity of these recurrent patterns was identified, nor were its consequences clearly distinguished from those of shared ancestry, so these speculations had little impact.

To assess the role of universals in the determination of form, the degree to which it is predictable, it is necessary to go beyond making an inventory of realized patterns and repetitions, although this may be suggestive. As Raup (1966) recognized, the larger theoretical spaces within which organic structures grow must be defined. To this end, he developed a well-known model which generates all the geometrically possible spiral shells that can be derived from changes in three fundamental variables. Analysis of the forms of real organisms in this context shows that higher taxa have limited, characteristic ranges and that many conceivable geometries occur little if at all in nature. Raup (1966, 1967) has shown that much of the "open" space is in fact forbidden, either by other features of particular growth patterns or because the forms are dysfunctional. In short, there are irrational morphologies, generated by the model in the discontinuities between what Galton or Thom (1975) would characterize as "regions of stability".

Much recent literature addresses the role of constraints on developmental processes and hence on evolution (e.g., Kauffman 1983; Thomson 1988; Goodwin and Saunders 1989). Most of this work focuses on the means by which information is called up and translated in a particular developmental context to organize material into patterns, as morphogenetic processes build living structure. We have shown elsewhere how growth processes, adaptation, and the contingencies of evolutionary history act together in the genesis of individual structures and taxa (Reif 1975; Thomas 1979; Reif et al. 1985). This integrative approach to the explanation of organic form, revived by Seilacher (1970; see also Vogel this Vol.), has become widely known as constructional morphology. It provides the larger evolutionary context for a generalization of Raup's argument that yields what we call the skeleton space. In this paper, we define the skeleton space and assess the extent to which it has been filled up. The skeleton space is based on inherent geometric and material constraints that define the universe in which animal skeletons grow, function, and evolve. If it is relatively fully occupied, we may infer that these constraints have had a large role in prescribing the opportunities open to evolving organisms. Alternatively, if the skeleton space is sparsely or very unevenly populated, then the role played by historical contingency may be inferred to be correspondingly great.

2 The Skeleton Space

Prompted by Seilacher's (1970) emphasis on the influence of growth processes on organic form, Reif and Robinson (1975) outlined a conceptual framework for the comparative analysis of the structures that are employed in animal skeletons. They listed alternative characteristics of skeletal growth processes and patterns under six headings. Some combinations of characteristics cannot occur, they noted, while others such as frameworks, spiral cones, and lever skeletons appear repeatedly in unrelated taxa. These concepts are revised and extended here, to define a multivariate "skeleton space" within which the forms of animal skeletons or particular structural complexes are specified. It is an empirical scheme, neither unique nor complete, but substantially representative of the dimensions of skeletal growth and form.

The skeleton space is based on the alternative character states of seven variables, listed in Table 1. Internal and external skeletons are distinguished conventionally, in relation to their secreting tissues, since the difference between growth on an ectodermal surface and growth within mesodermal tissue has major geometrical consequences for skeletal form. The simple characterization of materials as rigid or flexible is more fundamental than the long list of compositions and fabrics that underlie these properties. It serves to distinguish between those that are brittle, which resist breaking stress, and those that accommodate stress in deformation, until they reach breaking strain (Wainwright 1980). Single-element and multi-element structures represent fundamentally different levels of integration; two-element structures are distinguished largely to accommodate the extraordinary variety of bivalved shells. The shape categories are simply based on dimensionality. Rods are elongated structures, generally involved in support of the soft tissues. Structures secreted over surfaces enclose and partition space and provide support for soft tissues. Here, plates are distinguished as planar elements of a continuous surface, while cones are curved surfaces with open boundaries. Solids are largely machine parts, such as teeth and articulating bones. Among

Table 1. Alternative features of animal skeletons. The range of potential forms of animal skeletons or their subunits is defined in terms of seven essential properties, each with two to four possible states, yielding a total of 21 variables

1. Situation	**5. Growth**
Internal	Accretionary
External	Serial units ± branching
2. Material	Replacement/molting
Rigid	Remodeling
Flexible	**6. Assembly**
3. Number	Growth in place
One element	Prefabrication
Two elements	**7. Interplay of elements**
>two elements	No contact
4. Shape	Imbricate
Rods	Jointed
Plates	Sutured or fused
Cones	
Solids	

growth processes, accretion of material to a surface is distinguished from remodeling by resorption and redeposition, although it is recognized that most real structures are molded by both processes. Discontinuous growth of skeletons that fully enclose the soft tissues occurs by molting and replacement, or by the addition of serial units, which often involves branching, notably in colonies. Most skeletal structures grow where they are to be employed, but some are prefabricated and moved into functional position, as where teeth erupt. Alternatively, prefabricated materials may be obtained from the external environment and incorporated into the skeleton, as in the carrier shell, *Xenophora*, and agglutinated foraminifera. Skeletal elements interact more or less closely within a structural complex. At the extremes, they may have no contact, as in holothurian sclerites and some sponge spicules, or they may be rigidly fused together. Imbricate structures have the potential for sliding articulation, while joints allow rotation about a particular point or axis.

A couple of examples show that the skeleton space is a useful heuristic device, drawing attention to fundamental similarities and differences of structures that are compared in this context. The leg of a horse has an internal skeleton of jointed rods, composed of a relatively rigid material. Its bones are remodeled as they grow, in place. The limb of an insect has an external skeleton of jointed rods (cylinders in this case), also composed of a relatively rigid material. This skeleton is secreted in place, following periodic molts. The convergent employment of jointed rods made of materials that are remarkably similar in stiffness (Wainwright et al. 1976) in these disparate organisms rests on the physical specifications that must be met by any lever system. However, the bones of the horse are articulated by muscles attached to their exterior surfaces, while the muscles of the insect must be attached inside the cuticle, with generally less mechanical advantage. The fixed joints of the horse's leg can grow and function continuously, as they are remodeled. In contrast, the flexible cuticle of the insect joint allows free rotation that must be restrained by articulating pegs and sockets (Currey 1970).

Balanomorph barnacles and many articulate brachiopods, all obligate sessile suspension feeders, attach to hard surfaces, often in crowded clusters. The shells of both groups are external, composed of rigid elements that grow in place by marginal accretion. But the brachiopod shell consists of two articulated spiral cones, while the barnacle has up to eight sutured wall plates and four opercular plates with a sliding articulation. The growth pattern of the brachiopod shell requires it to retain its regular shape and not depart too far from bilateral symmetry. Most present-day brachiopods live attached by their pedicles to hard substrates and they are unable to reattach if they are dislodged (C. W. Thayer, in Vermeij 1978, p. 105). In contrast, the shape of the barnacle shell can be adjusted to fit available space and in some species these animals gain additional stability from becoming cemented together. This suggests the hypothesis that balanomorph barnacles are superior competitors on shallow-water hard substrates where epifaunal organisms are most vulnerable to disturbance and dislocation. Hence the striking Cenozoic evolutionary radiation of the barnacles may have taken place at the expense of surviving groups of articulate brachiopods, where both groups had the opportunity to take over the ecospace left vacant by the extinction of other epifaunal brachiopods.

Attempts to place individual structures within the skeleton space may also raise interesting questions. Is the belemnite guard an internal or external skeletal element? It is a solid, bullet-shaped object that grows concentrically outward by accretion and is usually described as being "internal" since it was evidently surrounded by body tissues. However, the guard was presumably secreted by infolded pallial epithelium, like the cuttlebone in *Sepia,* in which case it is an anatomically external structure. The belemnite guard is a straight, solid spiral cone, a molluscan shell turned outside in, as it were.

The larger value of the skeleton space is that it reaches beyond the descriptive classification of known skeletal structures. Its 1536 combinations of seven design characteristics represent paradigms for patterns of skeletal growth and form that may or may not be exploited in nature. It is not practical, nor would it be very interesting due to the large amount of redundancy, to examine every one of these combinations separately. However, it is instructive to consider the occurrence of all possible pairs of characteristics that can be drawn from any two of the variables. In Table 2, we have compiled our unabashedly subjective estimates of the frequency with which each character pair occurs among actual skeletal structures of living or extinct organisms. These assessments are supported by tables, included in the Appendix, which list organisms and structures that exhibit each character combination. Most assignments are obvious, but some raise intriguing questions, like the belemnite guard discussed above. Our tabulation is not exhaustive and if it is in places provocative, it serves its purpose.

It becomes apparent that not all the 186 pairs are in fact possible. Four pairs are disallowed because their specifications are mutually contradictory. A single skeletal element cannot be imbricate, jointed, sutured, or repeated in series. At least eight combinations are dysfunctional. Internal and sutured skeletons cannot be molted. Cones and solids cannot form an imbricate structure. Flexible struc-

Table 2. The Skeleton Space. Occurrences of the various character combinations in nature are recorded as being abundant (A), common (C), or rare (R). Combinations with mutually contradictory definitions and those that are clearly functionally inviable are disallowed (F). Taxonomic data from which these inferences are derived are provided in the Appendix

	Internal	External	Rigid	Flexible	One element	Two elements	> two elements	Rods	Plates	Cones	Solids	Accretionary	Unit/serial	Replace/molt	Remodeling	In place	Prefabrication	No contact	Imbricate	Jointed	Fused, sutured
1. Internal	–																				
External	–	–																			
2. Rigid	A	A	–																		
Flexible	A	A	–	–																	
3. One element	R	A	A	C	–																
Two elements	R	A	A	R	–	–															
> two elements	A	A	A	C	–	–	–														
4. Rods	A	A	A	R	R	R	A	–													
Plates	A	C	A	F	R	F	A	–	–												
Cones	R	A	A	R	A	A	R	–	–	–											
Solids	C	C	A	R	R	R	A	–	–	–	–										
5. Accretionary	A	A	A	R	A	A	A	A	A	A	C	–									
Unit/serial	A	A	A	C	F	R	A	A	A	C	C	–	–								
Replace/molt	F	A	A	A	R	R	A	A	A	C	A	–	–	–							
Remodeling	A	C	A	C	R	R	A	C	C	C	R	–	–	–	–						
6. In place	A	A	A	A	A	A	A	A	A	A	A	A	A	A	A	–					
Prefabrication	R	C	A	R	R	R	A	R	R	R	A	R	C	C	F	–	–				
7. No contact	C	R	C	R	A	R	C	R	R	R	R	A	C	R	R	C	C	–			
Imbricate	R	A	A	R	F	R	A	R	A	F	F	C	C	C	R	A	R	–	–		
Jointed	A	A	A	A	F	A	A	A	A	A	R	A	A	A	A	A	R	–	–	–	
Fused, sutured	C	R	A	F	F	R	A	C	A	R	R	A	A	F	C	A	R	–	–	–	–

Total combinations: $2 \times 2 \times 3 \times 4 \times 4 \times 2 \times 4 = 1536$
Total pairs (one character from each of any two groups) = 186
Frequencies: A = 94, C = 30, R = 50, F = 12

tures may be constructed with flexible materials, or with articulating rigid elements; flexible plates and sutured flexible structures are mechanically self-defeating designs. A structure composed of two plates cannot grow and enclose space unless these plates are cones. Finally, a structure that is prefabricated and moved into its functional position could only be remodeled by a process distinct from that by which it was grown, which is highly improbable.

In the skeleton space as a whole, all combinations of characteristics that include the mismatched pairs are, of course, inviable. This reduces the number of design combinations that are really available for skeleton construction to less than a thousand. Some further reduction may result if specific combinations of three or more characteristics are inviable, but there are relatively few of these that are not excluded already.

3 Conclusion

These abstractions lead to two important conclusions. First, the number of viable design paradigms prescribed by the skeleton space is not so very large. Second, the skeleton space is rather well filled by actual, known organisms. Fully half of the total number of character pairs occur abundantly and we have characterized less than a third of the viable pairs as being rare. These conclusions are consistent with the observation that convergence in organic skeletal design is ubiquitous. They affirm that the organizational themes of skeletal form are determined by the possibilities of structure and that those possibilities have been rather fully exploited.

This analysis of a scenario that prescribes possible animal skeletons does not deny important roles for chance and historical constraint in evolution. Rather, it confirms that the comparative analysis of morphology, in the structuralist tradition that extends back to the work of Cuvier (Webster and Goodwin 1982), is complementary to NeoDarwinian studies of population genetics and adaptation in arriving at a comprehensive understanding of the causes of evolutionary change and the patterns to which it gives rise. Wake and Larson (1987) have analyzed the roles of several developmental constraints on the evolution of salamander limb bones in these terms, but the constraints they have identified are deeply embedded in a preexisting historical context. In this study, we have taken up their challenge to predict from first principles "a finite set of possible forms that organismal systems can produce", apart from any particular history.

References

Currey J (1970) Animal skeletons. Arnold, London, 52 pp
Galton F (1869) Hereditary genius. Macmillan, London, 390 pp
Galton F (1892) Finger prints. Macmillan, London, 216 pp
Goodwin B, Saunders P (eds) (1989) Theoretical biology: epigenetic and evolutionary order from complex systems. Univ Press, Edinburgh, 230 pp
Gould SJ (1989) Wonderful life. Norton, New York, 347 pp
Kauffman SA (1983) Developmental constraints: internal factors in evolution. In: Goodwin BC, Holder N, Wylie CC (eds) Development and evolution. Br Soc Dev Biol Symp 6:195–225
Raup DM (1966) Geometric analysis of shell coiling: general problems. J Paleontol 40:1178–1190
Raup DM (1967) Geometric analysis of shell coiling: coiling in ammonoids. J Paleontol 41:43–65
Reif W-E (1975) Lenkende und limitierende Faktoren in der Evolution. Acta Biotheor 24:136–162
Reif W-E, Robinson JA (1975) Geometrical relationships and the form-function complex: animal skeletons. N Jahrb Geol Paleontol Mh 1975:184–191
Reif W-E, Thomas RDK, Fischer MS (1985) Constructional morphology: the analysis of constraints in evolution. Acta Biotheor 34:233–248
Seilacher A (1970) Arbeitskonzept zur Konstruktions-Morphologie. Lethaia 3:393–396
Thom R (1975) Structural stability and morphogenesis; an outline of a general theory of models. Benjamin, Reading, Mass, 348 pp
Thomas RDK (1979) Constructional morphology. In: Fairbridge RW, Jablonski D (eds) The encyclopedia of paleontology. Dowden, Hutchinson & Ross, Stroudsberg, Penn, pp 482–487
Thomas RDK (1988) Evolutionary convergence of bivalved shells: a comparative analysis of constructional constraints on their morphology. Am Zool 28:267–276
Thompson DW (1917) On growth and form. Univ Press, Cambridge, UK, 793 pp
Thomson KS (1988) Morphogenesis and evolution. Oxford Univ Press, New York, Oxford, 154 pp

Vavilov NI (1922) The law of homologous series in variation. J Genet 12:47–89
Vermeij GJ (1978) Biogeography and adaptation: patterns of marine life. Harvard Univ Press, Cambridge, Mass, 332 pp
Wainwright SA (1980) Adaptive materials: a view from the organism. Symp Soc Exp Biol 34:437–453
Wainwright SA, Biggs WD, Currey JD, Gosline JM (1976) Mechanical design in organisms. John Wiley & Sons, New York, 423 pp
Wake DB, Larson A (1987) Multidimensional analysis of an evolving lineage. Science 238:42–48
Webster G, Goodwin BC (1982) The origin of species: a structuralist approach. J Soc Biol Struct 5:15–47

Appendix

Catalog of organisms with skeletons or skeletal subunits exhibiting pairs of character states defined in the Skeleton Space. Estimates of abundance or variety listed in this Appendix constitute the data upon which Table 2 is based.

	1. Internal	1. External
2. Rigid	(A) Vertebrate bones, echinoderm plates	(A) Mollusc and brachiopod shells
Flexible	(A) Vertebrate cartilage, squid gladius	(A) Unmineralized arthropod cuticles, bivalve ligament, inarticulate brachiopods
3. One element	(R) Notocord of *Branchiostoma*, perforate foraminifera	(A) Gastropods, various other molluscs, serpulid worms, solitary corals
Two elements	(R) Not disallowed	(A) Bivalves, brachiopods, ostracodes
> two elements	(A) Vertebrates, echinoderms	(A) Arthropods, chitons
4. Rods	(A) Vertebrate limb bones = solid rods, sponge and other spicules, echinoid spines	(A) Arthropod limbs = cylinders, hollow rods, mollusc and brachiopod shell spines
Plates	(A) Echinoderms, bony plates of turtles, placodonts, crocodiles, human skull	(C) Barnacles, chitons, turtle scutes, dinoflagellate armor
Cones	(R) Internal shell of *Spirula*	(A) Mollusc and brachiopod shells, solitary corals, rhinoceros horns
Solids	(C) Vertebrae, ankle bones, otoliths	(C) Vertebrate teeth, shark denticles, belemnites, cuttlebone of *Sepia*
5. Accretionary	(A) Echinoderm plates, inter-vertebral disks, otoliths, bone of lower vertebrates	(A) Molluscs, brachiopods, corals, etc.
Unit/serial	(A) Echinoderm plates, graptolites	(A) Bryozoans, colonial corals, diverse chambered shells
Replace/molt	(F) Functionally implausible	(A) Arthropod exoskeletons, vertebrate teeth, cervid antlers
Remodeling	(A) Mammalian bone, echinoderms	(C) Molluscan shells (limited)
6. In place	(A) Vertebrates, echinoderms	(A) Most exoskeletons
Prefabrication	(R) Extruded sponge spicules, coccolith plates	(C) Vertebrate teeth, agglutinated tubes and shells of worms, molluscs, *Laternula*

	1. Internal	1. External
7. No contact	(C) Some sponge spicules, holothurian sclerites	(R) Shark scales (but collagen is continuous between them)
Imbricate	(R) Helicoplacoids	(A) Trilobite exoskeleton, fish scales, tail of pangolin, gastropod operculum
Jointed	(A) Vertebrates, echinoderms	(A) Arthropods, bivalves
Fused, sutured	(C) Vertebrate skull, echinoid tests	(R) Turtle scute plates, dinoflagellates

	2. Rigid	2. Flexible
3. One element	(A) Gastropods, scaphopods, solitary corals	(C) Notocord of *Branchiostoma*, cuticle of *Peripatus*, gorgonians. Monograptids?
Two elements	(A) Bivalves, brachiopods, ostracodes, Devonian rugose coral *Calceola*	(R) Bivalve *Pinna*, conchostrachans, modern phyllocarids
> two elements	(A) Vertebrates, echinoderms, many arthropods, chitons	(C) Chondrichthyan skeletons, elements of some arthropod skeletons
4. Rods	(A) Vertebrate and arthropod limbs. Sponge spicules	(R) Some arthropod limbs
Plates	(A) Echinoderms, ostracoderm fishes, placodont reptiles, turtles, barnacles	(F) Functionally implausible
Cones	(A) Molluscs, articulate brachiopods, solitary corals, mammalian horns	(R) Some bivalves, e.g., *Pinna*, *Solemya*; conchostracans, acrotretid brachiopods
Solids	(A) Vertebrate teeth, belemnites	(R) Intervertebral discs (for shock absorption)
5. Accretionary	(A) Molluscs, brachiopods, barnacles, stony corals	(R) Some bivalves, many inarticulate brachiopods
Unit/serial	(A) Echinoderm plates, bryozoans, colonial corals	(C) Gorgonians, graptolites
Replace/molt	(A) Most arthropod exoskeletons	(A) Some arthropod exoskeletons, cuticle of *Peripatus*
Remodeling	(A) Mammalian bone, limited in mollusc shells	(C) Cartilage in chondrichthyans and other vertebrates
6. In place	(A) Vertebrates, echinoderms, molluscs	(A) Chondrichthyans, some arthropods
Prefabrication	(A) Vertebrate teeth, agglutinated worm tubes+foraminifera, gastropod *Xenophora*	(R) Bivalve ligament
7. No contact	(C) Some sponge spicules, holothurian sclerites	(R) Some sponge spicules, Demospongia
Imbricate	(A) Trilobite exoskeleton, fish scales, tail of pangolin	(R) Bird feathers
Jointed	(A) Vertebrates, many arthropods, most bivalves, crinoids	(A) Chondrichthyans, some arthropods, conchostracans, some bivalves
Fused, sutured	(A) Echinoid tests, adult primate skull, turtles, placodont reptiles, dinoflagellates	(F) Functionally implausible

	3. One element	3. Two elements	3. > two elements
4. Rods	(R) Squid gladius, notocord of *Branchiostoma*	(R) Not disallowed	(A) Lever skeletons of vertebrates and arthropods
Plates	(R) Spherical foraminifera	(F) Two growing plates enclosing space are cones	(A) Echinoderms, turtles placoderms, ostracoderms
Cones	(A) Gastropods, cephalopods solitary corals, serpulids	(A) All bivalved shells, white rhino, bovid horns	(R) Cambrian chiton *Matthevia*
Solids	(R) Belemnites	(R) Cervid antlers	(A) Vertebrate teeth, vertebrae and discs
5. Accretionary	(A) Gastropods, scaphopods, belemnites, black rhino horn	(A) Bivalved molluscs, brachiopods, paired horns	(A) Echinoderms, lower vertebrate bones, barnacles
Unit/serial	(F) Logically excluded. One element cannot be serial	(R) Not disallowed	(A) Corals, bryozoans, echinoderm plates and arms, chambered shells
Replace/molt	(R) Not disallowed. Single horn, spherical shells?	(C) Ostracodes, phyllocarids	(A) Arthropods in general
Remodeling	(R) Gastropods, especially Nerineidae	(R) Limited in bivalves	(A) Vertebrates, especially mammals, echinoderms
6. In place	(A) Monoplacophorans, gastropods, cephalopods, notocord of *Branchiostoma*	(A) Bivalvia, brachiopods, white rhino, bovid horns	Vertebrates, most echinoderm plates
Prefabrication	(R) Bivalve ligament	(R) Not disallowed	(A) Teeth in vertebrate jaw, radula, shark's denticles, spines of bivalve *Laternula*
7. No contact	(A) Univalve molluscan shells: gastropods etc.	(R) Not disallowed	(C) Holothurian + tunicate sclerites, sponge spicules
Imbricate	(F) Logically excluded	(R) Radiolitid + hippuritid rudists, sliding articulation	(A) Trilobite exoskeleton, pangolin, helicoplacoids
Jointed	(F) Logically excluded	(A) All bivalved shells	(A) Vertebrates, arthropods
Fused, sutured	(F) Logically excluded	(R) Rostroconchs?	(A) Echinoid tests, primate skull, turtle shell plates

	4. Rods	4. Plates	4. Cones	4. Solids
5. Accretionary	(A) Long bones of lower vertebrates, echinoid spines, porcupine spines	(A) Echinoderms, vertebrate plate skeletons, barnacles	(A) Molluscs, brachiopods, solitary corals, vertebrate horns	(C) Belemnites, some gastropod opercula, intervertebral discs
Unit/serial	(A) Gorgonians, bryozoa, graptolites	(A) Echinoderms	(C) Colonial corals, trepostome bryozoa	(C) Vertebrate teeth, shark scales
Replace/molt	(A) Arthropod limb skeletal elements	(A) Arthropod tergite skeletal elements	(C) Bivalved arthropods	(A) Teeth of lower vertebrates, sharks Gastropod radula, antlers of cervids
Remodeling	(C) Mammalian long bones	(C) Echinoderm plates (limited), nerineid gastropods	(C) Mollusc shells, varying extent, mostly limited	(R) Surface/ volume ratio mitigates against it
6. In place	(A) Vertebrate and arthropod limbs	(A) Echinoderms, vertebrate plate skeletons	(A) Molluscs, brachiopods, solitary corals, horns	(A) Belemnites, antlers of cervids, intervertebral discs
Prefabrication	(R) Extruded sponge spicules	(R) Coccolith plates, *Thecamoeba*. Any animal examples?	(R) Not disallowed	(A) Vertebrate teeth, gastropod radula teeth
7. No contact	(R) Sponge spicules	(R) Holothurian sclerites	(R) Cambrian chiton *Matthevia*, rhinoceros horns	(R) Shark scales, tunicate sclerites
Imbricate	(R) Sponge spicules, octocorals	(A) Heliocoplacoids, edrioasteroids, fish scales, pangolin tail	(F) Functionally implausible	(F) Functionally implausible
Jointed	(A) Lever skeletons of vertebrates and arthropods, echinoid spines	(A) Echinoderms, especially echinoid tests	(A) Bivalved shells in general	(R) Vertebrate ankle bones, phalanges, intervertebral discs
Fused, sutured	(C) Sponge spicule lattices, calcified tendons	(A) Primate skull, vertebrate plate skeletons, echinoids	(R) Rostroconchs?	(R) Tooth plates of the eagle ray, *Myliobatis*

	5. Accretionary	5. Unit/serial	5. Replace/molt	5. Remodeling
6. In place	(A) Mollusc shells, belemnites, intervertebral discs	(A) Gorgonians, bryozoans, echinoid plates, graptolites	(A) Arthropod exoskeletons, cervid antlers	(A) Mammalian bone, echinoderm stereom
Prefabrication	(R) Bivalve ligament, spines of *Laternula*	(C) Vertebrate teeth, agglutinated carrier shell *Xenophora*	(C) Shark teeth, teeth of gastropod radula	(F) Functionally implausible
7. No contact	(A) All univalved molluscs, belemnites	(C) Sponge spicules, holothurian sclerites	(R) Not disallowed	(R) Not disallowed
Imbricate	(C) Fish scales, armadillo plates, helicoplacoids?	(C) Brittle star and crinoid arms	(C) Dorsal elements of skeletons of enrolling arthropods	(R) Not disallowed
Jointed	(A) Bivalved shells in general	(A) Echinoderm arms, serial limbs of arthropods	(A) Arthropods in general	(A) Mammalian bone, limited in shells of Bivalvia
Fused, sutured	(A) Echinoid tests, turtles, placodonts, barnacles	(A) Bryozoans, colonial corals, graptolites	(F) Functionally implausible	(C) Plates in tests of echinoids, primate skull

	6. In place	6. Prefabrication
7. No contact	(C) Sponge spicules, holothurian sclerites	(C) Vertebrate teeth, gastropod radula
Imbricate	(A) Helicoplacoids, edrioasteroids, fish scales, pangolin tail	(R) Coccolith plates, *Thecamoeba*. Any animal examples?
Jointed	(A) Lever skeletons vertebrates, arthropods and echinoderms	(R) Bivalve ligament
Fused, sutured	(A) Echinoid tests, turtles, placodonts, primate skulls	(R) Spines secreted, moved, cemented to shell of bivalve *Laternula*

A Theoretical Morphological Approach to Tooth Replacement in Lower Vertebrates

DAVID B. WEISHAMPEL [1]

Summary

The regulation of the pattern of tooth replacement in lower vertebrates is explored from a theoretical morphological perspective. When inhibition zones are postulated as controls on the vertical and horizontal spacing of tooth germs, all replacement geometries of extant and fossil lower vertebrates can be simulated. Further work on the existence of inhibition zones is necessary, but should they prove to be roughly as has been suggested here, then local controls at the level of tooth germs are sufficient to explain the global geometries of whole definitions.

1 Introduction

Function and phylogeny have long been seen as necessary parts of the explanation of biological form (Rudwick 1964; Cracraft 1981; Mayr 1983; Cheverud et al. 1985). However, in isolation neither provides a sufficient understanding of form. Seilacher (1970) argued instead that organic form could be understood only as the result of three different aspects: ecologic-adaptive factors, historic-phylogenetic factors, and bautechnische factors (roughly architecture). The impact of these three factors comprised Seilacher's Konstruktionsmorphologie research program for the analysis and understanding of biological form (see also Reif 1975, 1982; Thomas 1979; Thomas and Reif 1983; Reif et al. 1985; but contrast usage with Nachtigall this Vol.; Gutmann this Vol.).

As noted by Reif et al. (1985), Bautechnik recognizes that geometry, natural materials, and growth processes regulate morphologic patterns. Bautechnische factors are ahistorical elements that express the biological possibilities and limits on evolutionary change that stem from the physical and chemical properties of available materials (e.g., strength and failure, elasticity, adhesion, viscosity, etc.; Wainwright et al. 1976; Vincent 1982), cybernetic controls on development, maintenance, and repair (i.e., homeorhesis; Waddington 1968), and, finally, the geometry of pattern formation and self-organization (Wolpert and Stein 1984; Dhouailly 1984). The particular materials, growth programs, and regulatory systems are naturally acquired as specific historical events, hence the transfer of ahistorical factors to phylogenetic clades and lineages. Likewise, how these ahistorical factors

[1] Department of Cell Biology and Anatomy, The Johns Hopkins University, School of Medicine, Baltimore, Maryland 21205, USA

N. Schmidt-Kittler and K. Vogel (Eds):
Constructional Morphology and Evolution

are reflected in evolutionary fitness also shifts such influences to the unique historical nexus of adaptation.

Among bautechnische approaches to the study of biological form, none has been more important than theoretical morphology (Raup and Michelson 1965; see also Reif and Weishampel in press). This research program seeks to understand the ways in which morphogenetic "rules" impact on ontogeny and in phylogeny. To date, most theoretical morphological studies have been based on the geometry of logarithmic spiraling in shelled organisms (Raup and Michelson 1965; Raup 1966, 1967; Bayer 1978; McGhee 1980; Okamoto 1988a, b, c), with a few studies on trace fossil foraging patterns (Raup and Seilacher 1969; Papentin and Röder 1975), branching patterns in plants (Niklas 1978, 1986), dermal squamation in sharks (Reif 1980), and echinoid plate growth (Raup 1968). The present chapter extends these theoretical morphological approaches to the pattern of tooth replacement in lower vertebrates.

2 Tooth Replacement as Theoretical Morphology

The teeth of lower vertebrates are often arranged in a regular or near-regular alternating wave pattern of tooth replacement (Fig. 1). Edmund (1960) was the first to provide extensive documentation of this alternation pattern in fossil and living animals. In conjunction with the empirical evaluation of this alternation,

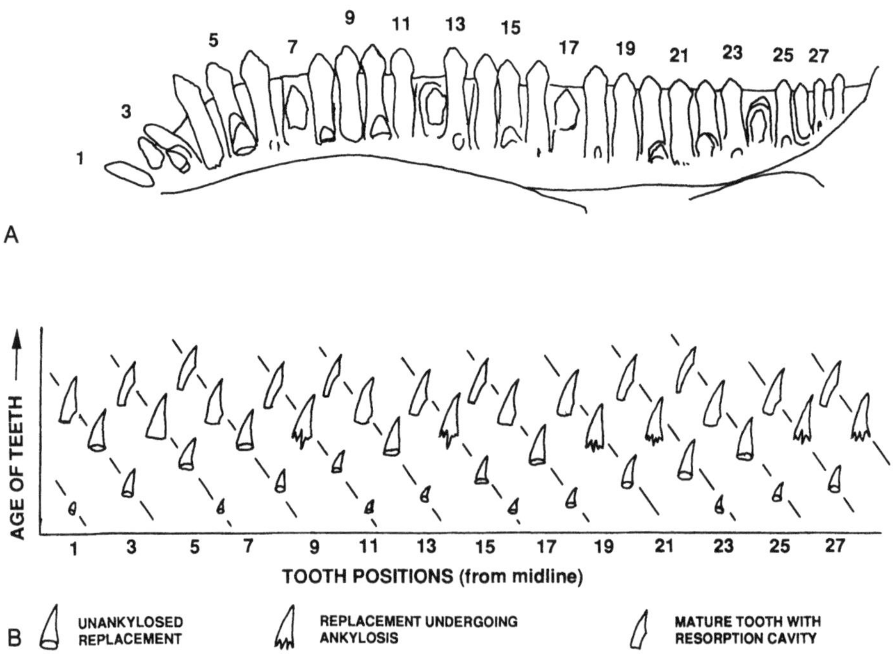

Fig. 1. Regular alternating tooth replacement pattern as seen in *Iguana iguana*. **A** Right upper dentition. **B** Graphic representation of (A). Diagonal lines represent Zahnreihen (After Edmund 1969)

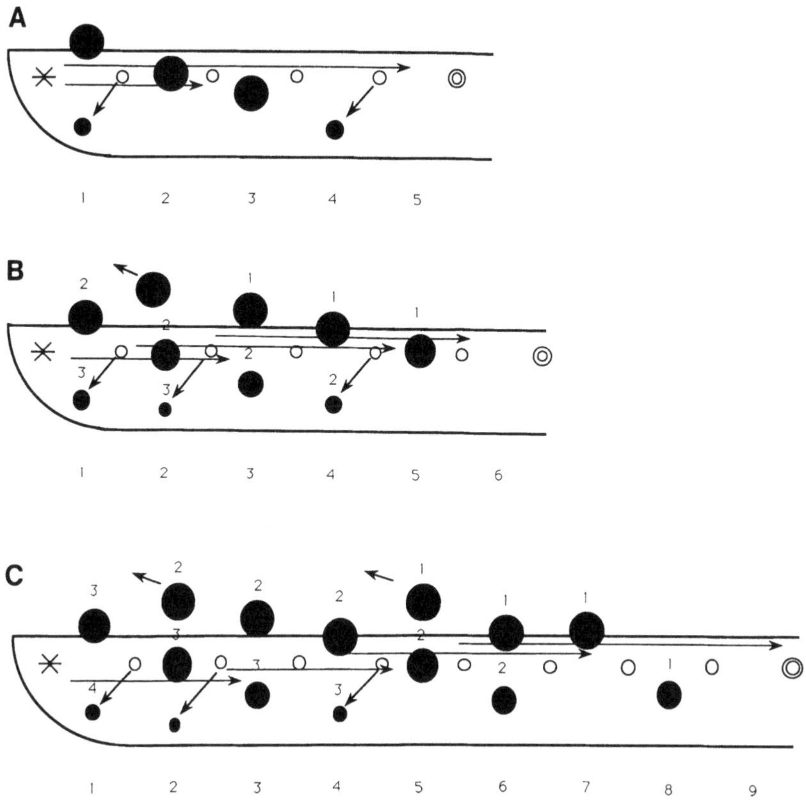

Fig. 2. Diagrammatic representation of the Zahnreihe theory. **A** An early phase of tooth replacement. **B** Successively older dentition. **C** A yet older dentition. In **A**, **B**, and **C**, the *asterix* in the front of the jaw (*left*) is the signal transmitter, the *open circles* are the signal receptors, the *double circles* are newly formed signal receptors at the back of the jaw, the *arrows* are the signals themselves, and the *solid circles* are maturing tooth germs. The *numbers assigned to each wave* of replacement represents the Zahnreihe arrangement of the developing dentition. The *numbers beneath the jaw* represent the tooth families within the jaw (After Osborn 1984)

Edmund (1960, 1969) also advanced some ideas about the morphogenetic basis for this alternation. As presently understood, this Zahnreihe theory postulates the existence of some sort of stimulus generator at the rostral tip of the jaw that sends chemical signals backwards through the jaw at regular intervals of time (Fig. 2). These signals are postulated to be received by a series of receptors that "turn on" the tooth primordia at each potential tooth position. In this way, the regular alternating waves of tooth initiation (viz., the diagonal lines or Zahnreihen in Fig. 1) seen in many lower vertebrate groups is achieved by global or field controls.

Subsequent work on Zahnreihen has involved not only the documentation of other patterns of tooth replacement beyond regular alternation in both recent and fossil vertebrates, but also a number of other studies on the theoretical distribution of Zahnreihe patterns as a function of the quality of the dentition, tooth

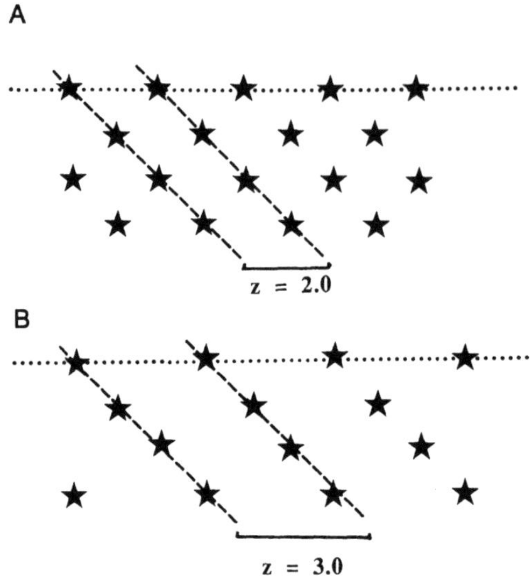

Fig. 3. Z-spacing. **A** 2.0 z-spacing. **B** 3.0 z-spacing. In both **A** and **B** the *dotted line* is the occlusal surface. *Stars along this line* are functional teeth. The *remaining stars* are progressively more mature tooth germs within a replacement sequence

packing, and growth (DeMar 1972, 1973; DeMar and Bolt 1981; Bolt and DeMar 1975, 1983, 1986).

In particular, DeMar and Bolt have described a host of parameters that characterize developing dentitions, most important among them the concept of z-spacing (see also Ziegler 1969; Hopson 1971, 1980; de Ricqlès and Bolt 1983; among others, for similar treatments). These z-spacings represent the distance, measured in number of tooth families (i.e., vertical series of replacement teeth), between the cycles of replacement down the dentition (Fig. 3). The alternating waves so common among lower vertebrates yield z-spacings around 2.0. Other z-spacings are apparently much rarer. For example, a 3.0 z-spacing is known in heterodontosaurid dinosaurs (Gow 1975; Hopson 1980), 5.5 is found in the archosauromorph *Trilophosaurus* (DeMar and Bolt 1981), and higher values can be found in a few vertebrate taxa (?8.0 in *Diademodon*; Hopson 1971). Finally, simultaneous replacement (z-spacing = 1.0, which is also equal to ∞) is found in piranhas (*Serrasalmus*; Berkovitz and Shellis 1978) and in some microsaurian amphibians (*Cardiocephalus* and *Euryodus*; Bolt and DeMar 1983).

All dentitions can be parameterized in terms of their z-spacing. This information provides not only a relatively simple description of the geometry of tooth replacement across lower vertebrates, but also the possibility of theoretically exploring all possible z-space geometries. That said, however, the mechanism by which z-spacing geometries come into being (i.e., the signals and receptors of the Zahnreihe theory) has been called into question.

Despite its ability to describe dentitional geometry through z-spacing, the Zahnreihe theory has come under fire principally from embryologists. In a series of papers, Osborn (1970, 1971, 1974, 1984) noted that the founder teeth in embryological sequences rarely are destined to contribute the first tooth family in mature dentitions as postulated by the Zahnreihe theory (see baseline study of

Lacerta vivipara; Osborn 1971). More recently, Westergaard and Ferguson (1986, 1987) pointed out that the observed initiation pattern of embryonic teeth does not automatically create the alternating initiation and eruption waves seen in many mature dentitions. Other patterns may be intercalated within the developmental sequence, again at odds with a model involving a single signal transmitter and sequentially placed receptors.

In place of the Zahnreihe theory, Osborn proposed the Clade theory of tooth replacement (Fig. 4; Osborn 1971, 1973, 1977, 1978). The Clade theory is based on local controls of tooth initiation, in which the tooth germ primordia are preceded by a clade of identical ectomesenchymal cells along the early developing dental lamina. At some critical size, this cellular clade becomes competent to form a founder tooth germ. Once formed, this founder tooth establishes an inhibition zone around it, thereby preventing tooth initiation anywhere along the dental lamina within the confines of the zone.

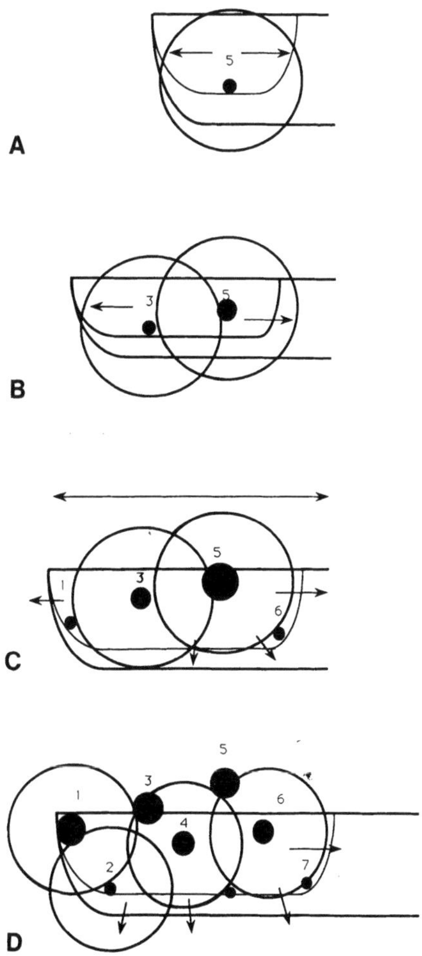

Fig. 4. The Clade theory of tooth replacement. In **A**, an initial tooth germ (*solid circle*) is formed along the free edge of the dental lamina within the developing jaw. Around each germ is a circular inhibition zone. The number *5* indicates the eventual tooth family in which this germ is destined to reside. In **B** through **D**, adjacent germs are initiated as tooth germs and inhibition zones migrate occlusally and as the dental lamina grows longitudinally (After Osborn 1984)

Unlike the Zahnreihe theory, subsequent teeth may be initiated in front of and behind the first tooth once it and its accompanying inhibition zone have migrated occlusally or the dental lamina has grown beyond its confines. Given these relationships, the local inhibition controls postulated by the Clade theory can easily produce the common alternating waving of tooth replacement (z-spacing of 2.0). However, how the theory accounts for z-spacings other than 2.0 is less certain. To gain insights along these lines, the relationship between inhibition zones and the array of empirical distribution of wave patterns expressed as z-spacing were explored using computer modeling.

3 The Model

A model of tooth replacement was set up in order to explore the ways in which local controls might produce the geometries of whole dentitions. Emphasis was placed on understanding the conflicts between whole dentition geometry as expressed in z-spacing and the morphogenetic basis of tooth formation postulated by Osborn, and later Westergaard and Ferguson.

The model (Fig. 5) is based on five parameters related to the environment and process of tooth formation:

1. The model is initially based on a single founder tooth germ established on the free edge of the dental lamina. This germ is not given a specific position in the jaw and hence may well end up being a member of any of the final tooth families in the jaw. It was referred to by Osborn (1971) and Westergaard (1980; see also Westergaard and Ferguson 1986, 1987) as the dental determinant; Reif (1984) termed this first germ the initiator tooth.
2. Surrounding the germ is an inhibition zone of varying size and shape. This inhibition zone is the same postulated by Osborn (1971 and elsewhere) and supported by Westergaard and Ferguson (1986, 1987) in their embryological studies of tooth replacement in *Lacerta vivipara* and *Alligator mississippiensis*, respectively.

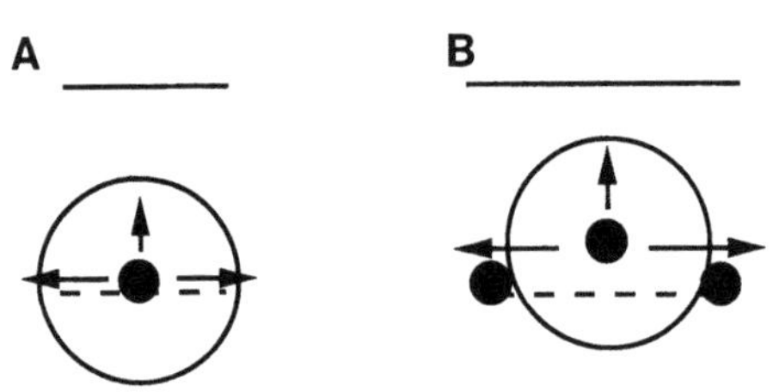

Fig. 5. A model for tooth replacement based on tooth germ, inhibition zone, and dental lamina interactions. In **A** and **B**, the tooth germ (*solid circle*) is formed along the free edge of the dental lamina (*dashed line*). Around the germ is an inhibition zone (*large open circle*). The dental lamina grows longitudinally (*horizontal arrows*), while the germ and inhibition zone migrate occlusally (*vertical arrow*). The *solid line* above the tooth germ complex is the occlusal surface of the jaw. The figure on the *left* represents an early stage in tooth development, while the *right* figure is a more mature stage in which the initial tooth has migrated sufficiently occlusally and the dental lamina has migrated sufficiently longitudinally for subsequent germs to have formed beyond the periphery of the first inhibition zone

The zone of inhibition can arise along two different courses. It may be produced by some sort of regulatory chemical substance that diffuses through extracellular matrix (chemical prepatterning models; Meinhardt 1982; Meinhardt and Klingler 1987). Or it may result from the local recruitment of cells around the initiator tooth (mechanochemical models; Oster et al. 1983, 1985). In either case, the inhibition zone is modeled as either a circle or an ellipse in order to reflect the most reasonable gradient geometries or diffusion patterns around the point source (that is, the initial tooth germ). The inhibition zone can vary in size and shape.

3. Once formed, the tooth germ and inhibition zone migrate occlusally from the free edge of the dental lamina to its ultimate functional position (occlusal migrational rates and distances).
4. At the same time, the dental lamina grows in both a mesial and distal direction at either equivalent or varying rates (relative dental lamina growth).
5. Successive tooth germs form along the free edge of the dental lamina that extends beyond the confines of the inhibition zone of prior tooth germs. This germ formation results from either the occlusal migration of inhibition fields or longitudinal growth of the dental lamina. A further implication of this inhibition zone-dental lamina interaction is that all portions of the free edge of the dental lamina are equally competent for the formation of tooth germs; there is no a priori subdivision into tooth families or other dental units.

The interaction of these parameters is reducible to a series of algebraic expressions that constitute the basis for a computer model of tooth replacement. By varying such parameters as size and shape of the inhibition zone, relative growth of the dental lamina, and relative migrational distances of the tooth germ, this model is capable of synthesizing the "universe" of tooth replacement patterns within whole dentitions. In another form, the model is capable of analyzing existing patterns of tooth replacement, treating them as preset configurations to which the parameters must conform. That is, this approach is directed at discovering the inhibition zone parameters that it would take to control the patterns of existing dentitions.

It is probably prudent at this point to note that this modeling consists of assertions about the kinds of morphogenetic "rules" governing the organization of whole dentitions. Whether these "rules" turn out to be true or false, at the very least they constitute predictions about the existence and form of inhibition zones and how these zones might interact with the dental lamina (see discussion below). As predictions about the real world of tooth replacement made on the basis of biologically sensible models, these exercises can be seen as comprising the general workings of model-based morphogenesis and functional biology (Weishampel in press).

4 Application

The first analytical modeling involves a z-spacing of 2.0, the commonest geometry found among extant and fossil lower vertebrates. This 2.0 z-spacing is shown in Fig. 6. As can be seen, the inhibition zone necessary to produce such a denti-

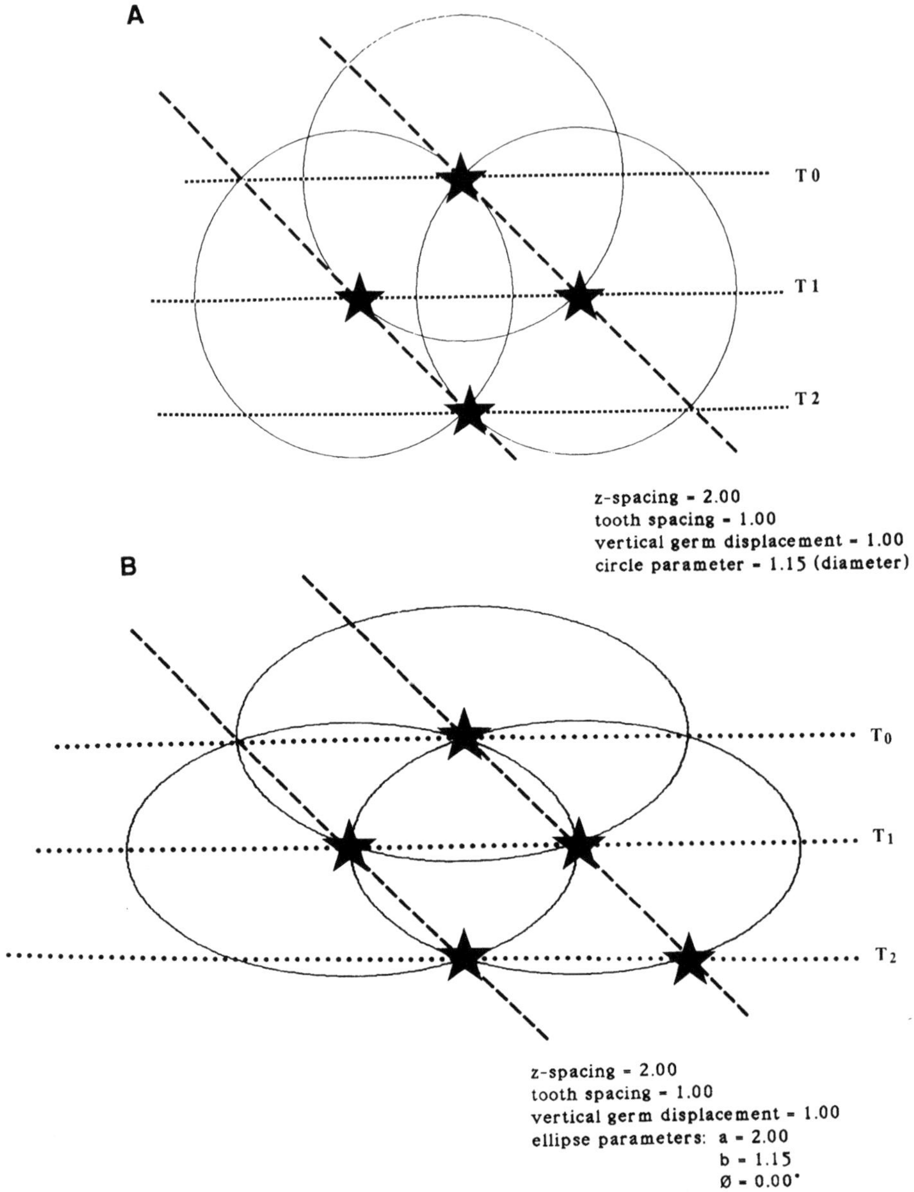

Fig. 6. Model application of 2.0 z-spacing. **A** Circular inhibition zone. **B** Elliptical inhibition zone. In this and successive figures, T_0, T_1, T_2, and T_3 refer to successive positions of the free edge of the dental lamina relative to the migration to new positions of previous tooth germs

Fig. 7. Examples of inhibition zone geometry based on dentitions with equal distances between tooth families (= 1.0 unit) and decreasing migrational distances. **A** Migrational (vertical germ displacement) distance = 0.67, z-spacing = 3.0. **B** Migrational (vertical germ displacement) distance = 0.50, z-spacing = 4.0. **C** Migrational (vertical germ displacement) distance = 0.20, z-spacing = 5.0

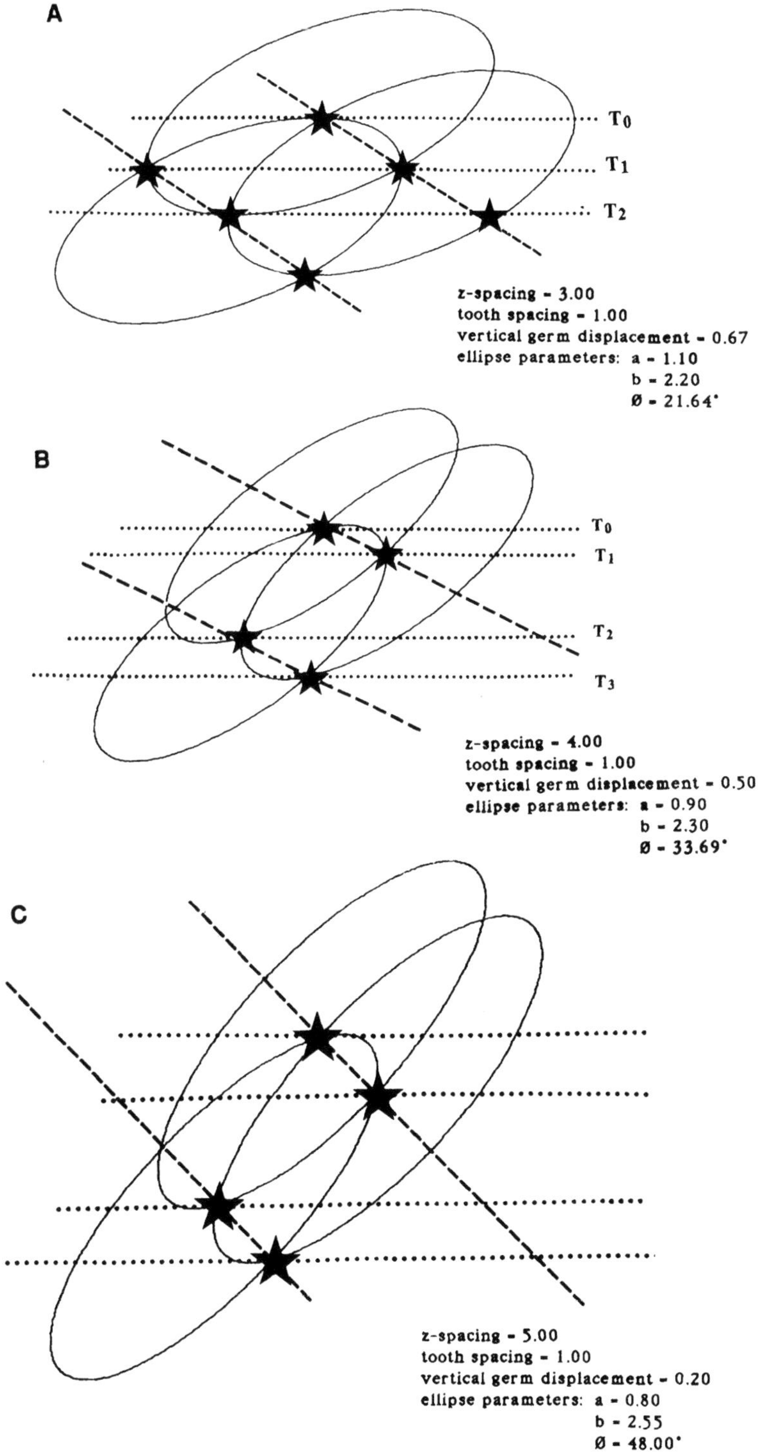
A
T0
T1
T2
z-spacing = 3.00
tooth spacing = 1.00
vertical germ displacement = 0.67
ellipse parameters: a = 1.10
b = 2.20
Ø = 21.64°
B
T0
T1
T2
T3
z-spacing = 4.00
tooth spacing = 1.00
vertical germ displacement = 0.50
ellipse parameters: a = 0.90
b = 2.30
Ø = 33.69°
C
z-spacing = 5.00
tooth spacing = 1.00
vertical germ displacement = 0.20
ellipse parameters: a = 0.80
b = 2.55
Ø = 48.00°

tional geometry is either circular or elliptical. If an ellipse, the axes must be symmetrical with respect to those of the dental lamina (i.e., major or minor axis parallel to the long axis of the dental lamina). Regardless of the size of the inhibition zone, so long as it is a circle or a symmetrically oriented ellipse, tooth replacement waves will always be alternating, hence a z-spacing of 2.0.

For z-spacings not equal to 2.0, two other analyses were performed. In the first analysis, z-spacings of 3.0, 4.0, and 5.0 were evaluated for inhibition zone configuration. Tooth family spacing was maintained at 1.0 units, but migrational distances were decreased from 0.67 to 0.40 with increasing z-spacing (Fig. 7). The inhibition zones necessary to account for these changes always involve an asymmetrically positioned ellipse. This ellipse is progressively steepened (from approximately 22 to 42°) and moderately deformed (i.e., lengthening of the major axis by approximately 10% and narrowing of the minor axis by 20%) from one z-spacing to the next.

Likewise, the inhibition zone in the last set of analyses is an asymmetrical ellipse. Here, both tooth family spacing and migrational distances are held constant at 1.0 units apiece (Fig. 8). The resulting ellipses are inclined distally by 45°. The minor axes of these ellipses are always equal (in this example, 2.8 units). In contrast, however, the major axes are lengthened with increasing z-spacing. Hence, each ellipse becomes more drawn out with higher z-values (a 10% increase in major axis length with increasing z-spacing).

5 Discussion

From the foregoing, it appears that inhibition zones can easily account for alternating waves of tooth replacement (i.e., z-spacing of 2.0) among lower vertebrates. However, in Osborn's formulation, there are problems in dealing with z-spacings other than 2.0. By allowing some flexibility in the parameters of inhibition zones, all of the z-spacings seen in real dentitions can be simulated.

From the point of view of tooth formation, the commonness of 2.0 z-spacings among lower vertebrates may well be due to the efficacy of symmetrically-oriented inhibition zones around founder and subsequent teeth within the jaws. If true, then slight deviations around 2.0 often seen in many real dentitions are likely due to slight asymmetries in elliptical inhibition zones, either in terms of mesodistal size and/or shape differences or a slight angulation in the long axis of the inhibition zone with respect to the dental lamina (Fig. 9). However, in order to account for widely varying z-spacings, it is necessary to invoke distinctly inclined elliptical inhibition zones. For instance, the two examples of 3.0 z-spacings presented here have elliptical inhibition zones, one at a relatively low angle and having a narrow width, or the other at a 45° angle and extended length (cf. Figs. 7a and 8a). Other possibilities for inhibition zone parameters yielding a z-space of 3.0 surely exist

Fig. 8. Examples of inhibition zone geometry based on dentitions with equivalent tooth family and migrational distances (= 1.0 unit). **A** Z-spacing = 3.0. **B** Z-spacing = 4.0. **C** Z-spacing = 5.0

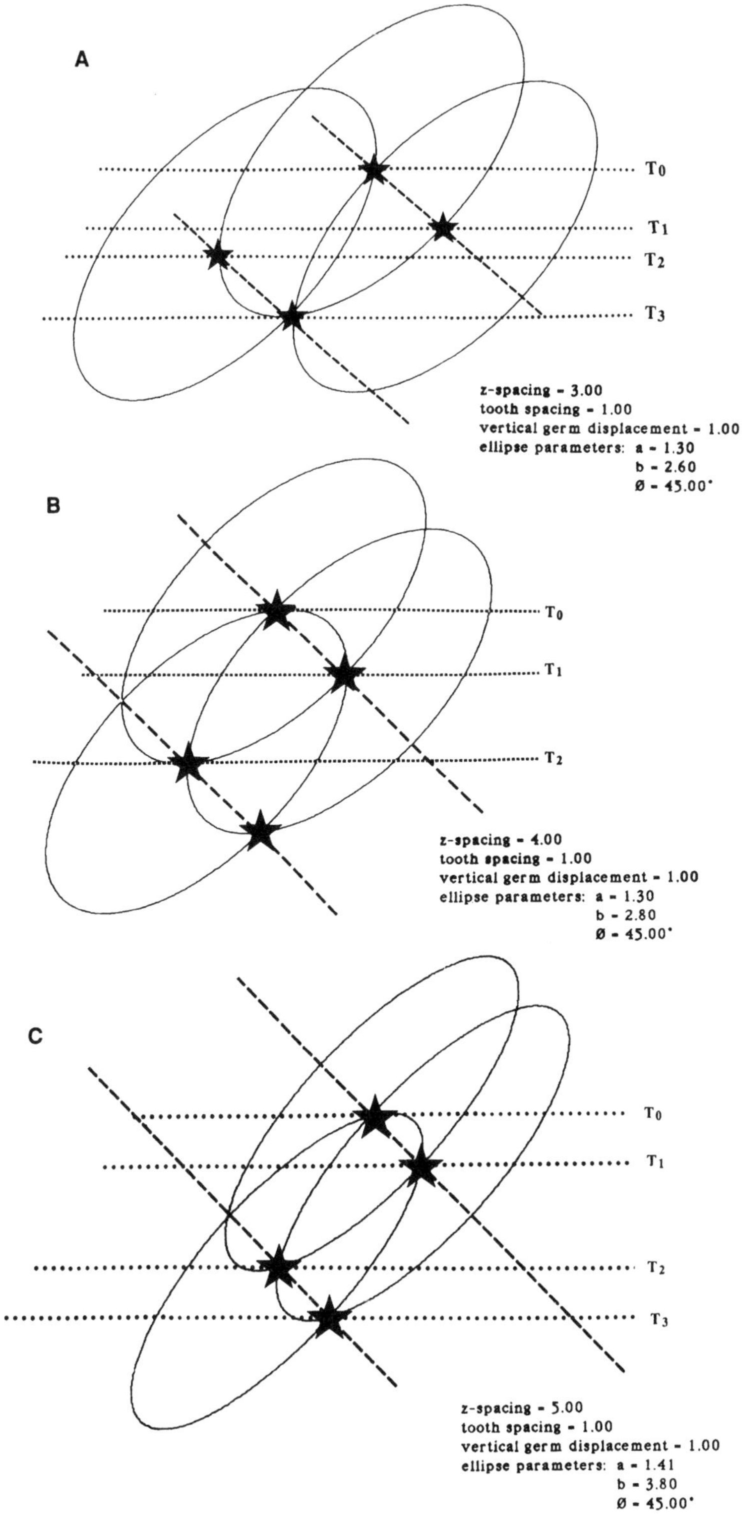
A
T0
T1
T2
T3
z-spacing - 3.00
tooth spacing - 1.00
vertical germ displacement - 1.00
ellipse parameters: a - 1.30
b - 2.60
Ø - 45.00°
B
T0
T1
T2
z-spacing - 4.00
tooth spacing - 1.00
vertical germ displacement - 1.00
ellipse parameters: a - 1.30
b - 2.80
Ø - 45.00°
C
T0
T1
T2
T3
z-spacing - 5.00
tooth spacing - 1.00
vertical germ displacement - 1.00
ellipse parameters: a - 1.41
b - 3.80
Ø - 45.00°

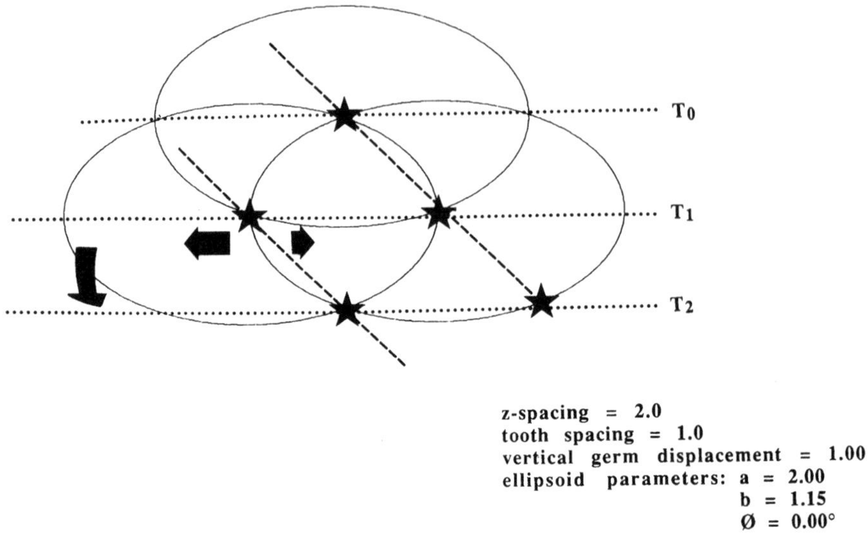

Fig. 9. Small variations in an elliptical inhibition zone that may alter a 2.0 z-spacing. In this example, the long axis of the ellipse may be slightly rotated (*curved arrow to the left*) or asymmetrically shortened or elongated (*horizontal arrows*)

as well, but all of these will be asymmetric with respect to the long axis of the dental lamina, as shown in these two examples.

A significant consequence of the differences among inhibition zone parameters is the relative space available for tooth maturation (Fig. 10). In the simplest case, where the inhibition zone is a circle or a symmetrically oriented ellipse, an increase in inhibition zone size has the effect of pushing successive tooth germs farther from the founder tooth, either in the vertical or horizontal direction. This extra space may well accommodate higher or wider mature teeth. For an asymmetrical, elliptical inhibition zone, both size and orientation of the zone play a part in providing for greater space available to a growing tooth germ. Again as an example, the rather narrow and slightly deformed elliptical inhibition zone used to produce a dentition with a z-spacing of 4.0 yields 2.0 migrational units between successional teeth (Fig. 10a). In contrast, the wide and long elliptical inhibition zone producing a z-spacing of 4.0 has 4.0 migrational units between successional teeth (Fig. 10b). Treating these units as available vertical space to accommodate mature teeth, the second condition provides for considerably higher teeth than the first.

The results of such preliminary modeling can be seen as portraying the sorts of morphogenetic "rules" that govern the organization of whole dentitions. Or they

Fig. 10. A Increasing the vertical and/or horizontal size of a circular inhibition zone. **B** Differences in migrational (vertical germ displacement) distance between 0.50 and 1.00 of asymmetrical, elliptical inhibition zones. Both have the effect of pushing successive tooth germs farther from the founder tooth in vertical or horizontal directions

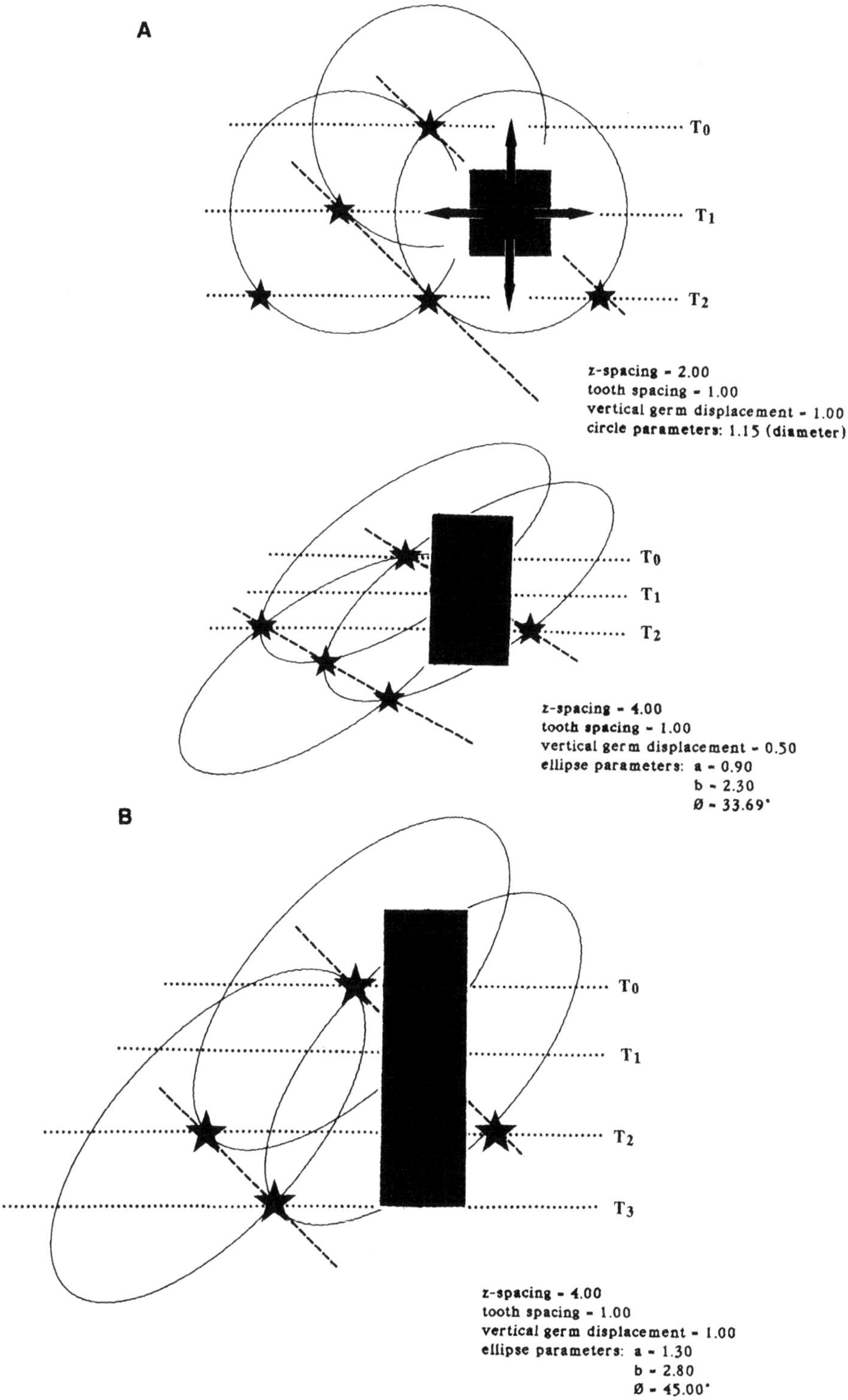
A
T_0
T_1
T_2
z-spacing = 2.00
tooth spacing = 1.00
vertical germ displacement = 1.00
circle parameters: 1.15 (diameter)
T_0
T_1
T_2
z-spacing = 4.00
tooth spacing = 1.00
vertical germ displacement = 0.50
ellipse parameters: a = 0.90
b = 2.30
Ø = 33.69°
B
T_0
T_1
T_2
T_3
z-spacing = 4.00
tooth spacing = 1.00
vertical germ displacement = 1.00
ellipse parameters: a = 1.30
b = 2.80
Ø = 45.00°

may be more profitable viewed as predictions about inhibition zone-dental lamina interactions, whether they exist at all and, if so, what the zone parameters might be. For example, a dentition with a z-spacing of 3.0, and equal tooth family spacing and vertical migrational distance should involve an elliptical inhibition zone inclined distally by 45°. Hence, this information amounts to specific predictions about the local environment surrounding the initial tooth germ, in particular about the presence and form of either cell recruitment or chemical patterning. Furthermore, it makes predictions about the ways in which successive tooth germs interact with migrating germs along the dental lamina. All may be used as statements about the geometric relationships of germs and lamina that can be used to experimentally test the model. Such tests could include the identification of inhibition factors (chemical regulation, local recruitment and depletion) and its manipulation to yield different inhibition (and hence dental) geometries. Other tests might fall under the realm of natural experiments, such as teratological modifications of dental replacement pattern and its relation to tooth formation processes. Should it pass these and/or other tests, the model implicates the control of global organization by local events; that is, tooth-germ-based regulation of the organization of whole dentitions. As such, the modeling of inhibition fields combined with the empirics of Zahnreihen will have been useful in outlining some of the morphogenetic "rules" in the self-organization of vertebrate dentitions.

Acknowledgments. This work was begun while I was a visiting professor in the Institut für Geologie und Paläontologie of the Universität Tübingen, Federal Republic of Germany, as part of Sonderforschungsbereich 230 (Natürliche Konstruktion – Leichtbau in Architektur und Natur). I thank W.-E. Reif for his generous support and direction, L. M. Witmer and A. Seilacher for their comments on the manuscript, and N. Schmidt-Kittler and K. Vogel for providing me with the opportunity to participate in the Constructional Morphology and Biomechanics Symposium in Bad Homburg.

References

Bayer U (1978) Morphogenetic programs, instabilities, and evolution – a theoretical study. N Jahrb Geol Paläontol 156:226–261

Berkovitz BKB, Shellis RP (1978) A longitudinal study of tooth succession in piranhas (Pisces: Characidae), with an analysis of the tooth replacement cycle. J Zool 184:545–561

Bolt JR, DeMar R (1975) An explanatory model of the evolution of multiple rows of teeth in *Captorhinus aguti.* J Paleontol 49:814–832

Bolt JR, DeMar R (1983) Simultaneous tooth replacement in *Euryodus* and *Cardiocephalus* (Amphibia: Microsauria). J Paleontol 57:911–923

Bolt JR, DeMar R (1986) Computer simulation of tooth replacement with growth in lower tetrapods. J Vert Paleontol 6:233–250

Cheverud JM, Dow MM, Leutenegger W (1985) The quantitative assessment of phylogenetic constraints in comparative analyses: sexual dimorphism in body weight among primates. Evolution 39:1335–1351

Cracraft J (1981) The use of functional and adaptive criteria in phylogenetic systematics. Am Zool 21:21–36

DeMar R (1972) Evolutionary implications of Zahnreihen. Evolution 26:435–450

DeMar R (1973) The functional implication of the geometrical organization of dentitions. J Paleontol 47:452–461

DeMar R, Bolt JR (1981) Dentitional organization and function in a Triassic reptile. J Paleontol 55:967–984

Dhouailly D (1984) Specification of feather and scale patterns. In: Malacinski GM, Bryant SV (eds) Pattern formation. Macmillan, New York, pp 581–601
Edmund AG (1960) Tooth replacement phenomena in the lower vertebrates. Contrib Life Sci R Ontario Mus 56:1–90
Edmund AG (1969) Dentition. In: Gans C, Bellairs Ad'A, Parsons TS (eds) Biology of the Reptilia, morphology A. Academic Press, New York London, pp 117–200
Gow CE (1975) A new heterodontosaurid from the Redbeds of South Africa showing clear evidence of tooth replacement. Zool J Linnean Soc 57:335–339
Hopson JA (1971) Postcanine replacement in the gomphodont cynodont *Diademodon*. In: Kermack DM, Kermack KA (eds) Early mammals. Academic Press, New York London, pp 1–21
Hopson JA (1980) Tooth function and replacement in early Mesozoic ornithishian dinosaurs: implications for aestivation. Lethaia 13:93–105
Mayr E (1983) How to carry out the adaptationist program. Am Nat 121:324–334
McGhee GR Jr (1980) Shell form in the biconvex articulate Brachiopoda: a geometric analysis. Paleobiology 6:57–76
Meinhardt H (1982) Models of biological pattern formation. Academic Press, New York London
Meinhardt H, Klingler H (1787) A model for the formation of patterns on the shells of molluscs. J Theor Biol 126:63–90
Niklas KJ (1978) Branching patterns and mechanical design in Palaeozoic plants: a theoretical assessment. Ann Bot (London) 42:33–39
Niklas KJ (1986) Computer-simulated plant evolution. Sci Am 254:78–86
Okamoto T (1988a) Analysis of heteromorph ammonoids by differential geometry. Palaeontology 31:35–52
Okamoto T (1988b) Developmental regulation and morphological saltation in the heteromorph ammonite *Nipponites*. Paleobiology 14:272–286
Okamoto T (1988c) Changes in life orientation during the ontogeny of some heteromorph ammonoids. Palaeontology 31:281–294
Osborn JW (1970) New approach to Zahnreihen. Nature (London) 225:343–346
Osborn JW (1971) The ontogeny of tooth succession in *Lacerta vivipara* Jacquin (1987). Proc R Soc London Ser B 179:261–289
Osborn JW (1973) The evolution of dentitions. Am Sci 61:548–559
Osborn JW (1974) On the control of tooth replacement in reptiles and its relationship to growth. J Theor Biol 46:509–527
Osborn JW (1977) The interpretation of patterns in dentitions. Biol J Linnean Soc 9:217–229
Osborn JW (1978) Morphogenetic gradients: fields versus clones. In: Butler PM, Joysey KA (eds) Development, function and evolution of teeth. Academic Press, New York London, pp 171–201
Osborn JW (1984) From reptile to mammal: evolutionary considerations of the dentition with emphasis on tooth attachment. Symp Zool Soc London 52:549–574
Oster G, Murray J, Harris A (1983) Mechanical aspects of mesenchymal morphogenesis. J Embryol Exp Morphol 78:83–125
Oster G, Murray J, Maini P (1985) A model for chondrogenic condensations in the developing limb: the role of extracellular matrix and cell tractions. J Embryol Exp Morphol 89:93–112
Papentin F, Röder H (1975) Feeding patterns: the evolution of a problem and a problem of evolution. N Jahrb Geol Paläontol Mh 1975:191–194
Raup DM (1966) Geometric analysis of shell coiling: general problems. J Paleontol 40:1178–1190
Raup DM (1967) Geometric analysis of shell coiling: coiling in ammonoids. J Paleontol 41:43–65
Raup DM (1968) Theoretical morphology of echinoid growth. J Paleontol Mem 2:50–63
Raup DM, Michelson A (1965) Theoretical morphology of the coiled shell. Science 187:1294–1295
Raup DM, Seilacher A (1969) Fossil foraging behavior: computer simulation. Science 256:994–995
Reif W-E (1975) Lenkende und limitierende Faktoren in der Evolution. Acta Biotheor 24:136–162
Reif W-E (1980) A model of morphogenetic processes in the dermal skeleton of elasmobranchs. N Jahrb Geol Paläontol Abh 159:339–359
Reif W-E (1982) Functional morphology on the procrustean bed of the neutralism-selectionism debate. N Jahrb Geol Paläontol Abh 164:46–59
Reif W-E (1984) Pattern regulation in shark dentitions. In: Malacinski GM, Bryant SV (eds) Pattern formation. Macmillan, New York, pp 603–621

Reif W-E, Thomas RDK, Fischer MS (1985) Constructional morphology: the analysis of constraints in evolution. Acta Biotheor 34:233–248

Reif W-E, Weishampel DB (1989) Theoretische Morphologie: Wachstum *und* Funktion? In: Int Symp SFB 230, Natürliche Konstruktionen – Leichtbau in Architektur und Natur, Stuttgart, Teil 1, pp 253–262

Reif W-E, Weishampel DB (in press) Theoretical morphology. An annotated bibliography since 1960. Courier Forschungsinst Senckenberg, Frankfurt

Ricqlès A de, Bolt JR (1983) Jaw growth and tooth replacement in *Captorhinus aguti* (Reptilia: Captorhinomorpha): a morphological and histological analysis. J Vert Paleontol 3:7–24

Rudwick MJS (1964) The inference of function from structure in fossils. Br J Philos Sci 15:27–40

Seilacher A (1970) Arbeitskonzept zur Konstruktionsmorphologie. Lethaia 3:393–395

Thomas RDK (1979) Constructional morphology. In: Fairbridge RW, Jablonski D (eds) The encyclopedia of paleontology. Dowden, Hutchinson & Ross, Stroudsburg, Penn, pp 482–487

Thomas RDK, Reif W-E (1983) Opportunity and constraint in the evolution of organic form. Geol Soc Am Abstr Program 15:705

Vincent JFV (1982) Structural biomaterials. John Wiley, New York, 206 pp

Waddington CH (1968) The basic ideas of biology. In: Waddington CH (ed) Towards a theoretical biology. Univ Press, Edinburgh, pp 1–41

Wainwright SA, Biggs WD, Currie JD, Gosline JM (1976) Mechanical design in organisms. John Wiley & Sons, New York, 423 pp

Weishampel DB (in press) Modeling approaches to analysis of skull form and function. In: Hanken J, Hall BK (eds) The vertebrate skull. Univ Press, Chicago

Westergaard B (1980) Evolution of the mammalian dentition. Mem Soc Geol Fr 139:191–200

Westergaard B, Ferguson MWJ (1986) Development of the dentition in *Alligator mississippiensis*. Early embryonic development in the lower jaw. J Zool A 210:575–597

Westergaard B, Ferguson MWJ (1987) Development of the dentition in *Alligator mississippiensis*. Late development in the lower jaws of embryos, hatchlings and young juveniles. J Zool A 212:191–222

Wolpert L, Stein WD (1984) Positional information and pattern formation. In: Malacinski GM, Bryant SV (eds) Pattern formation. Macmillan, New York, pp 3–22

Ziegler AC (1969) A theoretical determination of tooth succession in the therapsid *Diademodon*. J Paleontol 43:771–778

Part IV Bioconstruction: Evolution

Evolution of Biological Constructions: Concessions, Limitations, and Pathways

P. DULLEMEIJER [1]

Summary

This chapter discusses the possible role of constraints in the evolution of biological constructions. Examples on bone structure, hydraulic systems, head structures of fish, and their transformations are utilized to demonstrate the usefulness of the concept constraint and its various modalities in morphological explanation.

The deductive-nomological method is necessary to discover and to define the operation of constraints in development and evolution. The use of the concept constraint in this method as such, as sometimes occurs, is redundant to the already common logical terms.

Yet, in transformational problems in evolution and morphogenesis, the concept constraint is useful in recognizing actual limitations which are introduced in the models as 'mutual' boundary conditions.

1 Introduction

It is theoretically possible to construct a great number of organisms given some fundamental characteristics of matter and the empirical evidence obtained from the study of recent organisms. These theoretical constructions would certainly outnumber the 20 million species supposed to occur on this globe today. This discrepancy between the numbers of theoretically possible and actually existing organisms could mean that many forms have not yet evolved. But it is far more likely that many of the theoretical possibilities are not feasible on account of incompatibilities and constraints. Not every hypothetical form is feasible. This raises the question of what kind of constraints play a part in the evolution of organisms, and which pathways therefore are open and which are closed.

In an interesting article, Alexander (1985) shows, in a simple, clear style, that the fundamental properties of matter and energy constitute *universal constraints* and that organisms can make use of only a limited part of this spectrum of properties of matter. "An ideal animal might have a skeleton made of some unbreakable material which has infinite strength but no mass. Its muscles would be capable of responding instantaneously to stimuli, exerting unlimited force and contracting at infinite speed, and so on", notes Alexander (1985), citing theoretical limits which real organisms cannot achieve. In addition to these universal constraints, there are *general* and *local* constraints. The former term refers to the

[1] Zoölogisch Laboratorium, Kaiserstraat 63, 2300 RA Leiden, The Netherlands

N. Schmidt-Kittler and K. Vogel (Eds):
Constructional Morphology and Evolution

kinds of material and the construction principles used in almost all living animals, such as the DNA structure, the chemical composition of muscle tissue and the nervous tissue. The latter term applies to constraints resulting from the normal interaction of structures, functional components and constructions in individual, particularly during their ontogeny.

That various kinds of factors set limits on organic structures and functions is not in doubt, but it is difficult to define these specific constraints, and attempts to demonstrate their roles have been controversial.

In a review article, Maynard Smith et al. (1985) sum up the basic problems encountered and the methods used to discover and explain constraints, and to establish their effects in the course of evolution.

The impression gained from this article is that:

a) apart from simple and obvious cases, little evidence of the operation of constraints has so far been presented;

b) the methods to demonstrate constraints are scanty;

c) the relation of constraints to natural selection is far from clear. It is questionable that, as these authors suggest, these phenomena are mutually exclusive.

In a recent article on constructional constraint and its ecomorphological implications, Barel et al. (1989) describe some interesting examples of the operation of constraints, and end with a theoretical speculation, which to them "is indicative of our present ignorance when it comes to quantification of constructional constraints on environmental exploitation".

In my opinion, the role of constraints can be established only by comparison of real animal structures with optimal models (Dullemeijer 1974), a method also employed by Alexander (1985). This is the nomological-deductive method of Hempel (1965; see also Hempel and Oppenheim 1953; Rudwick 1964; Dullemeijer 1974; Dullemeijer and Barel 1977; Bock this Vol.). For this reason I shall begin with an example where this method is used in order to obtain a functional explanation of organic structure.

2 Bone

Animals have evolved a wide variety of structural materials that resist forces to protect and frequently to hold the soft tissues in place, and to shape or support them. These materials generally consist of several components, including one for resisting tensile forces, one for resisting compressive forces and a third one which cements the two together. These components are typically arranged and interwoven in a very complex manner. This holds for the cuticle or exoskeleton of insects, the shell structures of molluscs, and various other invertebrate skeletons, as well as for the various types of bone in vertebrates. These structures vary considerably in the proportions of the basic components.

For instance, to obtain the maximum resistance to compression, a bony element should consist exclusively of closely packed crystals with a very high Young's modulus, somewhere on the order of 100 GPa. But even so a compromise must be struck between the mass to be carried along and resistance to compression (Alexander 1985).

The disadvantage of such a construction lies in the low resistance of these materials against tensile forces. Long polymers of protein or carbohydrate are much better suited for this latter purpose. Protein polymers probably constitute the best material available in nature although, as Alexander (1985) remarks: "It is rather surprising that collagen is so widely used as a structural material although its strength is only about 0.9 GPa". From the viewpoint of an engineer, this is indeed surprising and the more so as in plant materials we find much stronger fibers, which in wood may be 100 times stronger. Collagen has to be built by the organism. From the very beginning, life has relied almost completely on proteins for stability and processes; they have to be assembled from a great variety of amino acids. This *historical constraint* has canalized the possibilities of skeleton formation in animals. It is very likely that collagen, of all the proteins that could be made, had the best potential, as it has suitable mechanical properties, combined with relatively easy assembly from amino acids and thus also an easy turn-over, breaking down into subunits used in growth and differentiation.

It has a strong capacity for energy storage and, last but not least, it has a maximum damping capacity for vibrations exactly in the range of frequencies of nerve signals which stimulate muscle contraction (Ramaekers 1978). If we had to construct a material from amino acids that could meet all these functional demands, we would very probably have come up with collagen. Collagen forms a very good mesh for the deposition of apatite crystals. Of course, this could not have been a factor in the origin of collagen, but it has made possible the development of bony tissues in vertebrates. This key innovation (Liem 1973) was followed by an enormous radiation of structural types of bone.

In the course of evolution another property of collagen has certainly contributed to its wide use. The polymer forms in a specific clockwise helical structure, with subunits that are themselves helices. This considerably strengthens the fibers and, if they are woven like cloth, it is possible to obtain a structure that resists stress in various directions, making it suitable for movement and changing directions of support. In bone we have a special case of reinforcement. Being packed between crystals, the collagen fibers resist compression in a way they cannot in a free position. Because of its twisted string connections to the crystals it cannot be compressed, and so contributes to compressive strength. Also, all fibers perpendicular to the compressive stress are under tensile strain and so help indirectly to withstand the compression.

Let us now turn to bone shape. Bone's most important function in support or protection of vital soft tissues and organs is to resist tensile, compressive, bending and torsional forces. The most influential and dangerous of these are the bending and torsional forces. To resist compression or tension, the cross-sectional area of material perpendicular to the force direction must be proportional to the magnitude of the force. In the case of bending or torsion, material must be appropriately distributed in the plane of the force and the area of cross-section must be proportional to the third power of the force. If a force is applied perpendicular to the longitudinal axis of a beam, bending occurs with compression on one side and tension on the other. In the center of the beam there is a neutral axis that curves without strain. Compression and tension are maximal at the two surfaces, their magnitudes proportional to the bending moment.

To maximize resistance to bending stress there are two alternatives. The material can be entirely at the surface or it can be distributed in proportion to the length of the lever arm, with hardly any material in the center and increasingly more nearer to the surface.

In bone, the properties of the material itself can be adjusted, with more crystallization increasing the E-modulus on the side of the compression, and more fibers, increasing the tensile modulus on the other side (Dullemeijer and Fruitema 1982).

In man-made constructions, engineers distribute the material almost exclusively at the two surfaces, with a thin interconnecting bar between them. This bar serves only to keep the two surfaces apart and connected (T-bar or I-bar). Frequently, a tube is used in which the connecting bar is a part of the tube itself. This tubular construction has the advantage that forces from various directions can be resisted. Which of the alternative solutions is adopted depends, in engineering science, on the expected variability in direction and magnitude of the bending forces, the manufacturing and material composition of the bar, the amount of material needed or available for its production and the handling and fitting of the bar into the construction.

The same factors are involved in animal constructions. Many vertebrates have more or less tubular bones, but there are so many exceptions that one is led to ask which mechanical principles are of critical importance here.

Let us first see how much the various constructions cost the animal in terms of amount of material and energy.

The common assumption that the cylindrical or tubular structure is cheapest needs to be put in perspective. Whether or not this statement is correct depends on the force pattern. If the bending force remains in one plane or varies only a little in direction, elliptical, narrow rectangular, T-bar or I-bar designs require less material and so are cheaper than a cylinder. Among cylinders, however, those with the greatest radius require the least material for any given magnitude of resistance, as the further the surface lies from the center, the less material is needed. The same holds for rectangular or I-bar constructions, especially if the I-bar has spherical or cylindrical thickenings at each end, in which case it becomes much cheaper. When we also recognize that the organism can influence the modulus of the material, we conclude that there must almost invariably be several good solutions to the same problem (Fig. 1).

Thus bones can be constructed in various ways to accommodate the demands of varying mechanical requirements with different maximal solution.

Yet there seem to be many cases where the solution is not ideal in terms of the amount of material or energy used. For this, the following reasons can be given:

1. there are other mechanical, local and spatial constraints;
2. there are other, nonmechanical functions in which the structure is employed;
3. there may be a relatively high safety factor (Alexander 1981);
4. there may be ontogenetic and (or) fabricational constraints.

Consequently, mechanical factors, although they must be fully satisfied, do not alone provide a sufficient explanation of skeletal form. The four reasons are related.

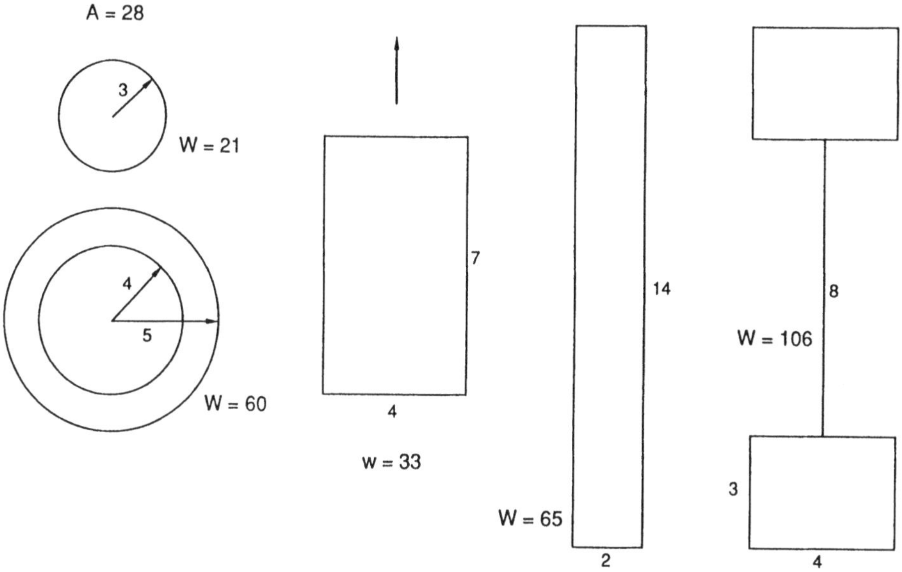

Fig. 1. Bars of different designs with equal cross-sectional areas ($A = 28$). These bars have different moments of resistance (W) against bending strain. The example shows that mechanical factors are necessary but not sufficient conditions to explain bone shape. There are "alternative solutions" available to the organism for the same mechanical demand

It is often said that *mechanical constraints* on skeleton design limit the maximum size of land vertebrates. The argument generally given is that the weight of an organism increases at the third power of its linear increase in size. This requires a relatively greater increase in bone diameter because resistance to compression depends on cross-sectional area, which increases only at the second power of the linear dimension. Bending stresses, as noted above, are much more demanding. The question has been raised several times whether in certain groups, such as elephants among the mammals and dinosaurs among the reptiles, the ultimate size has been reached. Alexander (1982) points to the usual quantitative relationship between the increase of weight to the third power, and the increase of muscle force only to the second power, of length. As muscle weight itself adds to body weight, this would imply that there comes a point when the total weight becomes too large to be supported. Many authors have pointed to the relatively slow locomotion of most large and heavy animals as an indication of a relative decrease in the ability to withstand the stresses generated in fast running. According to Alexander (1985), this is not so in the case of dinosaurs and elephants. Thus, it seems that the maximal size permitted by structural factors had not yet been reached.

Bones cannot be made excessively thin, with the mass too far from the center. Long bones must have sufficient bone cross-section to resist compression and tension, and very thin long bones are apt to buckle under very small compressional loads. There are a few other constraints. The advantage of stronger support is offset by the disadvantage of more energy consumption required to move heavy bones and by the need for stronger muscles to move them. These are functional constraints set by scaling laws and the interaction of several different mechanical factors.

Local and spatial constraints are well known. At any place where a soft tissue organ has to be protected the organization of the bone has to be adapted to fit around the structure and shape of the organ to be protected.

For instance, the brain and the eyes are supported by the base of the cranium, from which winglike, concave structures extend upward to contain and protect these organs. The amount of bone necessary would be far less than it is in mammals if the bone did not have to surround these organs, as can be seen in fishes, where the brain is small and narrow and the eye often protrudes from the skull. In fishes with reduced eyes, the bars which support the ethmoidal area and define the streamlined shape of the head run straight, take less space, and use less bony material.

Another good example is provided by the structure of the maxilla in viperid snakes (Fig. 2). The amount of bone needed to support the poison fang is relatively less in the common vipers than in the pitvipers. The pit requires an extra or stronger strut to provide mechanical support for the fang.

Muscle attachments, especially to long bones, sometimes require large areas and extensions of the bone which exist solely to provide adequate attachment surface area and mechanical advantage for muscles and tendons (for review see Van der Klaauw 1963). In shape and in the resistance moment of their cross-sections, these bones are overdesigned with respect to the load demand. It is clear, however, that the areas with extra bone cannot be considered as safety factors, because they generally do not occur over the entire length of the bone. These examples can be conceived as situations in which one functional requirement sets a boundary condition for the solution of other mechanical problems.

If high safety factors do occur, then this would imply that the structure is not constrained as to the amount of bone, but that on the contrary it is left free to add more bony material.

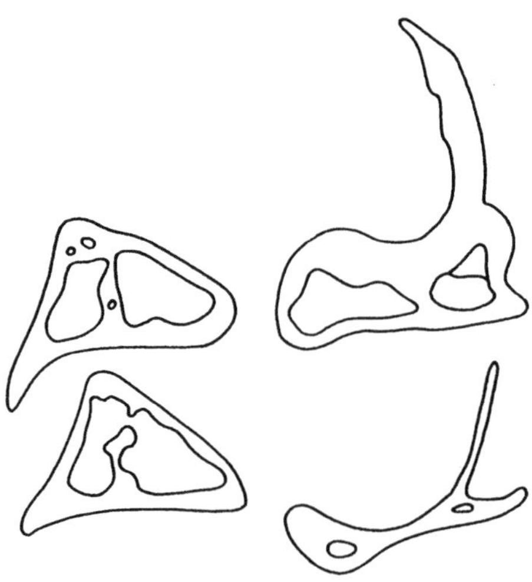

Fig. 2. Cross-sections of the maxillae of a common viper (*right*) and a pit viper (*left*) at the same relative level demonstrate clearly the effect of the position of a sense organ (the pit) on the solution of a mechanical problem (support for the poison fang) (After Dullemeijer 1959)

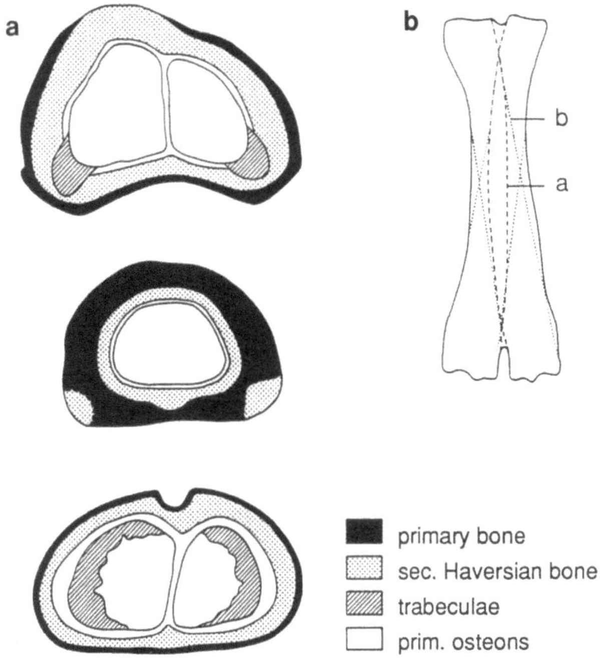

Fig. 3. a Three cross-sections through a cow metacarpal showing the difference in proportion and topography of bone types, each having different moduli of elasticity. **b** Two alternative integrations of the metacarpal bone: *a* when it is assumed that the central part is mainly affected by lateral bending; *b* when the metacarpal is affected by a differently directed compression (**b** after Dullemeijer 1989)

Determining safety factors is not easy. One must be sure to have included in the analysis all the functions of the bony elements over the entire bone. The bone density and local moduli must be known. The diameter of a bone may be large but it may consist of relatively weak bony tissue that for other functional reasons is spread over a broader and longer area. This we find in some long bones of recent and fossil cows (Fig. 3). Bone of various histological types, which clearly have different moduli, extends far along the bone. These differences are probably due to specific articular arrangements and to the fusion of some tarsals. Joints like these involve adjacent surfaces which roll over one another, reducing friction between them.

As during walking the relative positions of the bones change considerably, the working lines of the forces change too. To resist these stresses, struts exist within the long bone, related to these various directions of force. These struts generally fuse in the middle of the bone and diverge at either end. So, there may be three or four times more bone than is necessary to take up the strain developed in one position or in one phase of movement.

These examples suggest a more or less optimal adaptation of the structure and shape of the long bones to a combination of demands. With respect to any particular set of mechanical demands there are boundary conditions or constraints which make the simplest and otherwise most economic mechanical solution unfeasible. When we consider the organism as a whole, however, a viable compromise among conflicting demands is recognized.

Bony elements are fairly subordinate in shape and precise position relative to other structures such as muscles, sense organs, and brainparts. Their plasticity allows them to respond to varying demands while retaining their essential mechanical role. This may be accomplished in different ways. In response to increasing demands, bones react first by changing their molecular composition, then by rearrangement of their components, followed by a reorientation of the major bony bars and trabeculae, and finally they may change altogether the shape, position, and dimension (Dullemeijer and Fruitema 1982).

Some authors (Gutmann and Bonik 1981) claim that the shapes of tubular bones can be explained by their positional situation and fabrication during ontogeny. It is argued that bony material can only be deposited at the surfaces of hydraulic systems, at the cellular or higher level of organization. Indeed, bone is often deposited around spaces (not necessarily hydraulic) and it is translated with these spaces when they change in size and form (capsular matrix, Moss and Salentijn 1969). Yet bones do not always follow this pattern, and many other factors can be shown to influence bone position, shape, and structure (see Van der Klaauw 1963 for a review and, for the periosteal matrix concept, Moss 1968). In view of the rapid adaptation of bony tissue to changing demands, particularly mechanical ones, it is very unlikely that these fabricational factors constrain the functional shape to any substantial extent.

3 Methodology I

Let us now use this example to analyze the deductive procedure and discover whether or not the concept of "constraint" is necessary or useful.

In the preceding discussion, I followed the procedure of a designer, establishing the theoretical structure necessary for the performance of an activity. That structure (the bony element) had to be coordinated with the construction of a whole organism to perform a specific activity (supporting soft parts in a specific position). Consequently, a number of initial and boundary conditions were introduced and a structure was selected from those that could be constructed to serve the purposes of the organism. The initial conditions limit the number of possible design solutions and could therefore be called constraints, but the concept of initial condition is preferable because it makes it clear that we must take the historical relationships of the organism into account.

Another set of constraints arises from the kind of material that is available for fabrication of the structure. This constraint may arise as an initial condition, but generally it is considered as a separate factor. Little is gained, however, by assigning it to a category of constraint.

A unique logical position is taken by the formulas. On one hand, they express universal constraints in the sense of Alexander (1985). On the other hand, they constitute the explaining theory. Whereas the former aspect is always taken for granted and the second is expressed more clearly by "explaining theory", there is no obvious advantage associated with calling them constraints. In conclusion, it seems clear that in the procedure of nomological-deductive explanation there is no need for a concept of constraint.

4 Constructions

The situation is different when transformational problems and morphogenetic questions are considered and causal explanations are required.

A nice example of a transformation problem can be derived from the work of Otten (1983, 1985). The bony and muscular construction used in the uptake of food by fishes consists of many bars connected by joints and ligaments, and by muscles (Fig. 4). *Haplochromis elegans* is an insect feeder that swallows

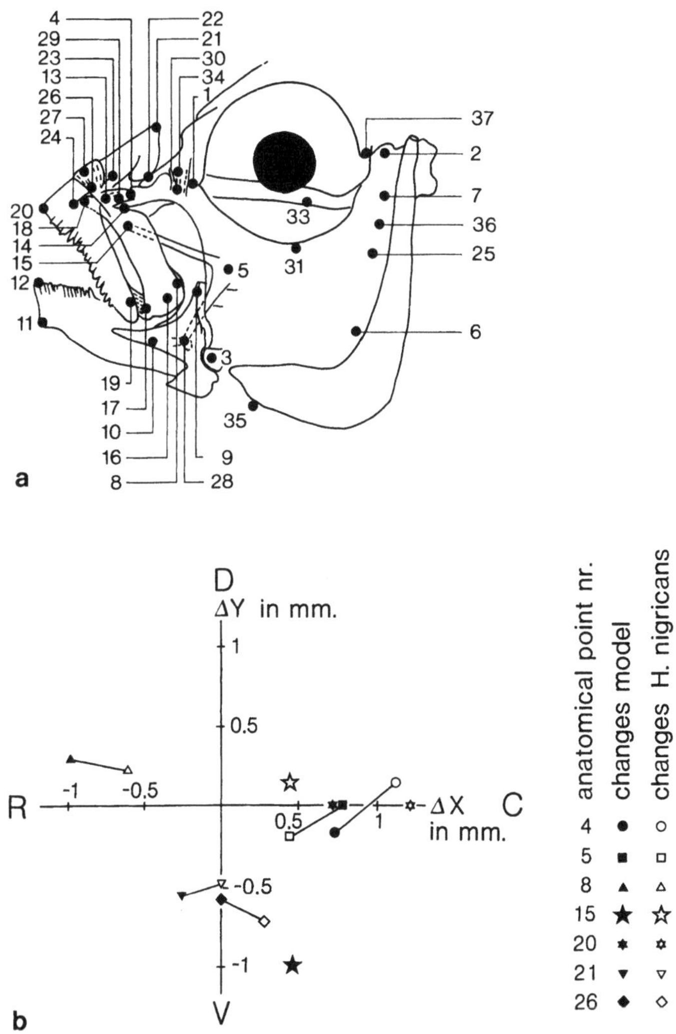

Fig. 4. a Points selected and perturbed describing the jaw mechanism of *Haplochromis*. **b** Displacements of seven of these anatomical points relative to adult *H. elegans* as a result of the iterative perturbation procedure demanding improved biting output (*filled symbols*), as found in *H. nigricans* (*open symbols*). Differences of 0.5 mm or less between the model and the actual specimen are considered to be similar (After Otten 1983)

its prey whole. Thus, it is in principle a suction feeder that can retain its prey in the mouth with small teeth. Otten selected 38 points connected by 22 bars in the system for a computer simulation. He determined that the number of degrees of freedom in the entire system was low, 4–5. The number of degrees of freedom could be reduced to one in various situations by the action of five muscles.

The computer model was now perturbed, in an iterative procedure, to simulate a progressively increasing output of biting force. The changes required in each bar and finally the optimal solution for all the bars together were determined for the stronger biting force. Otten found that of 38 connecting points only seven had to be changed. All the others remained the same.

Comparing an actual fish that bites algae from rocks, *Haplochromis nigricans*, with this model, six of these seven points were found to show similar changes. One point behaved quite differently. Otten suggested that this point could not be changed due to constraints placed on it by surrounding constructions: either it had to move into the opercular area, or the adductor muscle had to invade the orbita, neither of which was feasible.

If this interpretation is correct, it implies that the predicted change at this point in the actual fish is impossible, and consequently that the fish is not as good a biter as it might otherwise have been. Otten did not yet have experimental data on the biting forces in these fishes. It may be that other fishes will be found to fit the predictions of the model more closely.

This example shows a possible need for the concept of constraint, in cases where the predictions of a model and actual structure do not concur and where the deviation can be attributed to an unspecified boundary condition. Once recognized and defined, it can be called a boundary condition.

To determine these mutual limitations, in complex constructions, a necessarily biased selection of parameters must be made to develop a qualitative model. Thus it is the selection of a limited part of the organism or construction that leads to the introduction of the constraint concept.

5 Methodology II

One recognizes many kinds of explanations in biology (Jeuken 1969; Dullemeijer 1972; Bock 1988, 1989) such as historical, functional, causal explanations. As an explanation is always a relationship between a known parameter or feature and a parameter or feature to be explained by a theory (Hempel 1965), the various types of explanation must differ in the theory if the explanation concerns the same features.

For example, if I have explained the shape of a bony element by relating it to its known parameter by means of physical theory, it does not yet imply that the process or the cause is defined. For the latter another physical or chemical theory is required. As said, I have applied the procedure of the designer, i.e., I have given a functional explanation of structures and constructions. In the discussion of the transformation of constructions, we have refrained from introducing morphogenetic and causal explanations.

If we seek such explanations we have to distinguish at least two kinds of causal factors: (1) mutual interactions of the parts involved in the transformations, and (2) factors providing the material, energy, and stimuli for the process. Such factors could be called constraints.

It is usual to call the first set of factors conditions and the second set causes. If the former are limiting conditions, they may be called constraints.

6 General Constraint

Notwithstanding the arguments presented above that the concept of constraint is in general theoretically redundant, it is useful to refer to the limitations that occur in real ontogenetic and evolutionary development as constraints.

An example of the operating of a general constraint is presented in the many publications on the hydraulic system by Gutmann (1972, 1977, 1985 this Vol.). Gutmann starts from the generally accepted concept that a living structure can only be maintained if it is contained in a space, set off from the environment to provide a specific "milieu intérieur" (Cl. Bernard). The simplest form of such a space is a sphere surrounded by a membrane and filled with "fluid". Thus the hydraulic system consists of at least two components, both able to differentiate and to change their mechanical and other properties. The fluid exerts a small pressure on the membrane. The membrane is thus under tension and retains its spherical form. This well-known feature of all cells, particularly the first cleavage cells, is soon disrupted by internal or external forces. Within this protected area, many structures and processes occur, making the internal fluid inhomogeneous. The important question for the present discussion is: can the organism escape the basic hydraulic principle, and if so, how far and at what moment? In other words, to what degree does the hydraulic system constrain structural development and evolution?

Although it is generally understood that the cell, and many of its organelles, are basically hydraulic structures, the materials for these constructions are assembled along membranes which need not be hydraulic. Many structures within the cell are likewise not hydraulic. Thus, many structures arise in hydraulic surroundings without being hydraulic themselves. This is particularly likely when external stimuli or forces act on the membrane, as Bereiter-Hahn (1985 this Vol.) has shown. The membrane reacts, producing local changes and "skeletal" structures, and as a consequence it changes the direction and magnitude of forces inside the cell.

The basic cellular unit can be extended in various ways: (1) by adding similar constructions, by mitosis for instance, as in many protozoans; (2) by producing more cells and forming hydraulic systems at a higher level, as in *Volvox*, or nonhydraulic systems such as linear colonies of protozoans; (3) by reducing the cells to their outer membranes and producing basal membranes and connective tissue layers, from which the larger compartments of hydraulic systems arise.

Skeletal structures are then added to these hydraulic systems, within or surrounding the membranes, as in molluscs, insects, and vertebrates. At this point the system becomes independent of hydraulic forces, as a whole or in part. Production of a large variety of additional structures then becomes feasible (for protozoans see Gutmann and Vogel 1988).

It is clear that the hydraulic system must be a basic condition (= initial condition) ontogenetically and phylogenetically, that has both opened up and restricted the pathways of evolution. But alone it does not account for the enormous variety of animal forms and constructions. Many nonhydraulic structures (proteins, connective tissue components etc.) can be combined in various ways within or around the hydraulic system so that the constraint is released or broken.

Secondly, and even more important, the hydraulic system is what engineers call a multiple statically indeterminate structure, and consequently it can never fully determine new forms. There must be at least one additional factor to explain the direction and complexity of further development. The hydraulic system is one of the many necessary conditions for the shaping and evolution of animal constructions, but – though important – it is not a sufficient condition. All constructions that are not hydraulic have succeeded in breaking the constraint set by hydraulic systems that occur prior to them in development or evolution.

7 Local Constraint

The following example is based on a study by Barel and his co-workers (Barel 1985; Barel et al. 1989). Eyes in fishes differ in size, shape, and position. The size is an important parameter for vision: the bigger the eye, the more sensitive it can be or the better it can discriminate. These faculties depend on the number of retinal receptors, their packing, and their sensitivity to specific spectral lines.

The position of the eye is important in relation to orientation. It differs in fishes which mainly look forward, downward, or upward. In general the shape is spherical, but there are exceptions: there are oval eyes and eyes that are flattened in the lateral-medial direction.

All these features are adaptations to provide the kind of sight needed to find food and to see companions and enemies. In general, relatively small eyes are found in herbivorous fishes, such as algae scrapers, plankton sievers, and molluscs eaters. Zooplanktivorous fishes and small animal eaters have bigger eyes, and very big eyes occur in fish eaters, especially those which feed on fast-swimming prey. Relatively small eyes leave ample room for the bony suspensorium and the jaw adductor muscles. Consequently, the biters among those fishes, such as scrapers and mollusc crushers, which need only small eyes, possess heavy adductors. All these components must fit in a space determined by the outer shape of the head, which is related to the kind of movement of the fish.

In a broad quantitative comparison of the shapes and sizes of these three components among various related species, Barel found that fast-swimming predators have a long, low head, big eyes and small adductors. These are prey suckers, which move with great speed and generate a strong inward sucking force (cf. Osse and Muller 1980). In contrast, the algae scrapers and molluscivorous species, have relatively small eyes which occupy little space in a relatively high and short head, and heavy muscles attached to a large, strong suspensorium. The insectivorous and zooplanktivorous species are intermediate in form. As far as is known, all

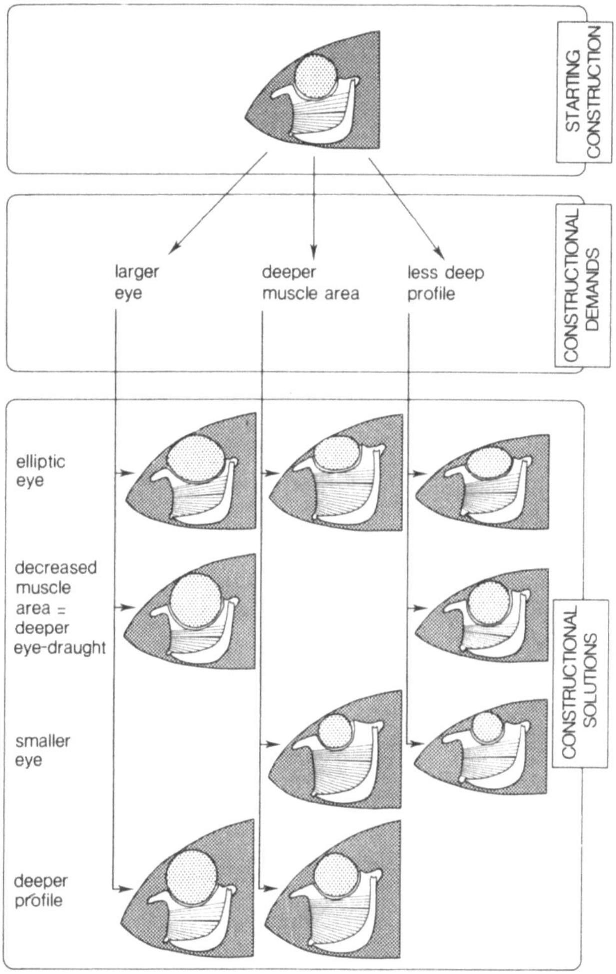

Fig. 5. Schematic representation of geometric transformation in response to three constructional demands (depth of head profile, depth of muscle area, size of eye) showing the incompatibility of low depth, large muscle area, and big eyes (After Barel 1985)

species follow this relationship, with only a few exceptions. When the eye becomes very big it tends to become elliptical in shape, reducing the dorso-ventral space it occupies. In fishes with big eyes the adductor muscle develops special structures that stretch around the eye and prevent the eye from being pushed outwards when the muscle contracts.

From these relationships it follows that some combinations are impossible (Fig. 5). Big eyes and large muscles are incompatible, so, although the eye is dominant in many respects, at a certain size it becomes elliptical or flat, or it moves to a position slightly detrimental to streamlining. The suspensorium seems to be almost completely subordinate to the eye and muscles, but the hyomandibular always maintains a minimum width necessary to fulfil its role in suspension. It cannot move into the opercular area. The outer shape is fairly dominant, in fast swimmers. In "slow"-moving fishes the smooth contour is sometimes broken.

These conclusions were reached in a comparative study based on a model designed to investigate the range of possible adjustments that could occur in response to changes in size and position of the eyes. Straightforward comparison would have led to recognition of correlations among the various quantitative parameters, but not to the conclusions reviewed here.

8 Conclusion

There is no doubt that constraints are very common at all levels of biological organization. Nor is there any question about the very important role constraints play in the pathways of the evolution (Reif et al. 1985). There is, however, very little definitive evidence of these constraints and the means by which they have their effects. The main reason for this lack of insight is that only a few investigations have applied a deductive methodology, using optimal models to study transformations that occur in ontogeny and evolution. This is the only proper tool to discover the effects of constraints.

Constraints can conveniently be classified in three groups: universal, general and local constraints. Universal constraints generally remained somewhat outside the scope of the investigations of the biologist, because they refer to the general physical properties of matter and energy. These properties have simply been assumed in most biological research, but they play a crucial role as Alexander (1985) has elegantly shown. Often, "there may simply be only one possible solution to a given structural problem due to the laws of geometry, physics and chemistry" to quote Reif et al. (1985).

The interesting problem is why the organism cannot or does not utilize the full band width of physical properties. We encounter here the consequences of general and local constraints that reflect the complex interaction of many different structures and demands.

Most general constraints arise from an early phylogenetic or ontogenetic "choice" of material, based on the need to meet the requirements of many functions simultaneously, as shown here in the case of collagen, or they arise from a "starting" construction as in the case of many hydraulic systems. Thus, what are in the first instance general constraints or universals become *historical constraints*. Local constraints are very important in understanding transformation during ontogeny and phylogeny. All the functional components which come together in one space to shape the organism's structure are in fact each other's boundary conditions or constraints. If they can be integrated in such a way that each functions optimally, then they operate only under universal and general constraints (Dullemeijer 1974, 1980; cf. also Reif et al. 1985).

More commonly, compromises are struck among the functional components and each function is carried out suboptimally compared to the ideal situation (*constructional constraints*). (Barel et al. 1989). For the organism it provides a "good enough" solution (Reif et al. 1985; Gans 1989) or a relatively "best" solution considering all the conflicting and constraining effects. To recognize these compromises, it is necessary to make models of combinations of two or more functional components and compare these with the real situation. Examples can be found in Seilacher (1970), Dullemeijer (1974), Wake (1982), Zweers (1985), Reif et al. (1985), Barel et al. (1989) and Lombard and Wake (1976, 1977).

The importance of compromises for these transformations lies in their broad spectrum. Between two components there can be a wide degree of compromise which can be shifted one way or the other as one functional component becomes more important for the survival and fitness of the animal than the other (*functional constraints*). I anticipate that the evolution of many lineages could be interpreted in this way.

Constraints may have a particularly strong influence on the pathway of evolution in two specific ways. One is where the presence of some functional component is essential. Without it the organism cannot survive. Then the line for further development in an alternate direction is closed off. The other is where the dominance of one component is so strong or becomes so strong during transformation that a compromise with the other components can no longer be made. This leads to incompatibility of structures and functions. Future research should investigate the extent of these effects in particular cases. When an organism is precluded from further improving its structural design. we may ask why it cannot break the constraint. Barel et al. (1989) point to three major factors which may block such a release:

1. Environmental factors impose conflicting demands on form, thus a specific compromise is essential.
2. Integration of several forms is already itself a dominant factor for high fitness.
3. The constraint has not yet been released; that is, the next key adaptation has not yet been discovered.

To these factors I add the following:

4. There is no suitable alternative in terms of materials or other necessary conditions (e.g., membranes, DNA, contractive tissue, etc.).

Yet a number of pathways remain open for new development for the following reasons:

a) Every functional component or construction has the potential to serve purposes other than the function for whcih it is "designed".
b) Every construction or structure has so-called irrelevant properties which are available for other purpose (Dullemeijer 1974). These are properties which come along with any form, not relevant for the existing task, but which can be called upon to serve a new purpose (exaptation of Gould and Vrba 1982).
c) Most constructions are statically indeterminate and thus can carry out their functions in many different ways, permitting change in behavior.
d) Change in proportions during growth, or in the relative rates of development of different structures, can lead to sudden new structural contacts, which may lead to the emergence of new potentials (catastrophe theory, Thom 1975; see also Gould 1977 and Kingsolver and Koehl 1985).

New "key innovations" can arise in any of these ways. However, prediction of such new developments remains hazardous, as the number of degrees of freedom is very large and slight changes may have great, unexpected effects.

Acknowledgments. I wish to thank the organisers of the symposium for their kind invitation, and Dr. R. Thomas for the critical reading and correcting of the manuscript.

References

Alexander R McN (1981) Factors of safety in the structure of animals. Sci Prog Oxford 67:109–130
Alexander R McN (1982) Optima for animals. Arnold, London, 112 pp
Alexander R McN (1985) The ideal and the feasible: physical constraints on evolution. Biol J Linnean Soc 26:345–358
Barel CDN (1985) A matter of space. Constructional morphology of cichlid fishes. Thesis, Univ Leiden
Barel CDN, Anker G Ch, Witte F, Hoogerhoud RJC, Goldschmidt PT (1989) Constructional constraints and its ecomorphological implications. Acta Morphol Neerl Scand 27:1–2, 83–110
Bereiter-Hahn J (1985) Architecture of tissue cells. The structural basis which determines shape and location of cells. Acta Biotheor 34:139–148
Bock WJ (1988) The nature of explanation in morphology. Am Zool 28:205–215
Bock WJ (1989) Principles of biological comparison. Acta Morphol Neerl Scand 27:17–32
Dullemeijer P (1959) A comparative functional-anatomical study of the heads of some Viperidae. Morph. Jahrb 99:880–985
Dullemeijer P (1972) Explanation in morphology. Acta Biotheor 21:260–273
Dullemeijer P (1974) Concepts and approaches in animal morphology. Van Gorcum, Assen
Dullemeijer P (1980) Functional morphology and evolutionary biology. Acta Biotheor 29:151–250
Dullemeijer P (1989) On the concept of integration in animal morphology. Progr Zool 35:1–18
Dullemeijer P, Barel CDN (1977) Functional morphology and evolution. In: Hecht MK, Goody PC, Hecht RM (eds) Major patterns in vertebrate evolution. Nato Adv Stud Inst Ser A 14:83–117. Plenum, New York
Dullemeijer P, Fruitema F (1982) The relation between osteon orientation and shear modulus. Neth J Zool 32:300–306
Gans C (1989) On phylogenetic constraints. Acta Morphol Neerl Scand 27:133–138
Gould SJ (1977) Ontogeny and phylogeny. Harvard Univ Press, Cambridge, MA
Gould SJ, Vrba ES (1982) Exaptation – a missing term in the science of form. Paleobiology 8:4–15
Gutmann WF (1972) Die Hydroskelett-Theorie. Aufs Reden Senckenb Naturforsch Ges 21:91 pp
Gutmann WF (1977) Phylogenetic reconstruction: theory, methodology, and application to chordate evolution. In: Hecht MK, Goody PC, Hecht RM (eds) Major patterns in vertebrate evolution. Nato Adv Stud Inst Ser A 14:645–669. Plenum, New York
Gutmann WF, Bonik K (1981) Kritische Evolutionstheorie. Gerstenberg, Hildesheim, 227 pp
Gutmann WF, Vogel K (1988) Protist skeletons: biomechanical preconditions and constructional utilization. Senckenb Lethaea 69:171–188
Hempel CG (1965) Aspects of scientific explanation. The Free Press, New York London
Hempel CG, Oppenheim P (1953) The logic of explanation. In: Feigl H, Brodbeck M (eds) Readings in the philosophy of science. Appleton Century-Crafts, New York, pp 319–352
Huson A, Spoor CW,, Verbout AJ (1989) A model of the human knee, derived from kinematic principles and its relevance for endoprosthesis designs. Acta Morphol Neerl Scand 27, 1–2:45–62
Jeuken M (1969) A note on models and explanation in biology. Acta Biotheor 18:284–290
Kingsolver JG, Koehl MAR (1985) Aerodynamics, thermoregulation, and the evolution of insect wings: differential scaling and evolutionary change. Evolution 39:488–504
Klaauw CJ van der (1963) Projections, deepenings and undulations of the surface of the skull in relation to the attachment of muscles. Verh Kon Ned Akad Wet Amsterdam 2, 55:247 pp
Liem KF (1973) Evolutionary strategies and morphological innovations: cichlid pharyngeal jaws. Syst Zool 22:425–441
Lombard RE, Wake DB (1976) Tongue evolution in the lungless salamanders, family Plethodontidae. I. Introduction. Theory and a general model of dynamics. J Morphol 148:265–286
Lombard RE, Wake DB (1977) Tongue evolution in the lungless salamanders, family Plethodontidae, II. Function and evolutionary diversity. J Morphol 153:39–80
Maynard Smith J, Burian R, Kauffman S, Alberch P, Campbell J, Goodwin B, Lande P, Raup D, Wolpert L (1985) Development, constraints and evolution. Q Rev Biol 60, 3:265–287
Moss ML (1968) A theoretical analysis of the functional matrix. Acta Biotheor 18:195–204
Moss ML, Salentijn L (1969) The capsular matrix. Am J Orthod 56:474–490
Osse JWM, Muller M (1980) A model of suction feeding in teleostean fishes with some implications for ventilation. In: Ali MA (ed) Environ Physiol Fishes. Plenum, New York, pp 335–352

Otten E (1983) Vision and jaw mechanism during growth of the cichlid fish, *Haplochromis elegans*. Thesis, Univ Leiden

Otten E (1985) Proportions of the jaw mechanism of cichlid fishes: changes and their meaning. Acta Biotheor 34:207–217

Ramaekers JGM (1978) The rheological behaviour of cortical bone and cartilage of *Bos taurus*. Acta Morphol Neerl Scand 16:55–67

Reif WE, Thomas RD, Fischer MS (1985) Constructional morphology: the analysis of constraints in evolution. Acta Biotheor 34:233–248

Rudwick MJS (1964) The inference of function from structure in fossils. Br J Philos Sci 15:27–40

Seilacher A (1970) Arbeitskonzept zur Konstruktions-Morphologie. Lethaia 3:393–395

Thom R (1975) Structural stability and morphogenesis: an outline of a general theory of models. Benjamin, Reading, MA, 348 pp

Wake DB (1982) Functional and developmental constraints and opportunities in the evolution of feeding systems in urodeles. In: Mossakowski D, Roth G (eds) Environmental adaptation and evolution. Gustav Fischer, Heidelberg, pp 51–60

Zweers GA (1985) Greek classicism in living structures. Some deductive pathways in animal morphology. Acta Biotheor 34:249–275

Constructional and Ecological Prerequisites for the Evolution of Homeothermy

HANS-RAINER DUNCKER[1]

Summary

For a full understanding of the evolution of homeothermy, the phylogenetic development of the respiratory system in lung-breathing vertebrates with its two independent lines towards birds and mammals is initially discussed. Secondly, the coordinated evolution of the cardiovascular system from reptiles to birds and mammals is described, which is directly functionally interrelated to lung evolution. Thirdly, the evolution of the locomotor system with its integration with the respiratory apparatus and its functional dependency on the high performance rates of both respiratory and cardiovascular systems is explained. Fourthly, the evolution of energy metabolism and its coupling to the system characteristics of the respiratory and cardiovascular apparatuses is discussed. Then, homeothermy is explained as a resulting phenomenon of the evolution of the locomotor, cardiovascular and respiratory systems. Finally, as a necessary environmental condition, the co-evolution of angiosperms and their high primary production is mentioned, which was actually the prerequisite for the high basal metabolism in homeotherms. This example of an analysis of the multiple interrelationships of organ systems in organisms and between organisms and ecosystems during their evolution towards higher integration and advanced functional levels is given to demonstrate that constructional morphology presents an adequate methodology by virtue of its comparative methods to analyze the complexity of the functional structure of organisms.

1 Introduction

By using the evolution of homeothermy in lung-breathing vertebrates as an example, the manner in which the evolution of the respiratory system is interconnected with that of the cardiovascular system is demonstrated, as well as how both are integrated into the evolution of the locomotor system. The evolution of the latter from reptiles to birds and mammals is characterized by a strong increase in the aerobic capacity, and thereby an increase of the sustainable maximal velocity of locomotion. The required strong increase in maximal exercise metabolism was enabled by the evolution of the respiratory and cardiovascular systems. Their "system characteristics" are responsible for the dependency of basal metabolic rate on

[1] Institut für Anatomie und Zytobiologie der Justus-Liebig-Universität Gießen, Aulweg 123, D-6300 Gießen, FRG

N. Schmidt-Kittler and K. Vogel (Eds):
Constructional Morphology and Evolution

the height of maximal exercise metabolism. At high basal metabolic rate it was advantageous to develop a constant control of high body temperature. This secondarily gained homeothermy, once developed, was the basis for coping with climatic changes and attaining a broad structural-functional and ecological dispersion as well as a broad geographical distribution.

However, it was only possible for this evolution to take place together with the co-evolution of angiosperms due to their high primary production, which was essential for the high food and metabolic requirements of homeotherms. The respiratory, cardiovascular, and locomotor systems in homeotherms evolved in birds and in mammals in two totally independent, parallel lines, starting from different constructional conditions and leading to physically and physiologically highly different constructional solutions, which have in common only a few general physiological characteristics, namely homeothermy.

Thus, for a sufficient understanding of the evolution of homeothermy, an excellent example is available by analyzing the construction of the respiratory, cardiovascular, and locomotor systems of birds and mammals. This analysis should always start with a typological analysis of structures and/or functions, which constitute the basic structural and functional systems. In the next step a functional analysis should attempt to reveal the functional meaning of structural components or organ systems or to evaluate the structural basis of investigated functional systems. In this way the structure-function relationships for the individually investigated parts of the organism can best be analyzed. In the third and final step, a system analysis must be undertaken to study the integration of different functional or organ systems into one entire organism and the organism into the ambient ecological conditions. To a certain extent these three steps of analysis describe in their sequence also the historical development of morphological research.

In all three steps of analysis special emphasis is made to incorporate (1) maximal divergence of taxonomic positions of the investigated species and thus their historical dependencies and their organ systems, (2) maximal differences in the functional specializations of their organ systems and ecological roles, and (3) maximal range of body size differences. Thus, it will be possible to analyze the principles of organismal constructions from their single components, covering organ and functional systems, and leading to the overall form and functional behavior of animals in their habitat and their role in the biosphere. All these steps of analysis depend on a comparative methodology. Experiments and anatomical descriptions deliver only knowledge of the basic elements of structural/functional systems. All three steps of this comparative analysis are based on the concept of evolution and the synthetic evolutionary theory enhanced by the ideas of systemization of functional systems to integrated complex organismic functions, constituting new hierarchical levels of qualitatively new functions, exemplified here for the development of homeothermy.

2 The Evolution of the Respiratory Apparatus in Reptiles

In reptiles, large differences in the structure of the lungs are found. The simplest type are the paired unicameral lungs, which are lying cranially and dorsally in the nondivided pleuroperitoneal cavity, where they are freely movable, being attached only via a dorsal mesopneumonium. The gas exchange surface is represented by the walls of small air chambers, ediculae, which cover the inner surface of this unicameral lung. The ediculae are wide and low in tuataras, whereas the majority of lizards demonstrate an increase of the gas exchange surface in their unicameral lungs by having narrower and higher ediculae. These structures are developed into very high and narrow, honeycombed faveoli in large tejus and in snakes. The height of the faveoli is mostly a few mm, never exceeding 8 to 10 mm even in large snakes (Duncker 1978a, 1981).

In a number of lizard species with special behavioral traits, such as chuckwallas, the lungs possess caudal dilatations with a strongly reduced exchange surface; the dilatations lack ediculae. In all lizards with an elongated body form, e.g., burrowing skinks, the asymmetrically elongated lungs possess caudal nonrespiratory dilatations. The largest nonrespiratory dilatations are found in snakes, which, with the exception of boas and pythons, have a reduced left lung. These dilatations enable the animals to maintain a certain body diameter, responsible for locomotor functions or as reserve volume for engulfing prey. The large dilatations in snakes lack a pulmonary vascular supply, and are fused with the body wall. The loss of the pleuroperitoneal cleft prevents the dilatations from being totally compressed or shifted craniad by body or intestinal movements (Duncker 1978b, 1979). The volume of the inflated lung without dilatations makes up 7% of the total body volume. This relative lung volume increases with the development of dilatations, reaching more than 30% in snakes.

Unicameral lungs are ventilated via volume changes during respiratory movements of the rib cage wall. Generally, the rib cage is developed as far caudally as the lungs extend. Respiration at rest is performed by the movement of a few cranial ribs, only during exercise-respiration are all ribs recruited for ventilatory movements. The expansibility of the lungs and thorax, the compliance, is measured as volume change per pressure unit change, and it depends on the structure of the lungs and the thorax wall, on which the respiratory muscles act to generate a pressure difference causing inflation or deflation of the lungs. The compliance of unicameral lungs is determined by the density of the exchange structures in the lung wall; it decreases strongly with the development of high and narrow faveoli. Accordingly, large tejus possess a posthepatic septum, which prevents the intestines from being sucked into the cranial pleuro-peritoneal cavity during strong inspiratory movements. Compliance increases dramatically with the development of the thin-walled caudal dilatations. Thus, the animals can change during exercise from a respiratory pattern characterized by small amplitude and high frequency (without dilatations) to a pattern with large amplitude and low frequency (with dilatations) (Perry and Duncker 1978, 1980). However, the dilatations are inflated only under special situations, such as defense display, or they act as reserve volume, e.g., for swallowing large prey. Regularly, the dilatations, even when they are air-filled, are not ventilated: the respiratory movements are restricted to the cranial parts of the lung which possess high faveoli.

In unicameral lungs gas in the central lumen is exchanged via ventilatory movements; gas in the high faveoli of the lung wall can only be exchanged via diffusion. Therefore, the height of the faveoli never exceeds 8 to 10 mm: their further elongation would be functionally inefficient. This demonstrates the functional limitation of the unicameral lung, because with an increase in body size or metabolism, the mass of body tissue increases in relation to the third power of body length, whereas with a limited height of the faveoli the exchange surface of unicameral lungs can only increase in relation to the second power of body length. Therefore unicameral lungs are only found in lizards with limited body size and in snakes because of their limited metabolism (Duncker 1978a).

This functional limitation of the unicameral lung is overcome by the septation of the lung into several chambers (Fig. 1), the walls of which enlarge the gas exchange surface to such an extent that these multicameral lungs increase their exchange surface in a linear relationship to their increase in volume and thereby parallel to the increase in body tissue mass. A few septa are found in lungs of chameleons and in a few species of agamids and iguanids. Multicameral lungs are mandatory for all turtles and tortoises, monitor lizards, and crocodiles; and it can be also assumed that all extinct large ichthyosaurs, archosaurs, and especially the dinosaurs, as well as the therapsids, possessed multicameral lungs. Even in the extant reptiles multicameral lungs can be very highly differentiated with a very large exchange capacity, as can be demonstrated in sea-water turtles or monitor lizards (Duncker 1978a, 1981; Perry 1989).

The basic type of multicameral lungs is found in snapper turtles, which possess only a body armor, but cannot retract legs and head into the shell. Their lungs contain 12 chambers of similar size with fairly large ediculae, which originate from a central main bronchus, the former central lumen. These lungs are attached to the dorsal body wall in the peritoneal body cavity via dorsal mesopneunomium, like all simple unicameral lungs. In most other turtles, with the exception of sea turtles and shoft-shelled turtles, and in all tortoises the lungs possess large ventral and caudal nonrespiratory dilatations, which greatly enlarge the lung volume. Due to this construction, the respiratory parts of the lung are situated dorsally in the lung around the intrapulmonary main bronchus in the form of small partitioned walls of the dorsal lung chambers, whereas the ventral and lateral chambers are dilated, at least their most ventral and caudal parts. These multicameral lungs are thus heterogeneously subdivided (Fig. 1). The lungs in monitor lizards and in crocodiles are differentiated in a way similar to that in turtles. They also possess multicameral lungs, which are heterogeneously subdivided with small partitioned dorsal parts around the intrapulmonary main bronchus for respiratory purposes and into ventro-caudal, large dilatations, which are only inflated in defense display, walking, or swimming (Duncker 1978a, 1981).

It is characteristic for all heterogeneously partitioned lungs, that they are separated from the intestines by a postpulmonary septum, which divides the body cavity into a pleural and a peritoneal cavity (Fig. 1). All heterogeneously partitioned reptilian lungs also lose the pleural cleft during embryological development, fusing with the body walls and with the postpulmonary septum. Thus, these lungs are pre-inflated to a certain extent, and cannot collapse totally. In monitor lizards and crocodiles the lungs fill the entire cranial part of the body cavity, not only the dorsal regions as in turtles. Thereby, the lungs displace the entire intestines together with the liver into the caudal part of the body cavity, including the

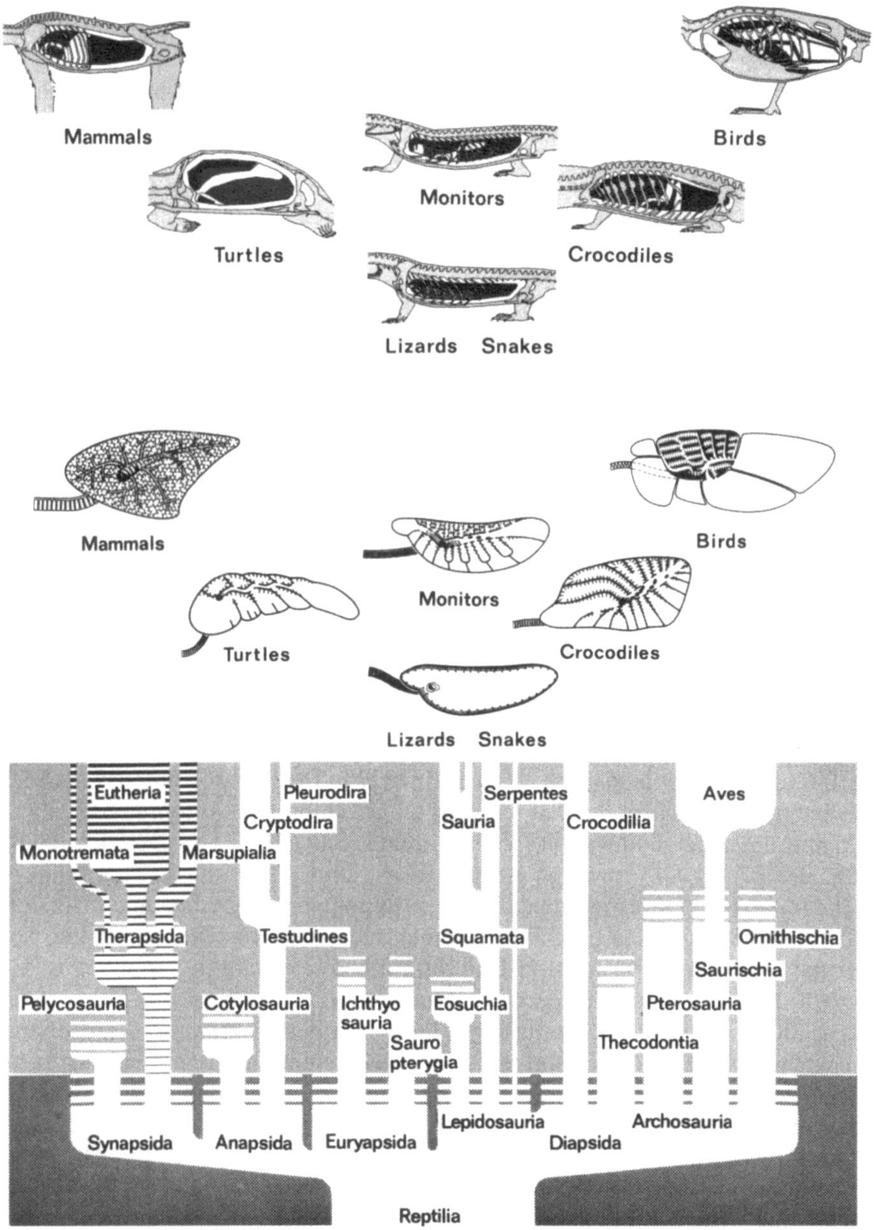

Fig. 1. Phylogenetic tree indicating the relationship between the different orders of amniotic vertebrates and types of their lungs as well as of their body cavity subdivisions (After Duncker 1978 d)

postpulmonary septum, which comes to lie perpendicular in the middle of the trunk between the pleural and peritoneal cavity. By this caudal shift of the postpulmonary septum, the pericardium, which is embryologically connected with the developing postpulmonary septum, has also undergone a descent to the middle of the body, lying at the caudal end of the sternum (Duncker 1978b, 1979).

The fusion of the heterogeneously partitioned lungs with the body wall and postpulmonary septum is functionally indispensable. The moist boundary layer between air and tissue in the lungs exhibits a certain surface tension, which in all vertebrate lungs is lowered by the surfactant material, consisting of phospholipids. Nevertheless, the remaining surface tension depends on the curvature of the surface structure, being reciprocally correlated to the radius of the curvature. Therefore, in heterogeneously subdivided lungs the surface tension is much greater in the small-partitioned dorsal lung chambers around the main bronchus than in the ventro-caudal dilatations. Because of their low surface tension, these ventral dilatations would become overly inflated in a freely movable lung, whereas the small-partitioned dorsal respiratory parts would maintain in a collapsed state. This is prevented by the fusion of the lungs with the body wall and the postpulmonary septum. By these differentiations the respiratory parts of these heterogeneously partitioned lungs are always pre-inflated to a certain extent, and cannot collapse totally. They are ventilated by slight respiratory movements of the body wall or the extremities in turtles (Duncker 1978a, 1981).

The ventro-caudal dilatations of these heterogeneously subdivided lungs are usually inflated only by stronger respiratory movements, especially in defense display in monitor lizards and crocodiles. In turtles, and especially in tortoises, the inflation of the dilatations is required for swimming or walking with extended legs, their volume is needed for the retraction of head and legs into the shell: this "turtle principle" depends on the availability of the rather large disposable volume, but also for the deposition of fat and water in the shell for hibernation. All these heterogeneously subdivided reptilian lungs exhibit a large compliance, similar to the unicameral lungs with caudal dilatations (Perry and Duncker 1978, 1980).

The few existing investigations of the gas exchange surface area and diffusion capacity of these reptilian lungs demonstrate that each lung type, taking into account the limitation of the lung types to certain body size ranges, supplies the animal with the necessary diffusion capacity according to its metabolic needs. This is especially striking when comparing a large teju with a monitor lizard of the same body mass. The unicameral teju lungs possess only half the surface area of the multicameral monitor lungs, whereas the teju lungs have an air-blood tissue barrier of only half the harmonic mean thickness compared to the monitor lung; thus both species possess the same available diffusion capacity (Perry 1983, 1989). This demonstrates best that the teju lung is the ultimate differentiation of the unicameral lung structure, not allowing a further increase in body mass, whereas the monitor lung represents a lung type that allows an enormous increase in body size.

3 The Evolution of the Avian Respiratory System

The evolution of the respiratory apparatus in birds and mammals took place in two totally different directions. The evolution of the avian apparatus is characterized by the following features. The lungs are separated from the intestines by a postpulmonary septum and fuse during their embryological development with the walls of the pleural cavity and the septum, which is identical to heterogeneously subdivided reptilian lungs. The lung chambers of these reptilian lungs originate in three rows from the main bronchus: the avian lung is similarly constructed, in which these chambers are differentiated into secondary bronchi, which originate in three rows from the main bronchus. In the avian lung the secondary bronchi are arranged at the ventral and the dorso-lateral lung surface. From the entire internal surfaces of both sets of secondary bronchi the parabronchi, which are typical for the avian lung, originate side by side and anastomose in a plane intermediate between ventro- and dorsobronchi. Thus, the ventro- and dorsobronchi are directly interconnected via the numerous parallel-running and interconnected parabronchi (Duncker 1971).

During the embryological development of the avian lung, the anlagen of the air sacs sprout from the secondary bronchi into the postpulmonary septum, dividing it into the horizontal septum, which maintains its position directly beneath the lung, and into the oblique septum. The thoracic air sacs expand between both septa, while the clavicular and cervical air sacs extend cranial to the pericardium into the upper thoracic region. These air sacs occupy a totally new part of the body cavity in the avian body, the subpulmonary cavity. In the kiwi, the end of the primary bronchus also penetrates into this cavity; in all other birds it penetrates through the dorso-caudal part of the nondivided postpulmonary septum and runs dorsally in the peritoneal cavity as the abdominal air sac, a thin walled, freely movable sac fixed only to the dorsal peritoneal wall (Duncker 1978c, 1979). As a consequence of this development, the avian respiratory system is differentiated into a gas exchange system, e.g., the lungs in the dorsal thoracic cavity, and into the ventilatory apparatus, the air sacs as bellows, located ventral and caudal to the lungs (Fig. 1).

The avian thoracic locomotor apparatus underwent a very special evolution. The broad sternum and the totally ossified sternal ribs, forming a hinge with the sternum, constitute, together with the vertebral ribs and vertebral column, a very rigid thoracic cage, which is not compressible in its lateral diameter. Via respiratory movements only the dorso-ventral diameter is changed, increasing from cranial to the caudal end of the sternum. In this laterally fixed thoracic cage the extremely dorsally positioned pleural cavity is ventrally separated off by the rigid horizontal septum, which contains small costo-septal muscles in its lateral margin. These costo-septal muscles relax during inspiration and contract during expiration, thus compensating for the small volume changes of the pleural cavity, which occur during the respiratory movements of the rib cage. Thereby, the avian lung remains practically volume-constant during the respiratory cycles, whereas the volume changes by the respiratory movements act entirely on the air sacs, which ventilate, like bellows, the rigid tubular system of the lung (Duncker 1971, 1983, 1989a).

The ventilation of the rigid tubular system of the avian parabronchi is aerodynamically directed. This type of ventilation is implemented by (1) the arrangement of the sites of origin of the secondary bronchi from the primary bronchus, (2) their angles of origin and the changes in diameter along their lengths, and (3) the connections of the cranial air sacs to the ventrobronchi as well as the connections of the caudal air sacs to the posterior primary bronchus. The parabronchi of the main part of the avian lung are ventilated unidirectionally during inspiration and expiration (Scheid 1982; Scheid and Piiper 1971, 1989). This unidirectional ventilation of the parabronchi is remarkable and guarantees that they are ventilated during the entire respiratory cycle by fresh air, which continuously exchanges with the venous blood along the length of the parabronchi. Thereby, a cross-current-exchange system is established, which is responsible for high gas exchange efficiency characteristic of the avian lung. This system enables birds such a high metabolic capacity, which is necessary for sustained flight, performed in some species even at heights above 8000 m. The large volume of the air sacs and their large compliance allows the respiratory system large ventilatory amplitudes, another feature of the large gas-exchange capacity in the avian lung.

The volume constancy of the avian lung, which is dependent upon the rigid structure of the avian thorax and upon the subdivision by the horizontal septum of the dorsally located pleural cavity, was an indispensible prerequisite for the evolution of air capillaries in the parabronchial walls. During embryological development, air capillaries sprout from the parabronchial lumen between and around the blood capillaries, which are developing in the thick mantle of the parabronchial wall. The blood capillaries are arranged radially in the parabronchial wall, running from the outer border straight towards the lumen of the parabronchi, where they are collected by small venules. The air capillaries form a three-dimensional, interconnected network around these radial blood capillaries demonstrating a similar diameter as the blood capillaries (5 to 14 μm depending on the species) (Duncker 1978a; Duncker and Güntert 1985). These air capillaries possess an excellent surfactant film, which lowers the surface tension; however, the remaining surface tension, depending on the very small curvature of the air capillaries, is so high that the air capillaries would collapse in a volume-variable lung and could not be inflated by respiratory movements. Thus first the volume constancy of the lung enabled the evolution of the air capillaries, which equipped the comparably small avian lung with the extremely large gas exchange surface in the blood-air capillary network (Duncker 1971). The surfactant is responsible for preventing the suction of fluid out of the blood capillaries into the air capillaries. Due to the extremely thin tissue-exchange barrier, the avian lung possesses the highest gas diffusion capacity of all vertebrate lungs (Duncker and Güntert 1985; Maina 1989).

The rigid, volume-constant avian lung cannot be inflated or expanded during the transition time from intra-amniotic development to aerobic life. The functional ability of the avian lung is attained by an overlap of gas exchange by the chorioallantoic membrane and breathing of air, which takes place during the last phase of incubation. During this phase the parabronchi are aerated by air-breathing, and the remaining pulmonary/amniotic fluid is removed by resorption from the growing air capillaries. Gradually, the lung then assumes the function of gas exchange. Thus, the overlap of these two functions by a rigid, nonexpandable lung is essential for the transition from a fluid-filled amnion to air. Therefore, birds had to retain "obligatory oviparity", inherited from their ancestors (Duncker 1971, 1978c).

4 The Evolution of the Mammalian Respiratory Apparatus

The avian respiratory system is the highest differentiated vertebrate gas exchange system, starting, in evolution, from a heterogeneously subdivided reptilian lung. The very efficient and also very highly differentiated mammalian lung is based on the principle of a homogeneous subdivision of the lung into the broncho-alveolar tree, obviously derived from a homogeneously partitioned reptilian lung (Fig. 1). Similar to the arrangement of three rows of lung chambers in reptilian lungs, the basic structural type of the mammalian bronchial tree is the origin of three rows of smaller secondary bronchi from the main stem (Nakakuki 1979; Valerius 1983). This system of branching is continued several times in identical fashion for subdivision of the bronchial tree. The frequently described dichotomous ramification of the bronchial tree is a derived mode of branching in lungs of larger mammals. Alveoli develop in the distal parts of the bronchial tree, singly or as groups of alveoli more proximally in the respiratory bronchioles, whereas in the terminal ducts the wall is entirely differentiated into alveoli. Depending on body size, the alveoli possess diameters ranging from 35 μm in shrews and bats to 500 μm in large mammals, such as elephants (Tenney and Remmers 1963; Weibel and Taylor 1981). Their well-developed surfactant allows not only volume changes of the alveoli during respiratory movements (by which they are ventilated) but also the inflation of collapsed alveoli. This is the principle used in deep-diving mammals, such as seals and whales, whose alveoli are totally compressed at greater depths but immediately reinflated upon surfacing. This is rendered possible by the cartilage-supported bronchioles.

The broncho-alveolar mammalian lung possesses a bronchial tree and alveoli of identical dimension throughout all parts of the lung. In all these parts the same retractive force is present, which is responsible for lung collapse after opening of the thorax wall. Two thirds of this retractive force are made up by the surface tension of the alveoli, the remaining third comes from the elastic fiber systems of the interalveolar septa. This retractive force is identical in all parts of the lung and allows the maintenance of a pleural cleft into adulthood. Thereby the mammalian lung is able to undergo large volume changes throughout the respiratory cycle. The subdivision into a large number of very small alveoli delivers a large exchange surface with a rather thin diffusion barrier, which is responsible for the high diffusion capacity (Gehr et al. 1978, 1981; Weibel and Taylor 1981; Weibel 1984). The large volume changes of the lung guarantee, especially during exercise, high exchange rates, which are necessary for high metabolism. The partitioning of the mammalian lung into a large number of very small alveoli drastically reduced its compliance. Probably early in the phylogeny of mammals the existing postpulmonary septum was supported by a strong layer of skeletal muscles. In this way the mammalian diaphragm evolved to assist forcefully the inspiratory action of the rib cage musculature. As in monitor lizards and crocodiles, the mammalian lung occupies the entire cranial body cavity. Thereby the lungs displace the entire intestines caudally, including the diaphragm. Because the pericardium is developmentally directly connected with the diaphragm, a descent of the pericardium occurs (Duncker 1978b, 1979).

The alveoli develop by a bulging of the proliferating epithelium into the meshes of a network of longitudinal and circular fibers consisting of collagen and elastin, which have been produced by the mesenchymal cells around the sprouting terminal ramifications of the bronchial tree during the late phase of ontogeny. Primarily, each growing alveolus is surrounded by a capillary net, which is interconnected with that of the neighboring alveoli. In their further growth these walls of the neighboring alveoli develop into the interalveolar septa, which also occur between the alveoli of adjacent bronchial branches. Through this growth of the interalveolar septa, the primarily double capillary nets are extended and transformed to the typical single capillary net of mammalian alveoli (Burri 1988; Duncker in press); by virtue of this development, mammalian alveoli are distinguished from reptilian ediculae and faveoli. The mammalian lung acquires its functional competence during embryological development by adherence of the capillaries to the epithelium of the terminal bronchial sprouts, and by a process of thinning of these epithelial cells. Being correlated with the mode of ontogeny, this takes place in embryos of altricial mammals in the walls of the bronchial sprouts before the formation of alveoli, and in embryos of precocious mammals in the already developing alveoli (Duncker in press). By virtue of its flexibility, i.e., not adhering to the walls of the pleural cavity, the mammalian lung can undergo large volume changes and therefore it can be forcefully inflated directly after birth, so that the exchange surfaces can be instantaneously expanded and immediately assume the function of gas exchange, while the remaining pulmonary/amniotic fluid in the terminal bronchial sprouts is being absorbed during first few hours of air breathing. This immediate transition from the gas exchange by the placenta to that of the lung takes place independently of the general developmental stage that the mammalian lung has reached at the time of birth. This was the developmental condition that allowed mammals to evolve "obligatory viviparity".

5 The Evolution of the Cardiovascular System from Reptiles to Birds and Mammals

The vascular system of the reptilian lung is perfused to a highly variable extent, dependent on whether the animal is breathing or in an apnoeic phase. This perfusion is centrally regulated and performed under vagal control by dilation or contraction of the pulmonary arteries, being functionally adjusted to a high perfusion rate whilst breathing and to a minimal perfusion rate during apnoe. Thereby, during a breathing phase gas exchange takes place very rapidly with a high oxygen saturation rate of the blood, whereas during apnoe the oxygen content of the air in the lung can be optimally utilized (White 1976, 1978). According to this functional specialization of the reptilian lung and its vascular system, the reptilian heart displays the most specialized structural construction of vertebrate hearts, allowing central shunting between the pulmonary and systemic circulation. This heart, best investigated in lizards and turtles, possesses in its ventricle a cavum arteriosum, which receives its blood via the left atrium from the lung, and a cavum venosum and a cavum pulmonale, which receive their blood from the systemic circulation via the right atrium. Both aortic arches originate from the cavum venosum, the pulmonary trunk arising from the cavum pulmonale. The cavum arteriosum is connected to the cavum venosum by the interventricular foramen in the ventricular septum. The cavum venosum and the cavum pulmonale are separated from each other only by a muscular ridge, a longitudinal upfolding of the ventricular septum (White 1976, 1978).

In the apnoeic phase, during which the pulmonary arteries are contracted, the systolic contraction of the ventricle starts from the left ventricular musculature. The ventricular pressure increases rapidly, due to the high resistance in the pulmonary vasculature, causing the muscular ridge to approach the ventricular wall, thereby occluding the cavum pulmonale from the cavum venosum. With increasing pressure the small amount of blood from the cavum pulmonale is ejected into the pulmonary trunk. At the same time the pressure of the blood in the cavum venosum exceeds the pressure in the systemic circulation, which opens the valves in the aortic arches and the blood is ejected into the aortas. Increasing pressure causes the blood from the contracting cavum arteriosum to flow through the interventricular foramen into the cavum venosum, and this blood is also ejected into the aortic arches. Thus, the majority of ventricular outflow is directed into the systemic circulation (White 1976, 1978).

During the breathing phase, while the pulmonary arteries are dilated, a reverse distribution of blood takes place, the contraction of the ventricular musculature starting from its right portion. At low pulmonary vasculature resistance most of the blood from the cavum pulmonale and cavum venosum is ejected into the pulmonary trunk, before the muscular ridge contacts the ventricular wall and occludes the cavum pulmonale. With increasing ventricular pressure the remainder of the blood in the cavum venosum and the blood from the cavum arteriosum are then ejected into the systemic circulation. Thus, the greater amount of total ventricular outflow perfuses the pulmonary vessels, enabling an efficient gas exchange during the short breathing phase.

This regulation of the overall blood distribution between the systemic and the pulmonary circulation is highly adapted to the specific construction and function of the reptilian pulmonary and respiratory systems. Not only the pulmonary arteries, but also the special mode of ventricular contraction are regulated by vagal innervation. In this way both the structure and function of the lung and the heart as well as their vasculature are closely integrated, not only with each other, but also with the structure and function of the systemic vascular system (White 1976, 1978). Limited perfusion of body tissues, even during long apnoeic phases with a reduced oxygen saturation, is sufficient due to the rather generally low metabolism of tissue, especially the low basal metabolic rate. The overall regulation of the perfusion of the lung is adapted to lung structure. There is mostly a homogeneous distribution of the gas partial pressure, so that the overall ventilation/perfusion regulation is functionally efficient. The design of the reptilian cardiovascular system is directly adjusted to these structural-functional relationships of the lung and to the limited metabolic needs of the reptilian body tissue.

In contrast, avian as well as the mammalian lungs, which are subdivided and display extremely fine-partitioned exchange structures, show large differences in ventilation of different parts of the lung and thus do not show a homogeneous distribution of gas partial pressure, especially for respiration at rest. Therefore, a general regulation of the perfusion/ventilation ratio is no longer functionally efficient. To adapt the perfusion to these inhomogeneities of ventilation and gas partial pressures, perfusion has to be regulated locally: this occurred by the development of local reactions of the smooth muscles of arterioles and perhaps also of venules to the oxygen partial pressure surrounding them in the form of a hypoxic contraction, which shuts off perfusion at low oxygen content (Klocke 1988; Weir and Reeves 1989). Thus, perfusion is always locally adapted to the existing partial pressures and the overall efficiency of the perfusion of the lung is maintained. Because the avian and mammalian body with its high metabolic rate requires a constant supply of oxygen even at rest, larger variations in lung perfusion and oxygen uptake below a rather high minimal level cannot be tolerated by the organism. A constant overall perfusion of the lung, guaranteeing a minimal level, but also a higher level of increase for exercise, does not tolerate a central shunting of blood between pulmonary and systemic circulation in the heart of birds and mammals as in reptiles.

According to this situation, the ventricles of avian and mammalian hearts are entirely subdivided into the right and left chambers, totally separating the pulmonary from the systemic circulation. The regulation of heart action concentrates on adapting the function of the heart to the general metabolic needs of the organism, from basal level to metabolic levels during exercise and vice versa. This is the functional correlate of the structurally subdivided heart in birds and mammals. A total subdivision of the ventricle has already been developed in crocodiles. They possess only a foramen between the aortic arch and the pulmonary trunk, the last functional passage for shifting blood between the two circulatory systems.

6 The Evolution of the Locomotor System and its Integration with the Respiratory System

The evolution of the respiratory system demonstrates, beyond its relationships to the metabolic needs of the organisms, another important functional aspect, i.e., its mandatory integration into the functions of the locomotor system. The development of the rib cage of land-dwelling tetrapods had its primary function in the maintenance of body form for terrestrial periods, when the buoyancy of the water no longer supports the body. The rib cage, with its ability to lower the intracoelomic pressure, assumed the function of ventilation for the lungs: thereby the very efficient aspirative ventilation could be substituted for the former buccal ventilation of the lungs, which further exists in rib-less lung-breathing vertebrates (such as lungfishes and amphibians).

However, this aspirative ventilation of the lungs by the movements of the rib cage functions in competition with the rib cage as a part of the locomotor system for body movements especially as site of origin for muscles of the shoulder girdle and forelegs. As has been demonstrated by Carrier (1987a, b), lizards can either breathe with the rib cage or alternatively use it for locomotor functions. The same is true in turtles and tortoises: they can walk or swim, or alternatively they can breathe, which they perform principally via actions of their legs and adjoining rests of body walls musculature. This alternative use of the locomotor apparatus for breathing or for body movements is in accordance with the general level of the basal or exercise metabolism in reptiles. Its relatively low level allows longer apnoeic phases so that the animals can arrange this alternative use of the same apparatus for respiration or locomotion.

With the transitional development to birds or to mammals, which necessitated a drastic increase in the metabolic level, this alternative use of the locomotor apparatus is no longer possible. These forms are forced to breathe continuously, which has then to be integrated into the locomotory functions. According to the very different structure and function of their respiratory apparatuses, birds and mammals solved this problem in two entirely different ways. In birds, a broad sternum and ossified articulating sternal ribs and their specific connections with the vertebral ribs evolved constituting the rather rigid, laterally incompressible thoracic cage, the dorsal part of which is occupied by the rigid, volume-constant lung. The broad sternum together with the coracoid and the furcula including their membranes function as the site of origin of the large flight muscles (Duncker 1971). Thus, this design of the skeletal-muscular system requires an anatomical dissociation between flight and respiratory movements, which has be proven experimentally (Berger and Hart 1974; Bernstein 1987). The remaining movements of the cranial shoulder girdle, laterally with the downstroke and medially with the upstroke of the wings, are insolated against the rib cage by air sac diverticula (Duncker 1971). Thus, birds breathe and fly with different frequencies, both of which are dependent upon body size. Smaller birds regularly demonstrate a higher wing frequency than breathing frequency, larger birds a lower wing frequency. Both frequencies have the same dimension in birds of 300 g body mass, e.g., in pigeons and crows. The investigation of these birds led to the erroneous conclusion of a synchronization of wing and breathing movements (Berger and Hart 1974).

In contrast to birds, in mammals breathing and gait frequency are synchronized, as far as our knowledge of the complicated integrations of these two functions of the locomotor apparatus allows such a statement (Bramble 1989; Bramble and Jenkins 1989). This may also be connected with the fact that mammals usually do not move their forelegs synchronously, as birds do with their wings, but rather in an alternating fashion. The functional relationships between foreleg movements and respiratory actions in mammals should be investigated more thoroughly. Similar to birds, mammals also depend on continuous respiratory movements for their high metabolic demands, especially during exercise (e.g., running).

In addition to the above-mentioned necessary integration of respiratory and locomotor activities the locomotor apparatus has undergone an astonishing evolution from reptiles to birds and mammals. Bennett (1985, 1989) demonstrated that reptiles and birds or mammals are quite comparable with respect to the general biomechanical design of their locomotor apparatus, e.g., they are capable of reaching the same maximal velocity, which is strongly dependent on body size. The large difference between reptiles and birds or mammals is found in their endurance to sustain running or flying velocities. According to Bennett (1985, 1989), investigated living reptiles are capable of high short-term movement velocities. They perform this activity mainly with their anaerobic fast-twitch, white muscle fibers, which constitute the majority of the musculature. In contrast, the amount of aerobic slow or fast-twitch red muscle fibers in reptiles is very limited; therefore they can only maintain a movement velocity of approximately 5% of their maximal velocity for a longer time. This limitation is set by the aerobic capacity of their locomotor apparatus, which is determined by the percentage of aerobic red muscle fibers and by the capacity of the respiratory and cardiovascular systems. Thus, most reptiles are designed for very fast, but short-term movements.

The evolution towards birds and mammals is characterized by a step-by-step increase in aerobic capacity, which is a stepwise increase in the sustainable running speed. Each increase in speed, even a very small one, has a biological advantage, enabling the animal to catch its prey better or to escape from its predator more successfully. This increase in the sustainable running speed is performed by an increasing percentage of aerobic red fibers from the total muscle fiber mass. The increasing amount of aerobic fibers, which are highly capillarized, requires an increasing capacity of the cardiovascular system for a permanently high rate of perfusion, combined with an increasing gas-exchange rate in the further evolved lungs. By this stepwise evolution, birds and mammals have gained the ability to maintain 80% of their maximal running or flying velocity as sustained speed (Bennett 1985, 1989). This high aerobic capacity of the locomotor apparatus is based on a high percentage of aerobic fibers in the body musculature as well as on the highly differentiated cardiovascular and respiratory systems, which are characteristic for these homeothermic vertebrates.

7 The Evolution of Energy Metabolism and the System Characteristics of the Respiratory and Cardiovascular Apparatuses

The evolutionary driving force was set by the goal of increasing performance of the locomotor apparatus to reach higher sustainable speeds in running or flight (Bennett 1985, 1989). Thus, the selection pressure was directed towards an increase in the maximal aerobic capacity of the locomotor as well as the cardiovascular and respiratory systems (Table 1), independently of whether the maximal aerobic capacity could be used seldom, e.g., to escape a predator occasionally, or frequently or continuously during the life time of the animal. This performance of maximal aerobic capacity requires the animal to use the maximal aerobic energy metabolism of which it is capable. It is well established that the level of basal energy metabolism is directly connected with this level of maximal energy metabolism. In poikilotherms and in homeotherms the basal metabolic rate is one tenth of the maximal exercise metabolism (Schmidt-Nielsen 1984; Weibel 1984; Wieser 1986). To date, the question remains open as to which factors determine this relationship. Biologically important is the fact that the increasing maximal aerobic capacity and energy production of the locomotor system immediately increases the basal metabolic rate, which determines the required minimal food intake. Thus, mammals and birds possess a basal metabolic rate tenfold that of reptiles of comparable body size at the same body temperature of 37°C, which puzzles most physiologists. They often explain this phenomenon by the high "leakiness" of cell membranes of homeotherm, which consequently need high energy-dependent ion pumps to maintain homeostasis (Wieser 1986).

However, the functional characteristics of the respiratory and cardiovascular systems together with some obvious functional aspects of the cellular biochemistry determine the relationship between basal metabolism and exercise metabolism. As far as data are available, the obtainable maximal rate for enzymatic processing of substrate is tenfold that of the basal rate, as determined for cytochromoxidase (Wieser 1986). Comparing the increased frequency and enlargement of amplitude in the respiratory system, the transported volume of gas during exercise can maximally reach the tenfold amount of the resting value in any investigated species (Schmidt-Nielsen 1984; Weibel 1984; Wieser 1986). Also the heart and vascular system are limited to a certain maximal transport performance by their structure and functional characteristics. In a similar way, the heart cannot fall below a certain minimal minute volume that is necessary to maintain perfusion of the vascular system, which is essential for survival. Thus in the cardiovascular system the increase in transported volume of blood from rest to maximal exercise reaches six- to maximally eightfold that of the minimal resting volume (Schmidt-Nielsen 1984; Weibel 1984). In combination with an increased difference in arterio-venous oxygen partial pressure during exercise the oxygen transport capacity of the blood can maximally increase up to tenfold (Weibel 1984; Schmidt-Nielsen 1984; Wieser 1986). This means that when evolution selects for reaching a maximal transport capacity and maximal metabolic values, also the necessary minimal values are raised to higher levels, which are determined by the "system characteristics" of cellular biochemistry as well as of the respiratory and cardiovascular apparatuses (Table 1).

To obtain the biologically desired and selected ability for sustained high locomotor activity, a high maximal energy metabolism was required, with concomitant rise in basal metabolic rate. The price for this drastic increase in locomotor abilities was a high basal metabolic rate, compared to reptiles. This high basal metabolic rate, which increased from the reptile rate by a factor of ten, also requires a tenfold increase in food intake. Therefore, the evolution of high locomotor ability in birds and mammals depended on the availability of food for this increased requirement. Before discussing this point, a further relationship in this particular aspect of evolution should be dealt with.

8 The Resulting Homeothermy and the Necessary Environmental Conditions for its Evolution

The increasing aerobic capacity of the locomotor system for reaching higher sustainable speeds in running or flying caused a special physiological problem: the organism had to deal with heat production during exercise. Therefore, it was necessary to develop a thermoregulatory system to avoid overheating, because a rise in body temperature beyond 41 °C is lethal (Schmidt-Nielsen 1979; Barnas and Rautenberg 1987; Eckert et al. 1988). Even reptiles possess a behavioral thermoregulation, which they demonstrate at sun bathing. As discussed above, during evolution the increase in amount of aerobic red muscle fibers in the locomotor system is based on the co-evolution of the cardiovascular and respiratory systems. As investigated in a broad range of avian and mammalian species, thermoregulation during exercise is performed by the interaction of the cardiovascular and the

Table 1. Evolution of homeothermy

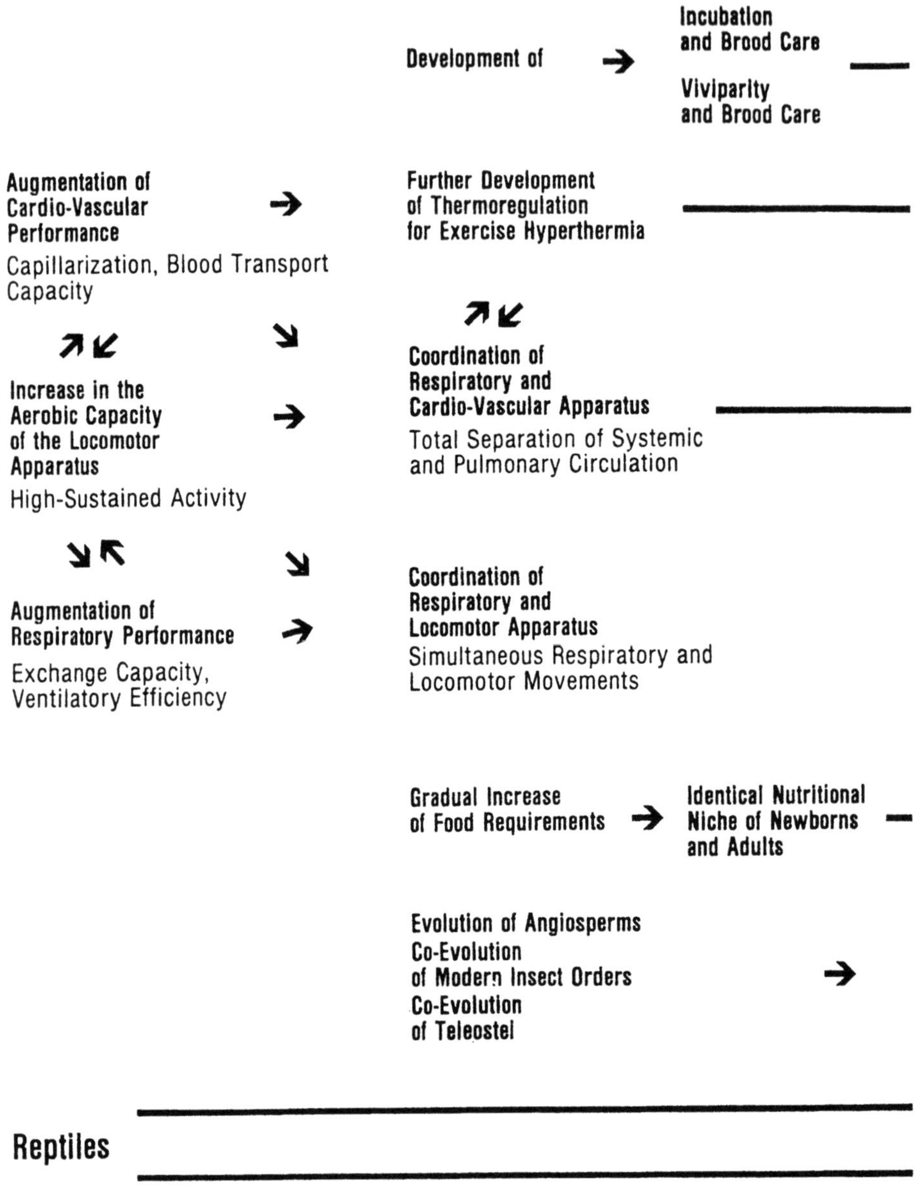

respiratory systems (Schmidt-Nielsen 1979; Barnas and Rautenberg 1987). Basically, this is a functional side effect of ventilation: water evaporates from the walls of the upper airways to humidify the inhaled air; by a hyperventilation of these upper airways, evaporation is used to give off heat, the known panting of birds and mammals. In this way blood is cooled down very effectively, getting rid of the metabolic heat of muscle work.

Table 1 (continued)

→ **Increase of Relative Brain Volume and Behavioral Performance** → **Complex Social Systems**

→ **Development of Thermoregulation for Control of Basal Temperature**

↗ ↙

→ **Increase of Exercise Metabolism and Concomitant Rise of Basal Metabolism** → **Homeothermy**

Systemic Laws of Cell Metabolism, Respiratory and Cardio-Vascular Systems

↘ ↖

Development of Insulation Mechanisms

Hair, Feathers, Panniculus adiposus

Homeothermy ↗ **Survival under Altered Climatic Conditions**

Homeothermy → **Invasion of New Ecological Niches**

Temperate to Boreal Zone

Continuous or Intermittent

Migrations

Homeothermy ↘ **Small Homeotherms under Extreme Conditions**

Torpor, Hibernation

Lowering of Body Temperature

Lowering of Basal Metabolism

→ **Development of New Nutritional Niches**

Granivores

Insectivores

Piscivores/Carnivores

Fruigivores

Herbivores

→ **Birds**

→ **Mammals**

When the evolving birds and mammals finally acquired a higher basal metabolism as an unavoidable side effect of the increasing upper limit of their aerobic capacity, the overall metabolic heat production dung rest also increased. At a certain level of the basal metabolism and basal heat production (probably at the level found in birds and mammals today) it became functionally opportune to maintain and regulate a constant basal temperature (Table 1). Together with the evolution of insulating structures such as feathers or hairs, it was functionally profitable to maintain a constant body temperature, which allows constant high reactivity of all sense organs, nervous system, and locomotor system. For the regulation of this constant body temperature, the already existing thermoregulatory systems were available, now regulating body temperature between the constant basal temperature and the only 2°C higher hyperthermia of exercise (Schmidt-Nielsen 1979; Eckert et al. 1988). Thus, homeothermy is a secondary result of the evolution of the aerobic capacity of the locomotor apparatus. In this way the evolution of homeothermy can be more readily understood: homeothermy is a "system characteristic" of highly developed vertebrates with specially evolved locomotor, cardiovascular and respiratory systems. Homeothermy is developed by a specific coupling of these functional systems together with those of thermoregulation and general regulation of metabolism. By means of this coupling this new quality "homeothermy" was able to emerge at all, delivering a new plateau for the further evolutionary diversification of the homeotherms.

The evolution of the aerobic capacity and its correlated rise in metabolic level, basal metabolism as well as exercise metabolism, depends on environmental preconditions. Especially the high basal metabolism, ten times as high as that of reptiles at the same body temperature, depends on the availability of food to maintain this metabolic level (Table 1). One astonishing feature in the evolution of birds and mammals is the fact that birds, with their characteristic constructional principle and obvious homeothermy, are known to have existed for at least for 120 million years, and mammalian fossil records are known for at least 190 million years (Starck 1978; Kemp 1982). But their radiation into the modern multiplicity of species and variability of form started at the end of the Cretaceous period, 70 million years ago. One possible explanation for this explosive development is the very similar time scale of evolution of angiosperms and the co-evolution of the modern orders of insects, and also the time scale of the evolution of teleosts.

The evolution of angiosperms and their diversification into the large variety of families and ecological forms started at the end of the Cretaceous period (Stanley 1986; Friis et al. 1987). Some basic characteristics of the angiosperms are (1) their high primary production rate, and as a consequence of this feature, (2) the often very short period of vegetation, so that the time from the seed and its growth to a fully development flowering plant and its production of new seeds is rapid, or even in perennials there is a very short growing and seed production period. By virtue of these features the angiosperms not only produced an immensely rich vegetation in humid tropical regions, but they were also capable of conquering the vast arid continental regions, which receive only very little rain during short periods of the year. Thus, the overall mass of primary production increased dramatically, delivering the necessary nutritional basis for the evolution of the metabolically demanding homeotherms. Most species feed only secondarily on plants, insects or fish being the primary food stables.

The modern orders of insects – the Coleoptera, Diptera, Lepidoptera, and Hymenoptera – developed in co-evolution with the angiosperms as pollinators or plant-feeding insects or insectivorous insects (Paulus 1978; Zwölfer 1978). Teleosts evolved in rivers, lakes, and the continental shelf area, subsisting mainly from the primary terrestrial production, which is washed to a certain extent into these water ecosystems. These primary and secondary productions were the basis for the evolution of homeotherms. Summarizing all these facts in a strict temporal setting (Table 1), it should be stated that the transition from the Cretaceous to the Tertiary period was characterized by an increase in the production and consumption of organic mass by a factor of ten. The modern biosphere, which is dominated by teleosts, modern insects, angiosperms and homeotherms, depends on this increase in production and consumption. Homeotherms with their specially evolved locomotor, cardiovascular and respiratory systems are an integral part of this biospheric evolution. The special design of their organ system not only depends on the different internal interrelationships, but also on the above-discussed external conditions.

In conclusion it should be emphasized that the achieved homeothermy enabled the homeotherms, with their newly acquired physiological potential, to radiate into different directions. They are able to invade cold areas in which only their constant body temperature allowed survival. They are able to migrate by their high locomotor capacity over large distances and utilize food resources that are only available seasonally. They are able to vary body size: especially toward very small organisms, which allowed specific strategies for utilizing strongly varying food production counteracted by high reproductive rates. In extreme cases, small organisms were even able to survive periods of extreme cold or nonavailability of food by means of hibernation or torpor (Schmidt-Nielsen 1979) (lowering their basal metabolism by regulated lowering of the body temperature). In general, the possibility for the evolution of homeothermic vertebrates is best illustrated by the development of brood care or viviparity in connection with highly developed social systems and an excessive cerebralization, which rendered birds and mammals capable of developing complex social systems and carrying out astonishingly accomplished behavioral achievements (Table 1).

The given example of the evolution of homeothermy possesses an excellent material basis to investigate two very different ways of achieving this functionally important goal by the parallel evolution of birds and mammals. The structure of the organ systems in birds and mammals is very different. This is especially true for the functional solutions that were evolved, especially with respect to the respiratory and locomotor systems, but is also reflected in the embryological development. An important step in the general analysis must be directed to the elaboration of the integration of these systems, which are quite differently interlocked with each other as demonstrated for (1) the structural integration of the respiratory system into the body cavity and with the locomotor apparatus, and for (2) the functional integration of respiration and locomotion, especially flight in birds and running in mammals. In contrast, the integration of the cardiovascular and respiratory systems seems to be identical in birds and mammals. However, we are just now starting to think about these questions of integration of organ systems. To date, investigations have delivered only preliminary observations. For the understanding of the constructional morphology of these organ systems it is not only necessary to analyze their structure and the functionally important parameters of structure, but also to analyze the systemic interrelationships with other organ systems of the investigated species, i.e., structural and functional integrations within the organisms, and also integration of the investigated functional systems with structures and functional processes in the environment of the organisms, to which special features of these organ systems are adapted or by which they are determined.

9 The Analysis of the Complexity of the Functional Structure of Organisms as the Central Goal of Constructional Morphology

Organisms are made up of functional systems, each of which serves a more or less specific function. These systems function as biochemical and biophysical entities, and fullfill manifold assignments including (1) constituting the basic chemical and physical properties of living organisms, (2) enabling metabolic exchange and growth and division of cells, (3) determining the mechanical behavior of the cells and tissues, (4) representing functional systems of organs for their regulation and interaction within the organism, (5) acting as functional systems of the locomotion and behavior of the organism in its environment, (6) facilitating as functional systems specific social structures and interactions, and (7) working as functional systems of the ecological interactions of the different groups of animals and plants in ecosystems. All these functional systems in organisms and in ecosystems are arranged in an hierarchical order, which represents the historical made of their formation during phylogenetic development of organisms and ecological systems.

The phylogeny of organisms and ecosystems depends on the evolution of these functional systems. Functional systems are differentiated during this evolution, often in such a way that the basic, functionally necessary system maintains and continues with its function, and alongside it evolves a second system with an altered function out of a duplication of the basic system. Most functional systems are already complicated in structure, thus they possess by virtue of their numerous elements several side effects in addition to their main functional goal. In the evolution of a functional system one or more of these side effects may become the main functional goal of the system. The vascular system primarily serves the metabolism of the perfused tissue, but also takes up and transports metabolic heat. Situated directly beneath the skin, the further differentiation of this part of the vascular system can be used for the elaboration of a heat exchanger for body temperature regulation (more examples are given in this chapter). One of the important events in phylogeny is the coupling of previously uncoupled functional systems to generate new functional systems with new functional qualities, as explained for (1) lung structure and body cavity septation in the evolution of heterogeneously subdivided lungs of reptiles and of the avian lung air sac system, or (2) for the integration of respiratory and locomotory movements in mammals and birds.

This integration of different functional systems into a new overriding system generates new functional possibilities for regulation or efficiency. Often qualitatively new functions generate a new level of functions in the organism, as demonstrated for the avian respiratory system or for homeothermy. Riedl (1975, 1977) coined the term systemization for this coupling or integration of previously uncoupled functional systems into new systems with the formation of a new functional level in the organism with qualitatively new functions. Via systemization the new functional system and level obtain the basic characteristics of a functional system with such components as heredity and its control by selection. In the course of phylogeny this evolution of new functional systems by the integration of previously uncoupled functional systems is responsible for the hierarchical order of functional systems within organisms (Riedl 1975, 1987). This hierarchical order of the functional systems and their multiple couplings and interconnections make up the complexity that is a basic characteristic of organisms and ecosystems.

After two centuries of scientific biology and many decades of molecular biology, it is now a central goal of morphology to analyze the complex structure of organisms. For the first time in the development of biology we are able to elaborate a coherent picture of organisms from the biochemical level of basic cellular functions to the level of the functional systems of organs and functional apparatuses, behavioral and social systems of animal communities and functional interrelationships within ecological systems. This coherent picture can also explain the multiple and different interactions by which these different functional systems are coupled with one another, often from different functional levels. Thereby such a coherent picture cannot only visualize the hierarchical structure of the functional systems, but also the complexity of their interactions and linkage. To understand why this specific group of organisms is constructed in a particular way and why it possesses this specific functional potentialities and limitations, we have to analyze the complexity of its historically developed hierarchical constructions.

The investigation of this complexity always has to start with the analysis of the functional systems that are integrated into the complex system. Regardless of which level of the functional system is to be analyzed, this analysis has to begin with a comparative elaboration of the structural and/or functional system (Riedl 1989). According to the level of the functional system to be analyzed, the investigated organism, organ system, cell but also behavior or ecosystem have to be compared with other organisms, organ systems, cells, behaviors, or ecosystems of related nature. Only by comparing species with (1) maximal differences in the functional specialization of the investigated functional system, (2) large differences in their ecological relationships, and (3) possible large differences in their body sizes, a basic knowledge of the functional system can be obtained. Only after the first often very vague concept of a new functional system or of some of its components and characteristics is it possible to determine and numerate some of the elements of the functional system. This first description has the inevitable requirement of investigating the interrelationships or interactions of these elements experimentally. It is often misunderstood by the experimental sciences that their planning of investigations depends on obtaining this initial knowledge of a hitherto unknown system, which can only be obtained by comparative investigation. This is the old, common scientific method, with which all biological investigations started.

These comparisons, as the basic morphological method, are necessary in order to elaborate the functional systems and their interrelationships, which make up organisms and ecosystems at all hierarchical levels. To prevent a misunderstanding: to detect homologies in the classical morphological meaning is only one way among a multiplicity of ways which have the goal of elaborating the knowledge of new functional systems, from biochemical to behavioral and ecological systems. Therefore comparisons are made not only with other organisms, organs, or cells, but also with known physiological, physical, chemical, and technical processes and systems, with models and machines, even with ideas and phenomena from quite different disciplines and fields of our daily experience. These comparisons are often made initially by thinking in analogies. The identification of the investigated functional system then proceeds in the manner of a hermeneutical spiral: with the results of the first comparison, more precise observations or measurements are possible, leading to a better understanding and to a more detailed comparison of the systems, followed by better observations and measurements, which result in a more precise formulation of the system and its elements, which are the basis for the planning of the first experiments to investigate the interactions of single elements. From this whole process causal interrelationships can be described, and with this better understanding and description of the functional system further experiments can be planned.

This is the basis of all experimental investigations, on which the fascinating results of all the modern experimental biological disciplines depend. But beyond this basis of all biological "sciences", the comparative approach is the only one which leads to the further elaboration of the functional systems and their integrations within the organism, and which allows the investigation of the hierarchical order of these functional systems. As a result of this elaboration, the multiple integration of the different functional systems and levels can also be investigated experimentally. These are made up of a broad multiplicity of mechanisms from molecular and physical to mechanical processes, from cellular and organic functions to neuronal or hormonal regulations, or from locomotor and behavioral events to ecological interactions.

However, a general problem arises during the attempt to obtain a better morphological insight into the structure of the hierarchical functional levels of organisms and into the integrations of their functional system of these different levels. Not only the special methods of observations, comparisons, and experiments are different, according to the characteristics of the level they are developed for, but also the terminology to describe the results and the general features of the investigated functional systems is highly different, each level having its own terminology. For this morphological approach of elaborating a coherent picture of organisms from the molecular to the ecological level, it will be necessary to find a way to overcome these terminological barriers. Therefore, a general rule for the description of the results of morphological investigations should be established: the results should be formulated in a very generalized way, expressing these findings, on the one hand, as made up by the elements of the underlying systems and their rules, laws, regulations, and interactions, and, on the other hand, explaining the importance of these findings for the following higher level, searching for those elements and systems that determine the functional characteristics of the higher level. The given example of the necessary prerequisites of the respiratory, vascular, and locomotor systems for the development of homeothermy attempts to demonstrate this methodological necessity as well as to explain the importance of homeothermy for the further evolution of birds and mammals, for their radiation into the multiplicity of forms and geographical and ecological types.

The development of a comprehensive picture of organisms will be the special task of morphologists in the general field of biological sciences. In this context

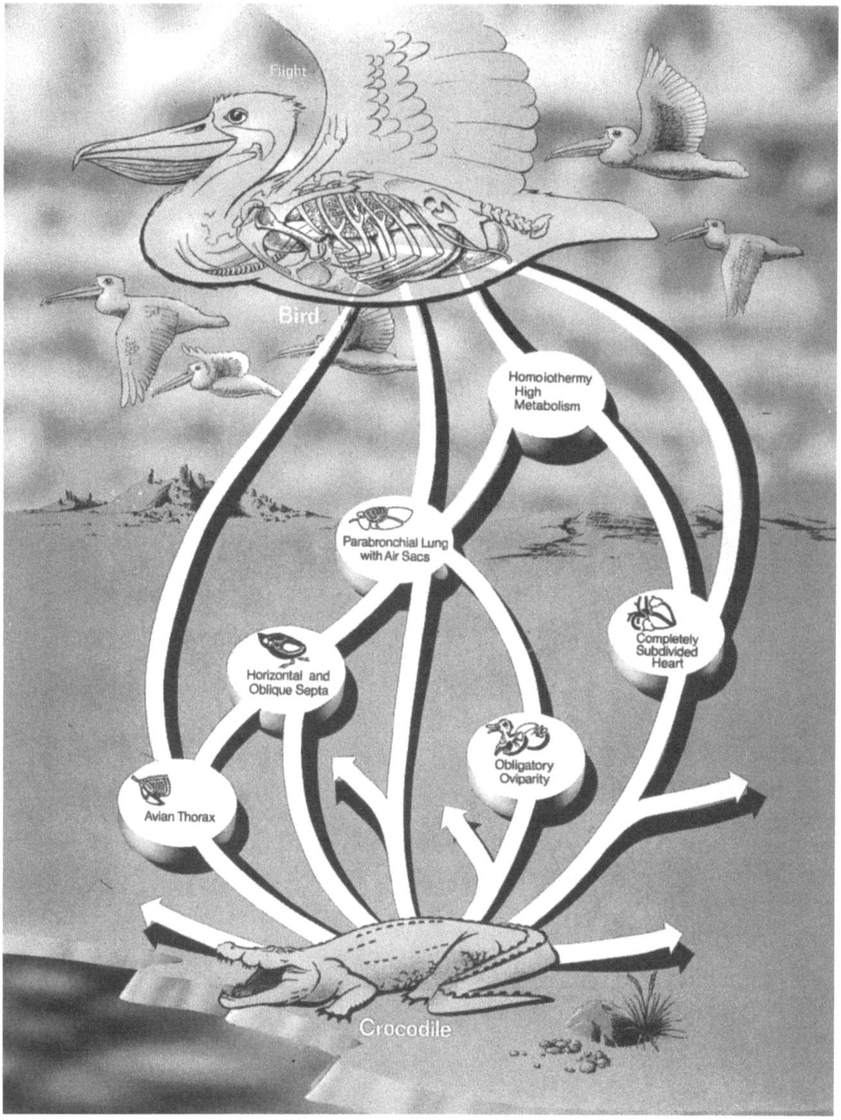

Fig. 2. Main functional components of the avian organism responsible for the evolution of homeothermy

they have to give such a general framework of the organism into which the results of the other disciplines can be incorporated. Here lies the special importance of morphological investigations and descriptions, to make the network of levels and system interactions mentally visible. For this purpose it will be very helpful to develop a special type of illustration to visualize the complex interrelationships of structural elements and functional systems which build up the different function-

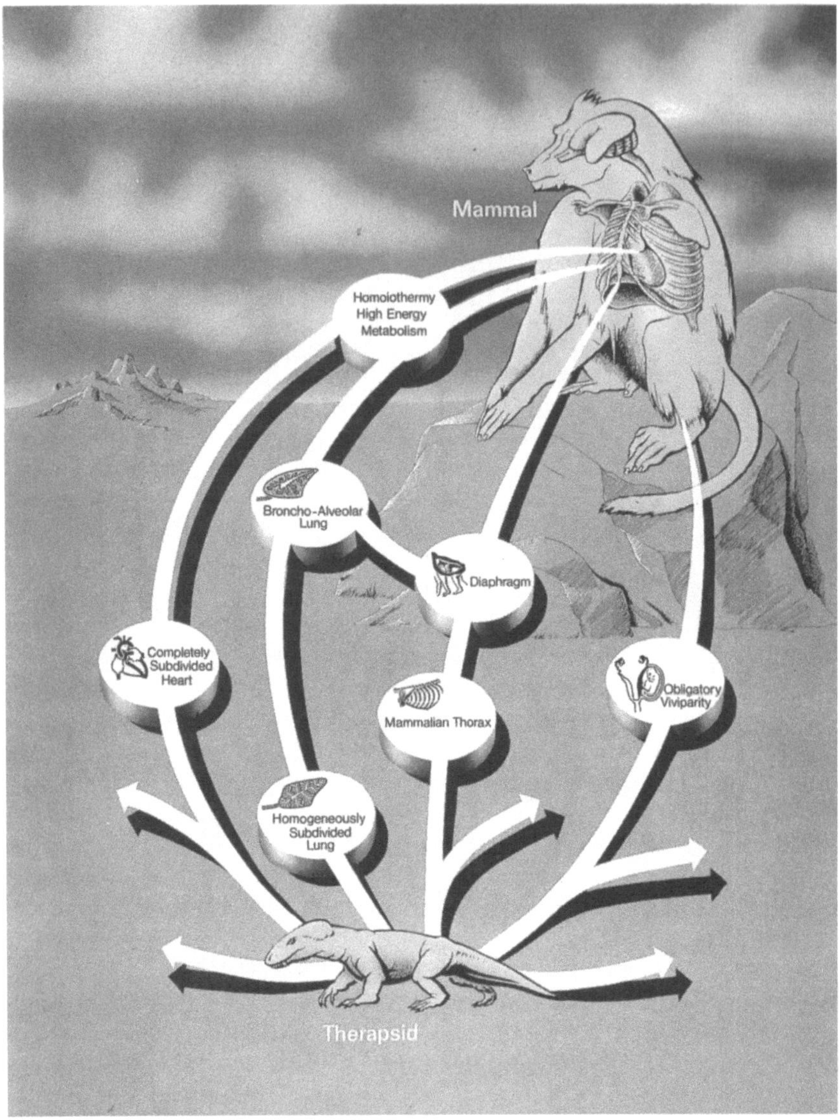

Fig. 3. Main functional components of the mammalia organism responsible for the evolution of homeothermy

al levels and their multiple interactions (Duncker 1989b). Two figures in this chapter are designed for this goal (Figs. 2, 3); they are part of a series of figures to illustrate the history of the respiratory apparatus in vertebrates from fish to man. Such illustrations utilize the ability of our visual system to perceive complex pictorial arrangements simultaneously and momentarily by pattern recognition. In contrast, the best description, on being forced to use sequential language, can only describe a single line of a complex pattern at one time, hoping that the sequence of different lines of this pattern are retained in the memory of the reader. One great advantage of the pictorial presentation of complex patterns is the momentary visual recognition of a multiplicity of single elements and interrelationships, which is so seldom used in biological sciences, even in morphology.

Morphological investigations start with the comparative elaboration of a functional system; they continue after the initial formulation of a new system and some of its elements to make further observations, measurements, or experiments, mainly to analyze the interactions of elements or interrelationships of different functional systems. In the next step the integration of the investigated functional system into the functional levels of the organism has to be investigated as well as the position and interrelationship of this functional level in the hierarchical system of the organism, its integration into the whole entity of the organism. Thus, the morphological investigation, starting from certain phenomena of the whole organism, after specific detailed analysis of specific structures and functions, returns to the totality of the entire organism or ecosystem. For morphology it is more than a fruitful heuristic principle, to return to the whole entity of organisms after analyzing some specific parts or functions. It is the central goal of morphology to obtain a better understanding of the organismic entity in its hierarchical structure from a detailed analysis of specific structures and functions in the organism and their interactions and importance for the entire organism. If we want to acquire not only more and more detailed information on different parts of organisms, this is the only way scientifically to overcome such common metaphysical thinking concerning organisms, as is so characteristically expressed by the fundamentalistic creationism.

References

Barnas GM, Rautenberg W (1987) Temperature control. In: Seller TJ (ed) Bird respiration, vol 1. CRC, Boca Raton, FL, pp 131–153

Bennett AF (1985) Chapter 10: Energetics and locomotion. In: Hildebrand M, Bramble DM, Liem KF, Wake DB (eds) Functional vertebrate morphology. Harvard Univ Press, Cambridge, MA, pp 173–184

Bennett AF (1989) Integrated studies of locomotor performance. In: Wake DB, Roth G (eds) Complex organismal functions: integration and evolution in vertebrates. Dahlem Worksh Rep. John Wiley & Sons, New York Chichester, pp 191–202

Berger M, Hart JS (1974) Physiology and energetics of flight. In: Farner DS, King JR (eds) Avian biology, vol 4. Academic Press, New York London, pp 415–477

Bernstein MH (1987) Respiration in flying birds. In: Seller IJ (ed) Bird respiration, vol 2. CRC, Boca Raton, FL, pp 43–73

Bramble DM (1989) Axial-appendicular dynamics and the integration of breathing and gait in mammals. Am Zool 29:171–186

Bramble DM, Jenkins FA Jr (1989) Structural and functional integration across the reptile-mammal boundary: the locomotor system. In: Wake DB, Roth G (eds) Complex organismal functions: integration and evolution in vertebrates. Dahlem Worksh Rep. John Wiley & Sons, New York Chichester, pp 133–146

Burri PH (1988) Development and regeneration of the lung. In: Fishman AP (ed) Pulmonary diseases and disorders, vol 1. McGraw-Hill, New York, pp 61–78

Carrier DR (1987a) Lung ventilation during walking and running in four species of lizards. Exp Biol 47:33–42

Carrier DR (1987b) The evolution of locomotor stamina in tetrapods: circumventing a mechanical constraint. Paleobiology 13:326–341
Duncker H-R (1971) The lung air sac system of birds. Adv Anat Embryol Cell Biol 45, 6:1–171
Duncker H-R (1978a) Funktionsmorphologie des Atemapparates und Coelomgliederung bei Reptilien, Vögeln und Säugern. Verh Dt Zool Ges 1978:99–132
Duncker H-R (1978b) Coelom-Gliederung der Wirbeltiere – Funktionelle Aspekte. Verh Anat Ges 72:91–112
Duncker H-R (1978c) Development of the avian respiratory and circulatory systems. In: Piiper J (ed) Respiratory function in birds, adult and embryonic. Proc Life Sci. Springer, Berlin Heidelberg New York, pp 260–273
Duncker H-R (1978d) General morphological principles of amniotic lungs. In: Piiper J (ed) Respiratory function in birds, adult and embryonic. Proc Life Sci. Springer, Berlin Heidelberg New York, pp 2–15
Duncker H-R (1979) Coelomic cavities. In: King AS, McLelland J (eds) Form and function in birds, vol 1. Academic Press, New York London, pp 39–67
Duncker H-R (1981) Stammesgeschichte der Struktur- und Funktionsprinzipien der Wirbeltierlungen. Verh Anat Ges 75:279–303
Duncker H-R (1983) Funktionelle Anatomie des Lungen-Luftsack-Systems. In: Mehner A, Hartfiel W (eds) Handbuch der Geflügelphysiologie, pt 1. Fischer, Jena, pp 436–485
Duncker H-R (1989a) Structural and functional integration across the reptile-bird transition: locomotor and respiratory systems. In: Wake DB, Roth G (eds) Complex organismal functions: integration and evolution in vertebrates. Dahlem Worksh Rep. John Wiley & Sons, New York Chichester, pp 147–169
Duncker H-R (1989b) The significance of morphology in the biological sciences: the necessity for pictorial illustrations of the complexity of morphological structures. In: Splechtna H, Hilgers H (eds) Trends in vertebrate morphology. Fortsch Zool 35:619–630
Duncker H-R (in press) 21 Respirationstrakt. In: Hinrichsen KV, Humanembryologie. Springer, Berlin Heidelberg New York
Duncker H-R, Güntert M (1985) The quantitative design of the avian respiratory system – from hummingbird to mute swan. In: Nachtigall W (ed) BIONA Rep 3. Bird flight. Fischer, Stuttgart, pp 361–378
Eckert R, Randall D, Augustine G (1988) Animal physiology. Mechanisms and adaptations, 3rd edn. Freeman, San Francisco New York
Friis EM, Chaloner WG, Crane PR (1987) The origins of angiosperms and their biological consequences. Univ Press, Cambridge
Gehr P, Backofen M, Weibel ER (1978) The normal human lung: Ultrastructure and morphometric estimation of diffusion capacity. Respirat Physiol 32:112–140
Gehr P, Mwangi DW, Ammann A, Maloiy GMO, Taylor CR, Weibel ER (1981) Design of the mammalian respiratory system. V. Scaling morphometric pulmonary diffusing capacity to body mass: Wild and domestic mammals. Respirat Physiol 44:61–86
Kemp TS (1982) Mammal-like reptiles and the origin of mammals. Academic Press, New York London
Klocke RA (1988) Ventilation, pulmonary blood flow, and gas exchange. In: Fishman AP (ed) Pulmonary diseases and disorders, vol 1. McGraw-Hill, New York, pp 185–198
Maina JN (1989) The morphometry of the avian lung. In: King AS, McLelland J (eds) Form and function in birds, vol 4. Academic Press, London New York, pp 307–368
Nakakuki S (1979) Comparative anatomical studies on the mammalian lung. Bull Fac Agric Tokyo Univ 21:1–74
Paulus HF (1978) Co-Evolution zwischen Blüten und ihren tierischen Bestäubern. Sonderbd Naturwiss Verl Hamburg 2:51–81
Perry SF (1983) Reptilian lungs. Functional anatomy and evolution. Adv Anat Embryol Cell Biol 79:1–81
Perry SF (1989) Morphometry of crocodilian lungs. In: Splechtna H, Hilgers H (eds) Trends in vertebrate morphology. Fortschr Zool 35:546–549
Perry SF, Duncker H-R (1978) Lung architecture, volume and static mechanics in five species of lizards. Respirat Physiol 34:61–81
Perry SF, Duncker H-R (1980) Interrelationship of static mechanical factors and anatomical structure in lung evolution. J Comp Physiol 138:321–334

Riedl R (1975) Die Ordnung des Lebendigen. Systembedingungen der Evolution. Parey, Hamburg
Riedl R (1977) A systems-analytical approach to macroevolutionary phenomena. Q Rev Biol 52:351–370
Riedl R (1987) Begriff und Welt. Biologische Grundlagen des Erkennens und Begreifens. Parey, Hamburg
Riedl R (1989) Opening address and introduction: the role of comparative anatomy in biology today. In: Splechtna H, Hilgers H (eds) Trends in vertebrate morphology. Fortschr Zool 35:VII–XVI
Scheid P (1982) Respiration and control of breathing. In: Farner DS, King JR, Parkes KC (eds) Avian biology, vol 6. Academic Press, New York London, pp 405–453
Scheid P, Piiper J (1971) Direct measurement of the pathway of respired gas in duck lungs. Respirat Physiol 11:308–314
Scheid P, Piiper J (1989) Respiratory mechanics and air flow in birds. In: King AS, McLelland J (eds) Form and function in birds, vol 4. Academic Press, London New York, pp 369–391
Schmidt-Nielsen K (1979) Animal physiology, 2nd edn. Cambridge Univ Press, London
Schmidt-Nielsen K (1984) Scaling. Why is animal size so important? Cambridge Univ Press, Cambridge, MA
Stanley SM (1986) Earth and life through time. Freeman, San Francisco New York
Starck D (1978) Vergleichende Anatomie der Wirbeltiere auf evolutionsbiologischer Grundlage, vol 1. Springer, Berlin Heidelberg New York
Tenney SM, Remmers IE (1963) Comparative quantitative morphology of the mammalian lung: diffusion area. Nature (London) 197:54–56
Valerius K-P (1983) Die postnatale Entwicklung des Bronchialbaumes der Ratte. Diplomarb Math Naturwiss Fak Univ Göttingen, pp 1–52
Weibel ER (1984) The pathway for oxygen. Structure and function of the mammalian respiratory system. Harvard Univ Press, Cambridge, MA
Weibel ER, Taylor CR (1981) Design of mammalian respiratory system. Respirat Physiol 44:1–164
Weir EK, Reeves JT (1989) Pulmonary vascular physiology and pathophysiology. In: Lenfant C (ed) Lung biology in health and disease, vol 38. Dekker, New York
White FN (1976) Circulation. In: Gans C, Dawson WR (eds) Biology of the reptilia, vol 5. Academic Press, London New York, pp 273–334
White FN (1978) Comparative aspects of vertebrate cardiorespiratory physiology. Annu Rev Physiol 40:471–499
Wieser W (1986) Bioenergetik. Energietransformation bei Organismen. Thieme, Stuttgart
Zwölfer H (1978) Mechanismen und Ergebnisse der Co-Evolution von phytophagen und entomophagen Insekten und höheren Pflanzen. Sonderbd Naturwiss Verl Hamburg 2:7–50

The Mechanical Constraints in Mollusc Constructions – the Function of the Shell, the Musculature, and the Connective Tissue

KARL EDLINGER[1]

Summary

The basic mollusc types and their development are presented as case studies for the reconstruction of phylogeny according to the principles of constructional morphology. It is shown that the sequence of transformation steps leading to solengastres constructions and several conchiferous types and the bifurcation in diverging lineages of the conchiferans are clearly constrained and determined by biomechanical principles and the energy converting capabilities of primitive molluscs.

1 Introduction

The application of the strict principles of biomechanics could have a strong impact on the understanding of morphology and the reconstruction of phylogeny. However, biomechanics in the sense of mechanical coherence and the hydraulic principle would have to be based on a theoretical concept of constructional morphology and would be required to be consistent with the law of conservation of energy.

Animals must be conceived as energy-converting systems (Fig. 1). They function by a step-by-step transformation and transduction of chemical energy, received from their input of matter, into a mechanical form of energy. The mechanical energy is transferred into the environment by motoric deformations of the organismic machine. The energy transduction process of living systems needs useful and highly ordered organismic, molecular, cytological, and histological structures and a morphological machinery which can work without difficulties for the efficiency of the organismic whole. A minimum of the energy is allowed to be set free by unrestrained dissipation, the restraints being provided by the morphological organization. In the mechanical framework of the organisms the contractile elements, which are working in an antagonistic way against a fluid-filling or skeletal structure, force energy into well-organized channels and over directing structures finally into the environment.

[1] Baseler Gasse 39, A-1232 Wien, Austria

N. Schmidt-Kittler and K. Vogel (Eds):
Constructional Morphology and Evolution

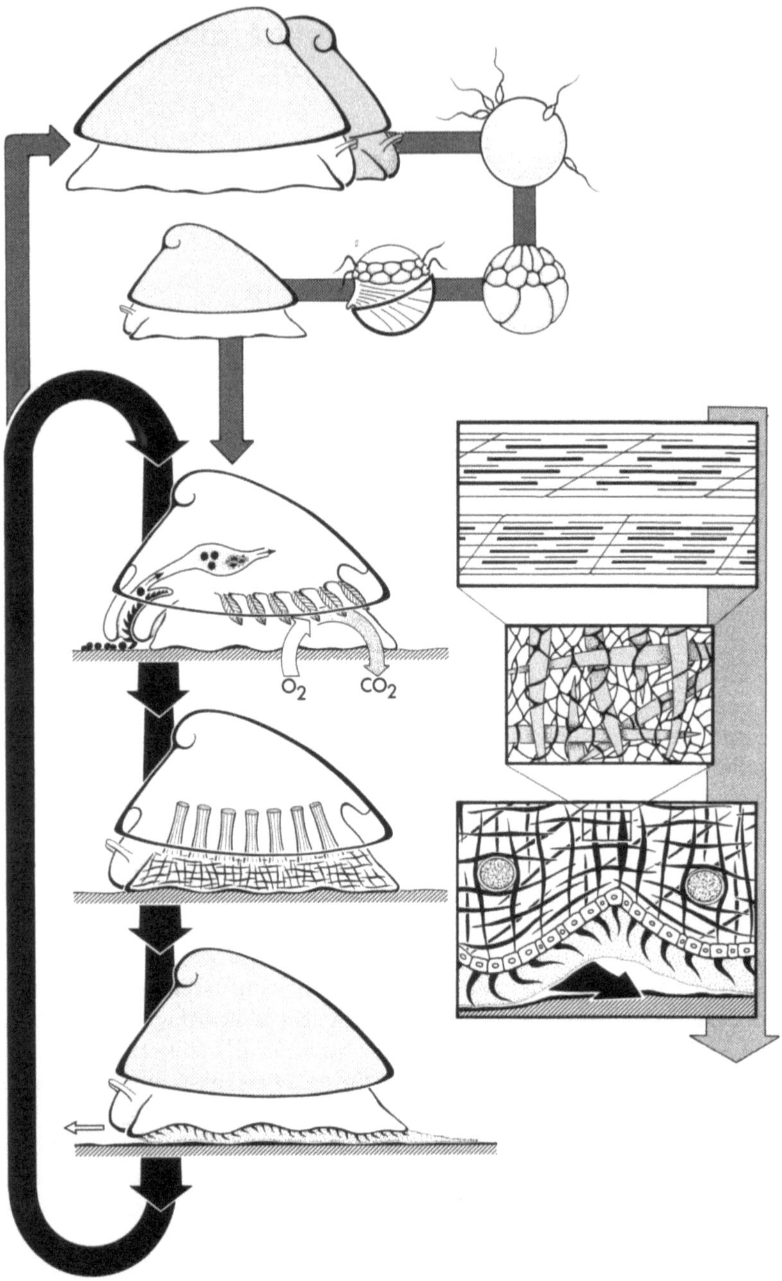
O2
CO2

2 The Basic Constructional Properties of Molluscs

Fluid-filled spaces of the organisms and the organismic machines as wholes must develop a tendency to assume a globular shape. Any other shape and specific form required for biological functions must be enforced by tethering fibers which are anchored in the enclosing membrane and form three-dimensional dense arrangements of tethering structures which underlie the surrounding membranes. This is very obvious in mollusc constructions in which the foot is braced by a three-dimensional muscle grid while the head portion is supported by a more tubelike system of muscles and connective structures.

Motoric deformations are actuated by contractions of fibers, especially by gliding of actin and myosin in the muscles. In order to become effective, muscles need to fulfill two constructional requirements. Muscle activity can cause motion only if force-transmitting structures are present which convey mechanical energy to other components of the living machine. Furthermore, muscle fibers can only shorten but are unable to regain their initial length. Restitution of the initial condition is only possible if the energy-transforming and contractile structures are integral elements in mechanically coherent systems in which contractile elements can effectively interact in such a way that an antagonistic relationship over the fluid filling is assured. For this reason all contractile and energy-canalizing structures require anchoring connective tissue within the context of the coherent apparatus.

All organisms are hydraulic systems, as the fluid or viscous filling with the chemical mechanisms of life is contained in closed but flexible membranes. According to the principle of hydraulics, all morphological structures on the macroscopic, the microscopic and the submicroscopic levels can be understood only as the expression of mechanical constraints, brought about by special arrangements of form-controlling and shape-enforcing structures within the context of mechanically coherent and permanently active hydraulic systems. Thus form, body architecture, motility, and behavioral activity are generated by energy conversion.

3 Ontogeny

The constructional principles just sketched are also valid for the ontogenetic processes and their stages. The alterations of ontogenetic forms are caused by expan-

---

Fig. 1. The bionomic circle of the organism-centered theory of evolution demonstrated by a conchiferan construction. The sexual interaction of two individuals generates the offspring. The process of ontogeny is energy-driven and consists in mechanical work of constructional units. The fully developed constructions work as machines and acquire matter and energy in the form of food. Primitive molluscs scrape living matter from a hard substrate. The ingested material is shifted in the construction by mechanical work and subsequently utilized by the metabolism; motoric activity is also responsible for the intake of oxygen. The energy obtained from the metabolism is transformed into mechanical energy which serves for propulsion. The motoric activity is utilized for renewed feeding activity, sexual behavior. The *inset on the right-hand side* indicates that energy conversion occurs on the macromolecular level by fiber sliding in acto-myosin structures; the energy is transmitted over histological and morphological structures and utilized in the mechanical work of the morphological apparatus

sive activity which is counteracted and restrained by the arrangement of form-enforcing structures. Form itself is not decisive, it is the arrangement of form-enforcing elements and structures in connection with the filling of the system that is relevant in constructional explanations. In the present chapter ontogenetic mechanisms are not further treated because we have to follow the course of fully developed forms and present examples for the reconstruction of gradual alterations of the construction types of some major mollusc taxa.

4 Evolution as a Process

Evolution has to be understood as the transformation process of such coherent energy-converting and working systems. The viability of the sequence of changes can only be reconstructed by the application of the mechanical principles just outlined. These are lawlike mechanisms and principles and allow the reliable reconstruction of possible phylogenetic pathways.

The understanding of the organisms as energy-converting and mechanically working constructions narrows down the potential transformation paths and does not allow free and unrestrained fantasy to gain control. It should, however, be emphazised that the explanations will only become comprehensible if form and body architecture are understood as manifestations of the above-mentioned constructional principles.

Because of the complexity of living constructions, there can be no doubt that evolutionary alterations can only occur gradually in a step-by-step manner. Selection favors those constructions which function in an increasingly economic way. The restraining mechanism of the transformation processes have to be elaborated and traced in models. In these it must be shown how antecedent constructions provided the determining factors for the subsequent modifications. In such a way phylogenetic reconstructions have to present sequences of functionally feasible stages.

Although the reconstructions show some narrative aspects they have to be based on abstract principles and lawlike mechanisms, on nomological-deductive explanations in the sense of Bock (this Vol.).

5 The Basic Mollusc Construction

In order to explain the establishment of the diverging mollusc taxa, we will start with a construction consisting of an elongated muscular locomotive organ, the head-foot construction, which is ventrally situated. On the dorsal side lies the visceral hump, which is usually covered by the calcified shell. As the rim of the shell extends laterally over the head-foot, it is able to mechanically stabilize the underlying mantle cavity with the gills. The "locomotive" parts of the head-foot consist of a three-dimensional muscular grid with fluid-filled interstitial spaces. Under the epidermis, the dense network of connecting tissue structures and muscles are effective in enforcing the form of the organ and in generating the locomotory movements.

As far as the histological structure is concerned, the musculature of the head-foot is of the obliquely striated or helicoid type. This pattern allows extreme contractions which are generated not only by gliding of actin and myosin but also by an alteration of the angle between the longitudinal axis and the z-zones of the muscle fiber. So the head-foot can be deformed in various ways and assume a great number of different shapes. This enables the creeping mollusc construction to actively adapt its sole to the substrate by deformation of the whole soft body construction in such a way that adhesion is continually ensured. In most cases the foot moves by peristaltic contraction waves which travel over the sole. However, other modes of muscle activity are also possible.

A considerable portion of the muscle grid of the foot is recruited from the dorsoventral musculature which inserts on the calcified shell and radiates into the muscle grid of the foot. The shell is a skeleton and functions as a rigid frame for the musculature. Because of its rigidity the shell can be understood as a special kind of connective tissue which forms an integral stabilizing element of the whole. It suppresses movements and deformations which might consume energy without contributing to the demands of the moving machine. The shell works as a deformation-suppressing system which influences the efficiency of the head-foot system in a positive way.

The shell can consist of seriated plates, as in chitons, or may form one single whole, as in conchifers. However, the loss of the shell by secondary reduction in some molluscan taxa can cause radical changes in the constructions and result in the development of secondary noncoelomic worms.

6 The Phylogenetic Problem

In the following it will be shown that the biomechanical and hydraulic principles of the mollusc Bauplan provide a reliable basis for the reconstruction of the transformation pathways leading to the different mollusc groups.

The mollusc model is based on the reconstruction deriving the molluscs from metameric predecessors (Gutmann 1974). It was shown that the formation of the foot by flattening of the ventral side required the dense bracing of the construction by anchoring of the dorsoventral muscles at developing dorsal plates (Fig. 2). Adhesive creeping allowed the formation of the radula and opened up the pathway into mollusc organization (Gutmann 1974). This constructional model has some affinity with unexplained hypotheses which derive the molluscs from annelids (Remane 1967; Götting 1980a, b), but stands at variance with attempts to conceive of a plathyhelminth ancestry (Vagvolgyi 1967; van Salvini-Plawen 1968, 1972). The basic model for the molluscs can be corroborated if a successful continuation into all other mollusc groups can be given.

7 The Archimollusc

The transition into the conchiferan construction can be explained if we start from an elongate chiton-like ancestor with high cambered shell plates and a massive

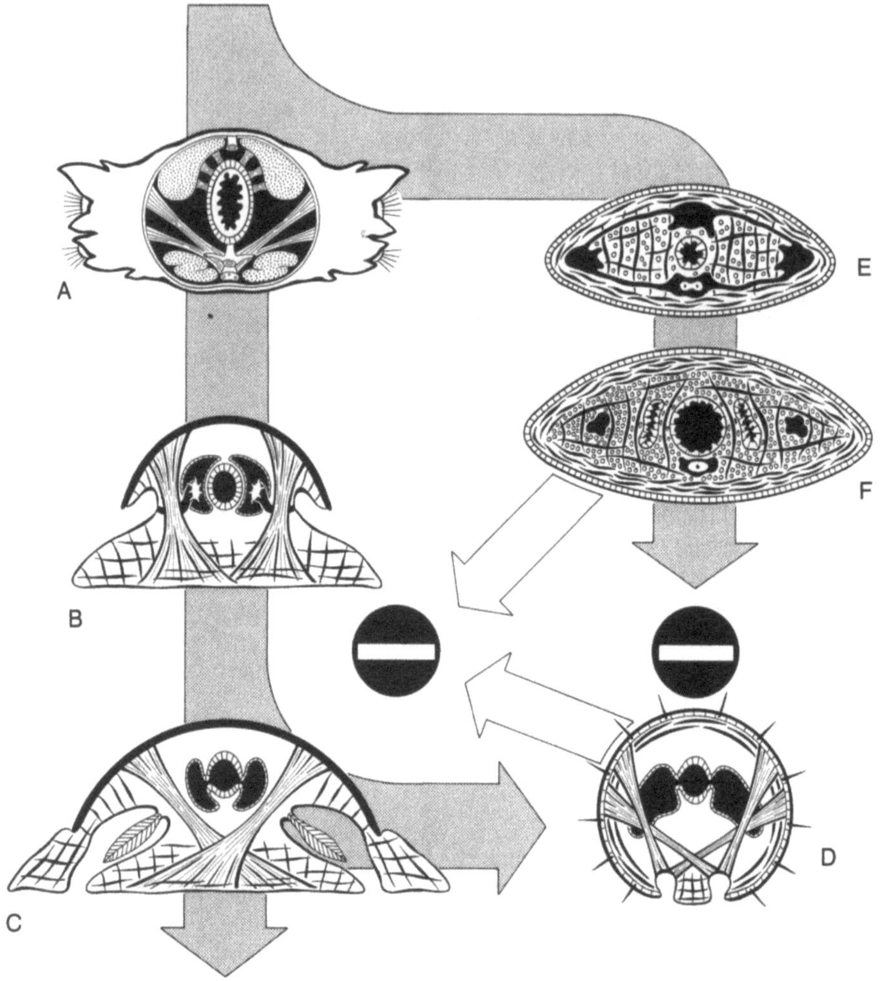

Fig. 2. Derivation of molluscs from annelid-like constructions, depicted in a sequence of cross-sections. **A** Annelid with tethering muscles determining the shape of the cross-section and allowing flattening of the ventral side for creeping on the substrate. **B** Formation of the muscle grid of the foot, allowing adhesive creeping. The shell structures are developed on the more or less immobile dorsal side. **C** Polyplacophoran construction with lateral girdle sealing the mantle cavity with the gills. **D** Solenogastric construction (ventroplicate) as a form derived from polyplacophoran ancestors. These organisms are secondary worms with specific differences from coelomatic constructions. **E, F** Derivation of plathyhelminths from coelomate constructions. Molluscs cannot be derived from plathyhelminths and from solenogastres as often suggested. The impossible morphoclines cannot be explained in a step-by-step form

head-foot for an adhesive way of creeping. The sole of the foot contained a layer of gland cells for the production of mucus which aided the peristaltic creeping movements. The radula was contained in a sac-like extension of the foregut. The conveyer belt of the radula ribbon with the rows of teeth was underlain by cushions of cartilage on which the scraping apparatus could be moved by the antagonistic activity of muscles. The function of the radula depended on the adhesive creeping of the foot and the independent movement of the head, which was able to swing sideways.

The visceral hump on the dorsal side enclosed the digestive apparatus, the kidneys, the gonads, and the pericardium with the heart. The digestive apparatus and the gonads were voluminous and turgescent organs and, as such, parts of the hydraulic apparatus in which they served as stabilizing structures and as a filling of the muscular apparatus which formed the frame of the whole construction.

The kidneys, in the form of metanephridia, were arranged in a metameric way. Each of the nephridia had a nephroduct which terminated in the mantle cavity. The gonads lay in vestigial cavities of the coelom and had connections with the pericardium. The pericardium surrounded the heart. Within the pericardium the heart could work autonomously. It had to transport the hemolymph into an aorta leading to the front and into the head-foot. The circular system was not closed. In the mantle cavity (mantle groove) the number of the seriated gills must have primarily corresponded to the arrangement of interspaces between the dorsoventral muscles. This conclusion is compeling, as the segmentation of the nephridia and the gills was enforced by the muscles, which left only narrow spaces for the passage of the ducts and the vessels for the hemolymph.

Under each of the skeletal plates the dorsoventral muscles were organized in pairs and were composed of a dorsoventral portion and a portion which was running obliquely to the opposite part of the foot. At the sides of the foot, extending to the rear end, was the mantle cavity with its great number of gills which were ventilated by the driving power of cilia.

8 The Radiation of the Mollusc Constructions

As soon as we understand the extreme constructional interdigitation of organs and the predominance of the mechanical elements of the construction, there can be no doubt that all phylogenetic pathways which lead away from the basic molluscs into the diverging constructions can only have passed through very narrow phylogenetic passage ways into transformed constructions. All of these have to be reconstructed according to biomechanical principles. None of the transformation paths leaves much freedom for arbitrary or fanciful considerations.

9 The Chitons

One of the basic evolutionary pathways resulted in flattened polyplacophoran constructions. Step by step the foot became transformed into an increasingly broadened adhesively active muscle grid with a flat sole. These changes allowed the organisms to live on rocky substrate. Beginning in the larval stages, the lateral parts of the skeletal plates extended far over the mantle cavity and formed a girdle of thickened muscle and connective tissue structures which gained contact with the ground.

On this basis, the mantle cavity could be closed by adhesion of the girdle on the substrate. This change permitted the animals to penetrate the tidal flats in which they were temporarily exposed to the air.

In the narrow mantle cavity of chitons the number of gills was growing, demonstrating their secondary nonrelatedness with the old metamerism of the body trunk with its muscles and skeletal plates. This independence was achieved when the lateral girdle developed and was supplied by independent blood vessels outside the series of adductor muscles. Thus the gills were decoupled from the ancestral blood vessels within the body trunk. The new vessels developed in conjunction with the girdle.

Independently of the conchiferan molluscs, most of the paired metanephridia were reduced in the polyplacophorans. This change was possible because one of the posterior pairs of nephridia formed glandlike excretory organs which extended through the whole anterior part of the construction, thereby replacing most of the segmented nephridia.

It is obvious now that the polyplacophorans are not in every respect representatives of the pre-conchiferan stage of mollusc evolution but were subjected to secondary modification. From the primitive molluscan construction only the segmented organization of the musculature and the shell plates was retained. In respect to gills, nephridia, and some aspects of the circulatory system, the polyplacophoran constructions are not representative for the predecessors of conchiferans.

10 The Solenogastres

Even the primitive molluscs with segmented shells could develop on diverging lineages. It is interesting to note that even the skeletal structure could become subject to reduction (Edlinger 1989b). Two lines diverged on the polyplacophoran level, one leading to the Ventroplicata, the other to the Caudofoveata (Fig. 3). Both lineages lost the shell rows of the chiton-like predecessors and became what might be called secondary worms.

The first group is characterized by a total reduction of the gills. In conjunction with the loss of the row of shells, the body assumed a circular cross-section. On the ventral side a groove was held retracted which contained a padlike foot rudiment. The structures holding the groove and the foot and preventing their bulging out were the just mentioned dorsoventral muscles, a heritage from the polycaphoran stage with the series of shells. The observable situation is what has to be expected if one considers the reduction of the skeletal plates. It goes without saying that a foot and the retracting muscles could never have formed directly in an ancestral coelomate worm.

The constructional changes and the high deformability of the body caused considerable alterations in the digestive apparatus. The massive midgut gland was reduced and the tube-shaped gut formed diverticles between the dorsoventral muscle bundles. The lateral pouches of the gut became essential component of the fluid filling of the hydraulic constructions. It was apparently impossible to achieve the de novo formation of a coelom as a hydraulic filling system. The high

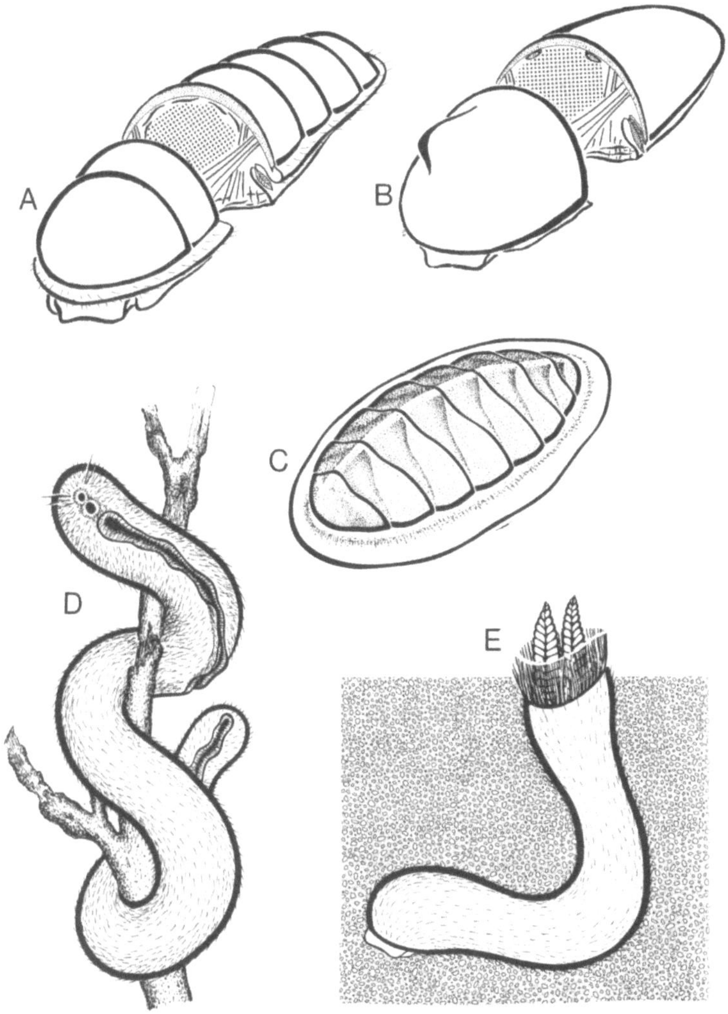

Fig. 3. Derivation of solenogastres from polyplacophoran-like forerunners. **A** Primitive mollusc with segmented shells and flexible body. **B** Preconchirefan construction with fused shells and segmented muscles. Basic construction from which all conchiferan groups must be derived (see Fig. 1). **C** Polyplacophoran construction which developed a girdle that seals of the mantle groove with the gills when the organisms penetrate the rocky beach. **D** Ventroplicate construction which, after the loss of the segmented shells, acquired the ability to climb on hydroid colonies. **E** Caudofeveate construction which in the course of shell reduction became capable of burrowing in the mud, sticking out the remaining caudal pair of gills

deformability and flexibility of the body allowed these organisms to climb on hydroid colonies and feed on the polyps.

The second group of chiton descendents, the Caudofoveata, also became wormlike and flexible constructions. The body became differentiated into an anteriorly situated active motoric portion and a more inactive rear part which was drawn along behind the front part of the organism when the organisms penetrated the sediment.

In a way similar to that in Ventroplicata, the midgut was widened to form diverticles which protruded into the space between the dorsoventral muscles which were retained as internal tethering structures controling the foot in the front part of the body. The reduction of the posterior dorsoventral muscles allowed the formation of a large expanded portion of the gut which could function as a stomach. A big glandy midgut sac filled the posterior part of the body and enclosed the pericardium and the tube-shaped hindgut, thereby forming the filling of the mechanical apparatus. Only the anterior part of the muscular foot persisted and functioned as a burrowing apparatus. As burrowing and tubicolous organisms, the Caudofoveata could retain only one pair of posterior gills, which are held exposed at the entrance of the burrow. In contrast to the Ventroplicata, the Caudofoveata evolved as organisms capable of burrowing in soft substrate.

It is obvious from a constructional understanding that the climbing and burrowing capabilities could only develop in the course of a reduction of the shell plates in polyplacophoran-like predecessor constructions. The de novo formation of the shell plates and serially arranged muscles was not possible in a flexible worm. The reduction of the typical skeletal plates was irreversible and accompanied by the loss of the typical creeping capabilities of primitive molluscs.

Suggestions that the solenogastres might have been the primitive molluscs (von Salvini-Plawen 1972, 1981) were never based on any kind of explanation. They have to be considered highly improbable, as the conditions for the formation of skeletal plates and corresponding muscles, extending into a ventral foot, cannot be based on a continuous biomechanical explanation.

11 The Transition to the Conchiferan Level

Most of the recent molluscs belong to the conchiferans, which are characterized by a unified and undivided shell which must have be formed by the fusion of the series of plates we have described in their predecessors. The shell scars of monoplacophorans are unequivocal evidence of the segmented organization of the ancestors.

The ancestors of the primitive conchiferans were certainly rather high cambered polyplacophoran-like constructions. However, neither the extant polyplacophorans nor the solenogastres can be considered representatives of the constructional situation preceding the establishment of the unified conchiferan shell. The above models explained the specific divergences in these groups. The conchiferan predecessors must have possessed and retained the nephridia which were integrated into the segmented pattern of the retractor muscles. The persistence of nephridia must have strictly precluded the fusion of the retractor muscles and enforced the metamerism of muscles and nephridia.

The essential change which occurred in the transition field between the polyplacophoran stage of mollusc evolution and monoplacophorans was the fusion of the skeletal plates. Such a transformation required the prior thickening of the muscle foot. This change was advantageous because it improved the capabilities to creep over irregularities of the substrate, and dispensed with the necessity of bending the whole body. In this situation the fusion of the dorsal shell plates became possible and allowed the formation of a more effective mechanical frame to which the retractor muscles could be anchored.

In the early stages of conchiferan organization, the dorsoventral musculature preserved its segmentation; the digestive apparatus, the nephridia, the gills, and also the basic organization of the foot remained virtually unchanged, giving rise to an organism basically identical with the extant Neopilina. From this basic group with the unified shell and an obvious tendency to abandon the remaining metamerism, the main molluscan groups can be derived over gradualistic reconstructions of the transformation process (Fig. 4). There can be no doubt that the present group of the Neopilina is a representative of the most conservative stock of the conchiferans. This organism certainly originated by a secondary flattening of the shell and the creeping foot. The dorsoventral muscles retained the serial organization because the persistence of most of the nephridia enforced this pattern. The spaces between the muscles are necessary for the dislocation of fluid to and from the lateral parts of the animals. Rearrangements of muscles and the formation of unsegmented retractor muscles by fusion in the narrowing waist of a developing conchiferan became possible only after most of the nephridia and many of the gills in the mantle groove were reduced.

The developmental trend of de-metamerization is indicated in the fusion of the anterior retractor muscles and the concomitant disappearance of the nephridia between these muscles in monoplacophorans. The enforcement of the segmented pattern of the retractor muscles by the nephridia remains obvious in the posterior portion of the same constructions. Apparently the fusion of the skeletal plates did not result in the loss of metamerism of muscles and nephridia.

Now the problem arises why most of the pairs of nephridia could be reduced in the conchiferans. The answer lies outside the realm of biomechanics. Apparently the formation of glandlike extensions of few nephridal ducts and the coupling of these newly formed organs with the blood flow allowed the loss of metameric nephridia. One or two pairs of transformed nephridal ducts with large glandular extensions could replace the great number of localized excretory organs.

As soon as we understand this situation, it is clear that the patellaceans, which were recently considered to be direct descendents of the monoplacophoran stock, cannot be derived from the basic monoplacophoran conchiferans, because they are forms which have undergone torsion and a shift of the posterior gills to the anterior part of the body (Fig. 5). Thus the suggestions of Haszprunar are certainly invalid.

12 Conchiferan Radiations

The space of this chapter is too limited to allow the strict reconstruction of all conchiferan groups. It should be mentioned that clarifying models and first approaches for the evolution of bivalves (Vogel and Gutmann 1980), cephalopods (Bonik et al. 1977), and gastropods (Edlinger 1988a, b) are given elsewhere. In this context we will focus on two examples which provide case studies for

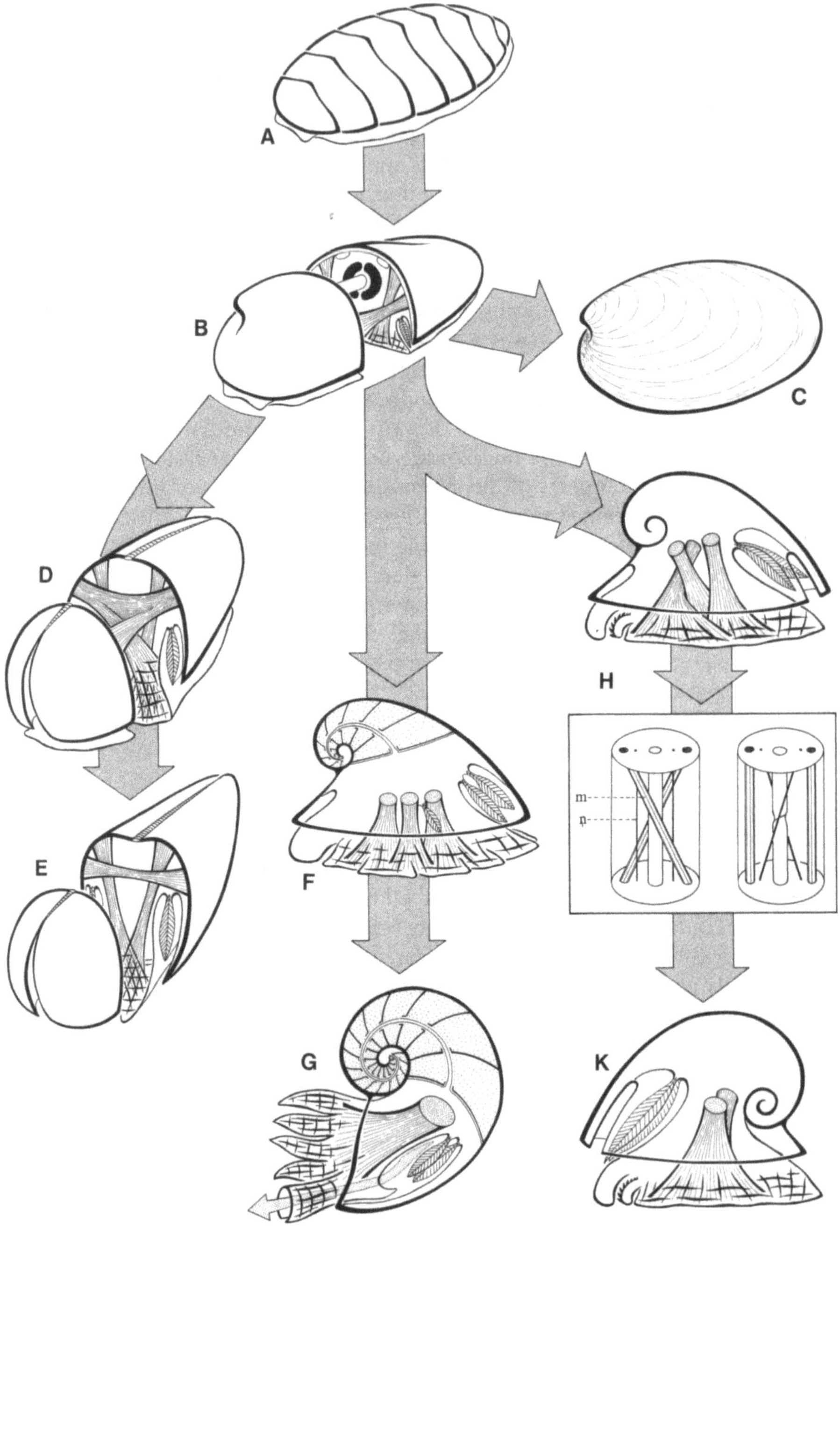
A
B
C
D
E
F
G
H
K
m
p

reconstructions, the interpretation of the patellaceans and the derivation of the scaphopods.

The gastropods, as the most diverse group of the conchiferans, have undergone a torsion of their pallial and visceral complex. This evolutionary alteration can be explained as the consequence of the high cambering of the caplike ancestral shell and the formation of an extremely narrow waist in the cephalopodium of the forerunners. It allowed the planispiral coiling of the shell. The formation of the waist enforced the reduction of most gill pairs with the exception of the most caudal ones. The remaining gills found their accommodation in the deepening caudal parts of the mantle cavity. The clearly demarcated hump could be easily turned in respect to the substrate-bound head-foot. Thereby it initiated the process of torsion, the key change for the formation of the gastropods. If this is kept in mind the derivation of the patellaceans, gastropods with falt caplike shells, cannot be directly derived from monoplacophorans, as suggested but not explained by some authors. Torsion of the nervous system and dislocation of the gills with the mantle cavity to the anterior part of the body could only occur in forerunners with a narrow waist which held the visceral hump in a cambered shell high over the ground. This intermediate stage is in conflict with the organization of possible monoplacophoran-like ancestors.

13 The Scaphopods

If one looks at the conchiferans with their narrow waist between the foot and the visceral hump high above the ground, the scaphopods stand out in a caricature-like way. The basic development of the conchiferans must have found an exaggerated expression in these animals. This situation is explainable if the scaphopods are derived from monoplacophoran forms with a posterior slit through which the water current left the mantle cavity. The whole construction was extended along the vertical axis. So the shell was held high over the mobile head-foot. As a result, the dorsoventral musculature became compacted in the waist and lost its metameric pattern. The remaining muscles were extremely elongated into the cone-shaped narrow shell. As a result, the foot was shortened in the longitudinal direction when the shell was stretched. The change of the shell into a conelike shape created a narrow mantle cavity in which the gills found no space and were

Fig. 4. Derivation of the main conchiferan groups. **A** Segmented primitive mollusc with a row of dorsal shells and a flexible body with a rather flat muscular foot. **B** Fusion of shells in conjunction with the formation of a thick muscular foot and separation of the visceral hump by a waist. **C** Monoplacophoran shell as a representation of the early conchiferan stage. **D, E** Development of the bivalve construction in forms which could pull the lateral parts of the (larval) shell together. The muscles became reorganized to form adductors and distinct muscles extending into the foot. **F, G** Formation of gas-filled chambers in the pre-cephalopods. The foot was transformed into the adhesively working arms which could grasp prey, and the funnel which directs the water current expeled from the mantle cavity for jet propulsion. **H, J, K** Development of the gastropod construction. The waist between the head-foot and the visceral hump became extremely pronounced. Only the caudal gills were conserved in the caudal region. Torsion of the visceral hump resulted in the shift of the gills to the anterior part and gave the head foot more space for retraction

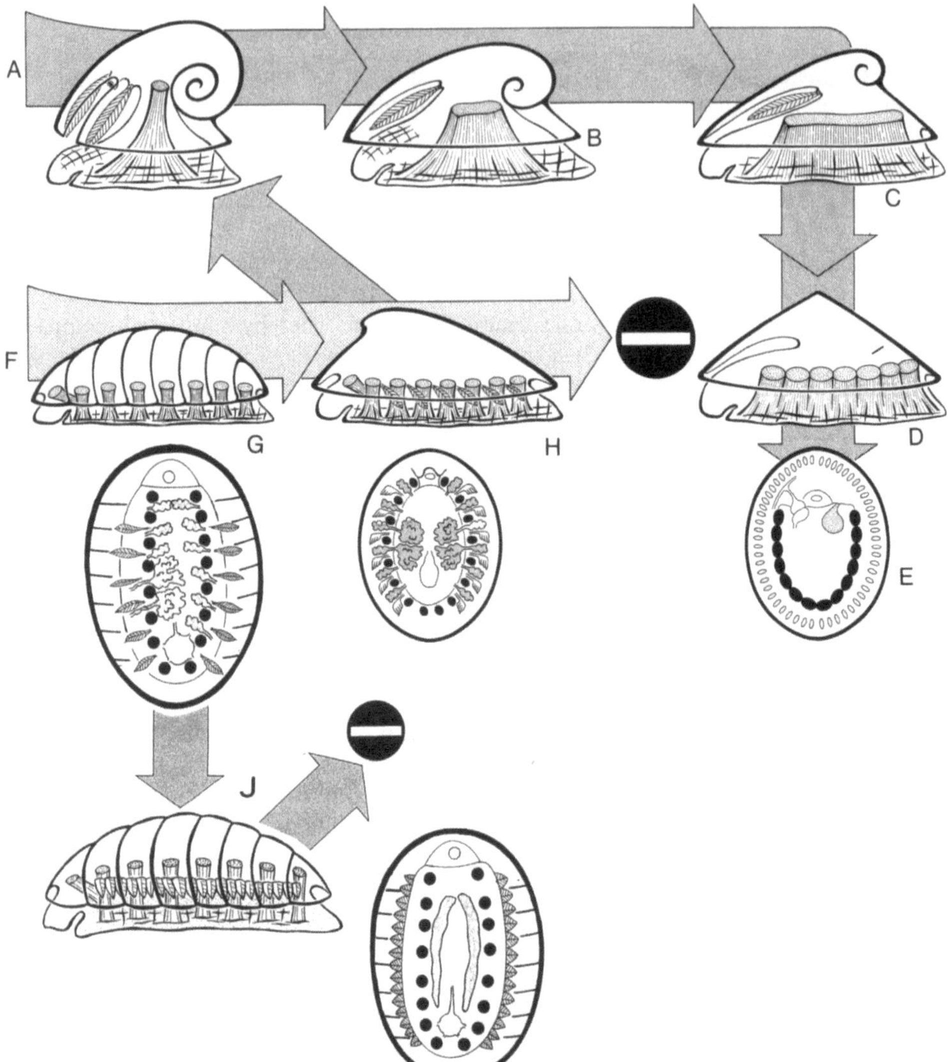

Fig. 5. Evolution of alternative pathways. **A–E** Development of the patellaceans from gastropods with torsion of the visceral hump. The constructional changes lead to a flattened shell and a U-shaped organization of the retractor muscles. The predecessors cannot have been monoplacophorans ancestors as they must have had a narrow waist and held the shell far from the ground. **F, G, H** The formation of the monoplacophorans from polyplacophoran ancestors. The gills and the nephridia determine the conservation of the metameric retractor muscles. As the constructional stages of the sequence never possessed a waist, they could not develop a torsion of the visceral hump, and therefore the continuation of the transformation series into the pattelaceans is not possible. **I** The formation of the recent chitons as described in the text. Constructional morphology can allow the convincing explanation of transformation sequences and reliably rule out alternative phylogenetic relationships (**H** to **D**, **J** to **H**)

reduced. A permanent water current used for respiration ventilates the mantle cavity, passing it through the slit which had become shifted to the apical point of the shell during the elongation process of the visceral hump. This slit is not the result of a lateral folding of a unified shell as often suggested. It can be easily derived from posterior slits in the shells of creeping forms. The derivation of the scaphopods from creeping forms is at variance with the totally unexplained hypothesis that they might have developed from a common lineage with pelecypods. The shell of the scaphopods has never been folded as the one of the bivalves, it is just an extremely distorted caplike shell of a monoplacophoran construction.

14 Concluding Remarks

As indicated in the presented models, molluscs provide paradigmatic examples for the importance of coherence, the hydraulic mechanisms, and the mechanical principles in the explanation of animal Bauplan constructions and for the reconstruction of their emergence. It should be added that the shape of the hydraulic body construction is determined by the tensile and contractile system of muscles and connective tissue complemented by some skeletal structures. All activity is generated by energy conversion and energy canalization in the mechanically coherent construction. Neither the working principles, the enforcement of form, nor the gradualistic transformation processes in the course of phylogeny can be explained or reconstructed on the basis of idealistic morphology and homology research.

One additional point has to be emphasized: as all explanations are based on the soft-body construction and the biomechanical principles ruling its architecture, none of the reconstructions can be directly deduced from the fossil record. The fossil record is, however, helpful, as it provides important evidence and documents as to when in the geologic history the respective constructional level was reached.

References

Bandel K (1982) Morphologie und Bildung der frühontogenetischen Gehäuse bei conchiferen Mollusken. Facies 7:1–98

Bandel K (1983) Wandel der Vorstellungen von der Frühevolution der Mollusken, besonders der Gastropoda und Cephalopoda. Paläontol Z 57 (3/4):271–284

Bjaly VI (1973) New early ordovician monoplacophorans of Siberia. Paleontol J 3:326–330

Bonik K, Grasshoff M, Gutmann WF, Klein-Rödder R (1977) Die Evolution der Tintenfische, ein Entwurf für das Schaumuseum. Nat Mus 107 (8):244–250

Crofts D (1937) The development of *Haliotis tuberculata*, with special reference to organogenesis during torsion. Philos Trans R Soc London Ser B:219–268

Crofts D (1955) Muscle morphogenesis in primitive gastropods and its relation to torsion. Proc Zool Soc London 125:711–750

Dullemeijer P (1989) Funktionelle Morphologie und Phylogenie. In: Edlinger K (ed) Form und Funktion – ihre stammesgeschichtlichen Grundlagen. Univ Press, Wien, pp 127–144

Edlinger K (1988a) Torsion in gastropods: a phylogenetical model. Malacol Rev Suppl 4:239–248

Edlinger K (1988b) Beiträge zur Torsion und Frühevolution der Gastropoden. Z Zool Syst Evol Forsch 26:27–50

Edlinger K (ed) (1989a) Form und Funktion – ihre stammesgeschichtlichen Grundlagen. Univ Verl, Wien, 196 pp

Edlinger K (1989b) Sekundäre Wurmkonstruktionen bei amphineuren Mollusken. In: Edinger K (ed) Form und Funktion – ihre stammesgeschichtlichen Grundlagen. Univ Verl, Wien, pp 35–54

Götting KJ (1980a) Origin and relationships of Mollusca. Z Zool Syst Evolutionsforsch 18:24–27

Götting KJ (1980b) Argumente für die Deszendenz der Mollusken von metameren Antezedenten. Zool Jahrb Anat 103:211–218

Gutmann WF (1972) Die Hydroskelett-Theorie. Aufs Reden Senckenb Naturforsch Ges 21:1–91

Gutmann WF (1974) Die Evolution der Molluskenkonstruktion: ein phylogenetisches Modell. Aufs Reden Senckenb Naturforsch Ges 25: pp 1–84

Gutmann WF (1988) The hydraulic principle. Am Zool 28:257–266

Gutmann WF (1989a) Konstruktive Vorbedingungen für die Eroberung von Lebensräumen. In: Edlinger K (ed) Form und Funktion – ihre stammesgeschichtlichen Grundlagen. Univ Verl, Wien, pp 5–20

Gutmann WF (1989b) Die Evolution hydraulischer Konstruktionen – Organismische Wandlung statt Anpassung. Kramer, Frankfurt/M, 201 pp

Haas W (1981) Evolution of calcareous hardparts in primitive molluscs. Malacologia 21 (1/2):403–418

Haszprunar G (1988) On the origin and evolution of major gastropod groups, with special reference to the Streptoneura. J Mollusc Stud 54:367–441

Lauterbach KE (1983) Erörterungen zur Stammesgeschichte der Mollusca, insbesondere der Conchifera. Z Zool Syst Evolutionsforsch 21 (3):201–216

Remane A (1967) Die Geschichte der Tiere. In: Heberer G (ed) Die Evolution der Organismen, vol 1 G Fischer, Stuttgart pp 589–677

Salvini-Plawen L von (1968) Die 'Funktions-Coelomtheorie' in der Evolution der Mollusken. Syst Zool 17:192–208

Salvini-Plawen L von (1972) Zur Morphologie und Phylogenie der Mollusken: die Beziehungen der Caudofoveata und der Solenogastres als Aculifera, als Mollusca und als Spiralia. Z Wiss Zool 184:205–394

Salvini-Plawen L von (1981) On the origin and evolution of the Mollusca. Atti Convegni Lincei 49:235–293

Stasek CR (1972) The molluscan framework. In: Florkin M, Scheer BT (eds) Chemical zoology. Academic Press, New York London, pp 1–44

Vagvolgyi J (1967) On the origin of molluscs, the coelom, and coelomic segmentation. Syst Zool 16:153–168

Vogel K (1983) Macht die biologische Evolution Sprünge? Sitzungsber Wiss Ges Goethe-Univ, Frankfurt/M 19 (4):3–19

Vogel K (1989a) Konstruktionsmorphologie und Rekonstruktion der Stammesgeschichte. In: Edlinger K (ed) Form und Funktion – ihre stammesgeschichtlichen Grundlagen. Univ Verl, Wien, pp 55–68

Vogel K (1989b) Constructional morphology and the reconstruction of phylogeny. Abh Naturwiss Verl, Hamburg 28:255–264

Vogel K, Gutmann WF (1980) The derivation of pelecypods: role of biomechanics, physiology and environment. Lethaia 13:269–275

Vogel K, Gutmann WF (1988) Protist skeletons: biomechanical preconditions and constructional utilization. Senckenb Lethaea 69 (3/4):171–188

Yonge CM (1947) The pallial organs in the aspidobranch Gastropoda and their evolution throughout the Mollusca. Philos Trans R Soc London Ser B 232:443–518

Morphogenetic Versus Environmental Cues for Adaptive Radiations

P. W. Skelton [1]

Summary

The linkage between adaptive divergence and taxonomic diversification is complex, but where a significant correlation of the taxonomic pattern with either an intrinsic (e.g. morphogenetic), or an environmental cue may be detected, an adaptive radiation may be identified as such. Both intrinsic and extrinsic factors are necessary for the radiation to occur, but in any given instance one can be shown to have been the effective cue, while the pre-existence of the other served as an enabling circumstance.

These relationships are illustrated with respect to the rudist bivalves of the Tethyan Cretaceous. The main radiation of the group was morphogenetically cued by the invagination of the ligament, which permitted uncoiled shell growth. Within that context, however, invasion of the recumbent adaptive zone, in two discrete phases (Aptian-Cenomanian and Campanian-Maastrichtian) was evidently environmentally cued. The adaptive radiation of the group thus shows a hierarchical pattern, with morphogenetic and environmental cues alternating at different levels.

1 Adaptation and Diversity

It would be grossly simplistic to suppose that there is a necessary correspondence between adaptive divergence and taxonomic diversification. The proliferation of taxa may proceed with little or no broadening of the basic adaptive repertoire of a clade for a variety of reasons. Measurements of variations in the gene products and chromosomes of natural populations in recent years have revealed large numbers of hitherto unsuspected sibling species (see examples in Levinton 1988), most no doubt arising as accidents of genetic divergence during allopatry (Mayr 1970). Likewise, Vermeij (1978) lists numerous pairs of adaptively similar, "cognate" molluscan species either side of the Panamanian Isthmus, derived from common ancestral populations split by the rise of the latter. The principle of species selection (Vrba and Gould 1986) may further account for the differential growth of some clades in a manner divorced from the adaptations of individual organisms. Conversely, Raup et al. (1973) have argued that the originations and extinctions of species are due to such myriads of particular causes that most pat-

[1] Department of Earth Sciences, The Open University, Milton Keynes, MK7 6AA, UK

terns of change in the diversity of clades might be effectively stochastic (though see criticisms of their model parameters by Stanley et al. 1981).

To identify genuine examples of adaptive radiation, then, is an onerous task. The hypothesis of a causal relationship between the broadening of adaptive possibilities, on the one hand, through the appearance of some key morphogenetic innovation(s) or a change in environmental circumstances, and, on the other, the taxonomic diversification of the clade(s) so affected must be tested against the alternative hypotheses for taxonomic proliferation, outlined above. Otherwise, the hypothesis of adaptive radiation risks amounting to no more than the "adaptationist storytelling" justly criticized by Gould and Lewontin (1979).

2 Identifying the Cues for Adaptive Radiations

It is often a vexed question as to whether such a radiation can be attributed to the influence of some intrinsic innovation, be it constructional, physiological or behavioral, or to the extrinsic stimulus of some environmental change. Confusion arises because both factors are required to permit an adaptive radiation: without innovation(s), the radiation would not entail adaptive divergence, while without a receptive environment, there would be no radiation. In seeking which factor might be deemed the effective "trigger" or "cue" for the radiation, one is attempting to establish which particular change co-incided precisely with the inception of the radiation and what, in contrast, were merely enabling circumstances. A couple of extreme examples will make the point clear.

The evolution of flight in pterodactyls and birds was presumably directly contingent upon the appearance of pre-adaptive innovations such as axillary membranes and feathers, respectively. There is nothing to suggest that the character of their environments changed in any way immediately prior to the times of origin of these groups.

The early Tertiary radiation of mammals involved the exploitation of many new adaptive zones, despite the long prior history of the group in the Mesozoic. There can be small doubt that this adaptive radiation, in contrast, was effectively launched by the clearing of ecological space associated with the mass extinction at the close of the Cretaceous. It is the co-incidence of the penetration of new adaptive zones by members of several already differentiated clades, in the episode following the mass extinction, which singles out the latter as the effective cue for the adaptive radiation. The various constructional attributes of members of the different clades, which allowed them to enter their new adaptive zones, can be thought of as the enabling circumstances in this instance.

The exercise I wish to explore in this chapter is the use of taxonomic and biostratigraphical data to resolve the respective roles of such intrinsic and extrinsic influences in adaptive radiations, and, in particular, to settle the question of constructional versus environmental cuing in particular instances.

I use the rudist bivalves by way of illustration, both because their expansive radiation throughout the latter half of the Mesozoic comprised several subsidiary radiations, and because many years of work upon them have furnished me with a tolerably serviceable data base of taxonomic and biostratigraphical data for testing evolutionary hypotheses.

3 The Adaptive Radiation of Rudist Bivalves

The rudists are an extinct superfamily (Hippuritacea) of bizarrely shaped sessile epifaunal bivalves which flourished on the shallow carbonate platforms of the Tethyan Ocean and contiguous regions in the late Jurassic and Cretaceous. Their main taxonomic diversification was during the Cretaceous. This has already been demonstrated to have been the direct consequence of a major morphogenetic innovation at the beginning of the period, which allowed exploitation of new adaptive zones, as I shall briefly summarise, from Skelton (1978, 1985).

The earliest rudists (paraphyletic family Diceratidae) all possessed a modified external parivincular ligament system, the exaggerated backward tangential growth of which was accompanied by extensive anterior splitting and shell incremental overgrowth of old ligament. Shell growth was thus constrained such that the umbones projected spirogyrally forward and out (Skelton 1978). Early cementation to the substratum of one of these provided an attachment platform, or passively implanted anchor from which the spiralling upward growth could maintain the feeding margin above the sediment surface (as in living chamids). Their growth habit was thus essentially that of spiral overgrowth of firm sediment, or hard surfaces. Later rudists in which the external ligament was retained (monophyletic family Requieniidae — attaching by the left valve) were also restricted to this growth habit. Their diversity remained relatively low.

It was among the diceratids attaching by the right valve that the major morphogenetic innovation occurred. This consisted of shortening and eventual invagination of the ligament, which had the effect of releasing the valves from the constructional constraint of having to grow spirogyrally. They now could, and did, grow with an uncoiled tubular (or, in the left valve, sometimes operculiform) morphology (Skelton 1978). From the time of its inception, the clade of "uncoiled" rudists underwent an initially exponential diversification of genera. On the basis of their palaeoecological and facies associations, these can be seen to have evolved a much broader adaptive repertoire than their spirogyrate sister taxa. That the radiation had its exponential focus on the inception of the first uncoiled genus, and that it was not matched by any such pattern among co-eval spirogyrate genera (which, served, so to speak, as the "experimental control group"), shows this to have been a genuine adaptive radiation cued by the pre-adaptive constructional innovation of ligamentary invagination (Skelton 1985).

It is on certain of the subsidiary radiations contained within this overall adaptive radiation of uncoiled taxa that I now wish to dwell, in order to resolve the subsequent contributions of morphogenetic and environmental cues. First, however, it will be necessary to provide more detailed, and objective characterizations of the adaptive zones to which reference will be made.

4 The Adaptive Zones of the Uncoiled Rudists

As sessile epibenthic suspension feeders, the rudists were only capable of adjusting to their situation in a habitat by shell growth. They therefore show a range of shell forms, the basic features of which can be closely linked with the character of the original substratum and ambient current regime. These may be classified into three broad morphotypes, on the basis of comparison with paradigms originally defined by Skelton (1979). However, these original definitions were only qualitative in character, and so the classification could not be used entirely consistently. A revised, parametized version of the scheme has been established by Skelton and Gili (in press), and is outlined here (Fig. 1).

1. Elevators. The entire growth margin of the attached valve was involved in upward growth. The elevation index, E, comprising the mean of the highest (α) and lowest (β) declination of the outer wall of the attached valve's growth margin therefore tended in life towards 90° (Fig. 1). Variation in E may have related to non-vertical orientation of the attached juvenile, secondary shell toppling (leading to geniculate growth) or obstruction to upward growth, but a lower limit for normal growth is defined at E = 45°. Stabilisation was achieved by passive implantation of the attached valve due to the accumulation of surrounding sediment, sometimes assisted by lateral attachment to neighbours. The morphotype was particularly associated with originally soft, usually muddy substrata, with positive net accumulation and only sporadic episodes of sediment removal during the animal's life. Rudists of this morphotype were therefore largely restricted to

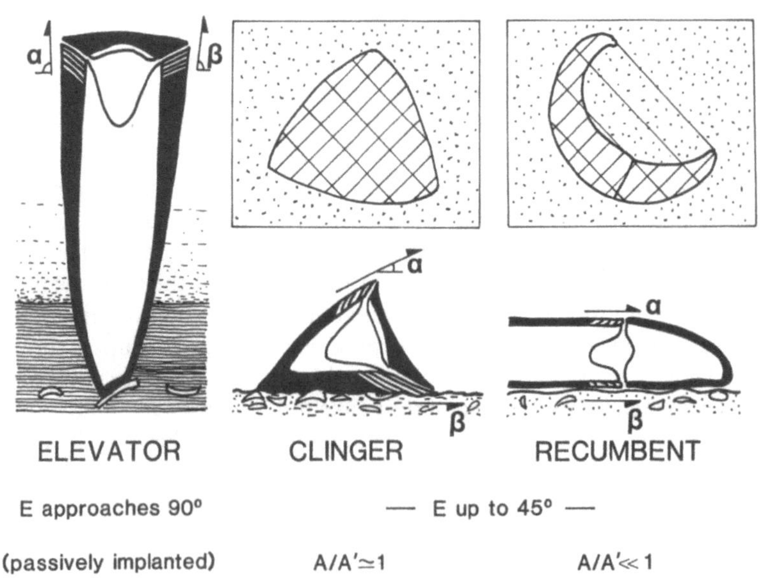

Fig. 1. Paradigms of palaeoecological morphotypes in rudist bivalves, modified from Skelton and Gili (in press), shown in vertical section and, for clinger and recumbent, in plan view (*above*). Close diagonal ornament, in the latter, shows A, and spaced diagonal ornament, A′. See text for explanation

low to moderate current energy regimes, with or without spasmodic (usually storm-induced) traction removal of sediment. Thickets of clustered individuals and boat-shaped forms, with the growth margins curving upwards from a reclining base, tended to dominate with lower levels of sediment accumulation.

2. *Clingers* (originally termed encrusters in Skelton 1979). Part of the growth margin of the attached valve directly overgrew the substratum, either cementing to it where it was hard (*attached clingers*) or growing closely over its surface (*frictional clingers*). Since β therefore tended to be about 0°, E is defined as having been $<45°$ in shells in life position. Stabilisation was achieved by the direct contact of the base of the shell (one or both valves) with the substratum. This basal contact surface area (A) tended therefore to be maximized, so approximating to the area of the convex polygon mapped onto the substratal surface from the peripheral outline of the shell (A′, or virtual area of support). Thus A/A′ tended to approach 1, and is defined as having been >0.75. The morphotype was associated with a range of substrata where net sedimentation was negligible to positive, though with only occasional influxes of sediment, during the animal's life. Sediment accumulation could be matched by upward stepping of the basal growth surface. Substratal surfaces ranged from being relatively stable to hard (including other shells). Associated current conditions might involve frequent or regular inhibition of settling, and even winnowing, of fine-grade sediment though with only rare traction removal of sand and coarser-grade sediment. However, attachment to shells and other exposed hard surfaces occurred in a complete spectrum of low to high current energy regimes.

3. *Recumbents.* Stabilization was again the key to this morphotype, with a zone of the (unattached) valve margins lying flush against the substratum, so as to produce a broad basal form to the entire shell. E was thus again $<45°$. Unlike the clinger, however, the area of contact with the (unstable) substratum, A, was of little importance for stabilization. Rather, the shell form resisted being flipped over by currents, or undermined and buried in a scour hollow, by having a widely dispersed arcuate or stellate basal form, with large effective diameters and, hence, maximised A′. Consequently A/A′ was much less than 1, and is defined as having been <0.75. This morphotype was associated with mobile lime sand to shell rubble substrata, where deflation by traction currents was frequent and net positive sedimentation negligible. It was the dominant morphology of rudists exploiting high current energy bedform settings with predominant bypassing of sediment, such as the top surfaces of current-swept banks and tidal delta lobes.

The morphotypes were by no means mutually exclusive in their distribution, but rather represent particular modes of association with substrata. Recumbents and clingers might co-occur on lime-sand banks, for example, with the former occupying the deflating crests and the latter, the prograding foresets, or indeed (as attached clingers) some of the shells of the recumbents. Some taxa were quite opportunistic, and could grow facultatively either as elevators or as recumbents, according to ambient sedimentation, although most elevators and recumbents were obligately so. Moreover, the juvenile shell in all rudists grew as an attached clinger (on a shell fragment or other hard surface) and so most rudists underwent an ontogenetic change in morphotype categorization in any case.

5 The "Golden Ages" of Recumbent Rudists

In their preliminary analysis of the distribution of rudist morphotypes through time, Skelton and Gili (in press) noted some distinct and intriguing evolutionary patterns (Fig. 2). Clingers and, from the advent of uncoiling at the beginning of the Cretaceous (p. 377), elevators persisted throughout the Cretaceous. Recumbent taxa, in contrast, were only noted from two discrete intervals, nicknamed golden ages. The first of these was from the Aptian to the Cenomanian, though with a significant depletion in the late Aptian, resulting from a mass extinction among rudists in the mid-Aptian (Masse 1989). The second comprised the late Campanian to early-Maastrichtian interval. Nowhere have any examples of recumbent rudists (as recognised by the criteria given above) been recorded from strata representing intervals outside of these two golden ages.

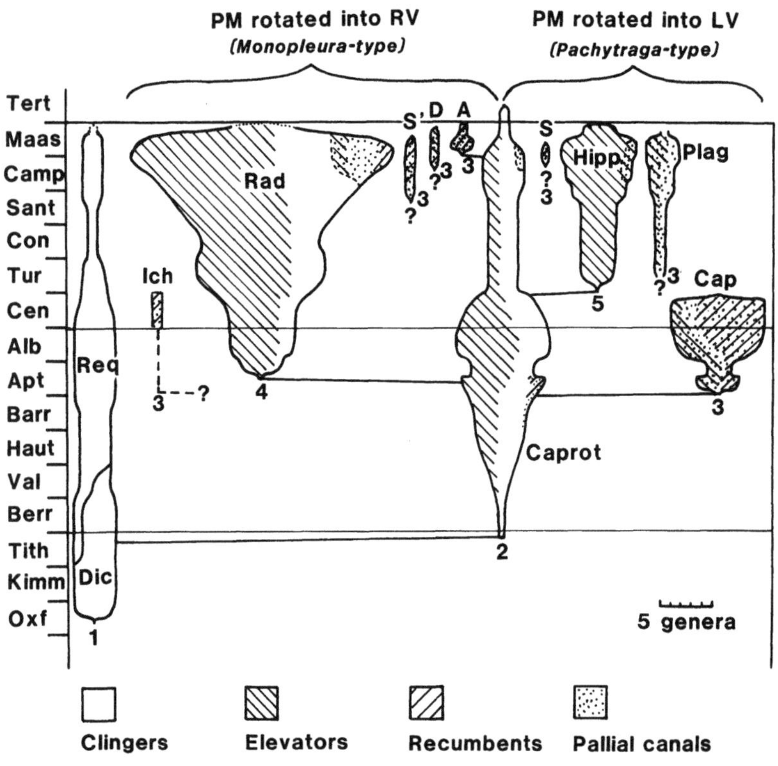

Fig. 2. Evolution of rudist morphotypes, modified from Skelton and Gili (in press). Spirogyrate rudists with an external ligament (*1*) comprise the Diceratidae (*Dic*) and Requieniidae (*Req*). Uncoiled rudists, with an invaginated ligament (*2*) comprise a paraphyletic stem group, Caprotinidae (*Caprot*) and several monophyletic offshoots, some having a monopleuriform posterior myophoral (*pm*) arrangement, and others a pachytragiform arrangement, associated with accessory cavities (see Fig. 5 for explanation). Pallial canals (*3*) evolved convergently in several clades: Ichthyosarcolitidae (*Ich*), Caprinidae (*Cap*), various *Sabinia* species groups (*S* and *S'*), Plagioptychidae (*Plag*), *Dictyoptychus* and allied forms (*D*), and Antillocaprinidae (*A*). The Radiolitidae (*Rad*) are characterized by sheet-cellular construction of the thickened outer (calcitic) shell layer (*4*). The Hippuritidae (*Hipp*) have an elaborate system of pores and canals in the left valve (*5*), originally associated with filter feeding

The reason for the termination of the first interval is relatively clear: it coincides with a major mass extinction associated with extensive carbonate platform drowning (Masse and Philip 1986). No recumbent taxa survived that event. What is of particular interest in the present context is the pattern of inception of this first interval. It certainly forms an integral part of the broader adaptive radiation of the uncoiled rudists, representing the opening of one major new adaptive zone. However, was it simply cued by the evolutionary appearance of features pre-adaptive for recumbent shell growth, or was there some environmental cue, which influenced the evolution of taxa already possessing such features? A similar question may be asked concerning the second golden age, following the extinction of all the earlier recumbent taxa.

To answer these questions, it is necessary to analyze the morphogenesis and the phylogenetic relationships of the recumbent taxa, and the latter entails combining some cladistic analysis with biostratigraphical data. If the first golden age was morphogenetically cued, then a single (monophyletic) radiation might be expected, immediately following the appearance of a species possessing some novel feature(s) characteristic of the construction of the ensuing recumbent taxa. If, on the other hand, the interval was environmentally cued, then a polyphyletic origin for the recumbents is more probable, with synchronous originations of the convergent taxa. Moreover, constructional features pre-adaptive for recumbent shell growth might be expected in taxa significantly pre-dating the interval in question.

Figure 3 shows a synoptic cladogram for all recumbent genera of the Aptian to Cenomanian interval, together with their stratigraphical ranges. Selected non-recumbent genera are also included in the cladogram to serve as outgroups, or because they were derived from recumbent clades. The stratigraphical ranges of the remaining non-recumbent genera are shown to the right, for comparison of their changes in diversity. Cladistic analysis of these has not been attempted, as further clarification of many of their character states is still needed. For the present purposes, however, mere recording of the changes in diversity of these genera suffices. Although it might, in theory, have been preferable to have conducted the analysis at the species level, this is impracticable at present since the species-level taxonomy of most of the genera needs much revision. Besides, in the rudists at least, it can be argued that genera give a better characterization of changes in the diversity of adaptive "types" than do species. The diversity of the latter may be more subject to factors other than adaptive divergence, as well as being more prone to preservational biases (Skelton 1985).

Figure 3 shows that the diversity of recumbent genera leapt from none recorded in the Barremian, to six in the early Aptian (= 'Bedoulian'), and that at least two, and perhaps three clades, independently derived from non-recumbent taxa, were involved. All have both valves extended, with one or both curved or loosely coiled, thereby providing the crescentic basal shape (when in prone life position) with $A \ll A'$, which diagnoses them as recumbents (Fig. 4e). Moreover, all, with the exception of *Pachytraga* (to be discussed shortly), have longitudinal "pallial" canals running through the originally aragonitic inner layers of the shell wall, in one or both valves (Fig. 4c, d). One biomechanical effect of this economical thickening of the valve walls would have been increased resistance to bending stresses (which might have induced brittle failure). This arrangement is analogous to the "sand-

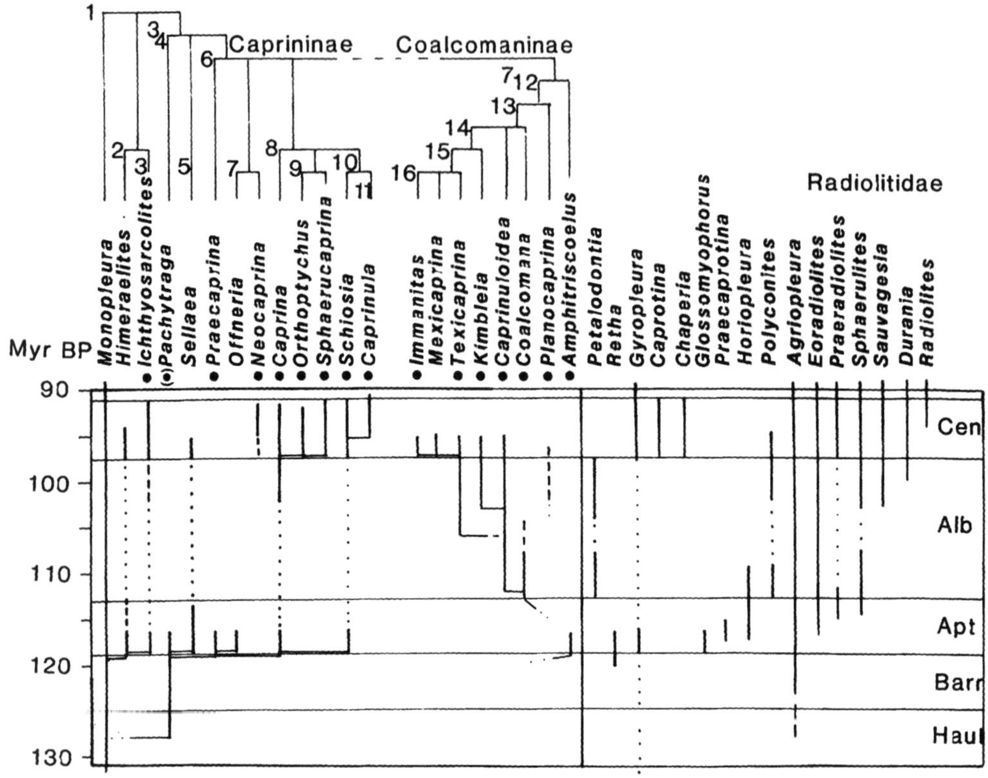

Fig. 3. Inferred phylogeny of the Aptian-Cenomanian recumbent rudists (marked by *black circles*), with ranges of other, non-recumbent uncoiled genera. *Solid lines* indicate sound records; *dashed lines* more dubious records; *dotted lines* inferred continuations of ranges where records are wanting. Cladogram for recumbents (plus selected other genera) has the following nodes (syn- or autapomorphies): *1* Ligamentary invagination (plesiomorphy for all genera shown here); *2* fine, capillary-form pallial canals; *3* extended left valve; *4* pachytragiform posterior myophores; *5* canals outside anterior myophore in right valve; *6* pallial canals outside both myophores in the left valve (may be convergent in Old and New World caprinid clades); *7* canals, as in *6*, in both valves (probably convergent); *8* pallial canals around entire ventral margin of left valve; 9 pallial canals, as in *8*, showing further subdivision by bifurcation of radial walls; *10* pallial canals, as in *8*, in both valves; *11* further complexification of canals, now of polygonal form; *12* broad, bean-shaped outline to posterior accessory cavity in right valve; *13* pallial canals extend around ventral margins of both valves; *14* complexification of canals by bifurcation of radial walls; *15* canals invade teeth; *16* canals throughout entire aragonitic inner shell. Systematic and stratigraphical data from numerous sources (details of which can be obtained from the author on request)

wich" of a corrugated layer between flat layers used to strengthen the walls of cardboard boxes, or, again to the "sandwich bone" of the human cranium, with its outer sheets of compact bone, surrounding cancellous bone (Wainwright et al. 1976). An additional biomechanical advantage of the canals, especially where these became complex, would have been inhibition of brittle fracture on impact, by prevention of crack propagation. Although other functions may readily be imagined for the canals, the co-incidence of their appearance in all the clades of re-

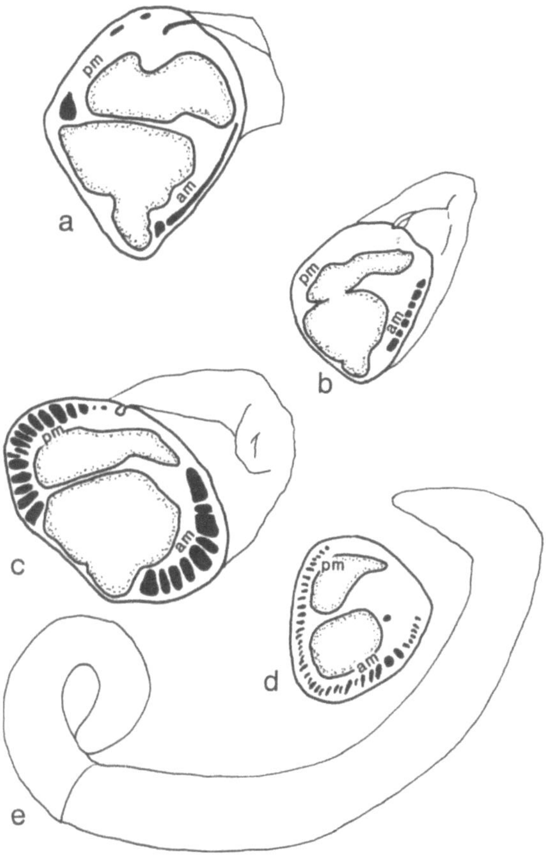

Fig. 4. Sections, close to the commissure, of left valves of *Pachytraga* (**a, b**), showing variable development and subdivision of the accessory cavities outside the anterior and posterior myophores (*am* and *pm*), *Praecaprina* (**c**) and *Caprina douvillei* Paquier (**d**), all from the Lower Aptian of France based on Paquier (1905); **e** plan view of shell form in recumbent *Pachytraga* sp. from the Lower Aptian of Ericeira, Portugal ($\times\frac{1}{3}$)

cumbent rudists, and the frequent association of the latter with relatively turbulent palaeoenvironments, strongly hints at some kind of strengthening role, of the sort suggested above – at least in the recumbent taxa.

In the Old World caprinids, a clear evolutionary derivation of the pallial canals from accessory cavities which separated the salient myophores from the valve walls (Fig. 4a–d) has been demonstrated by Paquier (1905). Some individual specimens of *Pachytraga* may show incipient subdivision of the accessory cavities (Fig. 4a, b), but only in *Praecaprina* (Fig. 4c) did this become a regular (and diagnostic) feature around the myophores of the extended left valve. Canals then invaded the ventral valve wall, as well, in *Caprina* (Fig. 4d). In terms of life habits, *Pachytraga* was usually an elevator, with a moderately extended left valve (Masse 1976 and pers. field observ.). However, specimens of a recumbent form of *Pachytraga* sp. (Fig. 4e) may be found in the Lower Aptian of the Cascais/Ericeira area, in Portugal (pers. observ.). It is unclear whether these represent a distinct species, or merely an ecophenotype of *P. paradoxa* (Pictet and Campiche). Nevertheless, they demonstrate that non-canaliculate taxa were perfectly capable of growing as recumbents. Pallial canals would seem, in this clade, to have been merely a secondary (strengthening) adaptation in already recumbent taxa. Moreover, both a degree of extension of the left valve, and the possession of accessory cavities characterized *Pachytraga* from its first appearance. Accessory cavities arose as a constructional by-product of uncoiled growth. Muscle-shell attachment in bivalves is mediated by specialised adhesive epithelium. Collagen fibres emerging from the epithelium form a dense mat over its outer surface, which is itself closely bonded to the shell myostracum by bundles of collagen fibres which penetrate the myostracal prisms (Nakahara und Bevelander 1970). This system allows muscles to "glide" tangentially across the inner shell wall with growth, by decoupling from the epithelium. However, the presence of the collagen sheath inhibits shell accretion normal to the myostracal surface. This is evident from the tendency of muscle scars on the insides of bivalve shells to be more or less indented, and especially so where there has been little tangential gliding, as with the anterior adductor in many forms. In spirogyrate rudists the posterior adductor muscles were supported on ledge-like myophores, along which the muscles glided with growth. With uncoiling, however, growth of the myophores became radially oriented (Fig. 5). For the gliding relationship of the muscles with the myostracal surfaces of the myophores to be retained, the latter had to become rotated with respect to the commissural plane. Otherwise, the inhibition of shell accretion normal to the

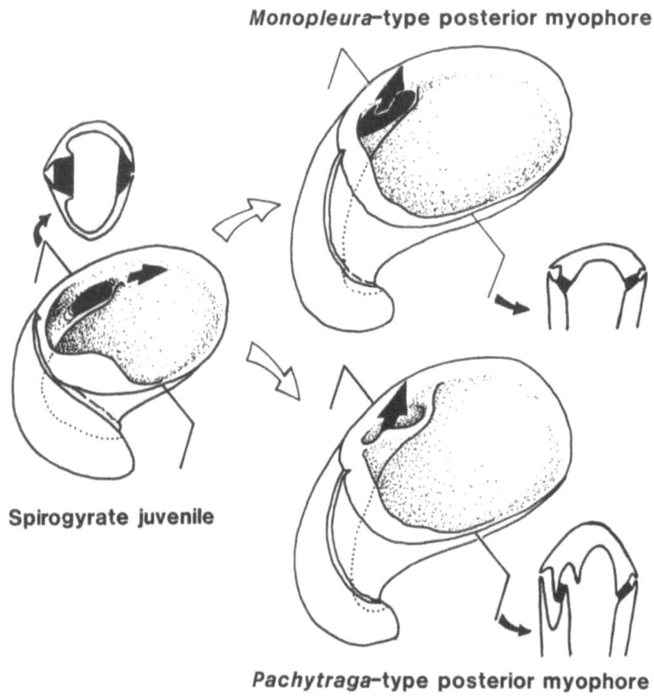

Fig. 5. Myophoral rotation in the uncoiled rudists. The *larger figures* show attached, right valves, and the *smaller figures*, diagrammatic sections across both valves, showing the arrangement of the adductor muscles and myophores, with neighbouring accessory cavities. *Large black arrows* indicate the growth paths of the myophores. Adductors (in the smaller figures) and their scars (in the larger figures) shown in *black*. See text for explanation

myostracal surface would have prevented the myophores from keeping up with growth at the valve margins. The two options for such myophoral rotation are shown in Fig. 5. The first option was for the posterior myophore of the (usually operculiform) left valve to rotate down into the right valve, such that the opposing myophore of the right valve retreated somewhat into the posterior wall (as in advanced *Monopleura* and *Agriopleura*). The second involved the upward rotation of the posterior muscle towards the domed left valve, so that a salient myophore from the left valve faced inwards, towards an erect myophoral lamina in the right valve (as in *Pachytraga*). In both cases, the anterior myophore of the left valve rotated down into the right valve. In many taxa, the rotation of the myophores to form erect, protruding structures thereby created gutters between the myophores themselves and the valve walls (Fig. 5). Where a valve became extended by growth, these gutters deepened, as distinct accessory cavities. Thus even the earliest species of *Pachytraga* (*P. tubiconcha* Astre, from the Hauterivian: Masse et al. 1989), in which the left valve was already somewhat extended, also possessed incipient accessory cavities, as an accident, so to speak, of its construction. So the constructional elements which in the Aptian proved pre-adaptive in this clade for a recumbent mode of growth already existed in the genus some 8 Ma previously. Yet during this interval, the genus had shown no detectable cladogenetic production of new genera, in contrast to the four new recumbent genera apparently arising from it, by successive branching events, within perhaps only a couple of million years in the early Aptian.

The origin of *Amphitriscoelus*, and hence of the clade of New World caprinids (Coalcomaninae of Coogan 1973), is less clear. Its simple canals, absent around the ventral margin, mean that *Praecaprina* is the only caprinid from which it might plausibly be derived (the derivation would not be the other way around, since *Amphitriscoelus* has canals in both valves, while *Praecaprina* only has them in the left valve). However, the distinctively bean-shaped outline of the posterior accessory cavity plus tooth socket in the right valve of *Amphitriscoelus*, which is a synapomorphy for the Coalcomininae (Coogan 1973), is quite unlike the narrow, elongate equivalent in both *Praecaprina* and *Pachytraga*. Its derivation from some other caprotinid is thus a distinct possibility.

Finally, *Ichthyosarcolites* is clearly of quite independent origin. It lacks the accessory cavities present in all caprinids (plus *Pachytraga*), possessing instead a modified monopleuriform myocardinal apparatus, with both the myophores of the left valve facing out onto the inner walls of the right valve (Douvillé 1887). On the other hand, its capillary-form pallial canals are a highly derived feature, unlike those in any of the caprinids, and indeed only shared, it seems, with *Himeraelites* (Camoin 1983), itself clearly an elevator. *Ichthyosarcolites* is usually considered an exclusively Cenomanian genus (e.g. Polšak 1967), but Paquier (1905) figured some entirely convincing fragments of the genus from the Aptian of Bulgaria.

The commencement of the first golden age of recumbents in the early Aptian, then, seems to have involved at least two and maybe three distinct clades. Moreover, one of these, that of the Old World caprinids, was evidently derived from a genus (*Pachytraga*) already pre-adapted for the condition in the Hauterivian. So, following the logic outlined at the beginning of this section, it is reasonable to conclude that this was an environmentally cued radiation.

This first diversification of recumbents suffered a severe setback in the late Aptian and Albian. Only the New World caprinids proliferated through the Albian, while the Old World genera either became extinct, or became confined to some as yet undiscovered refuges, to re-appear later in the record, mostly in the Cenomanian (Masse and Philip 1986). Only *Caprina choffati* Douvillé has a clearly established (Upper) Albian record.

The causes of this demise (see Masse 1989) are not the immediate concern of this chapter, but the pattern does highlight the apparently much greater sensitivity of the recumbents to environmental disturbance than was the case for elevators and clingers (Fig. 3). The recumbent taxa surviving from the early Aptian episode underwent secondary rapid cladogenesis when favourable conditions returned in the Cenomanian (though included at present in only one genus, several species of *Ichthyosarcolites* have also been recorded from this interval). The synchronised resurgence of these different recumbent genera again implies environmental cuing.

An analogous interpretation of the second golden age of recumbents can also be argued, though only the essential outlines will be given here (in part because much more systematic work on these taxa is required).

In the New World Caribbean region, the gigantic Maastrichtian *Titanosarcolites*, which was a recumbent par excellence (see Kauffman and Sohl 1974), appears to have been derived (via *Antillocaprina*) from some form with a monopleuriform myocardinal apparatus (MacGillavry 1937).

In the Old World, large, canal-bearing, crescentic recumbents are widely documented from Campanian to Maastrichtian platform margin facies (references in Skelton and Gili in press), and are generally assigned to the probably polyphyletic genus *Sabinia* (Skelton 1978). In addition to the *"Sabinia"* species, these platform margin facies typically also include some radiolitid genera of bizarre form, such as *Pseudopolyconites* and the biradiolitine, *Rajka* (see Sladić-Trifunović 1989), which may strictly be classified as recumbents, though of stellate, rather than the more usual crescentic form (Skelton and Gili in press).

Once again, then, it is evident that environmental change was the more probable primary cue for the second golden age of recumbents. The latter were synchronously derived from several different ancestors. It should also be noted that canaliculate forms did arise in the interim (Plagioptychidae), but remained as clingers and elevators.

6 Conclusions

Adaptive radiation – when it can be identified as such – requires both morphogenetic (or other intrinsic) innovation and a receptive environment. Yet the appearance of either may serve as an effective cue for the radiation, contingent upon the prior existence of the other, acting, so to speak as an enabling circumstance.

Phylogenetic analysis of the taxa involved allows one to distinguish which of the two possibilities is the more probable explanation in any given instance. Immediate cladogenesis from the first appearance of a single pre-adapted form, unmatched by any such pattern in other related taxa would argue in favour of morphogenetic cuing. A synchronous polyphyletic radiation from ancestral forms which may in some cases already have possessed pre-adaptations for some time, would militate, in contrast, for environmental cuing.

The adaptive radiation of the uncoiled rudist bivalves, as a whole, has already been established to have been morphogenetically cued (Skelton 1985). Within this radiation, however, there was clearly also some environmental cuing of some of its components, and in this paper it has been shown that the (repeated) rise of recumbent rudist taxa is best explained thus.

A hierarchical structure may thus be perceived in adaptive radiations where the contingent roles of morphogenetic innovation and environmental stimulus are reversed at different levels in the hierarchy. At one level, a morphogenetic innovation may cue a broad radiation, while components of that may in turn be cued, in due course, by environmental stimuli. But even within the latter level, different clades are manifestly more or less drawn into involvement according to their constructional pre-adaptations and constraints.

Acknowledgments. I thank the organisers, Drs. Schmidt-Kittler and Vogel, for inviting me to this stimulating meeting, and the Werner Reimers-Stiftung and the Stiftung Volkswagenwerk for their financial support. I am also grateful for the helpful verbal comments on the paper by participants at the meeting, as well as those of my rudistological comrade-in-arms, Jean-Pierre Masse, although I alone bear responsibility for any errors contained in it. Janet Dryden processed the successive drafts at the Open University.

References

Camoin G (1983) Plates-formes carbonatées et récifs à rudistes du Crétacé de Sicile. Thesis, Travaux du Laboratoire de géologie historique et de paléontologie (Offset), Univ Provence, Marseille, 244 pp, 24 pls

Coogan AH (1973) Nuevos rudistas del Albiano y Cenomaniano de Mexico y del sur de Texas. Rev Inst Mex Petrol 5:51–83, 9 pls

Douvillé H (1887) Sur quelques formes peu connues de la famille des chamidés. Bull Soc Geol Fr 3 (XV):756–802

Gould SJ, Lewontin RC (1979) The spandrels of San Marco and the Panglossian paradigm: a critique of the adaptationist programme. Proc R Soc London Ser B 205:147–164

Kauffman EG, Sohl N (1974) Structure and evolution of Antillean Cretaceous rudist frameworks. Verh Naturforsch Ges Basel 84(1):399–467

Levinton J (1988) Genetics, paleontology and macroevolution. Univ Press, Cambridge, 637 pp

MacGillavry HJ (1937) Geology of the Province of Camaguey, Cuba with revisional studies in rudist paleontology. Utrecht Rijks Univ, Geogr Geol Meded Phys Geol Reeks Amsterdam Diss 14, 168 pp

Masse J-P (1976) Les calcaires urgoniens de Provence (Valanginien-Aptien inférieur). Stratigraphie, paléontologie, les paléoenvironnements et leur évolution, 3 vols. Thesis, Univ Aix-Marseille 11, UER Sci Mer Environm, Marseille Offset CNRS No A 12390, 510 pp 60 pls

Masse, J-P (1989) Relations entre modifications biologiques et phénomènes géologiques sur les plates-formes carbonatées du domaine périméditerranéen au passage Bédoulien-Gargasien. Geobios Mem Spec 11:279–294

Masse J-P, Philip J (1986) L'évolution des rudistes au regard des principaux évènements géologiques du Crétacé. Bull Centres Rech Explor-Prod Elf-Aquitaine 10(2):437–456

Masse J-P, Conrad MA, Remane J (1989) Le "Calcaire à *Pachytraga tubiconcha*" (rudiste), épisode urgonien de l'Hauterivien carbonaté du Jura Franco-Suisse. Mem Soc Neuchâtel Sci Nat 11:73–80

Mayr E (1970) Populations, species, and evolution. Belknap Press of Harvard Univ Press, Cambridge, MA 453 pp

Nakahara H, Bevelander G (1970) An electron microscope study of the muscle attachment in the mollusc *Pinctada radiata.* Tex Rep Biol Med 28(3):279–286

Paquier V (1905) Les rudistes urgoniens, pt 2. Mem Soc Geol Fr Paleontol 29 (XIII):49–102

Polšak A (1967) Macrofauna crétacée de L'Istrie méridionale (Yougoslavie). Palaeontol Jug 8:219 pp

Raup DM, Gould SJ, Schopf TJM, Simberloff JD (1973) Stochastic models of phylogeny and the evolution of diversity. J Geol 81:525–542

Skelton PW (1978) The evolution of functional design in rudists (Hippuritacea) and its taxonomic implications. Evolutionary systematics of bivalve molluscs. Philos Trans R Soc London Ser B 284:305–318

Skelton PW (1979) Gregariousness and proto-cooperation in rudists (Bivalvia). In: Larwood G, Rosen BR (eds) Biology and systematics of colonial organisms. Syst Assoc Spec Vol 11, Academic Press, New York London, pp 257–279, 3 pls

Skelton PW (1985) Preadaptation and evolutionary innovation in rudist bivalves. In: Cope JCW, Skelton PW (eds) Evolutionary case histories from the fossil record. Spec Pap Palaeontol 33:159–173

Skelton PW, Gili E (in press) Palaeoecological classification of rudist morphotypes. In: Sladić-Trifunović M (ed) Proc 1st Int Conf Rudists, Oct 1988. Serbian Geol Soc

Sladić-Trifunović M (1989) Pironaea-pseudopolyconite senonian of the Apulian Plate: palaeobiogeographic correlations and biostratigraphy. In: Carulli GB, Cucchi FP, Pirini Radrizzani C (eds) Evolution of the karstic carbonate platform: relation with other peri-Adriatic carbonate platforms. Mem Soc Geol It 40

Stanley SM, Signor PW III, Lidgard S, Karr AF (1981) Natural clades differ from "random clades": simulation and analyses. Paleobiology 7:115–127

Vermeij GR (1978) Biogeography and adaptation. Patterns of marine life. Harvard Univ Press, Cambridge, MA, pp 1–332

Vrba ES, Gould SJ (1986) The hierarchical expansion of sorting and selection: sorting and selection cannot be equated. Paleobiology 12(2):217–228

Wainwright SA, Biggs WD, Currey JD, Gosline JM (1976) Mechanical design in organisms. Arnold, London, xii+423 pp

The Sequence of Body Plans and Locomotory Systems During the Precambrian-Cambrian Transition

JAMES W. VALENTINE[1]

Summary

Molecular data suggest that the body plans of the major phyla for which the data exist originated independently of one another, although a well-defined branching sequence of living phyla is indicated by 18S rRNA phylogenetic trees. It is plausible, then, that the branching occurred much earlier in the history of the metazoa than has been suspected, virtually among acoelomate-grade forms during the Late Precambrian. If so, the locomotory adaptations associated with the rise of the body plans of living phyla must each be traced as an independent evolutionary invention, chiefly in the Early Cambrian.

1 Introduction

The advent of molecular techniques of phylogenetic reconstruction (Zuckerkandl and Pauling 1965) has provided a large volume of new data on animal relationships. At the level of the phylum, the most useful molecule studied extensively is 16S or 18S ribosomal RNA (rRNA); this molecule has evolved relatively slowly and is thus especially useful in assessing relationships among taxa, such as kingdoms or phyla, that branched in the remote past (see Woese 1987). Two trees are available that have been based upon 18S rRNA sequences. Field et al. (1988; see also Ghiselin 1988) used a distance matrix method for treeing; Lake (1990) used a method (evolutionary parsimony, Lake 1987) that is less sensitive to evolutionary rate differences among the rRNAs of different taxa but that uses correspondingly less of the sequence data. In addition to these trees, there is molecular evidence of likely homologies in, and differences of expression among, some of the homeobox-bearing genes in annelids, arthropods, and chordates (Holland and Hogan 1988; Patel et al. 1989b). Finally, trees suggested by the molecular data are consistent with the admittedly spotty evidence on the sequence of body plans that is yielded by the fossil record (Valentine 1989, in press; Valentine et al. in press).

In this paper it is assumed that taken together, the molecular and fossil data indicate the branching order of many animal phyla. If true, then there are significant consequences for the order in which the architectural features of metazoan bodies are evolved, and in turn these inferences permit a reevaluation of their original adaptive significance.

[1] Department of Geological Sciences, University of California, Santa Barbara, CA 93106, USA

N. Schmidt-Kittler and K. Vogel (Eds):
Constructional Morphology and Evolution

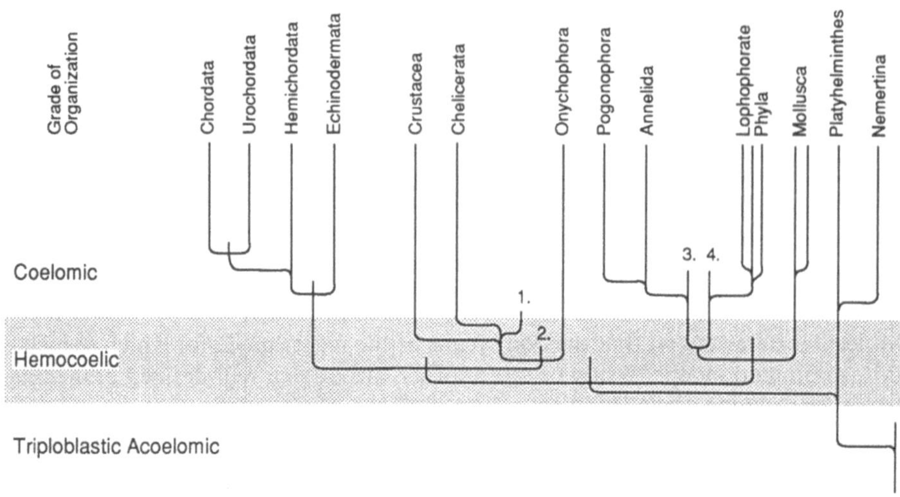

Fig. 1. Phylogenetic model of some metazoan phyla, based chiefly on the 18S rRNA tree of Lake (1989, 1990) and the fossil record as discussed in the text. *Horizontal bands* indicate the grades of organization at which the various phyla are believed to have originated. The relative vertical position of branches does not have temporal meaning except when they are in the same sequence; horizontal separations are merely for spacing on the page. Numbers indicate extinct body plans: *1* trilobites; *2* crawling vermiform segmented form; *3* narrow-bodied creeping semi-infaunal (?) segmented form; *4* narrow-bodied minute nonsegmented semi-infaunal form (After Valentine in press)

A possible phylogeny of some metazoan phyla is shown in Fig. 1. It is inferred chiefly from molecular data, although the routing of the branches that lead to deuterostomes and arthopodan phyla is not yet resolved by molecular techniques (see Lake 1990), but is inferred from what are believed to be synapomorphies within the deuterostomes. Figure 1 depicts the grade of origin of various phyla with reference to whether they began as hemocoelic or coelomic insofar as hydrostatic functions are concerned. By "beginning" is meant the evolution of those apomorphies that permit us to characterize the body plan as unique to the phylum-level taxon in question; an ancestor that has not yet advanced to this condition, even if it is an immediate and direct lineal ancestor, is excluded. For example, in Fig. 1 the deuterostomes are depicted as arising from an ancestral taxon that began with the hemocoel as the principal body cavity; this ancestor, however, does not belong among the living deuterostome phyla.

Clearly there are gaps in our knowledge of the evolutionary sequence of body plans. Probably many ancestral forms with characteristic body plans of their own have given rise to individual living phyla or underlie groups of allied phyla, but are now extinct. In the figure, the phyla are spaced out on the page simply to provide an uncluttered diagram, and in this presentation the "missing" ancestors are therefore represented chiefly by horizontal lines and the lower portions of vertical lines. It is precisely the implication of those portions of the lines that I wish to explore, first as to the architectures that they imply, and second as to the locomotory functions that may be associated with the rise of body plans.

2 Sequences of Bioarchitectural Features Implied by the Phylogeny

Body plans intermediate between those of living phyla may have been traversed rapidly in some cases, but as we shall see, they stand only a small chance of producing preservable fossil remains and some of them may have had long Phanerozoic histories before their eventual extinction. Such a possibility is underlined by the flatworms, a group that appears from all evidence to have an origin in the Late Proterozoic, but for which not a single body fossil is known.

That the earliest bilaterian architecture is likely to have been flatworm-like is a common postulate (see Willmer 1990). The evolution of a seriated architecture found in some large polyclads has long been postulated to be one response to increasing body size (review in Clark 1964). The advent of a circulatory system with hemocoelic vessels, channels, or sinuses can also be associated with size increases and perhaps with a concomitant deepening of the body. Nemertines possess such a system but have a coelomic-style body cavity (Turbeville and Ruppert 1985). Size increases in other forms may have been associated with an increase in the size and influence of the hemocoel, rather than with a secondary body cavity, and it is from sclerotized forms of this grade that the arthropod phyla appear to have been derived (Manton 1977); these phyla (and their ancestors) may never have employed a coelomic cavity as a hydrostatic skeleton in the adult (Valentine 1989). The adult arthropod "coelom" is composed of spaces that serve as ducts or organ receptacles and there is no special reason to regard the arthropods as coelomate, at least not in the sense that annelids or deuterostomes are coelomate. While arthropods retained a hemocoelic fluid skeleton, the annelids evolved their classic eucoelom, segmented extensively in coordination with bodywall segmentation and parapodial structures to produce their unique architecture. The segmental systems of arthropods and annelids are so anatomically distinctive that it seems doubtful that either evolved from the other, but rather that their last common ancestor was a lower invertebrate, perhaps of an acoelomate grade (Valentine 1989). From some annelidan lineage, or from a pre-annelidan form that lay on the ancestral annelidan line, have evolved several distinctive body plans that include brachiopods, molluscs and sipunculids (Field et al. 1988; Lake 1990).

The deuterostomes, it now appears, may have arisen from among hemocoelic worms, perhaps with internal seriation, for pattern-formation genes that become segmentation genes were evidently present (see Lewis 1989).

3 Sequences of Bioarchitectural Features Implied by the Fossil Record

As mentioned previously, we have no confirmed body fossils of flatworms in the Phanerozoic record, yet they (or organisms of their grade) must have been present throughout that eon. Therefore their absence in the fossil record of the Late Proterozoic should hardly be taken as evidence that they were not present then. Some Late Precambrian fossils from Vendian rocks have been classed as flatworms (Fedonkin 1985b) but this assignment is regarded as dubious. We certainly cannot be sure when flatworms originated. The earliest fossils that are likely to be bilaterians are Vendian (the Ediacaran fauna and its equivalents); they are trace fossils, rather meandering trails (Jenkins 1984) made by worms that are likely to have been at least vascularized acoelomate or perhaps nemertine-grade forms. The earliest body fossils that may represent bilaterians follow closely stratigraphically, and include what appear to be segmented forms – vendomiids and especially sprigginids (Glaessner 1976, 1984; Fedonkin 1985b). These forms show no trace of limbs (although there are good ventral impressions of sprigginids) and were probably at least lightly sclerotized (Seilacher 1984). If jointed limbs had been present, they should have required sclerotization as well. Perhaps in the sprigginids we have a representative of sclerotized, segmented, hemocoelic but pre-arthropodan forms (Valentine 1989).

Nonsclerotized benthic worms were likely to be represented only by traces, and these are not uncommon in Late Precambrian fossiliferous horizons, where they are chiefly horizontal trails or perhaps shallow horizontal burrows, some of which are back-filled (Fedonkin 1985a; Narbonne and Myrow 1988; Crimes 1989). Penetrating vertical burrows appear quite late in the Proterozoic and are small and rare, becoming progressively larger, deeper, and more common in successive Lower Cambrian stages (for examples see Droser and Bottjer 1988). The appearance and increasing importance of vertical burrowing suggests the origin and progressive success of coelomate burrowers such as annelids; if so, their origin follows after that of the segmented epifaunal forms. Well-defined arthropod traces appear in the Lowest Cambrian Tommotian Stage, and are joined by body fossils of trilobites and other segmented forms (anomalocarids, onychophoran-like forms, and numbers of arthropod-like organisms; see Hou 1987; Dzik and Lendzion 1988; Dzik and Krumbiegel 1989; Hou and Chen 1989) in the overlying Atdabanian Stage. Except for most trilobites, early segmented forms are not readily fossilized and are found as a result of exceptional preservation. The richness of the radiation that occurred chiefly in Early Cambrian time is indicated by the large number of arthropod-like forms, many with unique segment and appendage arrangements, observed in the marvelous Middle Cambrian Lagerstätten of the Burgess Shale and its equivalents (Briggs 1983; Whittington 1985; Briggs and Collins 1988).

The earliest undoubted coelomates are brachiopods, which appear at the base of the Tommotian (Pel'man 1979). Whether this appearance implies that the annelid body plan had appeared, or whether pre-annelid worms branched into forms that led to brachiopods first and then to annelids, is not certain. The oldest known molluscs also appear in Tommotian rocks, but again, whether their ancestry was among definitive annelids or among pre-annelidan worms is not certain.

The fossil record is thus not inconsistent with the phylogeny in Fig. 1. Flatworms are not found, but forms with apparent body-wall segmentation appear in the Ediacaran fauna. Arthropods may have arisen then also, or perhaps later; annelids seem to appear later still. In the Early Cambrian a variety of coelomates appear, both protostomes and deuterostomes, that are represented today by coelomate body plans. It is implicit in the phylogeny assumed here that the coelom is highly polyphyletic. It is noteworthy that over 90% of the orders that appear in late Precambrian and Cambrian time are extinct (Valentine et al. in press). It is possible that some of the orders that are assigned to extant groups on the basis of their skeletons may have lacked some of the characteristic architectural features associated with their living relations.

4 Sequences of Locomotory Mechanisms

Great general morphologists and anatomists of the late nineteeth and twentieth centuries have speculated upon the relationships among the organisms that they have known in such breadth. Similarly, generations of outstanding specialists have commented on the probable relationships of their special groups. The resulting collection of phylogenetic schemes, even those of relatively recent vintage, is contradictory and bewildering. It seems from this experience that while some of the gross relationships that morphologists have inferred appear to be valid, the power of comparative morphological studies to determine phylogenies at the scale of the phylum is quite limited. Many phylogenies have been inferred from hypothetical adaptive scenarios that in many cases seem quite plausible, and indeed are often ingenious and captivating, involving assumptions about homologies and about evolutionary trends of certain key features such as "the coelom" or "the larva". Ghiselin (1988) has critically reviewed a number of these scenarios in the light of molecular data. A major difficulty has been that the sequence of origin of the architectural features of body plans has not been known, and thus there has been no independent evaluation of which characters are ancestral and which descendant. It seems that such problems are now being solved. The adaptive scenarios of the past, even those that now appear to be incorrect, have nevertheless served to establish the distribution of adaptations through the phyla and are useful in approaching the questions raised by missing intermediates. These questions are very real. Although the intermediates that we construct may be hypothetical, they are models or at least proxies for what were certainly historical organisms – the organisms were very real, although our ideas about them may be flawed. Something existed in the gaps between the body plans that we know. Assuming for the sake of argument that the tree of Fig. 1 is approximately correct, how can a scenario be fashioned to explain the sequence of body plans in adaptive terms? Here I attempt only briefly to outline possible sequences of locomotory systems, drawing heavily upon previous work, in attempting to produce something similar to the sequence discussed above.

The earliest bilaterians were presumably ciliated benthic crawlers not unlike small flatworms. Size increases were evidently associated with a trend in the use of body wall muscles to form pedal waves (Clark 1964); sometimes flatworm points d'appui are near or at the lateral margins, at least among some terrestrial forms (Pantin 1950). It has been postulated that in some lineage(s), further size increases were accompanied by the evolution of serial organ systems. It is tempting to suppose that body-wall segmentation arose in elongate, hemocoelic forms that were already strongly seriated internally, so that the emerging segmental divisions became correlated with sets of organs, the relations among which became regularized as the segmental architecture became fully expressed (see Clark 1964).

Indeed, this may well have happened more than once, to produce several independently segmented lineages. Such a possibility is suggested by work on a homeobox gene, *engrailed*, the expression of which reflects the pattern of parasegment formation in arthropods (Patel et al. 1989a). In annelids, however, the expression of *engrailed* is rather different (Patel et al. 1989b). The gene *engrailed* is involved in neurogenesis, which was probably its primitive function at least within the higher metazoa (Patel et al. 1989b). Perhaps a previous employment of *engrailed* in regulating a series of transverse nerve connections in a ladder-like pattern in an elongate, seriated worm was an ancestral situation that led to the expression of *engrailed* during segment formation in arthropods.

At any rate, the immediate arthropod ancestor seems likely to have been a flattened, creeping form using pedal locomotion, with a hemocoelic hydrostatic skeleton, in which mobility and speed translated into fitness. Perhaps the lateral margins of segments became strongly involved in the pedal locomotory process. As segmentation evolved, sclerotization of the cuticle within segments provided mechanical support for the evolving system while retaining flexibility at segmental junctions for locomotory or other movements, and the vascular system continued to serve hydrostatic functions. Larval intramesodermal compartments were not exploited in the adult hydrostatic system. Gripping aids for purchase at points d'appui may have been provided by sclerotized segmental outgrowths which evolved into jointed appendages. A further ancestral locomotory technique may have been sinusoidal crawling – wriggling – for rapid or powerful movements (Valentine 1989), a technique that continued to be employed in modified form after arthropodization. The arthropodan groups radiated explosively as they exploited the many opportunities afforded to organisms of this general body plan. It is uncertain at what point in the rise of the arthropodan architecture the branching occurred that led to groups with the distinctive appendage systems described by Manton (1977 and references therein). Onychophorans appear to have produced an independent solution to locomotion; instead of sclerotizing they evolved the lobopodal system that depends upon the hemocoel as a hydrostatic skeleton. Perhaps they arose from narrow-bodied worms creeping on firm substrates that evolved pairs of accessory gripping structures, lateroventrally in this case, which became lobopods. If so, then the branch between arthropods and onychophorans lies so deeply within the ancestral groups that the living taxa should certainly be regarded as separate phyla.

A strong case that the segmented coelom of annelids served primitively for burrowing in marine substrates has been presented by Clark (1964). In these forms the hemocoel is largely restricted to vascular spaces with minimal mechanical functions. The near absence of penetrating vertical burrowing in the Precambrian suggests that the annelidan body plan was not fully evolved until the Early Cambrian, when penetrating burrows are not uncommon in some facies. To derive the annelids from among the hypothesized plexus of segmented but hemocoelic epifaunal worms, then, is to trace the invasion of the infaunal realm by a nonsclerotized ancestral form. A heterochronic shift of larval intramesodermal spaces into the adult in small-bodied semi-infaunal forms may have led to the eucoelom as the mesodermal spaces within the evolving segments came to surround the gut for penetration burrowing. Additional segments were then added to produce the classic elongate ancestral annelids, and radiation produced a variety of body plan modifications to suit distinctive habits. Body fossils that appear to be segmented worms are known from middle Lower Cambrian (Atdabanian) rocks (Hou and Chen 1989), and undoubted annelid body fossils are found in the middle Middle Cambrian Burgess Shale (Conway Morris 1979). The origins of the annelid body plan, then, may lie within a plexus of acoelomate worm lineages, and only there are we likely to find a common ancestor with arthropodan or onychophoran phyla.

Miniaturized annelids of the meiofauna display a number of architectural modifications that may provide clues to the derivation of the molluscs and other nonsegmented protostomes. Some annelid lineages that become miniaturized maintain segments and coelomic spaces, but in others both the eucoelom and segmentation are lacking, and the body plan becomes essentially that of an acoelomate (Rieger 1980; Fransen 1980). Thus an early meiofaunal (or at least small-bodied) offshoot of annelids, or more likely of proto-annelids, could furnish a stock from which a creeping epifaunal ancestral mollusc, still retaining aspects of organ seriation, could have been derived. The coelom in those protostomes that possess one has presumably evolved several times independently (as in brachiopods and in cephalopod molluscs).

Deuterostome coeloms have evidently arisen independently of the protostomes, and it is possible, though not certain, that the coelom is homologous throughout living deuterostome phyla. Presumably, deuterostomes also arose from creeping acoelomates, but their ancestral stock branched before the arthropod branch, at the acoelomate grade. Although there are clear homologies among homeobox genes in arthropods and vertebrates (Holland and Hogan 1988), there are also striking differences in their pattern of expression (Patel et al. 1989b). Presumably the homologies have common ancestors among acoelomates, while the differences may reflect independent origins of segmentation and other body plan features.

So far as we know, the early coelomate deuterostomes were oligomerous; deuterostomes may never have had segmented body walls. Organ series are common enough, as with the branchial openings and associated structures of hemichordates and gill arches of chordates. But the oligomerous coelom suggests an origin among tube-dwelling, suspension-feeding worms that used the trunk coelomic compartment (metacoel) for peristaltic burrowing and a more anterior compartment (mesocoel) as the hydrostatic skeleton of a tentaculate feeding system (Clark 1964). If the chordates are derived from such a form, then the early chordate body plan was evidently evolved in response to the requirements of wriggling through soft substrata and/or for swimming, in a microphagous clade. Assuming that cephalochordates are descended from a primitive deuterostome stock, one can postulate that creeping or any other variety of peristaltic locomotion was replaced entirely by sinuous, undulatory locomotion. Thickening of the dorsal margin of the gut may have served to antagonize the undulations created by body wall musculature, reducing the requirements for muscular activity and enhancing locomotion. Continued selection associated with this feature produced a notochord. Although it is possible that the notochord began as a larval adaptation, it seems at least as likely that it was useful throughout the life cycle from the first, and was evolved for adult as well as larval locomotion. The seriated muscle blocs (myomeres) which deform the cephalochordate and fish bodies are not homologous with any invertebrate segmentation, in this case.

5 Conclusions

It may be inferred from molecular evidence that the body plans of living phyla have been evolved quite independently. Even though some features – coeloms, segmentation of various sorts, and so on – occur in a number of phyla, they are likely to have arisen independently, and cannot be used to infer inter-phylum relationships. A corollary of this inference is that the branches that established the stocks within which the body plans of the extant phyla evolved occurred further back in metazoan history than has previously been thought. The fossil record establishes a rough timetable for those events. A round of cladogenesis among acoelomates in the Late Proterozoic, represented now chiefly by traces of surface trails in the Vendian, provided a variety of creeping worms with a number of distinctive architectural variations. Most were presumably flattened, some long and narrow, some relatively broad, some perhaps semi-infaunal and rounder in body cross-section. From various stocks within this plexus of forms arose the body plans of living phyla via anagenetic trends that chiefly occurred in latest Precambrian and Early Cambrian times. Molecular phylogenies may eventually produce an accurate picture of this branching pattern, but the entire sequence may lie among forms with little resemblance to the architectural frameworks and biomechanical systems of the living taxa.

If this is correct, then there were truly a large number of body plan sequences in higher invertebrate history, not merely two (protostomes and deuterostomes) or so. Locomotory innovations, often associated with changes in body size and, of course, with shifts in adaptive zones, seem now, as before, to be responsible for many of the major architectural features within these lineages. The techniques of interpreting this pattern can be found in the rich legacy of adaptive and biomechanical models that we already possess.

Acknowledgments. Work on which this chapter is based was supported by NSF Grants EAR 87-21192 and EAR 84-17011. I thank Tracy McLellan, University of California, Santa Barbara, and James A. Lake, University of California, Los Angeles, for assistance and advice and R. D. K. Thomas, Franklin and Marshall College, for manuscript review.

References

Anderson DT (1973) Embryology and phylogeny in annelids and arthropods. Pergamon, Oxford

Briggs DEG (1983) Affinities and early evolution of the Crustacea: the evidence of the Cambrian fossils. In: Schram FR (ed) Crustacean phylogeny. Balkema, Rotterdam, pp 1–22

Briggs DEG, Collins D (1988) A Middle Cambrian chelicerate from Mount Stephen, British Columbia. Palaeontology 31:779–798

Clark RB (1964) Dynamics in metazoan evolution. Clarendon, Oxford

Conway Morris S (1979) Middle Cambrian polychaetes from the Burgess Shale of British Columbia. Philos Trans R Soc London Ser B 285:227–274

Crimes TP (1989) Trace fossils. In: Cowie JW, Brasier MD (eds) The Precambrian-Cambrian boundary. Clarendon, Oxford, pp 166–185

Droser ML, Bottjer DH (1988) Trends in depth and extent of bioturbation in Cambrian carbonate marine environments, western United States. Geology 16:233–236

Dzik J, Krumbiegel G (1989) The oldest "onychophoran" *Xenusion*: a link connecting phyla? Lethaia 22:169–181

Dzik J, Lendzion K (1988) The oldest arthropods of the East European platform. Lethaia 21:29–38
Fedonkin MA (1985a) Paleoichnology of Vendian Metazoa. In: Sokolov BS, Ivanovich AB (eds) The vendian system, vol 1. Nauka, Moscow, pp 112–117 (in Russian)
Fedonkin MA (1985b) Systematic description of Vendian Metazoa. In: Sokolov BS, Ivanovich AB (eds) The vendian system, vol 1. Nauka, Moscow, pp 10–106 (in Russian)
Field KG, Olsen GJ, Lane DJ, Giovannoni SJ, Ghiselin MT, Raff EC, Pace NR, Raff RA (1988) Molecular phylogeny of the animal kingdom. Science 239:748–753
Fransen ME (1980) Ultrastructure of coelomic organization in annelids. I. Archiannelida and other small polychaetes. Zoomorphologie 95:41–84
Ghiselin MT (1988) The origin of mollusks in the light of molecular evidence. In: Harvey PH, Partidge L (eds) Oxford surveys in evolutionary biology, vol 5. Univ Press, Oxford, pp 66–95
Glaessner MF (1976) A new genus of late Precambrian polychaete worms from South Australia. Trans R Soc SO Aust 100:169–170
Glaessner MF (1984) The dawn of animal life. Univ Press, Cambridge, MA
Holland PHW, Hogan BLM (1988) Expression of homeo box genes during mouse development: a review. Genes Dev 2:773–782
Hou S (1987) Early Cambrian large bivalved arthropods from Chenjiang, Eastern Yunnan. Acta Palaeontol Sin 26:286–298 (in Chinese, English summary)
Hou X, Chen J (1989) Early Cambrian tentacled worm-like animals (*Fascivermis* gen. nov.) from Chengjiang, Yunnan. Acta Palaeontol Sin 28:32–41 (in Chinese, English summary)
Jenkins RJF (1984) Interpreting the oldest fossil cnidarians. Recent advances in the paleobiology and geology of the Cnidaria. Palaeontol Am 54:95–104
Lake JA (1982) A rate-independent technique for analysis of nucleic acid sequences: evolutionary parsimony. Molec Biol Evol 4:167–191
Lake JA (1990) Origin of the multicellular animals. Proc Natl Acad Sci USA 87:763–766
Lewis J (1989) Genes and segmentation. Nature (London) 341:382–383
Manton SM (1977) The arthropoda. Clarendon, Oxford
Narbonne GM, Myrow P (1988) Trace fossil biostratigraphy in the Precambrian-Cambrian boundary interval. In: Landing E, Marbonne GM, Myrow P (eds) Trace fossils, small shelly fossils and the Precambrian-Cambrian boundary. New York State Mus Bull 463
Pantin CFA (1950) Locomotion in British terrestrial nemertines and planarians: with a discussion on the identity of Rhyncodemus bilineatus (Mecznikow) in Britain and on the name *Fasciola terrestris* O. F. Muller. Proc Linnean Soc London 162:23–37
Patel NH, Kornberg TB, Goodman CS (1989a) Expression of *engrailed* during segmentation in grasshopper and crayfish. Development 107:201–212
Patel NH, Martin-Blanco E, Coleman KG, Poole SJ, Ellis MC, Kornberg TB, Goodman CS (1989b) Expression of *engrailed* proteins in arthropods, annelids, and chordates. Cell 58:955–968
Pel'man YuL (1979) Ancient brachiopod complexes. Sov Acad Sci Sib Div Trans Inst Geol Geophys Nauka 406:34–39
Rieger RM (1980) A new group of interstitial worms, Lobatocerebridae nov. fam. (Annelida) and its significance for metazoan phylogeny. Zoomorphologie 95:41–84
Seilacher A (1984) Late Precambrian and Early Cambrian Metazoa: preservational or real extinctions? In: Holland HD, Trendall AF (eds) Patterns of change in earth evolution. Springer, Berlin Heidelberg New York, pp 159–168
Turbeville JM, Ruppert EE (1985) Comparative ultrastructure and the evolution of nemertines. Am Zool 25:53–71
Valentine JW (1989) Bilaterians of the Precambrian-Cambrian transition and the annelid-arthropod relationship. Proc Natl Acad Sci USA 86:2272–2275
Valentine JW (in press) The macroevolution of phyla. In: Lipps JH, Signor PW (eds) Origins and early evolutionary history of the Metazoa. Plenum, New York
Valentine JW, Awramik SM, Signor PW, Sadler PM (in press) The biological explosion at the Precambrian-Cambrian boundary. In: Hecht MK, Wallace B, Prance GT (eds) Evolutionary biology.
Whittington HB (1985) The burgess shale. Yale Univ Press, New Haven
Willmer P (1990) Invertebrate relationships, patterns in animal evolution. Univ Press, Cambridge
Woese CR (1987) Bacterial evolution. Microbiol Rev 51:221–271
Zuckerkandl E, Pauling L (1965) Molecules as documents of evolutionary history. J Theor Biol 8:357–366

Subject Index

Index of Taxonomic Terms

MIX
Papier aus verantwortungsvollen Quellen
Paper from responsible sources
FSC® C105338

If you have any concerns about our products,
you can contact us on
ProductSafety@springernature.com

In case Publisher is established outside the EU,
the EU authorized representative is:
Springer Nature Customer Service Center GmbH
Europaplatz 3, 69115 Heidelberg, Germany

Printed by Libri Plureos GmbH
in Hamburg, Germany